AF788457

Monographs on Theoretical and Applied Genetics 17

Edited by
R. Frankel (Coordinating Editor), Bet-Dagan
M. Grossman, Urbana · H.F. Linskens, Nijmegen
P. Maliga, Piscataway · R. Riley, London

Monographs on Theoretical and Applied Genetics

Volume 1	**Meiotic Configurations** A Source of Information for Estimating Genetic Parameters By J. Sybenga (1975)
Volume 2	**Pollination Mechanisms, Reproduction and Plant Breeding** By R. Frankel and E. Galun (1977)
Volume 3	**Incompatibility in Angiosperms** By D. de Nettancourt (1977)
Volume 4	**Gene Interactions in Development** By L.I. Korochkin (1981)
Volume 5	**The Molecular Theory of Radiation Biology** By K.H. Chadwick and H.P. Leenhouts (1981)
Volume 6	**Heterosis** Reappraisal of Theory and Practice Editor: R. Frankel (1983)
Volume 7	**Induced Mutations in Plant Breeding** By W. Gottschalk and G. Wolff (1983)
Volume 8	**Protoplast Fusion** Genetic Engineering in Higher Plants By Y.Y. Gleba and K.M. Sytnik (1984)
Volume 9	**Petunia** Editor: K.C. Sink (1984)
Volume 10	**Male Sterility in Higher Plants** By M.L.H. Kaul (1988)
Volume 11	**Tree Breeding: Principles and Strategies** By G. Namkoong, H.C. Kang, and J.S. Brouard (1988)
Volume 12	**The Wheat Rusts – Breeding for Resistance** By D.R. Knott (1989)
Volume 13	**Population Dynamics and the Tribolism Model** Genetics and Demography By R.F. Costantino and R.A. Desharnais (1991)
Volume 14	**Genetic Improvement of Tomato** Editor: G. Kalloo (1991)
Volume 15	**Evolution and Function of Heterostyly** Editor: S.C.H. Barrett (1992)
Volume 16	**Distant Hybridization of Crop Plants** Editors: G. Kalloo and J.B. Chowdhury (1992)
Volume 17	**Cytogenetics in Plant Breeding** By J. Sybenga (1992)

J. Sybenga

Cytogenetics in Plant Breeding

With 74 Figures

Springer-Verlag
Berlin Heidelberg New York
London Paris Tokyo
Hong Kong Barcelona
Budapest

Prof. Dr. J. Sybenga
Department of Genetics
Wageningen Agricultural University
Dreijenlaan 2
6703 HA Wageningen
The Netherlands

ISBN-13:978-3-642-84085-2 e-ISBN-13:978-3-642-84083-8
DOI: 10.1007/978-3-642-84083-8

Library of Congress Cataloging-in-Publication Data.
Cytogenetics in plant breeding / [edited by] J. Sybenga. p. cm. — (Monographs on theoretical and applied genetics; 17)
Includes bibliographical references (p.) and index.
ISBN-13:978-3-642-84085-2

1. Plant breeding. 2. Plant cytogenetics. 3. Crops-Cytogenetics. I. Sybenga, J. (Jacob), 1926– . II. Series.
SB123.C98 1992 631.5'23—dc20 92-9632

This work is subject to copyright. All rights are reserved, whether the whole or part of the material is concerned, specifically the rights of translation, reprinting, reuse of illustrations, recitation, broadcasting, reproduction on microfilm or in any other way, and storage in data banks. Duplication of this publication or parts thereof is permitted only under the provisions of the German Copyright Law of September 9, 1965, in its current version, and permission for use must always be obtained from Springer-Verlag. Violations are liable for prosecution under the German Copyright Law.

© Springer-Verlag Berlin Heidelberg 1992
Softcover reprint of the hardcover 1st edition 1992

The use of general descriptive names, registered names, trademarks, etc. in this publication does not imply, even in the absence of a specific statement, that such names are exempt from the relevant protective laws and regulations and therefore free for general use.

Typesetting: Best-set Typesetter Ltd, Hong Kong

31/3145-5 4 3 2 1 0 – Printed on acid-free paper

Preface

The role of cytogenetics in plant breeding is varied. It provides methods for manipulating the composition of the genome and the genetic characteristics of the reproductive system. It also provides methods for extracting information from the material used in various stages of a breeding program, for instance the frequency of recombination between homologous and homoeologous genomes, the presence of abnormalities in normal, treated or hybrid material, the possible causes of irregularities in segregation, fertility etc. It also provides information of a more general nature on genetic transmission systems, general consequences of abnormalities etc.

Many of the techniques, approaches and concepts of cytogenetics have been available for a long time, but there has been continuous, gradual and occasionally rapid progress in the quantity and quality of techniques, knowledge and understanding of cytogenetics and their application to plant breeding. After a period of declining interest, mainly a consequence of the rapid development of cell biological and molecular techniques of manipulation, and the expectation that these would rapidly replace the more tedious, difficult and old-fashioned generative approaches, interest in cytogenetics is increasing again. This is due in part to cytogenetic developments in their own right and in part to the growing insight that the combination of cytogenetic and cell biological and molecular techniques can be very fruitful in both directions. In addition, experience is accumulating that molecular genetic manipulation after all does not produce results as rapidly and as easily as hoped earlier, and that cytogenetics might provide good alternatives. It should be noted, however, that cytogenetics, although often requiring less sophisticated techniques, is often conceptually more complicated than cell and molecular biology.

The long history of cytogenetic research (of which many of the reports are as relevant now as at the time of publication) and its varied potential for application in plant breeding have resulted in very extensive literature. It is not possible to be aware of, and even less possible to refer to, all relevant publications. The reader will find that the choices made in this book are personal and that many important references have been missed or for various other reasons not included. It is hoped that the most serious omissions will be brought to the attention of the author.

There are several reviews and proceedings of meetings dealing entirely or partly with the cytogenetics of different crop species, and cytogenetics in plant breeding in general, for example:

– *Cytogenetics of Crop Plants*, MS Swaminathan, PK Gupta, U Sinha (eds) (1983); MacMillan India Ltd., Delhi-Bombay;
– *Chromosome Engineering in Plants, Part A*. PK Gupta, T. Tsuchiya, eds. Elsevier, Amsterdam 1991;
– *Chromosome Engineering in Plants, Part B*. T. Tsuchiya, PK Gupta eds. Elsevier, Amsterdam 1991;
– *Proceedings of the International Wheat Genetics Symposia*, at 5 year intervals held at different locations;
– *Proceedings of the Barley Genetics Symposia*: published as Barley Genetics I, II, III etc. Held and published at varying intervals at different locations.
– Maize cytogenetics, Carlson W.R., in: *Corn and Corn Improvement*, G.F. Sprague and J.W. Dudley, editors; "Agronomy", Madison, WI, USA (1988). New editions will appear in the future.

Several chapters on papers from these books and proceedings have been referred to in the present text.

The approach to the subject and the decision which chapters to include are the choice of the author. For instance, a separate chapter on interspecific hybrids could have been included, but it was preferred to deal with them in the chapters where their use is discussed (especially Chap. 9 on genome analysis, Chap. 10 on gene transfer and Chap. 11 on manipulation of genome number). A general introduction to chromosomes and their behaviour is given, as it may be useful to be able to look up the most general principles in the book itself. However, for details the reader is referred to the literature. Such choices will not be argumented.

Many colleagues have contributed to this book by providing figures and special information. Without mentioning each personally, I wish to express my gratitude to all of them here. I am especially grateful to Dr. J.H. de Jong and Dr. J. N. de Vries, who took the trouble of going through much of the manuscript, giving many valuable comments and suggestions. I also thank Dr. de Jong and Ms. Jannie van Eden for drawing or plotting several of the figures.

Contents

Chapter 1 Cytogenetics in Genetics and Plant Breeding

Chapter 2 Chromosome Composition, Structure and Morphology

Chapter 3 The Mechanisms of Genetic Transmission

Chapter 11 Manipulation of Genome Composition. B. Gene Dose: Duplication, Polyploidy and Gametic Chromosome Number

Chapter 1

Cytogenetics in Genetics and Plant Breeding

In most of its stages, plant breeding makes use of auxilliary scientific disciplines. One of these disciplines is genetics, with its subdisciplines quantitative genetics, population genetics, cytogenetics, molecular genetics, etc. To understand the role of cytogenetics in plant breeding, it is useful to first give a brief review of the segment of genetics it covers, and what it is considered to include.

1.1 Cytogenetics as a Subdiscipline of Genetics

1.1.1 The Two Functions of Genetic Material

The science of genetics studies the composition and functions of the "genetic material" (Fig. 1.1).

The "genetic material" has two essential functions: (1) to bring about and regulate biological processes, and (2) to maintain, multiply and adjust itself.

It is composed of different substances, the most fundamental of which is DNA. The two functions of the genetic material reflect the two basic functions of DNA: *transcription* and *replication*. *Transcription* is the first step in a series of interactive and thoroughly controlled processes that ultimately constitute the vital functions of all living material. It is the assemblage of RNA on the DNA which serves as a template, with RNA polymerases as enzymes. Only part of the DNA, the coding DNA, is transcribed. The coding DNA occurs in functional units of roughly 1000 base pairs: the genes. Transcription is followed by *translation* of segments of the thus formed m(messenger)-RNA into polypeptides. This occurs in special structures, the ribosomes, consisting of ribosomal RNA and proteins. The polypeptides are combined with other polypeptides and, if applicable, other compounds into a great variety of enzymes, regulatory substances and structural elements. The result is a highly complex, strictly regulated system of interacting processes and structures. This, the domain of gene action and gene expression, will not be considered here to any significant extent.

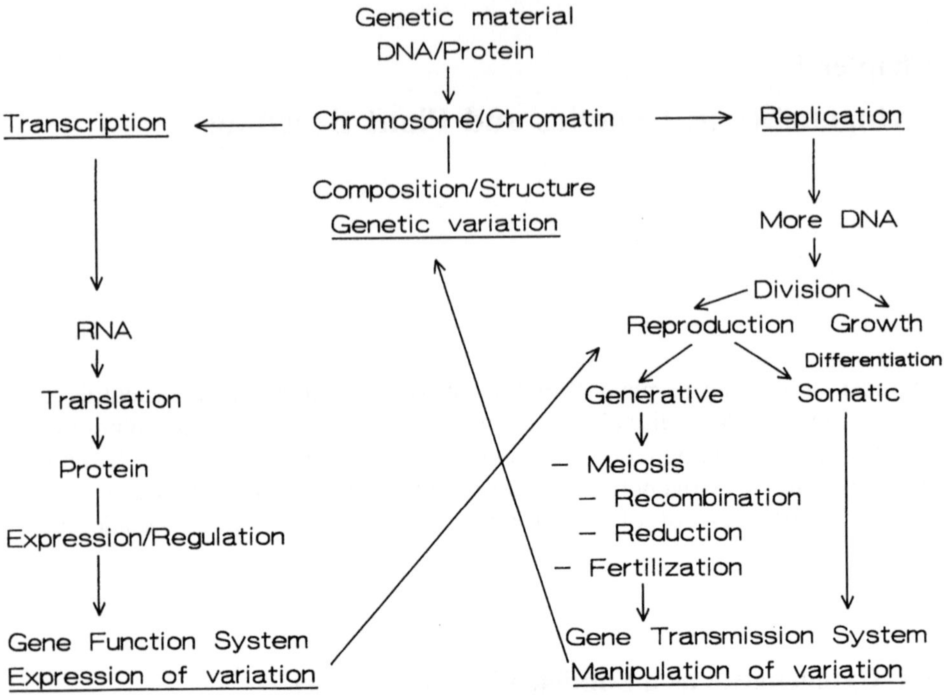

Fig. 1.1 The structure of genetics. Variation is an essential aspect of genetics. It can be induced as mutation or transformation of the genetic material, and fixed, eliminated or manipulated in the transmission systems

The second function of DNA, *replication*, is the first step in a series of highly regulated and interactive processes leading to systems of transmission of the genes from one generation of cells or individuals to the next, the domain of *gene transmission*. It involves interacting systems of structures and processes equally complex as the gene action and gene expression systems. Replication usually, although not without exception, involves all DNA of the nucleus and consequently results in complete duplication of the nuclear or organelle DNA with each round of replication. Usually, DNA replication of the nucleus is followed by nuclear and cell division leading to (somatic) growth and/or to multiplication.

The two functions of the genetic material (gene expression and gene transmission, Fig. 1.1) are biologically complementary and both necessary. The expression of a gene is of little consequence when it is not transmitted to a new generation. Life is finite, and without renewal by replication, the gene would be doomed to disappear. At first sight, replication and transmission seem to be of little use without function for the "genetic material". Yet, in many higher organisms, the larger part of the DNA is without apparent function and still is faithfully replicated with its associated proteins, and transmitted. It seems to exist primarily for its own sake: "selfish DNA". Part of this seemingly redundant DNA, however, may have a function in the genetic system by binding specific, regulating proteins. In addition, the total mass of DNA affects nucleus and cell size and as such has a pronounced effect on the organism.

The essence of the science of genetics is not simply in these two branches. The essence is *heritable variation* and its regulation. The origin of this variation is in the spontaneous, induced or introduced changes in the composition of the DNA and, to a limited extent, in its structure (epigenetics). These changes in the genetic material are either removed or consolidated during the reproductive cycle soon after their induction. Consolidated changes may (not necessarily always) be expressed in the phenotype of the organism in which they occur or in its progeny. If they are expressed, they are the basis of the heritable component of phenotypic variation. On this variation resides the potential of the members of a population to function, maintain and reproduce themselves in competition with other living forms in a specific habitat. In an agricultural environment the competition acts mainly through the grower, with respect to the crop as well as weeds, diseases and pests.

Competition in a limiting environment leads to selection, and selection eliminates genotypes and reduces genetic variation. Genetic variation has two components: (1) the differences between alleles of genes, and (2) the combination of specific allelic variants. New variation can arise simply by making new combinations of existing allelic variants: *recombination*. This is realized in meiosis and effected by fertilization during generative reproduction. It alters the composition of the genetic material, which ultimately results, through the gene function system, in new phenotypes. The transmission system, including the processes responsible for recombination, is itself under the control of the gene function (expression) system. Variation in the genes concerned will, therefore, be expressed as variation in transmission. This makes it possible, by selection, to optimize the recombination system.

From a biological point of view the two branches of genetics are completely interrelated and interdependent.

1.1.2 Cytogenetics Defined

Apart from parasitic elements, the *genetic material* is normally present in a number of different cell organelles: (1) the nucleus; (2) mitochondria; (3) plastids (in higher plants); and (4) plasmids (free in some lower organisms, in mitochondria and possibly other organelles in higher organisms). By far the majority of the *genetic material* in higher organisms is present in the nucleus. Here, the DNA is associated with different proteins (basic, neutral and acidic) and, to a lesser extent, with other substances, and aggregated in the form of *chromatin*. The chromatin of a *genome* (representing one full complement of genes and the accessory DNA) is divided over a number of *chromosomes*.

There is only one chromosome in a species of the nematode *Ascaris*, two in a number of species of the plant *Haplopappus*, three in some species of the plant *Crepis*, up to hundreds in some ferns.

Cytogenetics in the present context is the science of *chromosomes* in the broadest sense, including their light, electron microscopic and molecular *characteristics*, and especially the *function* of these characteristics; further, the *behaviour* of the chromosomes during *somatic* and *generative transmission*. Variation in characteristics, function and behaviour is an important aspect.

Chapter 2 deals with chromosome structure and Chapter 3 with somatic and generative transmission to provide a background for the chapters on the application of cytogenetics in plant breeding.

1.2 Cytogenetics in Plant Breeding

Cytogenetics can be applied in different phases of plant breeding, and in different ways. Plant breeding as an activity of manipulating genotypes requires

Table 1.1. The phases of a plant breeding program (Sybenga 1989)[a]

I	II	III	IV
a) Formulation of objectives	a) Collection of material (*)	a) Selection	a) Propagation/ maintenance (**)
b) Choice of approach in construction of genotype and reproductive system (*)	b) Testing and preselection of material (**)	b) Testing (**)	
c) Program design (*)	c) Induction of additional genetic variation (mutation, transformation) (**, ***)		
	d) Construction of special reproductive system (**, ***)		
	e) Combination (hybridization, etc.) and recombination (**, ***)		

[a] Asterisks indicate where which type of cytogenetic information or manipulation can be used: * information from literature; ** information to be collected; *** manipulation.

techniques and *information*. Cytogenetics provides both. The information provided can be of a *general* nature, valid for all materials and for several applications, or it can be *specific*, relating to the material and/or the problem at hand. The latter can be information on the structure of the starting material, on the material in intermediate stages, and on end products. It can concern the microscopic structure and number of chromosomes (*karyotype*) and their variation (Chap. 4), but also the *behaviour* of chromosomes during transmission, somatic or generative.

When special cytogenetic approaches (techniques) in genetic manipulation are used (Chap. 10, 11 and 12), again information is needed, both of a general nature, e.g. on the techniques to be employed (necessary for the design and implementation of the project), and of a specific nature on the material to be manipulated. Providing information, therefore, is an essential duty of cytogenetics in plant breeding.

Where cytogenetics plays a role in genetic manipulation, it is necessary to distinguish between the manipulation of the *genotype* (the specific genetic constitution of the genomes under consideration) and manipulation of the *genetic transmission* or *reproductive system*. Manipulation of the genotype may involve the introduction or replacement of specific genes, or the manipulation of the gene dose. Manipulation of the transmission system may, for instance, involve the introduction of limitiations to free segregation used in the fixation of heterozygosity.

Table 1.1 (cf. Sybenga 1983a, 1988) gives a review of the phases of plant breeding where cytogenetic information and manipulation are of potential interest.

Chapter 2

Chromosome Composition, Structure and Morphology

For a detailed review of the structure of the genetic material in the nucleus, the chromatin in its broadest sense, the reader is referred to the specialized literature (e.g., somewhat old but still useful for the present purpose: Bostock and Sumner 1978; more recent: Watson et al. 1987; Alberts et al. 1989). Most of the information relevant to the present context has been available since the late 1970s and the early 1980s. Modern developments mainly concern details that are of interest primarily to the specialist. In modern textbooks on cell biology most of the relevant information is given in a readily understandable form. Here, only the main issues will briefly be discussed.

2.1 DNA

2.1.1 The Chemical Basis

DNA consists of long chains of nucleotides, the monophosphate esters of nucleosides that contain one molecule of the pentose sugar deoxyribose (ribose in RNA), and one nitrogen base. The N-bases are the pyrimidines thymine and cytidine, and the purines adenine and guanine. These are the only four N-bases in DNA, but they may undergo functional modifications. The corresponding nucleosides are: thymidine (T), cytidine (C), adenosine (A) and guanosine (G). The composition of RNA is quite similar. In addition to differing from DNA in the character of the pentose sugar (oxyribose instead of deoxyribose), the only difference is that uracil takes the place of thymidine. The functional effects, however, are considerable.

During DNA synthesis (*replication*), the nucleotide to be built in is offered in the triphosphate form with the phosphate groups attached to the 5′ C-atom of the pentose. Two of the three phosphate groups are removed enzymatically and, at the same time, the remaining group is attached to the 3′ C-atom of the pentose of the nucleotide at the end of the chain already formed. The backbone of the DNA macromolecule is thus formed by alternating deoxyribose and phosphate groups with the N-bases sticking out at one side. The chain is not symmetric, one end being the 3′ C-atom of a ribose molecule, the other end a phosphate group on a 5′ C-atom of a ribose: the DNA molecule has *polarity*.

2.1.2 Heterogeneity of DNA; Unique and Repetitive DNA

Only a portion of the total DNA is transcribed into functional RNAs: the *coding DNA*. It contains the genes proper that are DNA segments of various lengths, but usually comprising roughly 1000 base pairs (1 kb), flanked by special sequences necessary for transcription and separated by spacer sequences. The exact base composition of many genes and their flanking regulating and spacer sequences in many plant and animal species has been established. Inside the genes, most eukaryotes have shorter or longer DNA segments that are transcribed but later removed from the messenger RNA by molecular excision before translation. These segments are called *introns*; the coding DNA outside the introns is composed of *exons*. For more details on possible origin and function of introns, see Watson et al. (1987).

Apart from typical gene DNA there is other *unique* DNA with special functions: the regulation of transcription (promoters, enhancers); replication (autonomous replication sequences: ARS), or other nuclear processes; or simply to serve as a spacer between different functional DNA segments. These are usually shorter than 1000 base pairs and have also been thoroughly analyzed. In most higher eukaryotes there is a considerable additional amount of unique DNA that does not seem to have any function other than to take up space and thus to affect the ultimate structure of the chromatin, and perhaps to bind specific regulating proteins.

In addition to unique DNA, most higher eukaryotes carry large amounts of *repetitive DNA*, i.e. DNA where a certain segment, usually not very long and usually not entirely faithfully preserved, is repeated a number of times. There is extreme variation in the degree of repetitiveness, the composition of the segment repeated and the total amount of repetitive nuclear DNA per organism. Variation in repetitive DNA is by far the most important cause of variation in nuclear and chromosome size between organisms. A representative example is rye (*Secale cereale*) with 7 chromosomes per genome, or 2n = 14 per diploid nucleus. It has 19 pg (picogram) DNA per 2C nucleus, i.e. per two genomes with unreplicated chromosomes (Bennett and Smith 1976). There is only slightly more than 1% coding DNA, about 19% other unique DNA and about 80% repetitive DNA of various kinds (Flavell et al. 1979). *Arabidopsis thaliana* (2n = 10), on the other hand, has only 0.5 pg DNA with hardly as much repetitive as unique DNA, and the latter is to a great extent coding DNA. Organisms with large chromosomes invariably have large amounts of repetitive DNA: e.g. *Allium cepa* (2n = 16) with 33 pg DNA or *Lilium* sp. (2n = 24) with over 80 pg DNA, all per 2C nucleus (Bennett and Smith 1976). In such large chromosomes the great majority of the DNA is not-unique.

Highly repetitive segments occur from 25 000 to 10^5 or occasionally even more than 10^6 times. When the frequency is very high, the unit segment is usually small, even less than 10 bp, and slightly variable. In *middle repetitive* DNA, the unit segment occurs from 500 to 25 000 times and with *low repetiviness*, the frequency is even lower than 500. Related families of repetitive DNA that have slightly different unit segments can be found.

There are different ways to analyze the characteristics of repetitive DNA. One way is to analyze the overall base composition.

Part of the repetitive DNA is species-specific. In rye, *Secale cereale*, for instance, 23% of the total of about 80% repetitive DNA in the genome is not found in closely related species such as *Secale montanum* (Bedbrook et al. 1980; Flavell et al. 1979), although the chromosomes pair quite well at meiosis. This indicates a very rapid evolution of this type of DNA. Whenever large quantities of highly repetitive DNA are found, this variation between related species is observed.

The distribution of repetitive DNA varies over the chromosomes, between locations within chromosomes, but also between species (Appels et al. 1978). In several species it is concentrated either near the centromere or at the ends. In other species it is found in bands at several, or occasionally many, locations in the chromosomes. Within species, some families of repetitive DNA may occupy specific positions, sometimes even specific bands in specific chromosomes. Other families of repetitive DNA may be more or less evenly distributed over the chromosomes (Flavell et al. 1979). There are different ways to make such specific DNA families visible in the chromosomes. The importance of repetitive DNA families as cytological markers of specific chromosomes or chromosome segments will be discussed in Section 4.2.4.2.

The function of the large amounts of repetitive DNA is not clear. Part may be "selfish DNA" existing for its own sake. Like non-coding unique DNA it may have the function to bind specific regulating proteins, or it may simply serve to fill space and thus affect nuclear and cell size. Genome size has been shown to affect the properties of te organism. Rayburn and Auger (1990), for instance, showed that the DNA amount of indigenous maize populations in the Southwestern USA increases with the adaptation to higher altitudes, confirming earlier observations on other species.

2.1.3 Replication

During replication, a new DNA chain is formed along an existing chain, which serves as a template. For stereochemical reasons, thymine is built into the existing chain in a position opposite adenine and vice versa, and guanine opposite cytosine. In this way the old chain serves as a template for the formation of a completely complementary new chain. The polarity of the new chain is reversed compared to that of the old chain. Slight modifications in the structure of the nucleotides do not prevent their acceptation as structural elements, but they do have effects on the function of the DNA. For instance, when offered in excess, bromodeoxyuracil (BUdR) will be built into DNA during replication and can then be used as a marker of specific processes or structures. For replication, the two strands must separate. Both single strands serve as templates for the formation of new (complementary) chains. Because each DNA double strand consists of one old and one new strand, the process is called *semiconservative replication*.

Under most conditions (there are important exceptions that will not be further considered here), DNA takes on the shape of a double helix with a diameter of 2 nm and 10 nucleotides per turn. This is too thin to be made directly visible even with the electron microscope. What one sees in EM photographs is DNA covered with protein, which can be shadowed, or stained, with heavy metals.

The entire period of DNA synthesis per nucleus is completed (during the synthesis or S-phase) in 6–8 h. Per chromosome, many segments are able to replicate independently and at least partly synchronously. These segments are called *replicons*. A mammalian nucleus is estimated to have at least 20 000 (Bostock and Sumner 1978). Replicons each have an ARS (*A*utonomous *R*eplication *S*equence) and may be activated separately, but tend to be regulated in groups. In fast dividing cells of *Drosophila* embryos, for instance, the unit size of replication is only 3–4 μm, whereas in slower dividing cells it may be 13 μm. Between groups, great differences in time and duration of replication can occur. This is related to the composition of the chromatin and therefore indirectly influenced by the composition of the DNA. Apart from this, replication is not affected by DNA sequences such as start or end of genes, regulating sequences etc.

2.2 Chromosome Structure; Histones and Other Chromosomal Proteins

The most important proteins in the chromatin are histones, simple basidic proteins, of which five primary types can be distinguished on the basis of order of elution after chromatographic separation. Histone fractions H3 and H4 are very conservative and practically identical for most organisms. H2A and H2B are somewhat more variable, while H1 is the most variable. H1 also undergoes the most extensive post-translational modification, necessary for its special role in chromosome condensation.

Per unit of weight there is approximately as much histone protein as there is DNA in the chromatin. The association between DNA and histones follows a characteristic pattern that is of great importance for the regulation of gene function and replication, as well as for the packing of the chromatin into chromosomes. Two molecules of each of H2A, H2B, H3 and H4 together form an octamer around which 146 base pairs of DNA are wound in slightly

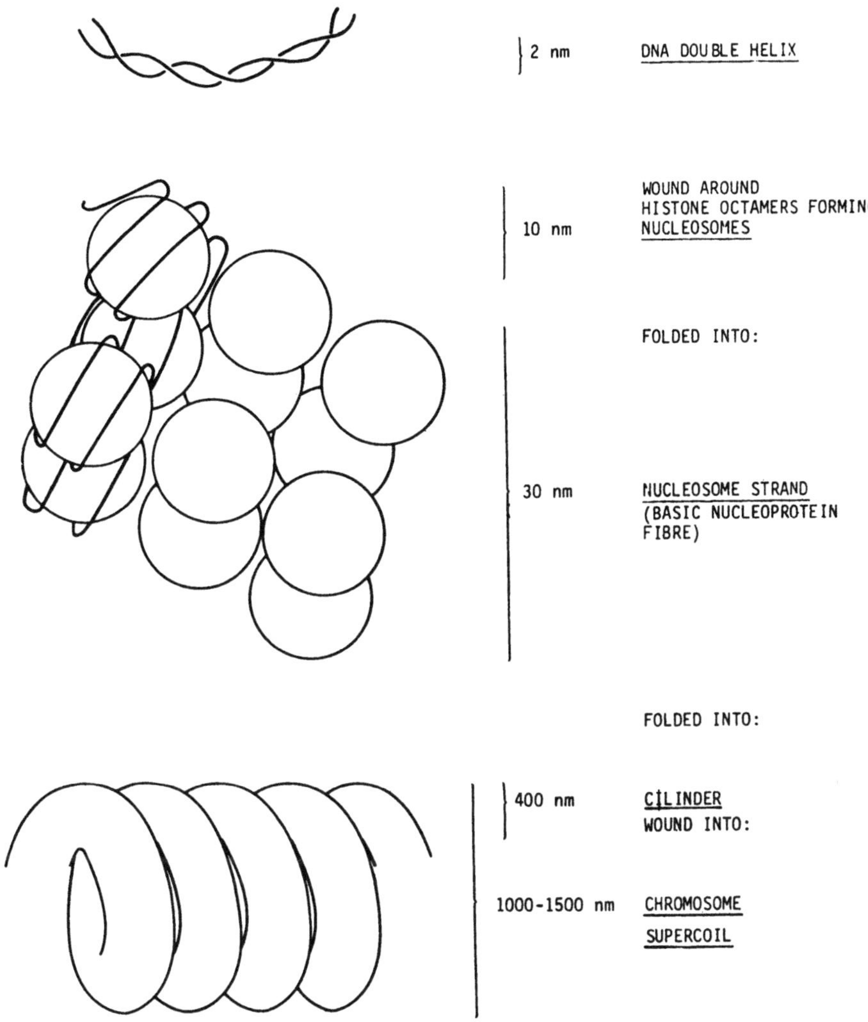

Fig. 2.1 Diagram of the structure of chromatin. **A** the DNA double helix measures 2 nm in diameter. **B** this double helix is wound around histone octamers consisting of two molecules each of H2A, H2B, H3 and H4, forming nucleosomes with a diameter of 10 nm. These nucleosomes are connected by the continuous DNA thread. By histone 1 (H1) they are folded into a 30-nm nucleosome strand, the basic nucleoprotein fibre. **C** This basic fibre is folded into a 400-nm cylinder, which in its turn is wound into the 100–1500 nm supercoil: the chromosome as visible in the light microscope. In the final condensation processes acidic proteins forming scaffolds play an essential role

less than two turns. The result is a disk of 10 nm in diameter and 5.5 nm in width. This is called the *nucleosome* or nu-body. It was first described by Olins and Olins (1974) in electron microscopic preparations of chromatin from which acidic proteins and H1 had been removed. Between two nucleosomes the DNA is not associated with histones over a length of 20 to 100 base pairs. Where the DNA enters and leaves the nucleosome, an H1 molecule is attached to the DNA, but it does not form part of the nucleosome; it plays a central role in the condensation of the chromatin (Kornberg and Klug 1981). Depending on local conditions (pH, salt concentrations etc.), but also on specific trigger substances, H1 molecules associate with varying strength and compactness, which makes the loose nucleosome strand of about 10 nm in diameter condense into a slightly wound 30-nm thread (Fig. 2.1). This is the basal strand of the chromosome, which already contains three superimposed helices: the DNA helix itself, the winding of the DNA around the nucleosomes and the final turns into the 30-nm thread. Its diameter is still below the resolution of the light microscope. The chromatin tends to remain in this condition except for slight modifications, such as incidental and local loosening of the DNA from the nucleosomes.

Further contraction is a consequence of phosphorylation and other modifications, mainly of H1, and is necessary as a preparation for mitosis. For the stability of the higher order structure, acidic proteins of a special nature are required. First, cylinders of about 400 nm are formed (Bak et al. 1979) that wind into the thick major coil (Fig. 2.1). At mitotic prometaphase this final structure attains such a smooth external surface that it appears to be a homogeneous mass. In the last stages of condensation, acidic proteins play a major role: they form a very stable *scaffold* onto which the basic 30-nm

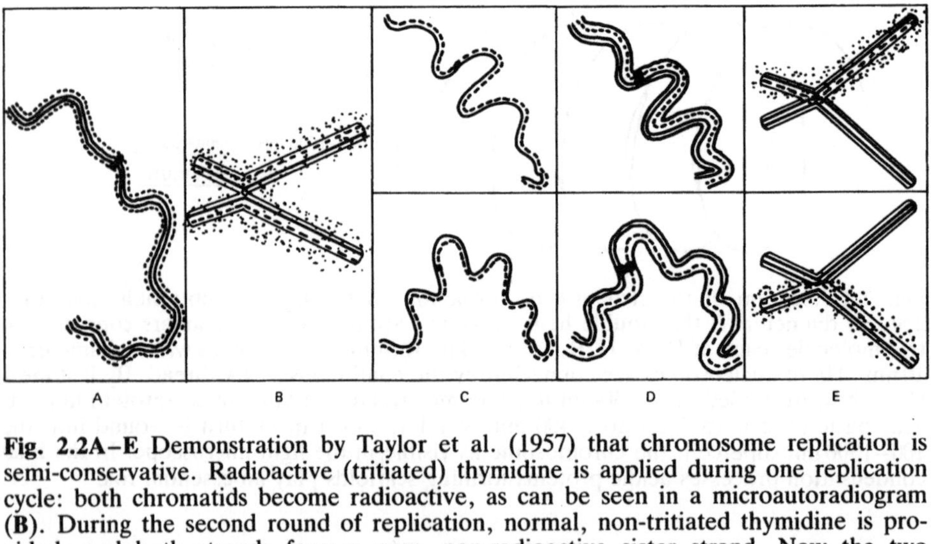

Fig. 2.2A–E Demonstration by Taylor et al. (1957) that chromosome replication is semi-conservative. Radioactive (tritiated) thymidine is applied during one replication cycle: both chromatids become radioactive, as can be seen in a microautoradiogram (**B**). During the second round of replication, normal, non-tritiated thymidine is provided, and both strands form a new, non-radioactive sister strand. Now the two chromatids are different: one parental strand was radioactive and the daughter double strand remains radioactive. The other strand is from the grandparental strand and, like the newest strand, is not radioactive: the entire chromatid is not radioactive (**E**)

chromatin thread is suspended. This makes it possible for the condensed chromosomes to resist the forces to which they are exposed during mitosis. When in a condensed chromosome some of the acidic proteins are gently removed, the 30-nm thread comes out in loops of 30000–90000 base pairs. When H1 is also removed, the 10-nm nucleosome thread appears in the same loops. When all histones are removed, naked DNA loops can be seen (by electron microscopy) attached to the scaffold.

Very shortly after DNA replication, the histones are replicated, but not in a semiconservative way.

Due to this complex but well-regulated process of chromatin replication, the complete linear structure of the entire DNA strand of each chromosome is maintained intact during the entire process of replication and subsequent separation of the two daughter units. Even the small chromosomes of *Drosophila melanogaster*, which are only a few μm long during mitotic division, each contain over 16 mm DNA, and large chromosomes contain several times as much. All this DNA will split and separate without error. The daughter chromosomes are called *chromatids* as long as they are still attached. They will ultimately replace the parent chromosome and are usually visible in light microscopical preparations from late prophase on.

The consequence of the process is that the chromosome, like the DNA it contains, must replicate semiconservatively, as was first shown by Taylor et al. in 1957 (Fig. 2.2).

2.3 Euchromatin, Heterochromatin

The manner in which, and the degree to which, DNA is compacted into chromatin is not homogeneous over the chromosome. Segments with a stronger or more resistant, but not necessarily denser, packing alternate with segments where the packing is less strong or resistant. The latter are the more common and follow the "normal" pattern of condensation and stainability in microscopic preparations: maximal stainability and density during the mitotic transport stages, and minimal stainability and condensation during interphase when transcription and replication take place. This is *euchromatin*.

Heterochromatin, on the other hand, also shows a compact and highly stainable structure during a large part of interphase and sometimes less condensation during the typical mitotic transport stages metaphase and anaphase. The term heterochromatin was first used by Heitz in 1929 and referred especially to stainability. It appears that specific highly repetitive DNA sequences can affect the histone and scaffold proteins such that chromatin packing is pronounced even at interphase. This is so with *constitutive* heterochromatin that has a permanent character. It contains practically no coding DNA. The repetitive DNA is often composed of several families that are not or not frequently found in euchromatin and that can even be specific for certain chromosome segments. Transcription is practically impossible and replication is late.

Facultative heterochromatin is chromatin which is not consistently heterochromatic. Entire series of genes are shut off from transcription without further subtle regulation. The cell progeny usually "inherits" this condition from the stem line cell in which it has been induced. Several instances are known where facultative heterochromatinization results in the appearance of sectors in the tissue where certain genes are not expressed. The most thoroughly studied case is the inactivation by heterochromatinization of one of the two X-chromosomes during early embryogenesis in mammals. When the two X-chromosomes are equivalent, either one can be inactivated; and in some sectors the alleles of one chromosome and in other sectors those of the other chromosome are expressed (Lyon 1963). Genes on autosomes, transferred to an X-chromosome by translocation, share this fate. These phenomena are not with certainty known to occur in plants.

Heterochromatin or heterochromatin-like segments in chromosomes, because of their special packing, can be made visible by special methods of preparation. Especially after cold treatment in some species of plants (*Trillium* spp., for instance), heterochromatin may be recognised as light-stained regions in otherwise dark-stained chromosomes (Darlington and LaCour 1940). Removal of the least strongly bound proteins followed by local denaturation and subsequent restabilisation of the chromosome structure can result in loss of stainability of the segments least compacted. The remaining compact and stainable segments can form a specific pattern. This is very helpful for the identification of specific chromosomes or even chromosome segments and will be further considered in Chapter 4, in which the karyotype is discussed.

2.4 Special Functional Elements in Chromosomes

Three elements with special functions and specific positions in the linear chromosome structure will be briefly considered: the *nucleolar organizing region* (NOR), the *centromere* and the *telomeres*. The centromere and the telomeres have functions that are restricted to and necessary for the chromosomes in which they occur and, therefore, are found on each chromosome. The nucleolar organizer has a function in the physiology of the cell as a whole and is often found in only one chromosome per genome, or at most in a few.

2.4.1 Nucleolus Organizing Region (NOR)

The NOR is a chromosome segment in which the multicopy gene (500–1000 copies) for the larger (18S and 28S) ribosomal RNA (r-RNA) fragments is located. Between the two sub-loci, a 5.6S segment is transcribed that is not included in the ribosomes. The gene coding for the third and much smaller RNA fragment of the ribosomes (5S-r-RNA) is also a multicopy gene, but is located elsewhere in the genome, either distributed over several loci or concentrated tandemly on one locus. The r-RNA is an essential part of the ribosomes, which are responsible for the assembly of amino acids into polypeptides with messenger-RNA as the template. The amount of RNA required is large and it must be present in all metabolically active cells. The r-DNA is not packed in nucleosomes, but occurs dispersed in special bodies, the nucleoli, where it is transcribed almost continuously during interphase before DNA replication. The dispersed state of the DNA in the nucleolus makes it invisible using most cytochemical methods, but very favourable for gene function studies.

Very often a heterochromatic segment is present near or almost at the NOR. During the mitotic contraction stages of the chromosomes, however, the NOR, if active during interphase, is often visible as a constriction. Then the nucleolus itself is normally not present: the DNA is taken up into the condensed chromosome body. The constriction is a convenient landmark in the chromosome in which the NOR occurs.

2.4.2 Centromere

The centromere plays a central role in the division of the chromosomes. In most organisms, each chromosome has one, but it has a compound structure. When active during nuclear division, the central part contains a proteinaceous element, the *kinetochore*. The kinetochore provides for the attachment and possibly some regulation of the microtubules that regulate or partly effect the movements of the chromosomes in mitosis and meiosis (see Sects. 3.1.1 and 3.2.2). Microtubules may have a variety of functions in the organism, especially in higher animals, but when attached to kinetochores, they exclusively serve to regulate the forces required for chromosomal movement. The typical kinetochore structure is present only during these stages and it is apparently attached to special DNA segments. These segments are repeats of relatively short sequences that may vary slightly between species and even within species between the centromeres of different chromosomes. This variability does not appear to reduce the functionality of the centromere.

Especially in yeast, the DNA sequence of the centromeres, mainly of chromosomes III and XI, have been analyzed (Fitzgerald-Hayes et al. 1982; Clarke and Carbon 1985). In yeast the most critical segment is only about 370 bp long with one very A+T-rich segment of about 80 bp. The DNA is not organized in nucleosomes. In higher organisms the base composition is comparable, but the centromere is very much larger. At least part of it may be associated with histones in nucleosomes. Its structure is morphologically symmetrical, as can be seen in prophase chromosomes where two or four chromomeres can be made visible (Lima-de-Faria 1956, 1986; Rattner and Lin 1987). Most eukaryote centromeres are large enough to be split into two parts that both are functional. A metacentric chromosome can thus be split into two acrocentric chromosomes, both having a functional centromere. In the analysis of eukaryote centromeres and kinetochores, effective use can be made of centromere-specific CREST auto-antibodies in the serum of skleroderma patients (Brinkley et al. 1985). These attach mainly to the peri-kinetochore elements of the centromeres.

There is some variability in the electron microscopic appearance of the active kinetochore of condensed metaphase chromosomes in higher organisms (see, e.g. Luykx 1970; Bostock and Sumner 1978). In some organisms it has a complex structure, one part having the shape of a ball to which the microtubules attach, the other the shape of a cup, associated with the DNA. In other organisms it has a trilaminar structure. There do not seem to be essential differences between animal and plant centromeres.

The centromere region contains more than the segment that forms the kinetochore. On both sides of the kinetochore-organizing region a short chromosome segment has the property of holding the two sister chromatids together even when in the rest of the chromosome they have begun separating. The importance of this function will be considered in Sections 3.1.1 and 3.2.2 on mitosis and meiosis. Sister chromatid cohesion is not affected by substances that inhibit microtubule polymerization (colchicine etc.) and that thereby prevent normal chromosome division. The chromatids separate normally when spindle development is inhibited but the separation is delayed. It appears that this separation is not entirely synchronous and that the centromeres have their specific properties in this respect (Vig 1983). Mutants affecting centromere cohesiveness can have drastic effects on the course of mitosis and meiosis (Lin and Church 1982; *ord* and other mutants in *Drosophila* and other organisms).

The centromere, especially the kinetochore region, is usually visible as a narrow constriction in the condensed mitotic chromosomes, especially when the cell has been cold-treated or treated with substances that prevent normal kinetochore function. Because the position of the centromere is characteristic for the chromosome, this *primary constriction* is a convenient morphological chromosome marker. The constriction of the chromosome associated with the nucleolus is called the *secondary constriction*.

As a result of chromosomal rearrangement, certain chromosomes may occasionally receive two centromeres. In most cases this results in irregularities during nuclear division because centromere action is not integrated and the orientation of the sister parts is independent. When close together, the kinetochores on the same chromatid may tend to orient to the same pole, which results in regular division. Sometimes one of the two centromeres is inactivated. When inactivation is complete, no constriction is visible and some specific centromere proteins are absent. When inactivation is incomplete, one of these proteins may be present and then the constriction may be visible, but no clear microtubule association is observed. Apparently, the formation of the

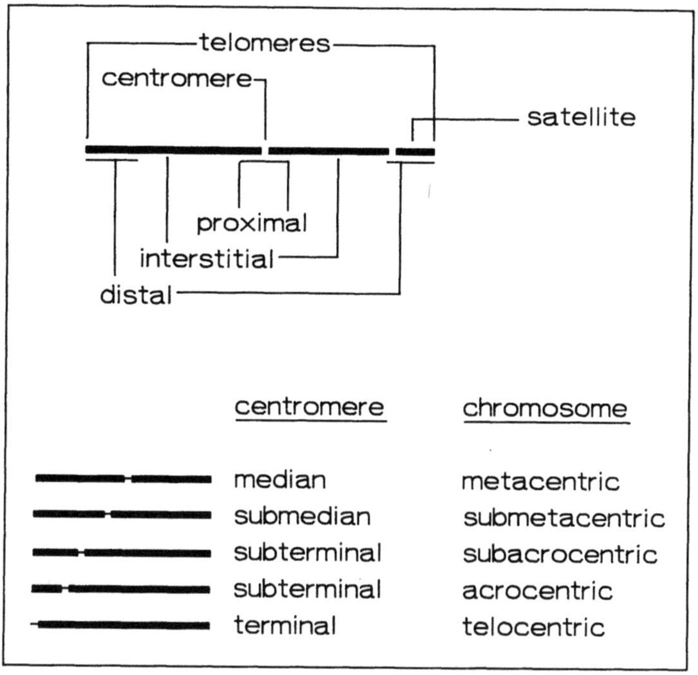

Fig. 2.3 Chromosome nomenclature. *Above* Terms used for the different segments of a chromosome in relation to the centromere. *Below* Terms used for centromeres and chromosomes in relation to the location of the centromere in the chromosome

constriction and other aspects of activity are regulated by different systems (Earnshaw and Migeon 1985).

Kinetic activity outside the normal centromere, usually in a distal position and without visible constriction, is found in special genotypes and favoured by special environmental conditions. These centres of secondary activity are called *neocentromeres* and have been described in several species, especially in meiosis of plants. In maize with an abnormal, heterochromatic chromosome 10, they may even become stronger during anaphase than the original centromeres (Rhoades 1952). In rye a special heterochromatic terminal segment in chromosome 4R similarly promotes neocentric activity (Kavander and Viinika 1987). Neocentric suppression in most normal genotypes may involve the same mechanisms as those suppressing additional normal centromeres in dicentric chromosomes. Neocentromeres may represent a remnant of an original holokinetic condition (Sybenga 1981).

Special terms are used for the position of the centromere: *median*, *submedian*, *subterminal*, *terminal*; and the chromosomes where the centromeres have these positions are called *metacentric*, *submetacentric*, *subacrocentric* and *acrocentric* respectively. *Telocentric* is a special case in which the centromere is at the chromosome end. These terms are illustrated in Figure 2.3.

The centromeric activity is not necessarily always confined to a strictly localized segment of the chromosomes. In several mostly primitive higher organisms, both plants and animals, the kinetic activity is distributed evenly over the chromosomes (*holokinetic* chromosomes) and then no primary constriction is visible. This is probably the original situation, and the localization of the kinetic activity is an evolutionary development that has taken place parallel in different taxonomic orders (Sybenga 1981). The possibility of reversal from monokinetic to holokinetic activity, however, cannot be excluded. The structure of the holokinetic kinetochore is necessarily somewhat different, but kinetochore plates and other components known to be associated with monokinetic kinetochores have also been observed in holokinetic chromosomes (Benavente 1982).

2.4.3 Telomeres

Telomeres are structures at the ends of the chromosomes that prevent the free ends of the DNA strands from being attacked by nucleases and from fusing with any other free DNA end that may occur in the nucleus, and they enable the terminal end of the chromosome to undergo replication. They apparently also play a role in temporarily associating non-homologous chromosome ends with each other (Wagenaar 1969) and in attaching chromosome ends to the nuclear membrane. The latter phenomenon is quite common and is assumed to play a role in the preparation for chromosome pairing at meiosis (Bostock and Sumner 1978). Because palindromic DNA sequences (inverted repeats) have often been found in telomeric heterochromatin, it has been postulated that the hairpin foldbacks such repeats can form would play a role in the many telomere functions (Cavalier-Smith 1983; Struhl 1983). Theoretically they could protect DNA ends and provide conditions for initiating replication. In artificial yeast chromosome ends, however, simple hairpins do not function. Moreover, in the most terminal regions of typical end segments, palindromes are not observed (Cavalier-Smith 1983; Richards and Ausubel 1988). It is now assumed that a template-independent terminal transferase adds a G-rich segment to the 3′ end that folds back upon itself to prime DNA replication of the C-rich strand. The terminal segment of the telomere in the cruciferous plant *Arabidopsis thaliana* appears to be very similar to that in lower eukaryotes, demonstrating an extreme evolutionary conservation (Richards and Ausubel 1988). There are up to 350 blocks of 7 bp (CCCTAAA) at each telomere of *Arabidopsis*, and maize and human telomeres are very similar. There is, however, considerable, heritable size differentiation between the telomeres of different chromosomes, even within species. This may be a consequence of incidental growth and diminution resulting from irregularities in the attachment of terminal segments during replication. In Trypanosomes this irregularity is systematic so that the chromosomes can grow during every replication cycle, but also shrink again (Bernards et al. 1983).

When chromosomes break, the broken ends do not have a telomere and are unstable. They may attach to other broken ends when available or they may degrade. Occasionally, however, the end is stabilized by the formation of a new telomere. This may find its origin in a telomere sequence present in low copy numbers in some locations in the chromosome. When occurring near the break, DNAse activity may remove the DNA up till such a sequence, and the telomere may, as in Trypanosomes, start growing during the next replication cycle, or even unscheduled, induced by the break. Healed ends have been observed in several places in the chromosomes, for instance after breakage of anaphase bridges resulting from dicentric chromosomes induced by radiation, or after exchange in inversion loops (cf. Sect. 5.3.4). In some tissues, for instance maize endosperm, such breaks do not heal, and when new chromatids form after the next DNA replication cycle, these fuse at the raw ends. A new bridge is formed in the next anaphase and the cycle starts over again: the breakage-fusion-bridge cycle (McClintock 1938).

Healed ends are observed most frequently after breakage in or near centromeres that have been torn apart as a result of bipolar orientation in meiosis (Darlington 1965). The result is a telocentric chromosome. Richards and Ausubel (1988) found that the new telomere near or at the centromere in a newly formed telocentric chromosome of *Arabidopsis thaliana* (see Koornneef and van der Veen 1983) had the same composition as the original telomeres, but with a smaller number of repeats. The potential of broken ends to heal instead of leading to a gradual degradation of the chromosome is of considerable importance for chromosome manipulation. Still more important is the potential of broken ends to associate with other broken ends, and even for merely damaged chromosome segments to associate with similar lesions in other chromosomes, ultimately resulting in chromosomal rearrangements. Its application will be considered in Chapter 10 on chromosome manipulation.

In several organisms, heterochromatin is associated with the telomeres of some or even all chromosomes. Its composition may be quite variable, although within a species, the same families of repetitive DNA may occur in all or most telomere-associated heterochromatin. There is no apparent relation to regular telomere functions.

2.5 Microscopic Chromosome Morphology: the Karyotype, Standard and Variations

The characteristics of the genome in terms of the number of chromosomes in which it is packed and the microscopic morphology of each of these chromosomes are called the *karyotype* (Chap. 4). The morphology of the chromosomes traditionally includes the length and the location of the primary constriction and the secondary constriction, if present. The karyotype of a species is remarkably constant and can therefore be a useful characteristic in taxonomy (cytotaxonomy). Yet occasionally, deviations from the standard type occur. Some are merely the result of random variation in contraction that can be dealt with by the proper statistical approach. These are not of interest in the present context. Other deviations are more fundamental. Some result from the presence or absence in certain individuals in a population of heterochromatic or otherwise dispensable segments. These *polymorphisms* are useful markers but otherwise usually of little consequence. Still other deviations may be the result of chromosomal *rearrangements* involving genetically essential chromosome segments. These too may occasionally "float" in the population as polymorphisms, but they then usually have specific consequences. They may also occur as the result of *accidents* or may be induced by ionizing radiations or certain mutagenic chemicals. These are usually rapidly eliminated from natural populations because of deleterious effects on the genetic balance or on meiosis (Chap. 5). In experiments they may be recognized and perpetuated if desired and if sufficiently balanced for transmission. At their origin, rearrangements may result in chromosomes with two centromeres or without centromeres, and these are readily eliminated in mitosis. Others result in somatically entirely functional chromosomes (Fig. 2.4).

Stable chromosomal rearrangements may affect chromosome morphology. The four main types are (cf. Chap. 5):

1. *Deficiencies* (also called deletions, and then often involving terminal segments) where a chromosome segment, either terminal or interstitial, has been removed. When large enough, the chromosome will be recognizably smaller.
2. *Duplications*, where a chromosome segment is present in more than one copy in a genome. It may be found in several locations: tandem, and then either in the same direction or reversed; in the same chromosome but removed from the original copy; in another chromosome in various orientations.
3. *Translocations*, where a chromosome segment is displaced. Several types exist. The most common is the *reciprocal translocation* or *interchange* where two chromosomes have interchanged a terminal segment (Fig. 2.4). When the segments are different in size, this may result in an observable change in chromosome morphology. A segment may also have changed position interstitially, either within a chromosome (*shift*) or towards another chromosome (simple or *interstitial translocation*). More complex types of translocation are possible.
4. The fourth type of rearrangement is the *inversion* (Fig. 2.4). Two types are distinguished: the *pericentric inversion*, which has the centromere inside the inverted segment, and the *paracentric inversion*, where the centromere is outside the inverted segment. The latter will not result in a change in length of the chromosome arm in which it occurs.

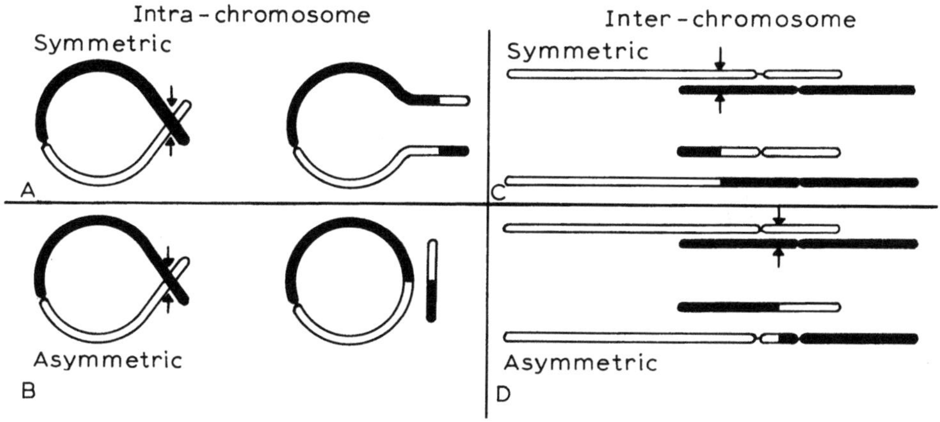

Fig. 2.4 Chromosomal rearrangements and their origin. **A, B** Breaks within one chromosome. **A** results in a pericentric inversion (cf. Fig. 5.4), **B** in an unstable ring chromosome and a non-transmissible acentric fragment. **C, D** Interacting breaks in two different chromosomes. **C** results in an interchange (reciprocal translocation, ct. Fig. 5.7), **D** in a dicentric, unstable chromosome and a non-transmissible acentric fragment

Some chromosomes are distinguished from normal chromosomes by having a special function. The most common, especially in animals, are the *sex chromosomes*. There are two sexes, one is the *homogametic* sex, which has two identical sex chromosomes, and the other is the *heterogametic* sex, which has one of these chromosomes and in addition a sex chromosome of a different type. The most common is the X-Y sex-determining system, where females have two X-chromosomes (the homogametic sex) and males have one X- and one Y-chromosome (the heterochromatic sex). In several insects the Y-chromosome is absent, and males have a single X-chromosome (X-O system). The homogametic sex makes one type of gamete: all with one X. The heterogametic sex makes two gametic types: one with an X and the other with a Y. The consequence is that half of the progeny have two X-chromosomes (females) and the other half, one X and one Y (males). In birds and a few more groups of animals the system is reversed: males are the homogametic sex and females the heterogametic sex. The sex chromosomes are here called W and Z.

Sex chromosomes are usually structurally somewhat different from normal chromosomes (called *autosomes*). Especially the Y-chromosome is largely heterochromatic and can be occasionally larger than the autosomes (*Drosophila*), but is usually considerably smaller. The X-chromosome often has a recognizable structure also, and is partly or largely heterochromatic. In female mammals one of the two X-chromosomes is heterochromatinized in half of the body cells, the other in the other half of the cells, giving rise to mosaicism for X-chromosome genes (Sect. 2.3).

In plants with sexual dimorphism (e.g. *Rumex*, spinach, hemp, asparagus), the sex chromosomes are usually much less differentiated and recognition in the karyotypes is often difficult.

A second category of special chromosomes are the *B-chromosomes* or *accessory chromosomes*. These have no apparently useful function for the organism, but appear to exist for themselves only. They usually contain genes that have an effect on the internal or morphological phenotype. They have a mechanism that allows them to accumulate during the sexual cycle or, in plants, during pollen or embryo sac mitosis (Sect. 3.1.4.1.2.3). This compensates for their lack of useful genes or even negative effects. Their morphology is usually sufficiently different from that of the autosomes (or A-chromosomes) so that they can be recognized in the karyotype.

This summarizes the main structural features of the chromosomes, of which the somatic and generative transmission will be considered in the following chapters.

Chapter 3
The Mechanisms of Genetic Transmission

3.1 The Somatic Cycle

3.1.1 Mitosis

Somatic cell multiplication is realized by cell division, resulting in two equivalent daughter cells. In higher organisms cell division normally is preceded by nuclear division, almost without exception following a strict pattern: *mitosis*. This is the process basically responsible for chromosome (and thus gene) transmission. The sequence of mitotic stages is presented diagrammatically in most textbooks on cytogenetics and general genetics, often combined with photomicrographs of the different stages. A well-known series of photographs of mitosis and meiosis in *Lilium* is that of McLeish and Snoad (1958 and later editions). In Fig. 3.1 the mitotic stages are diagrammatically represented on a horizontal line. When both daughter nuclei divide again, the line branches, which is not shown in the diagram. Often, mitotic division is represented as a cycle, which does not represent the actual situation.

The first prerequisite for somatic nuclear division is the replication of all chromosomal DNA in the nucleus, accomplished in the synthesis (S) phase. In rapidly dividing tissues the S-phase is short and follows soon after completion of the previous division. In more slowly proliferating tissues the S-phase is longer and there is a resting phase between divisions in which transcription can take place. The faster the DNA replicates, the more independent origins of replication are activated. The stages between DNA synthesis and mitosis in which very little can be seen to happen have been named "gap", abbreviated G. The stage between the previous mitosis and the S-phase is G1, and that between the S-phase and the following mitosis is G2. The transcription phase just following mitosis is often called G0. After a cellular "signal" to the nucleus that mitosis is to start (Fig. 3.1), G1 proper is initiated, which is automatically followed by S. G1 can thus be considered a preparatory stage for DNA synthesis.

Once DNA synthesis has started, it will continue to completion unless: (1) the necessary enzymes are blocked artificially; (2) the necessary precursors are not available; or (3) specific segments are protected from replication. At completion of DNA replication, the nucleus arrives in G2. During G2 the

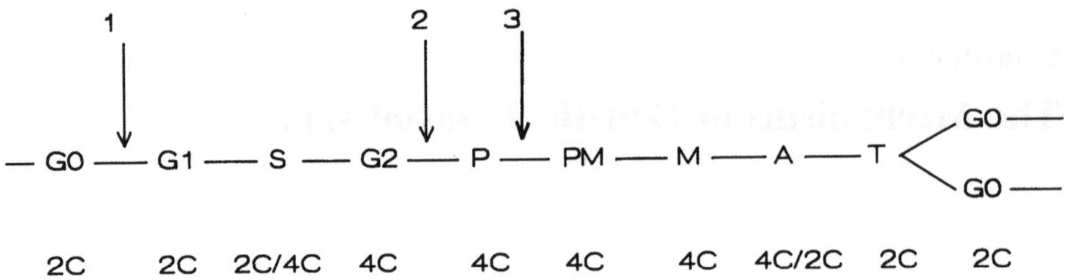

Fig. 3.1 The mitotic sequence line. *Arrows* indicate the signals necessary to continue mitosis. If the signal is not given, the cell remains in or reverts to interphase. If signal *1* fails, *GO* continues; if *2* is not given, endoreduplication follows; without *3* there is endomitosis. The latter two result in polyteny or polyploidy respectively

nucleus can again adopt a resting state, but the doubled DNA is not necessarily transcribed. Protein replication immediately follows DNA replication, so that during G2 two complete strands per chromosome are present. It is said that the nucleus now contains the "2C" amount of DNA per genome (Bennett and Smith 1976) in contrast to the 1C amount per genome between completion of mitosis and DNA replication. Since there are usually at least two genomes in somatic tissues (one from the mother and one from the father), a nucleus normally contains the 2C amount before DNA synthesis and a 4C amount after.

In normal tissues G2 is short and a second mitotic signal (Fig. 3.1) causes the chromosomes to condense. The two strands condense simultaneously but separately, although still remaining closely associated. They become *chromatids*, the new daughter chromosomes, as they are called after separation during the last stages of mitosis. Occasionally, however, the condensation signal is not given and after some time a new replication signal follows. Then there is an 8C amount of DNA in the G2 nucleus. The chromatin strands do not separate and when the process is repeated several times, multistranded, *polytene* chromosomes are formed by *endoreplication*. This has been observed in a number of glandular tissues of the Diptera. In *Drosphila* salivary glands, for instance, a ten-fold series of replications leads to a giant, uncondensed chromosome with 1024 strands that can be studied in great detail under the light microscope, especially because alternating dense and less dense chromosome segments form a banding pattern. Chromosome segments with transcribed genes are decondensed and consequently looser in structure; they become recognizable as *puffs*. Polytene chromosomes have been of great importance for gene physiology studies and cytogenetics.

In plants polytene chromosomes have been observed in specialized cells or tissues, e.g. in the embryo suspensors of *Phaseolus* beans and in synergids in embryo sacs. They are not as favourable for study as dipteran polytene chromosomes, but can be used for karyotype and gene physiology studies (Nagl 1969, 1978). After several cycles of endoreduplication, cell division is usually not possible: polytene nuclei are at a dead end. However, in some tissues, e.g. in leaf mesophyll, in which only one endoreduplication cycle has taken place, the cells remain capable of division, but only when it is artificially induced. In tissue culture such 8C leaf mesophyll cells may proliferate to polyploid callus or embryoids. During the first division the two sister chromosomes formed after endoreduplication, each containing two chromatids, can be seen to be connected: *duplochromosomes* (Pijnacker et al. 1986). At metaphase the four chromatids each have an active centromere, which can show complex types of orientation (Pijnacker and Ferwerda 1990), but subsequent segregation is usually into two sets of two chromosomes. Only a fraction of the proliferating mesophyll cells is polyploid. Polyploidy observed in cultured cells or callus is usually not due to endopolyploidy of the explant material, but to disturbance of mitosis during the callus phase.

By substituting bromodeoxyribose uridine (BUdR) or a similar base analogue for thymidine during the last two replication cycles and staining with Giemsa, it is possible to distinguish the chromatids having two new strands from those having one new and one old strand. The double-substituted strand stains more lightly than the single or non-substituted strands. It appears that, in duplochromosomes, the two chromatids with the two new strands are found on the outside. Chromatid separation apparently is not random (Pijnacker et al. 1986).

In normal meristems the signal for condensation follows automatically in G2. Histones are phosphorylated, particularly a special mitotic post-synthesis modification of H1 called H1m. Another H1 variant, H1s, plays a role in the regulation of transcription. H3 phosphorylation is also important for chromosome condensation, but not as much so. Condensation is regulated at the cellular level.

When an interphase cell with dispersed nuclear chromatin is fused artificially with a cell with condensed chromatin, the chromatin of the interphase nucleus condenses prematurely and very rapidly before the nucleus is actually ready for it (Johnson and Rao 1970). This "premature chromatin condensation" (PCC) results in fragmentation of the chromatin, especially during the S-phase, which can severely damage and disrupt the chromosomes. It has been observed in plants as well as in animals.

A nucleus that has received the signal to condense will normally start division, but not always. Instead, the chromosomes may wait in a condensed state for some time and finally decondense again. The nucleus may receive another signal to replicate its DNA and the process is repeated. In this case polyteny will not result because the chromatids have started to separate as a result of condensation. Because the chromosomes go through a condensation cycle, this phenomenon is called *endomitosis* and leads to *endopolyploidy*. In plants endopolyploidy has been observed, e.g. in endosperm tissue (Fig. 3.2) and in tapetum cells of several species, in which the resulting giant cells are very effective sustaining tissue for the developing embryo and the pollen mother cells respectively. They have only a short life. Endopolyploidy was

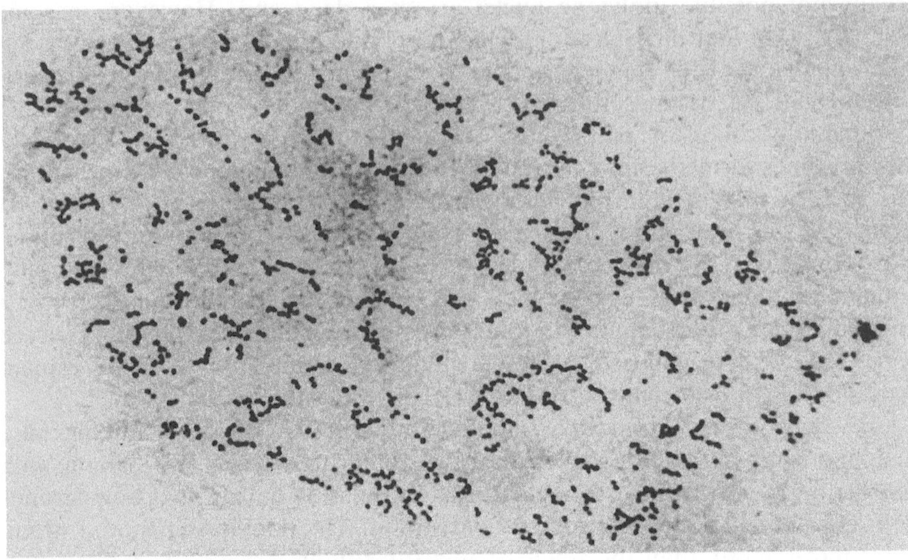

Fig. 3.2 High degree of endopolyploidy as a result of endomitosis in the short-lived endosperm of *Cucurbita pepo*. Note the rows of identical daughter chromosomes derived from the same mother chromosome in a series of replications not followed by nuclear division. (After Varghese 1971; cf. Sybenga 1972)

described earlier in several tissues of the waterstrider, *Gerris lateralis* (Geitler 1939).

In normal meristems, chromosome condensation is followed by a new signal (Fig. 3.1) that leads to completion of mitosis when no special measures are taken. During these stages of mitosis a large number of changes take place in the cell. The nuclear membrane disappears and the condensed chromosomes then lie freely in the cell. Because of their compactness and the (probable) presence of a special membrane ("matrix"), whose nature is not known, this is not deleterious. In some material it has been shown that microtubular material from the cytoskeleton is deposited as a band around the nucleus: the preprophase band. The position of this band determines the direction of the division axis and is, therefore, an important factor in development. Perpendicular to this band, a spindle-shaped structure develops with a pole at each end. At the poles a protein body, the centrosome, is formed in the cells of most animals, very exceptionally in those of plants. At its center a darkly stainable smaller body, the centriole, can sometimes be seen. It follows a special division cycle on the outside of the nuclear membrane before it organizes the centrosomes at the spindle poles; and it is the centriole from which the flagellae of the sperm cells develop during spermatid differentiation. Higher plants do without this. Between the poles, those with centrosomes or without, bundles of *microtubules* (or *microtubuli*) are formed, some of which run from pole to pole, others from pole to kinetochore. The mitochondria collect in bundles outside the spindle area and as a consequence the spindle has practically no metabolic activity. Through diffusion, its oxygen content is, therefore, higher than in the remainder of the cell. The pH rises somewhat and the free Ca^{2+} and Mg^{2+} ions are apparently removed: these are unfavourable for microtubular function. The amount of bound divalent cations increases. Some of these changes, such as that of the viscosity in the spindle area, were observed many years ago (cf. Swanson 1957).

The microtubules that bring about the movement of the chromosomes have been analyzed both biochemically and by the electron microscope (see, e.g. Bostock and Sumner 1978; Brinkley et al. 1985; McIntosh 1985). They are hollow cylinders of about 25 nm in cross-section after glutaraldehyde fixation, but of only 15 nm after osmium tetroxide fixation. The cylinders are composed mainly of 3–4 nm dimers of tubulin, arranged in 10–13 slightly slanting (10–20°) threads around a hollow center. Tubulin comes in two basic forms: α and β, each of about 55 kDa. They differ somewhat in amino acid composition and phosphorylation, and consequently in electrophoretic properties. The dimers are probably predominantly of α–β composition, less frequently of αα or ββ. The β-tubulin is very conservative and almost identical in all higher organisms, whereas α-tubulin is more variable.

Substances that specifically disturb spindle function, such as colchicine, vinblastine, benomyl and many more, bind at specific places on the tubulins or on smaller proteins that function in tubulin polymerization and thereby inhibit microtubule polymerization. Existing microtubules are usually not degraded by such agents. Tubulin polymerization and depolymerization are in balance under normal conditions: the microtubules are not permanent structures. An object, even a chromosome, can move straight through the spindle without disturbing it, and a mechanically or chemically disturbed spindle repairs itself effectively.

It still is not clear how the forces of chromosome movement are exerted, but it is probable that the microtubules serve primarily to regulate and direct these forces. Relatively large amounts of ATP are associated with the microtubules, but this ATP does not transfer appreciable quantities of energy-rich phosphate: chromosome movement requires little energy.

The mechanism of anchoring the microtubules to the endoplasmic reticulum around the poles is not understood. The attachment of the microtubules to the chromosomes is concentrated in the kinetochores in monokinetic chromosomes, but is distributed over the chromosomes in holokinetic chromosomes. When the microtubules attach to the kinetochores, the chromosomes are already double and each chromatid organizes its own complete kinetochore. This means that some spindle microtubules attach to one daughter kinetochore and others to the other. In the beginning the number of microtubules attached to the kinetochores is small; it increases during prometaphase-metaphase.

The first visible stage of mitosis (Fig. 3.1) is *prophase*, when the chromosomes begin condensing within the intact nuclear membrane. In some organisms the spindle may begin developing while the nuclear membrane is still intact; in some protists the entire nuclear division takes place within the membrane. These are exceptions. About the time the nuclear membrane disintegrates, the nucleolus disappears as a separate body and the nucleolar DNA is taken up in the chromosomal body; all that remains visible is a constriction in the condensed chromosome. With the onset of spindle activity and the start of movement, *prometaphase* begins. At the beginning of prometaphase the chromosomes appear to move rather haphazardly about the cell and the two halves of the centromeres are not yet coordinated.

When a moving kinetochore meets a counterforce, its movement and the orientation of this movement tend to be stabilized. This is an essential feature of mitotic stability, but the character of this stabilization is unknown. A counterforce is exerted on a moving kinetochore by its sister kinetochore when it also is pulled at by a polar force. There are two conditions: there must be some form of association between the two kinetochores and the forces must act in opposite directions. The association is realized by pericentromeric chromosome segments where the chromatids stick together more strongly than in the rest of the chromosome. Oppositely oriented forces can only come from opposite poles. When two sister kinetochores are associated with the same pole, they will not stabilize each other's orientation. Sooner or later the pulling forces will lapse or disappear as a consequence of microtubule depolymerization. When, by accident or otherwise, the microtubules attached to two sister kinetochores come from opposite poles, the forces will be stabilized. Even then, however, reorientation remains possible for a considerable length of time. One cause may be chance depolymerization when only a few microtubules are attached. Another may be the early approach of one of the kinetochores to one of the poles. This reduces the pulling force, because it is proportional to the distance between the kinetochore and the pole, i.e. the longer the microtubule bundle, the stronger the force exerted (Östergren 1951). As soon as the pull from one pole relaxes, the microtubules pulling the sister chromatid towards the other pole are also destabilized. Reorientation now becomes possible.

Gradually the centromeres of the chromosomes move to a position between the poles where the forces on the two kinetochores are equal: the centromeres are there at the *equator*. During this process, the number of microtubules attached in bundles to the kinetochores increases. When the centromeres of all chromosomes are at the equator, the stage is called *metaphase*. It may take many minutes, occasionally quite long, until the association between the chromatids is released, usually almost simultaneously for all chromosomes, by a sudden, biochemical change in the cell. The half-kinetochores then move to the poles, dragging the chromosome arms behind them: this is *anaphase*.

The delay of chromatid separation until all chromosomes have their sister kinetochores stably oriented to opposite poles guarantees a correct distribution of a balanced set of chromosomes to each pole and consequently the conservation of an intact genotype with each cycle of chromosome transmission. Mitosis is a thoroughly conservative process and seemingly of little interest for manipulation of the genotype. As will be shown later, there are important exceptions that permit genetic changes to be induced, selected and perpetuated.

Anaphase movement ends when the kinetochores reach the poles. Even then, chromosome movement does not stop. Unknown forces may cause the chromosomes to move slightly farther and pull the chromosome arms into the group without kinetochore action. The chromosomes thus collect into a roundish body and the contraction is gradually undone: this is *telophase*. A new nuclear membrane is formed and the nucleolus returns. Two new *interphase* nuclei are formed.

3.1.2 Duration of Mitosis

The entire duration of mitosis is quite variable and depends on the organism, the tissue and external conditions. There are three main methods for determining the duration of mitosis or its separate stages:

1. Visual observation of living material. This is in principle possible in isolated cells or small cell aggregates under phase contrast or other systems of microscopy that permit the observation of unstained material that is, as much as possible, undisturbed. There are only few plant tissues where this is possible, e.g. endosperm and pollen mitosis, although in the latter, observation is difficult and there is no guarantee that the process takes a natural course. Some cell and tissue cultures can also be observed in living condition under the microscope, but in general, animal material is more favourable.

2. Application of a label (e.g. tritiated thymidine) for a short time, followed by thorough washing (pulse labelling). If the label is taken up, the cell is necessarily in S-phase during application. By making micro-autoradiograms at specific intervals after application and scoring the mitotic stage the labelled cell is in, the time between labelling and that stage can be accurately determined (Fig. 3.3). The mitotic process can be followed for more than one cell generation. Cells do not divide synchronously, and divide at different rates, so some will reach a specific stage before others. There is, however, a large group that goes through mitosis at about the same rate and these cells will reach a defined stage more or less as a group. This results in a frequency peak after the average interval required to reach that stage. After some time these cells leave mitosis; later they will enter the second cycle and a new peak appears. The length of time between the peaks is the average length of a complete cycle, including interphase.

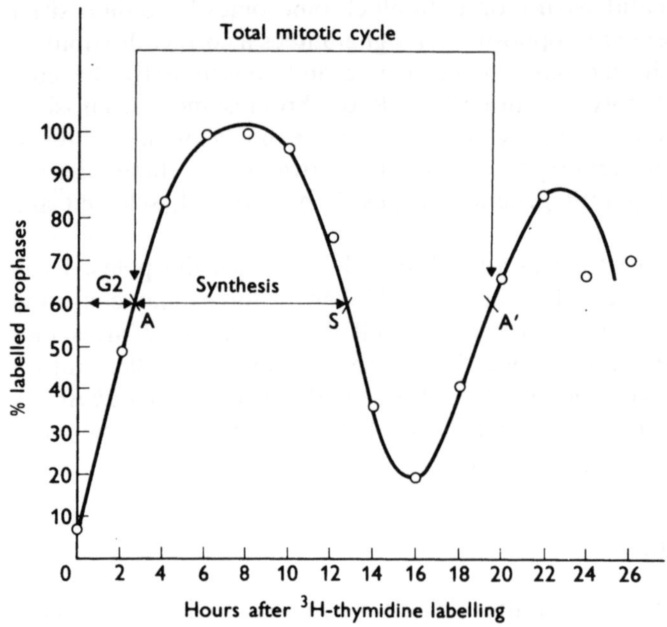

Fig. 3.3 Timing of the mitotic cycle in *Nigella damascena*. (Rees and Jones 1977)

3. Blocking mitosis at a specific stage, e.g. just before S-phase, by applying 5-amino-uracyl or hydroxyurea. Removal of the agent after an appropriate period of time permits the mitotic cycle to restart. By scoring the stage of the cycle at specified intervals after removal of the agent, the duration between S-phase and the observed stage can be determined.

It appears that there is a correlation between DNA content and the duration of the mitotic cycle. In addition, there is considerable genetically determined variation in mitotic cycle time, and internal (mainly tissue-determined) and external (temperature) factors play an important role. From prophase to telophase, a complete cycle in *Tradescantia* microspores at 30 °C takes about 30 h. In stamen hairs it takes only about 1 h, but at 10 °C, more than 2 h (Swanson 1957). It is not always easy to compare different observations, because the onset of prophase and the end of telophase may be difficult to determine, and different observers may have different criteria.

3.1.3 Order and Disorder in the Somatic Spindle and Nucleus

In principle, mitotic chromosome behaviour is very regular and leads to an exact replication of existing nuclei. This orderly and conservative behaviour is based on mechanisms that are very similar or practically identical in a wide range of higher organisms. The way in which the different forces operate during mitosis not only results in exact nuclear replication, but also appears to affect the arrangement of the individual chromosomes in the spindle, and therefore may have consequences for the order of the chromosomes in the newly formed nucleus. Slight variations in the mitotic processes will not normally have noticeable consequences. When mitosis is more seriously disturbed, the specific way individual chromosomes behave on the spindle and subsequently become arranged in the nucleus may have interesting consequences. For artificially manipulating chromosome transmission in mitosis, this potentially could be, but is not yet, widely exploited. Different types of systematic variation in the order of the chromosomes on the spindle and in the nucleus can be distinguished.

3.1.3.1 Relative Position of Large and Small Chromosomes in the Spindle and in the Nucleus. Hollow Spindle

In organisms with a combination of large and very small chromosomes (e.g. several birds and some species of insects), the large chromosomes are positioned on the periphery of the spindle and the small ones in the centre. In several insects with large chromosomes exclusively, the centre of the spindle is empty (hollow spindle, cf. Swanson 1957) except for occasional sex chromosomes. In both cases the centromeres point to the centre of the spindle and the arms seem to be pushed outside. The forces responsible have not been identified. This position is maintained during anaphase and telophase, and in the resulting nucleus, a similar relative position of the small and large chromosomes is maintained. Now the centromeres point to one pole and the arms to the other side of the nucleus.

3.1.3.2 Rabl Orientation

The telophase orientation of the chromosomes, with the centromeres pointing to one side of the nucleus and the telomeres to the other, is maintained during the entire interphase and can still be recognized in the next prophase nucleus when the chromosomes recondense: this is called *Rabl orientation*, named after the discoverer. When heterochromatin is concentrated in specific chromosome segments, e.g. near the centromeres and/or the telomeres, concentrations of heterochromatin are formed in specific areas of the nucleus, and these may fuse into *chromocentres* that dissolve again during prophase. The reason for heterochromatin fusion is not clear, but could possibly be a high degree of homology between highly repetitive DNA in the heterochromatin of different chromosomes. The fact that, most of the time, the heterochromatin in the resting nucleus is condensed, and only little of the DNA is exposed and even less is single-stranded, does not favour this hypothesis, however.

3.1.3.3 Nucleolar Fusion

Like heterochromatic segments, but perhaps easier to understand because of the dispersed nature of the repetitive DNA, nucleoli tend to fuse during interphase. An additional reason may simply be their relatively large size, which promotes contact and initial membrane fusion. As a result of nucleolar fusion, the nucleolar chromosomes tend to be closer together than other chromosomes. Their secondary constrictions are often entangled. This has frequently been observed in rye (own unpublished results) and is common in human somatic cells. It may occasionally lead to breakage.

3.1.3.4 Somatic Pairing

In several organisms the homologous chromosomes tend to be positioned nearer to each other than is expected with random distribution. In some organisms this can result in the close somatic pairing of homologues, as in dipterous insects. In most organisms homologous somatic pairing is weak or completely absent, and statistical methods are necessary to determine whether homologues are actually closer together than when randomly positioned. In some genotypes, homologous somatic pairing is closer than in others and this may have consequences for meiotic pairing (see, e.g. Brown and Stack 1968; Avivi et al. 1982).

3.1.3.5 Non-Homologous Chromosome Association. Nuclear Compartmentalization

In addition to association between homologous or even homoeologous chromosomes, specific non-homologous chromosomes have been suggested to be associated in somatic tissues, mostly on the basis of chromosome-arm length (Bennett 1982). Later (Callow 1985; Dorninger and Timischl 1987), the statistical significance of the deviation from random distribution was contested. At meiosis, the segregation of unpaired chromosomes on the basis of size can be demonstrated in *Drosophila* (Grell 1964), although apparently not in plants.

More clearly demonstrated is the separation of entire genomes (Finch et al. 1981). Especially in hybrids between species that are not closely related, the genomes appear to take special positions. In the hybrid between rye (*Secale cereale* L.) and barley (*Hordeum vulgare* L.) as studied by Finch et al. (1981), the chromosomes of rye can be distinguished from those of barley because they are larger. Banding was not possible in these unsquashed cells.

The nucleus was sectioned in ultrathin sections for electron microscopy, the chromosomes were measured and the nucleus was reconstructed using a computer program. The barley chromosomes appeared to be positioned in the center of the nucleus, the rye chromosomes in the periphery. It was was originally supposed that this was a result of differences in size, although slight size differences do not necessarily have an effect on chromosome position within genomes. Bennett (1988) reported that at prometaphase in the proembryo of a hybrid between *Hordeum vulgare* and *H. bulbosum*, the centromeres of the *vulgare* chromosomes started activity before those of the *bulbosum* chromosomes. According to Pohler and Claus (1985), the cause of genome separation in hybrids between *Hordeum* species and between *Hordeum* and *Secale* was a genomic difference in the timing of anaphase separation. The chromatids of *Secale* chromosomes separated later than those of *Hordeum* and were positioned at the periphery of the nucleus. Other irregularities occurred also. Differences in centromere separation have repeatedly been reported by Vig and coworkers (for a review, see Vig. 1983) and may well play a role in chromosome positioning in the resulting daughter nuclei.

The recent developments with confocal scanning laser microscopy greatly facilitate the exact localization of chromosomes at metaphase and occasionally at prophase in intact and even living nuclei of favourable material. Using this method, Oud et al. (1989) could show that in root tips of *Crepis* species the chromosomes were not positioned randomly in the nucleus. Specific systems of chromosome order were not discovered. Rickards (1988) ascribed apparent meiotic pairing difficulties in heterozygotes for an interchange in *Allium triquetrum* to the fact that the translocated homologous segments were forced into unusual positions in the nucleus. From here they had problems associating.

Bennett (1982) suggests that the position of the chromosomes relative to each other may have consequences for the interaction between their genes or gene products. As a result of Rabl orientation and other systematic arrangements of the chromosomes, specific genes are always found in each other's neighbourhood, which facilitates interaction. The position within the nucleus may also have effects: in human fibroblasts, chromosomes at the periphery of the nucleus replicated later than those in the center (Ockey 1969), and condensation was more pronounced. It is probable that this also affects transcription. When such genes fail to interact properly, either by translocation between non-homologous chromosome segments or by disturbance of the system of (relative) chromosome positions (for instance in species hybrids), this may have consequences for the physiology of the organism. The importance of such interactions is difficult to ascertain. Proximity to the nuclear membrane is apparently important for transcription.

The relative positions of genomes and chromosomes are of considerable importance, not only for cell physiology, but also for induced or spontaneous deviations from the normal course of mitosis (in vitro cultures, treatment with mitosis-disturbing substances) as well as for meiotic chromosome pairing. In the next section and in Chapters 10 and 11 on chromosome manipulation, the consequences of variation in the arrangement of chromosomes in the nucleus will be discussed further.

3.1.4 Mitotic Variants with Genetic Consequences

Mitotic variants can be spontaneous or induced. Only the first category (intact organism, in vitro cell, callus, tissue and organ culture) will be discussed in this chapter. Induced variants will be considered in later chapters. The genetic consequences of variant mitosis are expressed as *somatic segregation*: the sector of the organism formed by the progeny of one daughter cell has another genetic composition than the sector composed of the progeny of the other daughter cell. The result is *chimerism*.

3.1.4.1 Intact Organisms

Spontaneous variants in intact organisms occur in two forms: *accidents* and *systematic* (natural) *variants*.

3.1.4.1.1 Accidents

These occur continually as a result of imperfect control of the somatic cellular processes. Even in the most stable systems, occasional gene *mutations* and chromosome structural *rearrangements* are encountered. Except for a few chromosome structural rearrangements, gene mutations and chromosome rearrangements are supposed to result not from mitotic errors but from errors in DNA replication and repair. Infrequently, incorrect timing of chromatid separation or failure of spindle function under normal conditions result in the absence of a chromosome from one daughter cell and the presence of an extra chromosome in the other: this is called *non-disjunction*. Systematic *elimination* of specific chromosomes or genomes suggests a more active process, but it is usually based also on some form of non-disjunction. When anaphase separation fails for the entire cell, the result is *chromosome doubling*, as after endoreduplication and endomitosis.

Chromosomes, that tend to show parallel alignment or even close somatic pairing are liable to *somatic crossing-over*, particularly when they are homologous (Fig. 3.4; cf. Jones 1937). In heterozygotes this may lead to somatic segregation of genes. It is observed as a sector in which the recessive phenotype is expressed, in the middle of an otherwise normal tissue where the dominant allele is expressed. When the heterozygote has an intermediate gene expression, a *twin spot* may appear: one sector with the recessive and another,

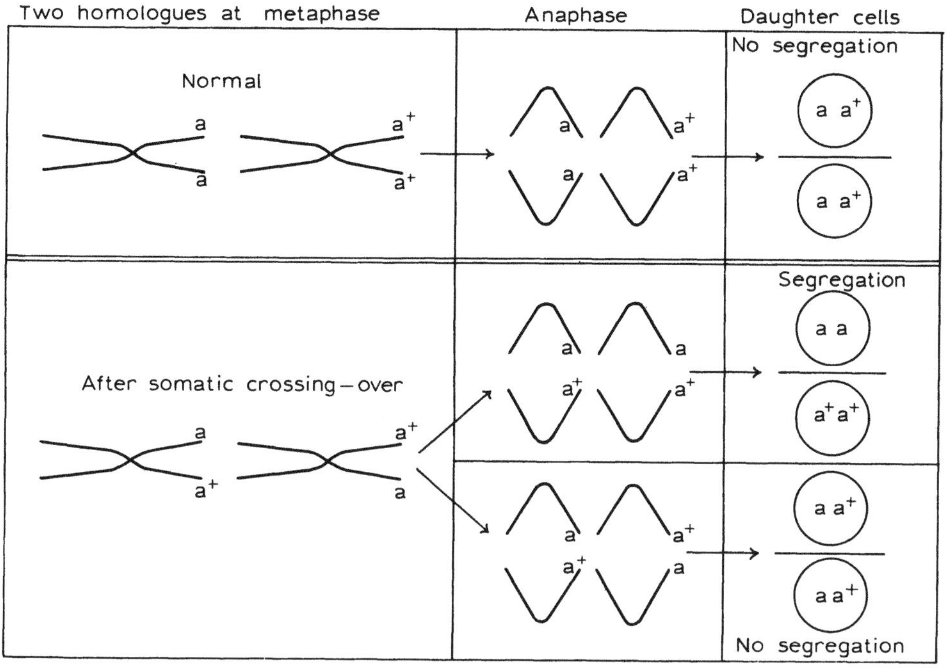

Fig. 3.4 Somatic crossing-over: there is no separation of entire chromosomes from bivalents as in meiosis. The chromatids of the recombined chromosomes separate independently. Segregation will be observed only when two chromatids with the same alleles move to the same pole. This is expected in 50% of the cases where somatic crossing-over has occurred

next to it, with the dominant expression. The twin spot is embedded in the heterozygous tissue where the gene expression is intermediate. Gene mutation, deficiency of a segment with the dominant allele of the gene and loss of the relevant chromosome may mimic somatic crossing-over, and may in fact be much more probable in many instances.

There are reasons for spontaneous irregularities to occur relatively frequently: species hybrids and inbred lines of outbreeders. Chromosome breakage, for instance, is more frequent in species hybrids and in inbred lines than in more balanced genotypes. In *Nicotiana* (tobacco and relatives), hybrids have been described that show distorted growth, or bear tumors with high frequencies of chromosome number and even chromosome structural deviations (Burk and Tso 1960; Yang 1965). In other *Nicotiana* hybrids, certain chromosome segments reduplicate several times, resulting in giant chromosomes (Gerstel and Burns 1966) that have mechanical problems in mitosis.

Chromosome elimination is observed in several hybrids, of which that between barley *Hordeum vulgare* and *H. bulbosum* is best known. It may even result in the loss during young embryonal stages of the entire *bulbosum* genome, so that a haploid barley plant remains (Kasha and Kao 1970; Lange 1971). The practical importance is considerable (Sect. 11.4.2). It is most probably a passive elimination resulting from differences in developmental speed between the chromosomes of the two species (see Sect. 3.1.3.5). Delayed metaphase congression and anaphase separation of the *bulbosum* chromatids, as has also been suggested for genome segregation without elimination, is apparently the cause (Bennett et al. 1976). The genotype of the parents is important for the expression: complete, partial or no elimination. The loss of single chromosomes in human lymphocytes leading to aneuploidy has been shown to result from the displacement of such chromosomes from the equator, for instance towards the hollow spindle (Ford and Correll 1989). It is not improbable that similar displacement may play a role in chromosome loss in cultured plant cells.

Sometimes one or a few genes may be responsible for deviant mitotic behaviour: *Bloom*'s disease is a serious disorder in humans, caused by a monofactorially inherited very high frequency of sister chromatid exchange. *Fanconi*'s anaemia and Ataxia telangiectasia (*Louis Bar* syndrome) are also associated with spontaneous chromosome breakage. Unlike *Bloom*'s syndrome, which primarily involves randomly distributed interchanges between sister chromatids, these diseases mainly involve specific non-homologous chromosomes and chromosome segments. This resembles radiation and chemically induced chromosome breakage, and may be associated with a systematic non-random distribution of chromosomes in the interphase nucleus (Bostock and Sumner 1978).

In addition to genetically conditioned mitotic disturbances, chromosomal abnormalities can also be induced by diseases. Human viruses like the measles virus can cause relatively high frequencies of chromosome aberrations. Neoplastic tumor tissues frequently have abnormal karyotypes. Plant tumors may be chromosomally abnormal, but in tumors induced by the soil-borne parasitic bacteria *Agrobacterium tumefaciens* and *A. rhizogenes*, no abnormal karyotypes are observed.

3.1.4.1.2 Systematic Deviations

These are part of special systems that may be deviant from a general point of view, but that are "normal" for the organism or tissue concerned.

3.1.4.1.2.1 Chromosome Doubling

Chromosome doubling by endomitosis or endoreduplication is found in many plant tissues with special functions: tapetum, endosperm, animals in the liver etc. This was mentioned earlier.

In certain parthenogenetically reproducing insects (*Carausius*, stick insects: Pijnacker 1966; Pijnacker and Ferwerda 1982) and in the parthenogenetic tetraploid plant species *Allium tuberosum* (Gohil and Kaul 1981), programmed doubling takes place just prior to meiosis by an extra premeiotic DNA replication cycle. This results in two daughter chromosomes so close together during meiotic prophase that they pair excluding the homologues. There is no recombination and the resulting diploid cells develop parthenogenetically without fertilization. This process is of considerable potential interest for the induction of parthenogenesis in plants (Sect. 12.5).

3.1.4.1.2.2 Elimination

The elimination of chromosomes, part of chromosomes or even entire genomes has been reported repeatedly as part of the reproductive cycle, occurring often shortly before or after meiosis and especially in insects and other invertebrates (Swanson 1957; White 1973).

Although plants tend to behave much more conventionally, it is good to be aware of the fact that in all living organisms numerous drastic variants of what is considered normal behaviour are possible and that plants in principle do not lack this potential. Under exceptional genetic or environmental conditions, such possibilities may be realized and exploited. An important example is the production of haploids from interspecific hybrids by the elimination of all chromosomes of one of the parental species (Sect. 11.4.2.1).

3.1.4.1.2.3 B-Chromosomes

In plants as well as animals, especially insects and even small mammals, *B-chromosomes* (also called supernumerary chromosomes, accessory chromosomes and accessory fragments) accumulate (Jones 1975; Jones and Rees 1982). Morphologically, they differ from species to species, but within a species usually only one type or, exceptionally, a few different types occur. Some B-chromosomes are mainly heterochromatic (maize, many insects), others partly (rye). In most species, B-chromosomes are about the size of normal chromosomes or somewhat smaller, about half the size of a normal chromosome. In others they are considerably smaller. Often, B-chromosomes have some effect on the host phenotype, especially the "endophenotype", such as chiasma frequency, and on less specific reproductive characteristics. In a complex way these effects depend on the number of B-chromosomes (Jones 1975).

In highly bred plant cultivars, B-chromosomes cannot maintain themselves, but in more primitive varieties they occur in varying frequencies: e.g. maize, rye, several grasses. This is apparently related to the effect of the genotype on the rate of transmission of B-chromosomes (Jones and Rees 1982; Romera et al. 1991).

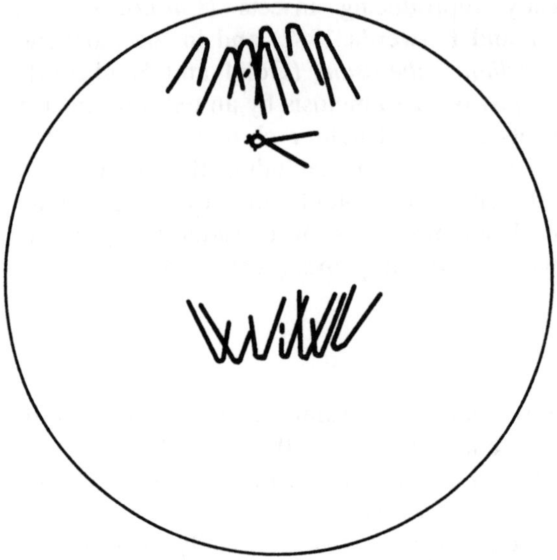

Fig. 3.5 Non-disjunction of a B-chromosome in the first pollen mitosis of rye. The two B-chromatids stay where the generative nucleus is formed. This nucleus will have two B-chromosomes, the vegetative nucleus none

The number of B-chromosomes per cell in field populations does not exceed four in most cases, and seldom is more than two. Even numbers are favoured because of their system of accumulation, but perhaps also because of differences in phenotypic effects between odd and even numbers. In experiments, up to 30 per cell in maize and 12 in rye have been reported. With such high frequencies, plant fertility is severely affected. B-chromosomes may disappear from the somatic cells in several insects, but this is not the rule in plants. Accumulation in insects usually takes place during meiosis, but in plants it is during the first (rye) or second (maize) pollen mitosis. In the first case the B-chromosome moves unsplit (non-disjunction) to the pole, where the generative cell is formed (Fig. 3.5) and both generative nuclei receive two chromatids instead of one. This implies a direct doubling in number. In maize, non-disjunction takes place at the second pollen mitosis so that one generative nucleus has two B-chromosomes and the other none. Accumulation then results because the cell with the two B-chromosomes fertilizes the egg cell and the other fertilizes the doubled polar nucleus: selective fertilization. These processes explain why an even number of B-chromosomes is usually found in these species. This is necessary also for functioning in meiosis where at least a bivalent must be formed for proper segregation at the first division. Nevertheless, accumulation does occasionally fail and single B-chromosomes result that encounter meiotic problems. In plant species in which single B-chromosomes are the rule (*Lilium* spp., *Trillium* spp.), unpaired B-univalents move preferentially to the functional pole of the embryo sac mother cells. Here, no accumulation occurs during pollen mitosis.

The origin of B-chromosomes is uncertain. Because of their occasional effect on sex determination, their occasional meiotic association with sex chromosomes and their heterochromatic appearance, it has been suggested that in animals they have been derived from sex chromosomes. In plants this is not possible in monoecious species. The parallel between their system of non-disjunction and the tendency to non-synchronous centromere split in some species hybrids might suggest that B-chromosomes may be altered alien chromosomes derived from interspecific introgression, followed by loss of active genes and accumulation of properties that favour their special behaviour.

The capacity to accumulate has been used for duplicating A-chromosome segments which first were translocated into B-chromosomes. This has been more promising for gene function and gene localization studies than for practical use (Sect. 11.2.2).

3.1.4.2 In Vitro Culture

Mitotic irregularities are much more frequent in in vitro culture than in intact organisms. There may be two reasons: (1) the artificial conditions disturb the intracellular regulating systems; (2) the artificial medium supports deviant cells more effectively than the organism. Differentiated tissues or organs are more stable in culture than free cells or callus. Whenever regeneration from cell or tissue culture is desired, as is not uncommon with plants, it is attempted to keep the cell or callus phase as short as possible, unless aberrations and mutations are desired. This *somaclonal variation* has in some instances been considered an interesting source of selectable variation (Larkin and Scowcroft 1981). The most common deviations are aneuploidy resulting from non-disjunction, chromosome doubling, gene mutations and chromosome structural aberrations (translocations, inversions, deficiencies, duplications). There are numerous publications on the subject. A detailed report on chromosomal abnormalities in in vitro culture is given, for instance, by Ashmore and Shapcott (1989) for *Haplopappus gracilis*. In animals, especially humans, spontaneous gene mutations and translocations in tissue culture have been very important for gene localization. In plants similar methods have been proposed, but their realization is not simple.

As in the intact organism, there is genetic variation in mitotic in vitro instability and, again, species hybrids, especially hybrids between remotely related species, are more unstable than balanced genotypes. Very wide hybrids that are not possible through generative hybridization can be made in vitro by cell fusion and these appear to be especially unstable. Their instability remains partly after regeneration to plants, in those cases where this is possible, for instance the hybrids between potato (*Solanum tuberosum*) and tomato

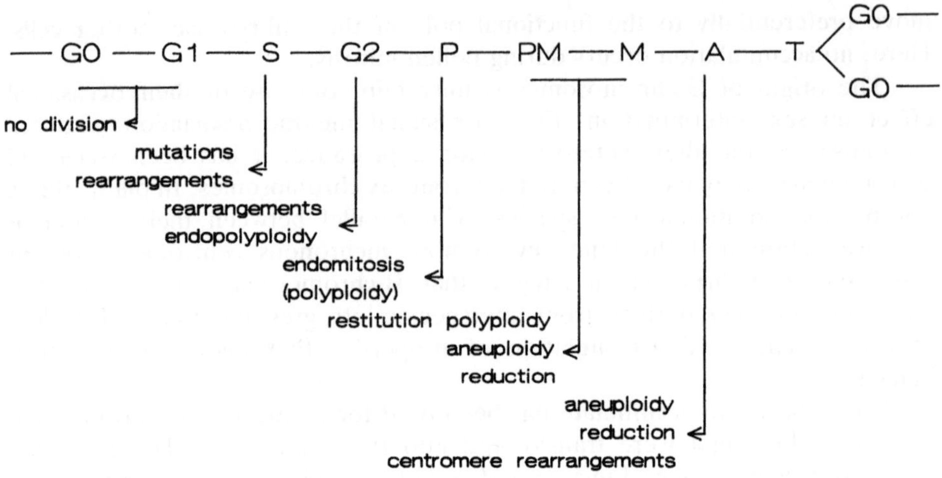

Fig. 3.6 The mitosis line with the moments indicated where the process is sensitive to disturbance due to abnormal conditions of in vitro culture. The expected consequences are indicated

(*Lycopersicum esculentum*) (Melchers et al. 1979). In cell fusion hybrids, as in the intact hybrid plants mentioned above, genome separation has been observed (Gleba et al. 1987). Although aberrations may be induced during stages that are not considered mitotic in the broad sense (which here is understood to include S-phase), it is clear that especially the mitotic stages are sensitive to disturbing factors.

When the subsequent stages of mitosis are shown on a horizontal line (Fig. 3.6; cf. Fig. 3.1), the first stage liable to disturbance is *S-phase*. DNA synthesis may be disturbed in different ways but it is not clear exactly how these play a role in causing somaclonal variation. It has been suggested that, instead of errors in DNA synthesis, defects in the repair system, which normally corrects spontaneous damage to DNA, are responsible. In view of the complexity of DNA synthesis and the importance of correct timing of the subsequent events, it is not improbable that both DNA synthesis errors and repair defects are the cause of the aberrations, at the gene as well as at the chromosome structural level.

In Section 3.1.4.1.1 the formation of giant chromosomes in *Nicotiana* hybrids by lengthwise reduplication of specific chromosome segments was mentioned (Gerstel and Burns 1966). A very similar phenomenon is observed in cell cultures of mammals (mouse, man and others) that are exposed to methotrexate (Cowell 1982). Only cells that for some reason have the capacity to repeatedly replicate the segment with the gene coding for the enzyme dehydrofoliase (which breaks down methotraxate) survive in high concentrations of this toxic substance. The segment apparently corresponds with a replicon, but its autonomous replicating segment (ARS) is deregulated. The newly formed DNA, including the gene and all necessary regulating sequences, is inserted next to the original segment in the chromosome. As in the *Nicotiana* hybrids mentioned, where no special selected gene product is associated with the phenomenon, giant chromosomes arise that have a very homogeneous appearance without observable G-bands in the segment involved.

An alternative is the appearance of large numbers of small, free chromosome segments containing the same segment, again capable of similar massive reduplication, not in lengthwise succession in the chromosome this time, but liberated after replication. The critical gene is transcribed in each separate segment, resulting in the required overdose of the dehydrofoliase. The segments lack a centromere, so their distribution in the daughter cells is irregular. Because of the presence of two chromatids in each segment at metaphase, the fragment appears to be double: *double minutes*. The phenomenon is not widespread, but not restricted to this particular segment with this gene. Manipulation of the genome using this system has not yet been successful in plants (Sect. 11.2.2).

Other forms of deregulation of S-phase and G2 may result in *endoreduplication* and *endomitosis*, but these processes may occur normally in several specialized plant tissues such as leaf mesophyl. Then, if polyploidy is recovered in regenerants from cells or protoplasts derived from such tissues, it may not be considered an abnormality. The relative frequency of the event can be evaluated by analyzing the first division after explantation (Pijnacker et al. 1986; cf. Sect. 3.1.1).

Chromosome condensation in normal cells in vitro (Fig. 3.6) is not expected to experience irregularities with genetic consequences. After fusion between cells in different mitotic stages that have different degrees of chromosome condensation, however, the least condensed chromosomes will condense unnaturally (PCC, see Sect. 3.1.1), which may result in fragmentation when this cell is at S-phase. This has been suggested as a means to liberate DNA that might subsequently be incorporated in intact chromosomes by a process of transformation.

The next stage liable to disturbance is *spindle formation*. When the spindle fails entirely, polyploid cells result. With abnormal organization of the spindle, for instance when the cytoskeleton is disturbed and the perinuclear preprophase band of microtubules is not developed or not oriented normally, various complications may result. The details of these abnormal phenomena have not been studied in sufficient detail, in spite of their considerable interest. Non-disjunction, resulting in aneuploidy, is a rather common result, and tissue cultures, especially those involving extended callus periods, often show drastic chromosome number variations. In spite of inevitable selection for viable genotypes, some very deviant karyotypes can maintain themselves. In both mammalian and plant cell and tissue cultures, it appears that a number of specific karyotypes are stably established after a period of instability. These may be more stable than other types, but more probably they have a selective advantage that is not readily equalled by new variants.

Lack of *synchronization* between *spindle* activity and *chromatid separation* may be one of the factors in the origin of aneuploidy, both when the chromatids separate too early and subsequently fail to co-orientate, and when they separate too late and undergo non-disjunction. In the latter case, the forces exerted on the chromosomes are so strong that they may break, especially in the centromeric region. Subsequent fusion with other breaks may result in translocations. These phenomena have been studied in more detail in meiosis and their occurrence in in vitro mitosis is insufficiently documented. Translocations and other chromosome-structural rearrangements are probably most frequently the result of S-phase disturbances and not of spindle and centromere irregularities.

Chromosome number variations can result from *multipolar spindles* that may be formed when cellular polarization is imperfect, perhaps as a consequence of premitotic cellular disorganization. Certain chemicals affecting spindle development can enhance this. Especially when there is a special order in chromosome position in the nucleus (genome separation, somatic pairing etc., Sect. 3.1.3), more or less systematic spindle abnormalities may have interesting consequences (Sect. 10.4.4.1).

3.2 Generative Transmission: Fertilization – Meiosis

Although somatic transmission is not as simple and straightforward as sometimes suggested, the possibilities of generative genetic transmission for manipulating the genotype are far better. This is clearly the reason for the development and maintenance in nature of such complex systems in practically all forms of life. Generative transmission has two complementary components. One is fertilization, the other is the formation of the gametes that perform fertilization. Gametogenesis requires that the chromosome number of the parental diplont (the sum of the chromosome numbers of the two parents in fertilization) be reduced in meiosis to the number originally present in the parental gametes. Use is made of meiosis to rearrange the genetic make-up of the chromosomes so as to create gametes with a genetic make-up different from that of the parental gametes: *recombination*. Through fertilization, genetically new progeny are formed. Both in the original and in new (or altered) environments, the most successful genotypes will contribute most to the next generation. This, very briefly, is the essence of the importance of generative transmission in nature.

Whereas somatic transmission offers opportunities for eliminating or fixing accidental changes in DNA sequences, in chromosome structure or in chromosome number, generative transmission has the capacity to combine and recombine existing changes in each successive generation. To plant breeders, it offers an array of opportunities for manipulating the genotype as yet not equalled by any other approach.

Although fertilization has several cytogenetic aspects of interest to the plant breeder, it is meiosis that offers the most important cytogenetic phenomena and the most interesting opportunities for manipulation.

3.2.1 Fertilization

Fertilization is the fusion of a female with a male nucleus, located in the *gametes*. In plants the male gamete is one of the generative nuclei in the pollen grain. The female gamete is the egg nucleus in the embryo sac. Both pollen grain and embryo sac have an organization and genetic individuality that justify their classification as individual organisms, the *gametophytes*, in spite of the fact that during most of their lives they are embedded in the parental *sporophyte*. The pollen grain is liberated from the parental host (anther) to enable the generative nuclei to perform their act of fertilization. The pollen grain contains either two or three cells at the time of maturity: bicellular (or binucleate) and tricellular (or trinucleate) pollen. These cells are separated by membranes and not by cell walls like normal plant cells. There is one vegetative ("somatic") cell and one or two generative cells in the mature male gametophyte (the pollen grain). When the pollen grain reaches the surface of the stigma of a plant with which it is compatible, a *pollen tube* is extruded from the grain through which the *generative nucleus* is transported through the pistil to the embryo sac. Stigma and pistil are diploid transporting tissues of the maternal sporophyte and may have a genotype different from that of the haploid egg. The generative cell of binucleate pollen divides during pollen-tube growth to produce two sperm cells.

The relative competitive abilities of different male gametophytes on the stigma, in the style, at the entry of the ovule and at the moment of fusion with the egg nucleus are to a large extent determined by their genotype. From a cytogenetic point of view it is important to note that chromosome numbers deviating from the norm (aneuploids with too many or too few chromosomes, polyploids with entire genomes in excess) tend to give the male gametophyte, and consequently the gamete, a competitive disadvantage. They are selected against in male transmission.

During fertilization, one male gamete fertilizes the egg nucleus, the other the secondary polar nucleus of the embryo sac. The two gametes that carry out this *double fertilization* usually derive from the same pollen grain and then have the same genotype. Occasionally, they may originate from different pollen grains and then their genotypes may differ.

The embryo sac develops from the meiotic end-product in a more complex way than the pollen grain, and there is considerable variation between taxa in embryo sac development. In most cases, the products of the first mitotic division of the embryo sac move to opposite poles of the cell and each nucleus divides two more times. One of the four resulting cells of each group (again not surrounded by rigid cell walls) moves to the centre of the embryo sac and the two fuse to form the diploid *secondary polar nucleus*. When this is fertilized by a haploid male gamete, a triploid nucleus is formed that is the initial cell of the *endosperm*. This is nutritive tissue necessary for initial, and in several species for long-term, embryo development. In several plant species (for instance many Gramineae), the endosperm forms the bulk of the seed. In other taxa it degenerates at an early stage and the bulk of the seed is formed by specialized parts of the embryo, mostly the thickened embryonic first leaves (cotyledons).

One of the three cells left at the micropylar end of the embryo sac becomes the *egg cell*. The other two are the *synergids*, which have only an accessory role in fertilization. The three cells at the opposite end of the embryo sac are the *antipodes*. The nuclei of the synergids and antipodes may occasionally become polytenic or endopolyploid and may have an abnormal nuclear phenotype. The synergids may block the passage of cytoplasmic organelles from the pollen tube into the egg cell.

In most cases the embryo is diploid and the endosperm triploid. In some plant species embryo sac development is somewhat different: the endosperm may, for instance, be pentaploid (five genomes). The interaction between embryo and endosperm is so delicate that ratios of the number of genomes of maternal and paternal derivation deviating from the norm are often not tolerated. Epigenetic "genomic imprinting" by the paternal and maternal tissues seems to play a role in this balance (Lin 1982, 1984) and special chromosome segments may be involved. When a diploid is fertilized by a tetraploid, for instance, the gametes are haploid and diploid, respectively, and the embryo is triploid, with two genomes derived from the mother. This in itself is no problem. The endosperm, however, now is tetraploid, with two genomes from the mother and two from the father. The maternal tissue is diploid. This may deviate too much from the standard relation to be tolerated. Endosperm abortion may occur, resulting in subsequent embryo abortion. The reciprocal cross, with different genomic combinations, may be more balanced. There is genetic variation not only between, but also within, species for the degree of embryo abortion resulting from such genomic interactions. Failure of triploid embryos to develop after hybridizing a diploid with a tetraploid is called the *triploid block*. It can play a negative role when triploids are desired, but it can be a positive asset when rare deviant ploidy gametes are selected (Sect. 11.3.1.2.1.2). Embryos that, because of failing interaction between embryonic, endosperm and maternal tissue (or for other genetic reasons as are common, for instance, in interspecific hybrids), will normally not develop, may be saved by in vitro culture (*embryo rescue*).

Although the processes involved in fertilization and their variants are of considerable importance for plant breeding, especially for hybridization, the subject will be further discussed in later chapters only when specific cytogenetic aspects are concerned.

3.2.2 Meiosis

Halving the chromosome number is an essential part of generative reproduction, as it enables a diploid generation (*diplophase*) resulting from fertilization to alternate with a haploid generation (*haplophase*) that forms the gametes. Generative reproduction as such, however, is of no real significance without recombination: the *primary function of meiosis* is the realization and regulation of *recombination*. Reduction, as the complement of fertilization, is no more than one of the means of effectuating recombination. For the plant breeder, recombination is one of the major tools in realizing his aims. In the following discussion, therefore, meiosis will be considered mainly from the viewpoint of recombination. The main features of meiosis are assumed to be known. An extensive review of meiosis was given by John (1990). Only a few special points will be referred to here.

Meiosis is derived from mitosis, the difference being that there are two nuclear divisions, but only one chromosome division. This results in a reduction of the chromosome number per nucleus. The way in which it is realized makes it possible to accomplish recombination in a very controlled fashion. Some apparently specific features of meiosis, such as chromosome pairing, delayed centromere split etc., are sometimes observed in mitosis, as seen earlier in this chapter.

The *duration* of meiosis varies greatly depending on the species and the environmental conditions, especially the temperature. It is important for the meiotic functioning of hybrids and allopolyploids that the duration of meiosis is not too different in the two parental species. One of the reasons the allopolyploid between wheat and rye (triticale) usually has a somewhat irregular meiosis is thought to lie in the differences in meiotic timing in the parental species. Reviews of the duration of meiosis in different species have been given by Bennett (1971) and Bennett and Kaltsikes (1973).

The basic meiotic processes are very similar in all eukaryotes, but in the morphology and in quantitative aspects, especially those related to recombination, striking differences may occur. The general principles are shown in the diagram of Figure 3.7. In species with holokinetic chromosomes some seemingly fundamental differences from the principles of the diagram are observed. On closer inspection, these differences appear not to involve central issues (Sybenga 1981).

3.2.2.1 Prophase I

The prophase of the first meiotic division is of crucial importance for all further meiotic processes and their consequences: here chromosome pairing and genetic exchange take place and some of the conditions for segregation are set. Before prophase proper starts, however, some important lines of development have been fixed. As indicated above, the chromosomes, also in somatic tis-

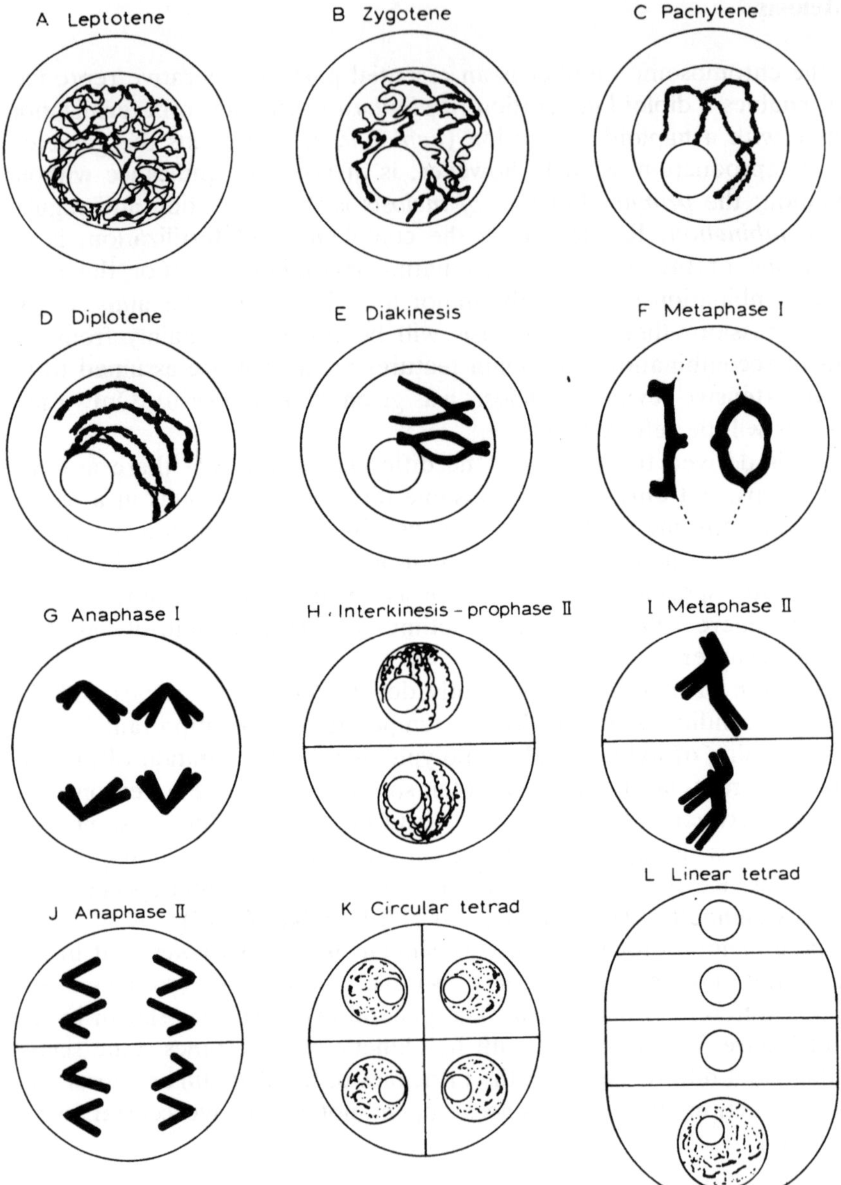

Fig. 3.7 The stages of meiosis

sues, are not always distributed at random over the nucleus. This affects the meiotic process. For instance, when homologues are somatically associated and remain continually in each other's vicinity throughout the mitotic cycle up to meiosis, meiotic chromosome pairing is merely an intensification of an

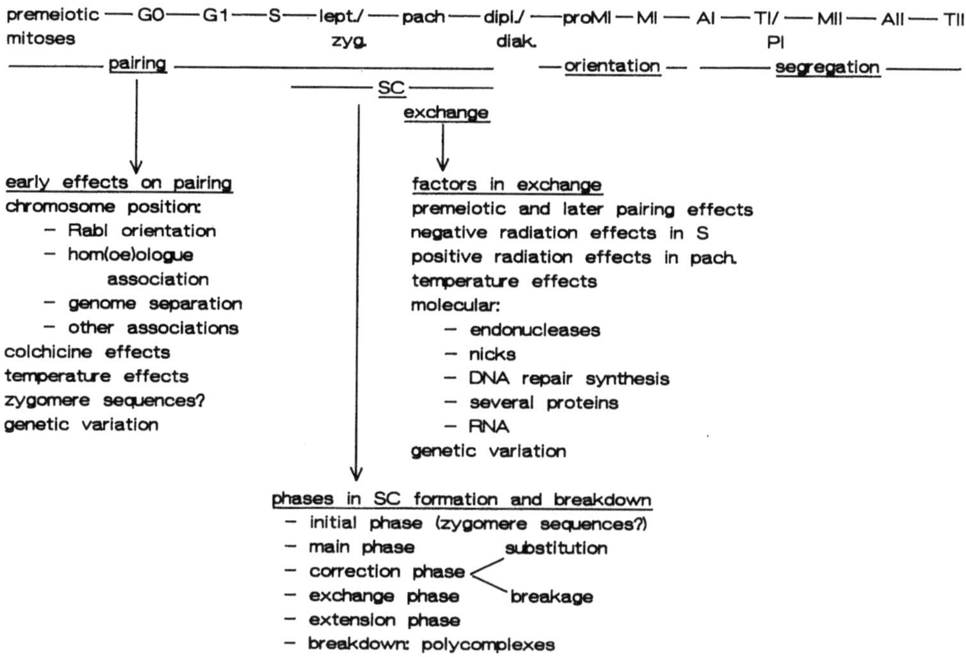

Fig. 3.8 The meiosis line: processes taking place, specific substances present and effects of external agents

existing pattern. On the other hand, in allopolyploids with related genomes, genome separation will decrease the probability that homoeologues pair, and the closer somatic association of homologues than of homoeologues will have a very similar effect.

The many factors involved in the initiation of meiotic pairing have been thoroughly reviewed by Loidl (1990); a few will be briefly considered here (Fig. 3.8). Some have an effect long before microscopically visible characteristics of meiosis appear, occasionally even during premeiotic mitoses. In situations where meiosis takes place synchronously in a number of closely associated cells [e.g. spermatocytes in cysts in several animals; pollen mother cells (PMCs) in anthers of many plant species], the premeiotic divisions stop when a certain number of cells are present and the cells accumulate in GO (Bennett 1976). The cell population then simultaneously shifts to G1 and subsequently to S-phase, sometimes, however, only after a considerable delay. When the cells are in G1, sensitivity to colchicine is observed, which suppresses meiotic pairing and consequently exchange. Somewhat later, sensitivity to high temperatures appears, especially in special genotypes (Riley 1966). During this stage microtubules may appear at the periphery of the nucleus ("intranuclear fibrillar material", Bennett et al. 1974), but also outside the nucleus (Sheldon et al. 1988). These are known to be sensitive to colchicine;

however, their actual role in pairing is not known. Puertas et al. (1984) observed multivalents in diploid pollen mother cells of rye after early colchicine application, which they ascribed to disturbance of the premeiotic alignment of the chromosomes. The result would be that homologues are farther apart than normal during the initiation of pairing and that duplicated segments in different chromosomes have an opportunity to find each other, which normally would not be possible. Homologous segments in different chromosomes that can pair and form chiasmata, resulting in unexpectedly large configurations, have also been found in haploids. On the other hand, the induction of translocations by colchicine cannot be entirely excluded. The accompanying phenomenon of reduced pairing after colchicine treatment was also observed and analyzed for instance by Driscoll et al. (1967). The resulting univalents were found to be in a higher frequency than corresponds with the reduced chiasma frequency alone, and were again explained by the disturbed spatial relation between the chromosomes during pairing. The manipulation of the course of meiosis by disturbing premeiotic processes is an interesting possibility for plant breeding but has not seriously been exploited.

Ionizing radiation applied during, or perhaps somewhat before, S-phase has a negative effect on genetic exchange (Lawrence 1961, 1965), possibly by disturbing the normal replication processes, of which special meiotic variants are essential for exchange. The actual close pairing may start before the end of S-phase (Grell et al. 1980), although not in chromosome segments that are actually replicating. In any case, pairing is initiated soon after S-phase (Oud and Reutlinger 1981). This can be considered the beginning of *zygotene*. Pairing is completed at *pachytene*. The term *leptotene* is merely descriptive and refers to the stage in which the chromosomes first become visible under the light microscope; a strict differentiation between S-phase, leptotene and zygotene therefore does not make much sense.

The processes involved in the primary attraction between chromosomes are not understood (Loidl 1990). As indicated above, the pairing of homologues is facilitated by several conditions, including Rabl orientation (Fussell 1987). At the beginning of pairing, the chromosome ends are usually attached to specific segments of the nuclear membrane. The centromeres of metacentric chromosomes tend to point in the other direction or, in acrocentric chromosomes, occupy a position in the same area. The pairing chromosomes together often attain the shape of a bouquet, the *bouquet stage*.

Shortly before they pair, the chromosomes form a special "scaffold" of acidic and neutral proteins, the *axial cores*, consisting of elements attached to specific chromosomal sites that interconnect to form a continuous thread. The chromatin between these elements comes out of the axial cores as long loops. In some organisms the axial cores are formed some time before they actually pair (rye), in other organisms they are formed just prior to close pairing (mouse). The axial cores of the pairing homologues approach one another in a parallel fashion, but remain at a certain, specific distance that does not vary much between species. Together they form the *synaptonemal complex* (SC).

For a general review, see von Wettstein et al. (1984); for a review of plant SCs, see Gillies (1984) and John (1990).

The axial cores, after having been integrated into the SC, form the *lateral elements* of the SC. Between them they form a thinner protein strand, the *central element* (Fig. 6.4). Pairing between the axial cores of the chromosomes proceeds zipperwise from the points of pairing initiation. Pairing and SC-formation initially stop at discontinuities in homology, but at the end of zygotene, pairing may extend over non-homologous segments, although it will not in those places enable genetic exchange to take place.

The reason why chromosomes do not get entangled during pairing is not entirely clear. *Interlocking bivalents* have been observed, infrequently in normal material (Darlington 1965) and more frequently in material treated with high temperatures at G1 or earlier, especially in specific genotypes (Yacobi et al. 1982). Parallel orientation during early meiotic stages (Rabl orientation, cf. Fussell 1987) and, in some cases, premeiotic pairing plays a role in avoiding interlocking, but this is not sufficient. It is also possible that, during pairing, a chromosome that has been captured between two pairing chromosomes is simply pressed away. When such a chromosome is itself engaged in pairing, this may be undone locally to permit the interlocked chromosome to be removed. An alternative (Holm and Rasmussen 1981) is that chromosomes actually break to free an interlocked chromosome, after which the broken ends fuse again. Scheduled chromosome breakage and reunion is not very uncommon. Topoisomerase 2, an important structural element of the cores with enzymatic characteristics (Moens and Earnshaw 1989), may play an important role in this process.

With homologous pairing in SCs, the DNA forms short loops into the central elements and it is assumed that the process of genetic exchange between completely corresponding DNA sequences takes place here. Small nodules are visible, in large numbers and rather randomly distributed, between the lateral elements at the beginning of zygotene (early nodules). Their role is not clear. It has been suggested that they represent sites actively in search of DNA homology (Loidl 1990), where enzymes necessary for recombinational strand exchange are involved (Roeder 1990). Their role in recombination would be limited. *Gene conversion*, where one allele of a heterozygote is converted into the other allele, could possibly take place in early nodules when sufficient homology is encountered. It is related to exchange recombination, but nonreciprocal instead of reciprocal, and is common in lower organisms, but difficult to detect and analyze in plants. Early nodules are later replaced by the somewhat larger recombination nodules, best visible at mid-pachytene and late pachytene. These are believed to contain the elaborate machinery necessary for exchange (*recombination nodules*, Fig. 6.4). Their number and position correspond to what is expected for exchange. They are not visible in the SCs of all material and no biochemical analysis has been possible so far. It is probable that a nodule starts as a condensation nucleus for enzymes and other specific recombination proteins available in low concentration on or around the chromosomes. As a result of such concentration, this material is not available for a second nodule in its immediate neighbourhood. This may be the basis of *interference*: the phenomenon where points of genetic exchange within chromosomes (sometimes even in different chromosomes) do not occur near each other. This is a hypothetical explanation of interference and not yet generally accepted.

The molecular processes involved in exchange in higher organisms, especially plants, are not fully understood, although many of the enzymes, and protecting as well as regulating proteins, involved have been isolated. After detailed studies in the 1970s, progress has slowed down. Much of this work has been done in *Lilium*, which has large anthers with many pollen mother cells and large chromosomes, and in the mouse (cf. Stern and Hotta 1987; John 1990). In microorganisms, including lower eukaryotes, where comparable but somewhat different systems operate, especially the biochemistry of and the enzymes involved in exchange at the level of the DNA have been analyzed in much more detail. Incision and repair are prerequisites for genetic exchange, but only a very small fraction of the nicks is actually involved in exchange. Whereas ionizing radiation applied during S-phase reduces genetic exchange, it has a stimulating effect when applied during zygotene-pachytene (Lawrence 1961, 1965), probably by inducing repair processes that are largely identical to those operating in genetic exchange. Mutant repair enzymes or other proteins involved in repair result in an increased sensitivity to ionizing radiation and at the same time reduce the capacity for genetic exchange. This has been studied in detail in bacteria (*rec* proteins). The first eukaryote in which these mutants were discovered was the fungus *Ustilago* (Holliday 1967), where UV-sensitive mutants had drastically reduced recombination frequencies. They were also found in *Drosophila* (Watson 1969). Although early reports on their presence in man and in higher plants were later contested, there is little reason to doubt that such mutants occur in all higher and lower organisms. The *rec* proteins of eukaryotes are not identical but related to those of prokaryotes.

Several models have been proposed for the process of exchange at the DNA level. It is certainly very similar in all living organisms, because many comparable enzymes, specific proteins and DNA repair systems are involved. All models include the resolution of heteroduplex (hybrid) DNA and repair processes as first proposed by Holliday (1964) and Whitehouse (1963). For reviews, see Alberts et al. (1989) and John (1990).

Light microscopically, the paired chromosomes (bivalents) at pachytene often give the impression of being completely fused. These bivalents are usually not smooth, but have a beaded appearance: more or less condensed segments alternate. The more condensed *chromomeres* follow a specific pattern and, especially when they are large and represent heterochromatic segments, they are useful landmarks for the recognition of chromosomes and even specific chromosome segments (Sects. 4.2.4.1 and 8.3.3.1.2).

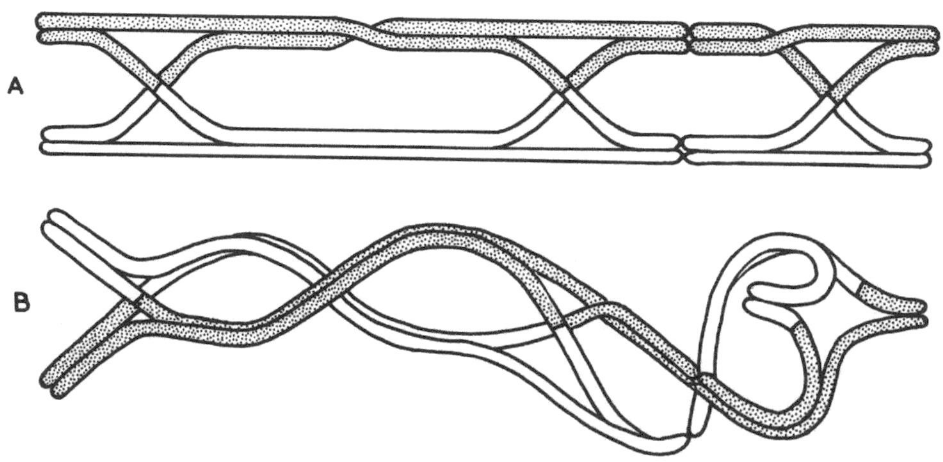

Fig. 3.9 Chiasmata at diplotene. **A** The diagram; **B** more realistic drawing: the chiasmata in the *middle region* maintain their shape, but the distal chiasmata change under the influence of strong forces resulting from contraction. The four chromosome segments around the chiasma try to establish equal angles (90°) between them. Bivalents with only one chiasma attain a cross shape. (After Sybenga 1975)

Pachytene tends to be a long stage. At its close, the lateral elements of the SCs can occasionally be seen under the electron microscope to be double. Gradually, the two chromatids of each chromosome become visible under the light microscope also, and the chromosomes separate, except for the points of exchange where the X-shaped structure can clearly be seen, at least in favourable material (Fig. 3.9). This structure has been given the name *chiasma* because it resembles the Greek character chi. This marks the beginning of *diplotene*, where condensation, which was slowed down during pachytene, is resumed. In several cases, in plants as well as in animals, the breakdown of the synaptonemal complex, in animals often accompanied by the formation of polycomplexes from the released proteins, is not immediately followed by the incorporation of new scaffold proteins necessary for further condensation. Then, instead of condensing, the chromosomes become diffuse again. In the oocytes of higher animals this is the start of a long "resting" period (*dictyotene*), in which transcription starts again on a large scale to provide the transcripts necessary for the large mass of yolk. Only when the egg matures is meiosis resumed and a new round of condensation starts. In large mammals, such as humans, dictyotene may last from the embryonic stage until the end of the fertility period and may involve several decades.

In the spermatocytes of some animals and in the pollen mother cells of some plants, a similar interruption of chromosome condensation at early diplotene is observed (*diffuse diplotene*), but without transcription activity

(Wilson 1925; Klasterska 1977; Oud et al. 1979). When the sequence of stages can not easily be recognized, this stage has often been mistaken for zygotene, especially when the chromosomes are weakly visible as fine threads and seem to be synapsing instead of separating. In the large oocytes of amphibia, at a stage comparable to dictyotene but morphologically different, the decondensed bivalents are stretched out in the greatly enlarged cell and can attain a total length of 500–1000 µm. These "*lampbrush*" chromosomes are not known in plants and will not be further discussed; for a review see Callan (1986).

Whatever the structure of the diplotene bivalents, normal prophase condensation, if not already begun during diplotene, begins or is resumed at its end. This then is the stage that is comparable to the beginning of somatic prophase and all that precedes it takes place in what would be interphase in somatic cells. In this respect, *meiosis is precocious* (Darlington 1965).

In all models, the classical observation that, of each chromosome, only one chromatid is involved at each genetic exchange is implicitly assumed. Through BUdR incorporation during two premeiotic DNA replication cycles, it is possible to mark the two chromatids of each chromosome involved in exchange. During the condensation stages following pachytene, the exact location of the point of exchange can then be seen (Jones 1987), provided the chromosome has not condensed too far.

Gradually, diplotene shifts to *diakinesis*. As a result of progressing condensation, the chromosomes become rigid and attempt to straighten. This is impossible where homologues are connected by chiasmata, and special shapes arise there to accommodate the forces operating on the associated chromosomes. With only one chiasma, the bivalent will attain the shape of a cross, or the shape of a rod when the chiasma is near the end. With two chiasmata, depending on their location in the bivalent, a ring of variable appearance will be formed (Figs. 3.10, 3.11). The chiasmata sometimes give the impression of slipping off towards the chromosome ends, especially when metaphase I is approached or during metaphase I. In most cases this is only seemingly so. In very condensed chromosomes the exact place of the chiasma is not well visible and the resolution of BUdR differentiation of the two chromatids is not sufficient. The chiasma may seem to be situated at the end (in *terminal* position), while actually it is only *subterminal*. When the chromosome ends can be marked by C-banding, it can often be seen that the terminal bands are situated not at the place where the chromosomes are connected by the chiasma, but to the sides of this point (Fig. 3.11). Chromosomal deformation makes it difficult to locate chiasmata exactly at highly condensed stages. In some insects the typical scaffold or core proteins can be accentuated and the chromatin outside the core destained or possibly partly removed (Rufas et al. 1987; Santos et al. 1987). The course of the core can then be followed and the location of the chiasmata exactly pinpointed, even in the highly condensed chromosomes of diakinesis and metaphase I. All such observations show that terminalization of chiasmata, believed earlier to be a

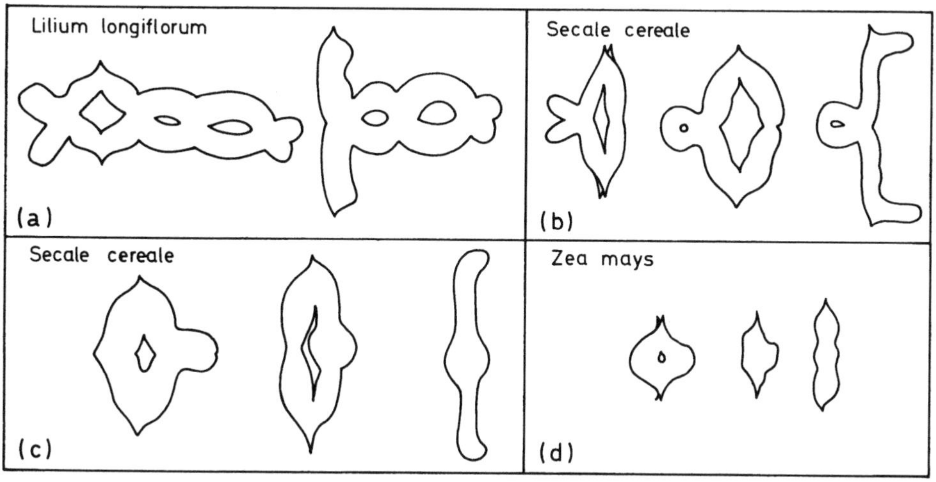

Fig. 3.10 Different bivalent shapes at metaphase I. In some species of *Lilium* (**a**) the number and location of chiasmata can be readily determined. In the first example of rye (*Secale*, **b**) this is also possible, but less readily. In other genotypes of *Secale* (**c**) the morphology of the chromosomes permits only approximate recognition of number and location of chiasmata in the arms. Even when the chiasmata are not clear, the specific shape of the bivalent can often be used to derive the number and approximate location of the chiasmata. Whether or not an arm has a chiasma, however, is always clear. In maize (*Zea*, **d**) details of the chiasmata are much more difficult to see, although here, too, considerable differences between genotypes exist. When the chromosomes are still smaller, this may become even worse, and in a number of species it is not even clear whether an arm has a chiasma or not, but this is not always the case. (After Sybenga 1975)

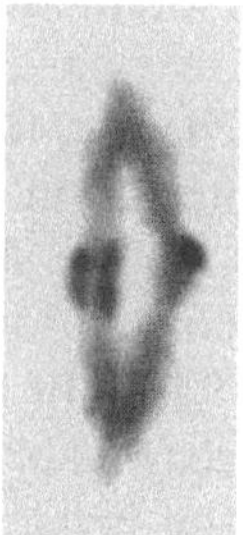

Fig. 3.11 Practically no chiasmata are formed in heterochromatic C-bands. In rye, the terminal C-bands are often seen to be positioned distal to the chiasma at metaphase I. Very distal chiasmata give the impression of being entirely terminal, but when C-banded, it is seen that here, too, the connection between the two chromosomes is not through the bands, but just proximal to them. The bands have been pushed into the chromosomal body at the sides of the chiasma, in principle like more proximal chiasmata. The bivalent shown has a seemingly terminal chiasma at the *left*, but the bands are clearly at the sides of the arms. There is C-band heteromorphism in the right arm. (Courtesy of J.H. de Jong)

general phenomenon, hardly plays a role, even at metaphase I where strong pulling forces are exerted on the bivalents.

The analysis of the frequency and location of chiasmata is of considerable practical interest, as it gives a measure of the recombination level and the possible presence of recombination-free chromosome segments containing specific gene combinations. There are no reports yet on the successful application of techniques for specifically staining diakinesis and metaphase I chromosome cores in plants, although several attempts are being made. This is important for organisms in which diplotene is not accessible for chiasma analysis, as in most plants.

3.2.2.2 Metaphase I

The transition from diakinesis to *metaphase I* (metaphase of the first meiotic division) is marked by the initiation of kinetic activity. As in mitosis, the chromosomes first move towards the equator (*congression*); this stage is designated *prometaphase I*. Unlike mitosis, in which the centromeres of the two chromatids have each other as orientation partners (*amphitelic orientation*), at meiotic prometaphase the centromeres of the two chromatids of each chromosome remain to function as one unit: *syntelic orientation*. Apparently, kinetochore separation starts much later than kinetochore activity, and the two united kinetochores have the double kinetochore of the homologue as their co-orientation partner. Because a counterforce is necessary for co-orientation, co-orientation suitable for regular segregation is possible only when the two chromosomes are connected mechanically. In mitosis this connection is formed by the pericentromeric region where the chromatids do not separate. This connection is present in meiosis also, but does not have a function at the first metaphase, because the sister chromatids must stay together during the first division and may not coorient. Chiasmata, the visible result of genetic exchange, fulfill the essential mechanical function of keeping chromosomes connected in meiotic metaphase I.

The shape of the metaphase I bivalents varies considerably, depending on chromosome size, location of the centromere and condensation pattern (Fig. 3.10). In favourable cases, the number and location of the chiasmata can be reasonably well observed directly or derived from the typical morphology of the bivalents (or other configurations). In other cases, the morphology of the bivalents does not permit more than concluding which arms are associated by one or more chiasmata, and which arms are not.

3.2.2.3 Anaphase I, Telophase I

After having congressed with their chiasmata positioned at the equator, the bivalents remain there for some time, apparently to ensure the most complete co-orientation attainable. This is a prerequisite for the subsequent segregation of the chromosomes to the two daughter cells. When chromosomes with specific functions and special constitution are present, their regular segregation may require extra time. Examples are sex chromosomes, especially when they occur as univalents, as in X-O males of several insects. At a certain moment, however, as in mitosis, chromatid cohesiveness is released, however, excluding the centromeric region. The chromatids of the arms outside the centromeric region separate actively as a result of their stiffness, and the chiasmata are released. At anaphase, the centromeres pull the chromosomes to the poles, dragging the chromosome arms behind them. The chromosomes still contain both chromatids.

The reason the centromeres of the sister chromatids do not separate at the first division of meiosis as they do in mitosis, but orient themselves to the same pole (syntely), is not clear. During metaphase I, the distance between the centromeres in crane fly spermatocytes appears to increase without affecting co-orientation (Janicke and LaFountain 1989). With low temperatures and special chemical treatments, the separation still follows the same pattern, but centromere activation is delayed until the sister centromeres have separated farther than normal. Now the bivalents fail to orient normally. When the sister chromatid centromeres in *Acroptera* B-chromosome univalents remain associated, the chromosomes lag at the equator because they have no co-orientation partner. When the sister centromeres tend to separate, they start amphitelic orientation (Suja et al. 1991). There are several genetically conditioned errors of meiotic disjunction that are of interest for the study of orientation and segregation, but these will not be discussed here.

With respect to the number of chromosomes, anaphase I segregation implies reduction, and therefore the first meiotic division is called *reductional*. This term, in addition to a numerical meaning, has also received a qualitative meaning and then refers to chromosome segments: those segments in which the sister chromatids remain together, as around the centromeres, are said to undergo *reductional segregation* at first anaphase. Those segments in which the sister chromatids separate as segments distal from a chiasma (Fig. 3.7) are said to segregate *equationally* at first anaphase.

When homologous chromosomes differ in structure (as a result of heterozygosity for insertion, duplication, deficiency or C-bands), the two chromatids attached to the same centromere may, as a result of exchange, differ in morphology. This permits an estimate of the frequency of genetic exchange between the centromere and the locus of the special structure in which the chromosomes differ. This subject will be discussed again in Chapter 8.

Telophase I is the stage in which the chromosomes, reduced in number but still double, decondense and form a nuclear membrane. The ensuing resting stage is *interkinesis* and is usually very short, and sometimes hardly present when *prophase II* follows immediately upon telophase I. There is no need for replication because two chromatids are already present and there is usually no transcription. The resulting two cells are often referred to as *dyads*. It is somewhat confusing that the same term dyad is occasionally also used for a chromosome that is visibly double and contains two chromatids at this stage.

3.2.2.4 Meiosis II

In prophase II (Fig. 3.7), the chromosomes become visible again under the light microscope and can be seen to contain two chromatids that are connected only in the centromeric region. The arms usually project in different directions. Condensation causes the arms to line up more or less parallel, but without close association. The chromatids are genetically different in the segments distal from the first chiasma from the centromere, as in anaphase I, but this is visible only when a sizable chromosomal marker is present. As in mitosis, the centromeres line up in the metaphase II equator after congression in prometaphase II and separate at anaphase II. Now the centromeres segregate equationally, but the segments distal from the first chiasma segregate reductionally because they are not sister chromatids. These segments undergo *post-reduction*. After telophase II, a *tetrad* of four haploid cells is formed, each of which has a specific genetic composition.

After meiosis in female animals, only one of the two products of the first meiotic division develops, as does only one of the two products of the second meiotic division. The result is that each meiosis results in only one gamete. In plants, all four end products are formed, but usually the female gametophyte originates from only one. Male meiosis in both animals and plants results in four functional end products. In plants these tend to remain together for a short time, forming the *tetrad*. The same term tetrad is occasionally used also for a bivalent consisting of four chromatids.

3.2.3 Systematic Variants of Meiotic Behaviour

3.2.3.1 Achiasmate Meiosis

In one of the two sexes of some organisms, usually the male, meiosis is *achiasmate* (chiasmata are absent). Synapsis as such leads to a sticky association between the homologues that replaces the chiasmata. The dipterous insects are the best known example for this unusual behaviour. Whereas chiasmata, even when distributed according to species-specific systems, can be formed at variable sites and often in the proximity of the chromosome ends, achiasmate association is restricted in location to the centromeric region. This is not of much consequence in normal material, but the meiotic consequences of chromosomal aberrations may be rather different in chiasmate individuals (i.e. the females in *Diptera*) than in achiasmate males (Vosselman 1981). Because achiasmate meiosis is exceptional in plants, (*Fritillaria amabilis*, Noda 1968), this will not be further considered here.

3.2.3.2 Holokinetic Chromosomes

More complications arise when the chromosomes are *holokinetic*, i.e. when the kinetic activity is not restricted to one specific single structure in each chromosome, but is distributed evenly over the chromosome body (cf. Swanson 1957; John and Lewis 1965; John 1990). Chiasmata may then have "kinetochores" on both sides and cannot be released properly at anaphase I. Different organisms have acquired different solutions to this problem. In the simplest case, as found in the males of several insect species, two or more chiasmata are distally located and centromeric activity is concentrated in the central region of the chromosomes. When only one chiasma is present, it may be located at any place in the chromosome and the kinetic activity is concentrated in one of the end segments. This then resembles a monokinetic chromosome and may in fact be a transition from more primitive holokinetic to more evolved monokinetic systems (Sybenga 1981). More primitive and more typical for holokinetic behaviour, chromatid cohesiveness and consequently the chiasmata are released before kinetic activity starts. The weakly paired sister chromatids separate, as in mitosis, after amphitelic coorientation. There must be renewed association between homologous chromatids at meiosis II to ensure proper segregation. The mechanism is poorly understood. It results in what is called *post-reduction* because, at the first meiotic division, a typical reduction division does not take place. The term is not quite correct because, as a result of genetic exchange, the separating chromatids are only partly sister chromatids. Similarly, the terms syntelic and amphitelic do not fully apply to these chromosomes. In many instances the chromosomes involved are so small that it is very difficult to decide exactly how they behave in meiosis.

Holokinetic chromosomes are common in many Heteroptera and they are found in several plant species. Well-known examples are the genera *Luzula* and *Carex* (LaCour 1952; Nordenskiöld 1962). Among the relatively primitive Ranales, the genus *Myristica*, including the nutmeg tree, *M. fragrans*, also has holokinetic chromosomes (Flach 1966). These were not observed in any of six other Ranales families studied (Flach, pers. comm.). Holokinetic chromosomes are not sufficiently frequent among commercial plant species to justify a more detailed treatment here.

3.2.3.3 Neocentric Activity

Perhaps as an atavistic trait left over from the development of monokinetic from holokinetic behaviour, normally monokinetic chromosomes may under certain conditions develop one or more extra centromeres during meiosis, usually in distal heterochromatic regions. The most striking example is that described in maize when an abnormal, highly heterochromatic chromosome 10 is present (Rhoades 1952). In bivalents where one chromosome has a heterochromatic knob and the other does not, the neocentric activity of only one chromosome can affect chromosome behaviour and even segregation. In rye and a few other species, weaker neocentromeres have been observed in inbred lines and populations under certain circumstances (Prakken and Müntzing 1942, Kavander and Viinika 1987). No serious practical use in plant breeding has yet been suggested for neocentromere activity.

3.2.3.4 Specialized Chromosomes

The mitotically deviant behaviour of *B-chromosomes* was referred to above (Sects. 2.5 and 3.1.4.1.2.3). In animals, their accumulation is usually due to their special meiotic behaviour. In plants, they tend to behave like any other chromosome in meiosis. *Sex chromosomes* are another category of specialized chromosomes that may show unusual meiotic behaviour. In addition to animals, where they are normally present, sex chromosomes are also found in a number of dioecious plant species, some of these cultured plants (spinach, hemp, asparagus). In a few instances the sex chromosomes of plants are sufficiently morphologically differentiated to be recognized. In most cases their behaviour is quite regular, and sex determination is simply a consequence of normal sex bivalent segregation and the fact that self-fertilization (within sexes) is not possible.

In a number of animal species, and even some plants, deviant systems involving sex chromosomes have been observed that are quite revealing as to what chromosomes are capable of when operating under conditions that deviate from what usually is considered normal. Much interesting information on normal and abnormal meiosis can be found in John and Lewis (1965) and John (1990). The subject will not be further discussed here.

3.2.4 Recombination

Recombination is essential for plant breeding. There is considerable variation in recombination, which can be manipulated to some extent. In some instances an increase in recombination is desired, in other instances genes must be kept together in specific allelic combinations. In this chapter a few principles will be briefly considered. Later, (Sects. 8.4 and 10.4.2) manipulation will be discussed in more detail.

3.2.4.1 Two Forms of Recombination

3.2.4.1.1 Chromosome Recombination

In addition to recombination as a result of genetic exchange between paired chromosomes (*exchange recombination*), recombination results from the in-

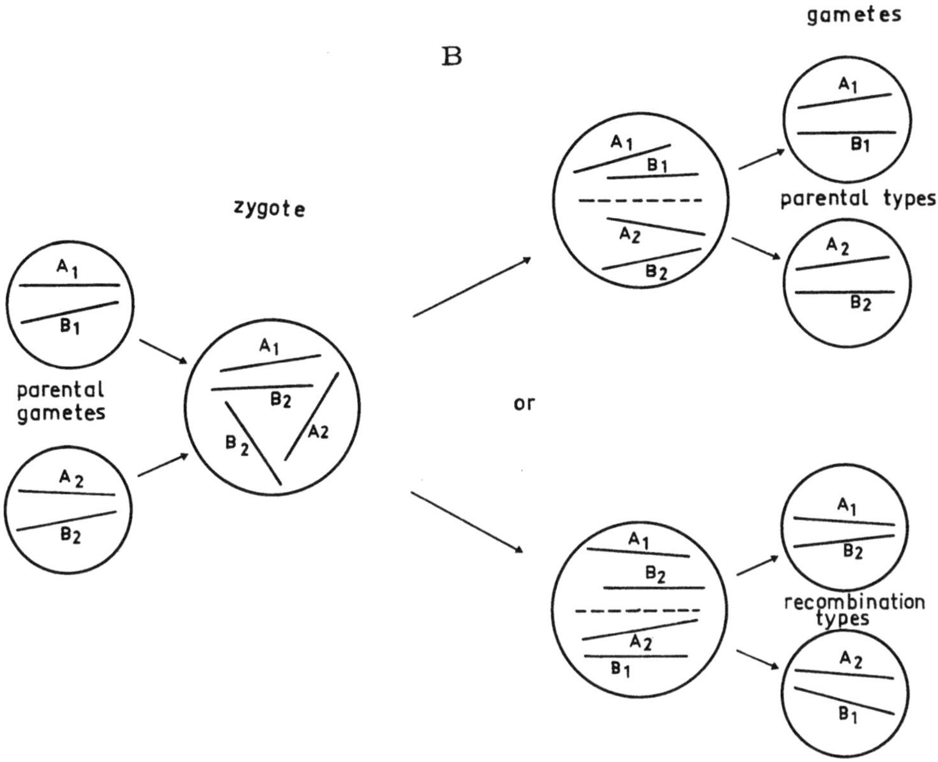

Fig. 3.12 Diagram of chromosome recombination. (After Sybenga 1975)

dependent orientation of different bivalents (Fig. 3.12): *chromosome recombination*. With only one chromosome, there is no chromosome recombination. The larger the number of chromosomes, the more chromosome recombination occurs. Chromosome recombination is normally 50% between genes in different chromosomes, unlike exchange recombination (crossing-over) between genes in the same chromosome, which depends on the frequency of exchange between the genes considered. For plant breeding, chromosome recombination is often quantitatively more important than exchange recombination, especially when the number of chromosomes is large. The lower the number of chromosomes, the more important exchange recombination becomes.

Normally, chromosome recombination is constant at 50%, but this can be changed by translocation (chromosome segments being moved from one chromosome to another). When the translocation is homozygous, chromosome recombination will be constant at 50% again, but now for other gene combinations as in the original chromosomes. In translocation heterozygotes, however, chromosome recombination is (almost completely) suppressed, at least as far as the translocated chromosomes are concerned. This can play an important role in breeding programs, in scientific experiments and in nature (Sects. 5.4.1.4 and 12.3).

When metacentric chromosomes are split in their centromeres (centric split, Robertsonian split), the two parts behave independently and exhibit chromosome recombination (Sect. 5.5.4). The opposite occurs when two acrocentrics fuse to form a metacentric chromosome. The effect of doubling the chromosome number to produce autopolyploids is different. The effect on recombination is complex, mainly because the entire genetic system is altered (Sect. 11.3.1.2.2).

There are exceptional cases in which, without a mechanism that is clearly understood, two apparently independent chromosomes do not segregate independently (e.g. "affinity", and the interdependent segregation of unpaired sex chromosomes and autosomes in some insects). The phenomenon is not of sufficient frequency in plants to be further discussed here.

3.2.4.1.2 Exchange Recombination

Exchange recombination results from the exchange between chromatids of paired chromosomes. A single exchange, resulting in a single chiasma, will recombine the genes on both its sides in two of the four chromatids. A single exchange, therefore, results in a maximum of 50% recombination. When, on an average, there is less than one exchange (or chiasma), recombination is reduced correspondingly. More than one chiasma between specific genes does not lead to an increase in recombination between the alleles of these particular genes. However, the overall level of recombination between all genes does increase with an increase in exchange frequency.

3.2.4.1.2.1 Two and More Chiasmata in One Chromosome

Crossing-over (exchange recombination) is observed in the segregation of genetic markers only when a single point of exchange is present between the markers. Two exchanges between the same two chromatids will cancel each other. In each exchange that results in a chiasma between chromosomes, only one chromatid of each chromosome participates. Each chromatid ends up in a separate spore. Thus, for each chiasma, only half the number of spores has a recombinant chromatid, and the number of cross-overs corresponds to half the number of chiasmata. When two chiasmata are formed between two markers (Fig. 3.13), the same two chromatids will be involved in some cases (*reciprocal chiasmata*); two different chromatids in other cases (*complementary chiasmata*); and in still other cases, one chromatid will be the same in the two chiasmata and the other will be different (*disparate chiasmata*). There are two different types for the latter case. Thus, four different combinations of two chiasmata are possible, each with equal probability if the chromatids take part randomly in the chiasmata.

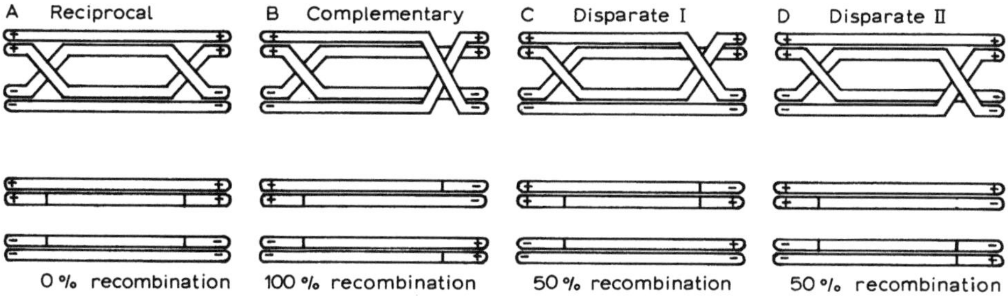

Fig. 3.13A–D The four combinations of two chiasmata. Loci at the ends of the chromosome show no recombination when the chiasmata are reciprocal; they show 100% recombination (both chromatids of each chromosome have recombined) with two complementary chiasmata, and 50% recombination (one chromatid has recombined in each chromosome) with the two disparate combinations of two chiasmata. Thus, there is a maximum of 50% recombination if all four types have an equal probability of occurring. (After Sybenga 1975)

Considering recombination chromatids, reciprocal chiasmata result in no recombination because the two exchanges cancel each other's effect. Complementary chiasmata will result in four recombination chromatids or 100% recombination. When the two have an equal probability, the average recombination is 50%. The two disparate types each result in 50% recombination chromatids and, as a consequence, the maximal recombination percentage with two chiasmata is always 50. It can easily be shown that, with more chiasmata, the same result is obtained. With a single chiasma, recombination is 50% also, and, therefore, the maximal recombination percentage is 50. Lower percentages occur only with, on an average, less than one chiasma in the segment considered. More than 50% recombination is possible only when complementary chiasmata are more frequent than reciprocal chiasmata. More than 50% recombination has been observed in some fungi and phages (Esser and Kuenen 1965), but only incidentally and probably not convincingly in higher organisms.

An excess of the combination of reciprocal and complementary chiasmata, as opposed to disparate, has been reported in grasshoppers, where the course of the chromatids can be followed (Hearne and Huskins 1935). The phenomenon has not been sufficiently studied to be certain that it is significant. In this case, there are no effects on recombination.

3.2.4.1.2.2 Genetic Variation in Exchange Recombination

3.2.4.1.2.2.1 Genetic Variation in Frequency of Exchange

True *asynapsis*, where no pairing takes place, can be genetically (mutants) or chromosomally (lack of homology) conditioned. It is possible that complete failure of the entire chromosome pairing system tends to result in a reversion from meiosis to mitosis, with the consequence being that no meiotic processes are observed. Even with apparently complete asynapsis the axial cores are usually formed, indicating that the system effectuating the initial steps in pairing is still operational. There are two main reasons why, in the absence of sufficient homology, pairing can be initiated (including SC formation) but is followed by very limited or almost no chiasmate association at later stages: (1) there is a sufficiently extensive relict of homology (in distant hybrids) or there are sufficiently extensive duplications present (in the haploid) to result in at least some true homologous pairing that may even be effective for genetic exchange; (2) the pairing system, although not finding sufficiently extensive homologous stretches to initiate true homologous pairing, finds it possible to align (after some delay) non-homologous chromosome segments and even form synaptonemal complexes.

There are several forms of *desynapsis*. Genetic desynapsis is common: many mutants are known to show chromosome pairing but no maintenance of the bivalents, either because no chiasmata are formed or, less frequently, because chiasmata are not maintained. It is not always simple to distinguish between these two possibilities, not only because it requires observation of chiasmata at early stages, but also because desynapsis is often variable in expression. This means that, when early diplotene chiasmata are observed in one group of cells but not at metaphase in a neighbouring population, the conclusion that chiasmata are formed but not maintained is not necessarily valid. The diplotene cells with chiasmata could have produced metaphase I cells with chiasmata, and the metaphase I cells without chiasmata could have resulted from diplotene cells without chiasmata that were not observed in the sample studied.

Inbred lines of outbreeders often show a reduced chiasma frequency that results in an increased number of univalents. This is a polygenically regulated partial desysnapsis.

For the many other effects of environmental (physical, chemical) and genetic conditions on recombination and random variation in recombination, the reader is referred to the literature (cf. Sybenga 1972; John 1990). In this text, such effects will only be referred to when relevant in context, for instance in Section 8.4.3.

It is clear that, when chiasmata are not formed because there is no chromatid exchange, there will be no genetic exchange and that this will be reflected in a low frequency of recombination in the few progeny that still can be formed. When pronounced desynapsis is followed by frequent first meiotic division restitution resulting in unreduced but viable gametes, the occurrence of exchange recombination can be tested. This requires analysis of the next generation in a half-tetrad analysis. Jongedijk et al. (1991a) report strong reduction of recombination between several markers in desynaptics of diploid potato (*Solanum tuberosum*) with high levels of restitution. In special regions, however, apparently due to changes in interference or pairing patterns, normally low recombination remained unchanged or even increased in the desynaptic. There may be other reasons for unaltered or increased rather than decreased recombination in desynaptics, especially in those that fail to show restitution and, as a consequence of irregular meiosis, have considerably reduced fertility: the few progeny plants formed are derived from a selection of meiotic cells with an almost normal level of recombination, because only cells that have a sufficiently high level of chiasmata will be able to go through a relatively normal anaphase and further meiotic stages. In view of the often irregular expression of desynapsis, this may be a reasonably large cell population. Recombination is then no indication of the type of desynapsis (operating before or after genetic exchange). Unless early diplotene can be studied in large and varied cell populations, it cannot be ruled out that chiasmata have formed but could not be maintained.

3.2.4.1.2.2.2 Genetic Variation in Pattern of Exchange

Exchange recombination does not take place randomly over the chromosomes. Especially the location of chiasmata has been studied in detail, because in some organisms it can be relatively easily determined. The pattern of chiasma localization is not constant, but can vary in nature as well as in the experiment. There are overall, usually genetic, effects and there are localized effects, often resulting from chromosomal rearrangements. In Section 8.3.4, this will briefly be discussed.

Chapter 4

The Somatic Chromosome Complement: Karyotype Analysis

4.1 The Karyotype

4.1.1 Characteristics of the Karyotype

The number of chromosomes over which the genome is distributed and the microscopically visible morphology of these chromosomes form the *karyotype* (Sect. 2.5). Chromosome morphology traditionally includes the length of the chromosomes, the location of the primary constriction (centromere) and, if present, the secondary constriction (near the NOR). Tertiary constrictions in somatic metaphase chromosomes are occasionally visible and can then be used to specify chromosome segments. They tend to coincide with heterochromatin.

Presently, a karyotype description includes chromosome segments with staining characteristics that are consistently different from the remainder of the chromosome. Depending on the staining technique used, such segments may appear as dark staining, light staining or as brightly fluorescing bands in the chromosomes. Usually such selectively stained bands are considered to be heterochromatic (Sect. 2.3), but whether or not this is the correct term for all such segments, there is no doubt that the structure of the chromatin in these segments is different from that of the remainder of the chromosome. The techniques producing differentially stained bands and the resulting karyotypes will be briefly discussed in Section 4.2.4.

The description of the karyotype can be further extended to include chromosomal characteristics that can not be made visible by mere staining. One is the total *amount of DNA* per somatic nucleus. This is known for many plant species (Bennett and Smith 1976). It is even possible to determine the amount of DNA and other components per chromosome (e.g. White and Rees 1987 for *Petunia*), but this is not yet a very common procedure. Other microscopically invisible chromosomal characteristics may be specific DNA sequences that do not affect the chromatin structure but that can be made visible by *molecular hybridization* with known homologous DNA or RNA probes. This may be high- or low-copy-number repetitive, nontranscribed DNA or even low-copy-number gene DNA. This too will be briefly considered in Section 4.2.4.

In mammals, especially in man, somatic cell genetic analysis has resulted in very detailed *gene maps* of chromosomes (Sect. 4.2.4.3) combined with a detailed karyotype description. These include variants of coding and non-coding, but unique, DNA segments known as restriction fragment length polymorphisms (RFLP). The first steps towards in vitro somatic cell genetics of plants have been set, but the detailed chromosome maps that can be obtained in animals are still a long-term objective for plants.

Recombination frequencies obtained from generative recombination analyses may be used as measures of distances between markers on chromosomes. When they involve microscopically visible chromosomal markers together with genes of which different alleles are available, or RFLPs, these genetic markers can also be included in the karyotype (Sect. 8.3.1). What results is a *genetic chromosome map*.

4.1.2 Applications of Karyotype Analysis

The plant breeder's interest in karyotype analysis is the *information* it provides, which can be used for a number of different purposes.

In *taxonomy*, karyotype descriptions are often used as a character in species descriptions (cytotaxonomy). Classical cytotaxonomy has had relatively little impact on taxonomical revisions (Grant 1984). It has at most been a useful additional factor in strengthening or weakening a conclusion based mainly on macromorphological characteristics. Yet in some cases, for instance in (allo)polyploid complexes, even simple chromosome number differences could well be a reason to separate different forms into distinct species, because chromosome numbers as such are discrete distinguishing characteristics, and chromosome number differences are usually effective genetic barriers for keeping closely related sympatric species evolutionarily separated. Such considerations have not been applied extensively in taxonomy. Yet, for understanding evolutionary processes and thereby indirectly for taxonomy, karyotype analysis and karyotype evolution can be of considerable importance (see, e.g. Ehrendorfer 1986). In addition, it may give important information to the plant breeder regarding barriers to the introduction of genes from related or more distant species.

Differences in chromosome number (other than polyploidy) and in chromosome morphology between wild species and cultivars suggest chromosomal differences that may disturb meiosis in the hybrid, endangering the proposed *gene transfer*. In addition, such differences, even if not directly affecting the transfer of specific genes because they involve other chromosomes, may be carried over to the cultivar if not detected in time. In later stages they may affect the breeding program.

An extended karyotype description including the location of known desired genes in relation to other markers (RFLPs, bands, recognizable rearrangements) contains useful information for the planning of gene transfer or chromosome segments. The *transfer* can be conveniently *monitored* by following the RFLPs by molecular methods, enzyme markers by biochemical methods and chromosome-morphologically marked segments by cytological methods.

When a genetic chromosome map (Sect. 8.3) is included in the karyotype description, with map distances as measures of *linkage* between *genes* and *markers*, the effort required to separate the desired gene from undesired genes or from recognizable chromosome segments can sometimes be estimated. When the map distances between these genes and markers are known, the recombinant may actually be already available. Such detailed extended karyotype descriptions are available for only a few species, among which are some economically important cultivated plants. The detail of genetic chromosome maps is rapidly increasing.

Chromosome morphological aberrations within species, between cultivars or incidentally arising within a cultivar, can cause unexpected, undesired complications. They may disturb, or even prevent, recombination and they may cause partial sterility. In some cases they are produced on purpose. When they alter the morphology of the chromosomes sufficiently, they can be recognized in the analysis of the karyotype and used for the classification of the rearrangement in segregating progenies. The same is true for *chromosome number variants*, occurring incidentally or induced. Such a classification of karyotypes, when only the known or expected variation is scored, is the simplest form of karyotype analysis but quite important in many fields of applied and theoretical cytogenetics.

4.2 Approaches to Karyotype Analysis: Chromosome Number and Morphology

The degree of refinement in karyotype analysis depends on the objectives of the analysis on one hand and on the characteristics of the material and the available input on the other. In several cases, the chromosome number or even the number of genomes per nucleus are accepted as the maximum of information. The reason may be that this serves the purpose. Another reason may be that the material does not yield more information (e.g. the chromosomes are very similar in size and do not respond to banding techniques etc.). The mere establishment of a chromosome number is the barest minimum of karyotype analysis, but is in many cases of considerable importance. Also, even though the material may be suitable for more sophisticated techniques and the extra information would be welcome, it is possible that the laboratory involved is not equipped or not prepared to spend sufficient effort to carry out these specialized techniques.

When planning an experiment or analysis, it is useful first to decide if one is capable of carrying out more than simple techniques and if so, whether it will pay to apply them in the situation at hand.

4.2.1 Number of Genomes Per Cell: Indirect Methods

The simplest characteristic of the karyotype is the chromosome number and the simplest interesting information in respect to chromosome number is the number of genomes. These are many instances where the determination of the number of genomes is important: comparing the ploidy level of related species or forms; checking material that has been treated for chromosome doubling; establishing the frequency of meiotic doubling; checking for endopolyploidy in explants from tissue culture; and testing the genetic stability of tissue, callus or cell culture.

For simply determining the number of genomes per cell, the simplest techniques can be used. Since the number of chromosomes determines the size of the nucleus and indirectly the size of the cell, it is often sufficient to determine the relative nucleus or cell size, or any attribute that depends directly on the number of genomes or indirectly on their relation to nuclear or cell size. Because of the variation in nuclear and cell size, and the indirect relation between nuclear and cellular characteristics, it is not possible to derive the number of genomes reliably from a single measurement. There are two main levels of variation that must be taken into account:

1. Nuclear and cell characteristics vary not only between species, but also between and sometimes within tissues of an individual; the least variable and most accessible tissue must be selected.

2. The number of genomes is usually not proportional to the quantitative measure of the characteristic analyzed and there is usually considerable variation here even in the most favourable circumstances.

It is necessary, therefore, first to establish the relation between the number of genomes and the characteristic measured, and then, on the basis of the variation between observations, to determine the number of measurements necessary to permit a reliable conclusion. In special cases, the results of the analysis are sufficiently accurate to conclude that incomplete genomes are in excess or deficient. Single chromosomes that are deficient or in excess, however, can almost never be traced.

The most direct measurement of the number of genomes (or, near the haploid level, even segments of genomes) is by measuring the amount of DNA per nucleus. Measuring UV absorption of unstained nuclei at a wavelength of about 260 nm or visible light absorption in Feulgen stained nuclei at 565 nm in microscopic cytophotometry is usually more laborious and not much more reliable for the purpose than more indirect techniques as discussed below.

A very useful direct DNA measurement for scoring genome numbers is the flow cytometry of isolated and fluorescent stained nuclei. Leaf segments, or parts of other organs with readily extractable nuclei, may be frozen in liquid nitrogen, ground in a mortar, suspended and shaken in an appropriate fixative, filtered to remove large debris, centrifuged to precipitate the isolated nuclei, resuspended in the fluorescent dye (e.g. DAPI: 4,6-diamidino-2-phenylindole) and run through the flow cytometer. Several variants are available. For pollen, slightly different methods are used (van Tuyl et al. 1989). The numbers of nuclei in each genome class are plotted. When the number of chromosomes per nucleus is relatively small, aneuploid individuals can sometimes be recognized by frequency peaks deviant from those of euploids. Because of the use of autofluorescence, the shape of the nucleus is much less critical than with absorption cytometry. It even appears possible to accurately determine and compare the amount of DNA in the generative and vegetative nuclei of pollen grains that have a very pronounced difference in DNA packing (van Tuyl et al. 1989). It is presently possible to have nuclear DNA amounts determined by flow cytometry on a commercial basis.

Other direct methods are those in which nuclear characteristics directly related to genome number are measured. As expected, the size of the mitotic metaphase plate is proportional, although not simply, with the number of chromosomes in the nucleus. Especially with small and numerous chromosomes, where direct counting, although not impossible, is laborious, measuring the metaphase plate may be used to establish the number of genomes. The accuracy is usually insufficient to detect aneuploidy, but is much better than measuring interphase nuclei in the same tissues. It is necessary first to estimate the variation between the measurements in order to establish the number of observations required to reach acceptable conclusions.

Another approach is the estimation of the number of "prochromosomes", conspicuous heterochromatic segments in interphase nuclei. When their number in the standard material is relatively small and when they are large enough and sufficiently well separated to be consistently recognized and reliably counted, they can occasionally present a simple way to establish genome numbers. Because of the fusion of heterochromatic bodies in interphase nuclei, in combination with possibly some somatic pairing between homologues, the increase of such bodies is not necessarily proportional to the increase in the number of genomes. In organisms with heterochromatic sex chromosomes (very rare among plants), these are good examples of countable heterochromatic bodies in interphase cells.

The number of nucleoli per nucleus depends directly on the number of genomes. It is a very general characteristic of nucleoli to fuse when more than one per nucleus are present, but nucleolar fusion is usually not complete, and in many cases the maximum number of nucleoli observed may be a good indication of the ploidy level. There are several complications. One is that occasional endopolyploid nuclei may give the impression that the entire organism has the ploidy level of those cells. A second complication is that some organisms have more than one nucleolus per genome. A well-known example of a crop species with two nucleoli per genome is barley. Tetraploid barley has

Table 4.1. Stoma characteristics in relation of ploidy level

A The number of chloroplasts in the stoma cells in plants with different ploidy levels (Gottschalk 1976)[a]

Species:	2x	3x	4x	5x	6x
Beta vulgaris	12–16	17–22	22–28		
Beta vulgaris	14.23	20.34	25.36	32.10	
Beta vulgaris	15.61	20.01	25.26		
Medicago sp.	9.2	9.1	12.8		15.2
Trifolium hybridum	8		14		
Brassica campestris (var. *rapa*)	11.4		19.3		

B The length of guard cells of stomata in plants with different ploidy levels, in percentage of diploid number (Gottschalk 1976)[a]

Species:	3x	4x	5x	6x	8x
Brassica oleracea		121			159
Plantago coronopus		141			190
Funaria hygrometrica	144	162			
Bromus inermis				131	165
Pirus malus	121	129	154		

[a] Note: aneuploidy may affect cell size and number of chloroplasts per guard cell in a specific way. This can disturb the analysis.

eight, but only very infrequently are all eight observed. When classifying diploids, triploids and tetraploids, it becomes necessary to study distributions per plant rather than maximum numbers, especially when incidental spontaneous (endo)polyploid cells occur.

Although these indirect nuclear methods are occasionally used for establishing genome numbers per nucleus, they are not much superior to indirect cellular methods and they all have the disadvantage that reasonably complicated preparations must be made.

Quantitative observations on cell characteristics, although not directly related to genome number, have had considerable application because they can often be made with very simple techniques. The most obvious is measurement of cell size, but cell size is quite variable within the individual, sometimes even within tissues, and some well accessible epidermis cells have shapes that hardly permit measurement. The most favourable material with respect to accessibility, lack of variation and ease of measuring are the cells, especially the guard cells, of the leaf or stem stomata. Their size and the determination of their number per unit leaf area have been in general use for determining ploidy number (Gottschalk 1976; Table 4.1). Simple techniques for preventing the underlying tissue from interfering with the measurements by making prints on tape have been developed (Sarvella et al. 1961). Even then, it is necessary first to establish the size distribution in order to know how many cells must be measured or counted per area for a reliable distinction between ploidy levels.

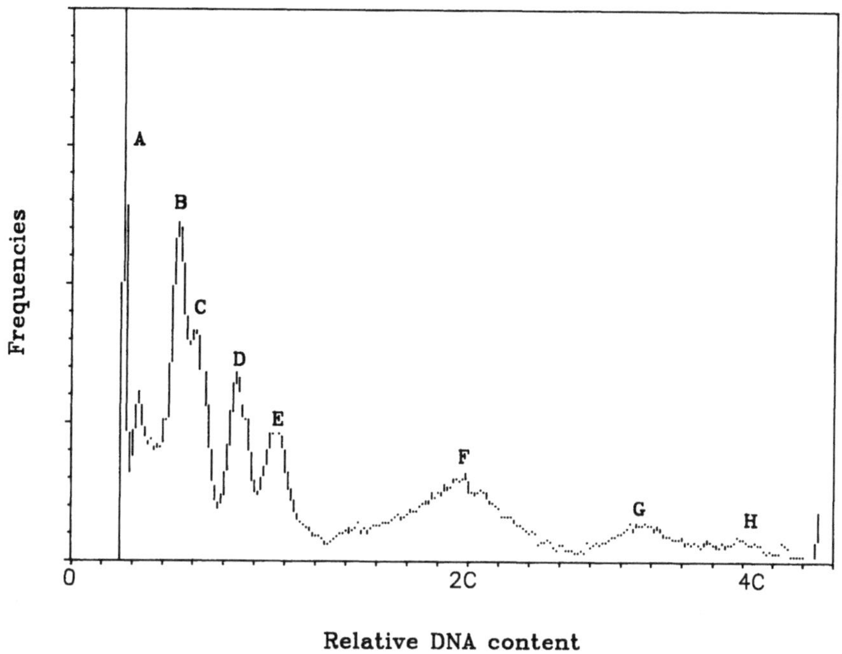

Fig. 4.1 Flow cytometry of the isolated metaphase chromosomes of a cell culture of *Haplopappus gracilis* (2n = 4). *A* Cellular debris; *B*, *C*, *D*, *E* the four chromosomes, modified during culture. Consequently, all four are different in size, but still in the original diploid number. Complete interphase nuclei with the 4C and 8C DNA level indicated on the *ordinate*. (After de Laat and Blaas 1984; courtesy of ITAL Wageningen)

Another frequently used attribute of stomatal guard cells is the number of chloroplasts per cell. This is an even less variable parameter and closely correlated to ploidy level (Gottschalk 1976; Table 4.1).

For the exact determination of chromosome number, especially when aneuploids are expected, these approaches are usually insufficient. When the number of chromosomes is relatively low, modern flow-cytometric DNA determinations are sufficiently accurate to detect the presence or absence of large single chromosomes. With larger numbers of chromosomes or when a small chromosome is involved, the chromosomes must be made visible individually. A variant is the flow-cytometric measurement of chromosomes separated from the cell. An example is given in Fig. 4.1 of a *Haplopappus gracilis* cell line cultured in vitro for many generations. It still has the diploid number of four, but with clear morphological abnormalities (de Laat and Blaas 1984). When this approach is insufficient or not available, light microscopy reparations are required of tissues with sufficient mitotic activity. When simple chromosome counts must be made, the quality requirements of the preparations are much less stringent than for a complete karyotype analysis.

The same is true when simple, well recognizable chromosome structural variants are scored.

4.2.2 Exact Chromosome Number; Chromosome Morphology

The traditional karyotype descriptions refer to chromosomes in the contracted state, i.e. they must be at prometaphase, metaphase or anaphase of mitosis, exceptionally of meiosis. This prerequisite restricts the material suitable for karyotype analysis to tissues with a high mitotic rate: root tips, very young leaves, shoot tips, pollen during pollen mitosis, rapidly growing tissue, callus or cell cultures. Except in in vitro cell cytogenetics, root tips are preferred because of (usually) easy accessibility, suitability for preparation and often high rate of cell division, which can be stimulated by cultural techniques. However, several methods have been developed for studying chromosomes in young leaves and vegetation points, which are generally the only material available from established plants, shrubs and trees in the field. These methods resemble those for root tips, with some modifications. The reader is referred to the specialized literature. Occasionally, endosperm mitosis in developing seed has been used for chromosome studies. In several instances pollen mitosis has been found to be favourable material, notably because of the haploid chromosome number. Yet its use has been limited, mainly because of insufficient availability of pollen at the right stage, technical complications and sometimes the compactness of the metaphase plate.

4.2.2.1 Techniques of Preparation

For details on techniques, the reader is referred to the specialized literature (e.g. Darlington and LaCour 1976; Dyer 1979; Sharma and Sharma 1980; or newer editions of these books), but research reports on chromosome studies of specific plant species are usually the best source of information on how to prepare mitotic cells for karyotype analysis. Here only a few main lines are indicated.

The choice of the starting material is important. Because of accessibility and ease of prepation, root tips are favoured when available. Techniques have been developed to optimize mitotic frequency in root tips. A very simple but also important prerequisite is that growing conditions for the plant are optimal, with ample air and often minimal, although sufficient, moisture around the roots. Pretreatments that suppress the spindle and that are primarily intended to spread the chromosomes throughout the cell, to contract and straighten them, also tend to accumulate metaphase cells because anaphase is suspended. Large numbers of mitoses in young seedlings can be obtained by choosing the optimal time after germination when waves of cell divisions occur (Wolff and Luippold 1956). Synchronization can be induced by several substances that block development reversibly, such as hydroxyurea (Oud et al.

1979) or 5-aminouracyl (Wagenaar 1966). Such techniques require experience before they can be applied effectively and many cytogeneticists simply accept the material as it comes.

The normal mitotic metaphase cell is usually not very suitable for detailed analysis of the chromosomes: the chromosomes are not straight, unless they are short by nature, and the equatorial plate is compact. Many treatments have been developed to prevent microtubule polymerization and consequently to prevent the spindle from being formed or maintained. This results in the scattering of the chromosomes over the cell and in continued contraction. As a consequence, after the proper treatment, the chromosomes come to lie straight and well separated in the cell when the tissue is flattened by squashing or spreading. A day in ice water (0 °C) will often contract metaphase chromosomes and spread them over the cell. Colchicine in water (e.g. 0.5%) is frequently used because of its usually good and reproducible result, but there are several cheaper and less toxic alternatives.

After the pretreatment, the cell is usually fixed in the mixture of acetic acid with alcohol (1:3), with or without chlorophorm added, but many variants, most of them developed at the end of the last centrury, are known. Many grass root tips are excellently fixed in 1N hydrochloric acid at 58–60 °C. This, at the same time, takes care of the next step: maceration.

Maceration is necessary to make the tissue soft enough for spreading out into a single layer of flat cells by squashing or other spreading techniques. It can be done simply in 1N HCl at 60 °C for 6–12 min. Variations for special material or special tastes are available. Too long a duration of maceration may destroy the material or greatly reduce its stainability. For some techniques, especially most banding techniques, a HCl-maceration strong enough to sufficiently soften the material is not compatible with the further steps in the procedure. Then, but also for many other situations, an excellent alternative is a treatment with a mixture of cellulase and pectinase (each 1–2%) at the prescribed temperature for a few hours. The softened tissue can be squashed on the slide or, often quite effectively, suspended in aceto-alcohol and dropped on a pretreated slide (Mouras et al. 1978; Pijnacker and Ferwerda 1990). This technique is especially suitable for C-banding and in situ hybridization (Sects. 4.2.4.1 and 4.2.4.2). Following maceration (for C-banding after spreading on the slide), the material is stained.

The cells must be flat, in a single layer, but still complete. The last requirement cannot often be realized when the first two have been met and this can present serious problems for a reliable karyotype analysis. Squashed or spread preparations can be made permanent by inclusion in resins that harden in the air. Several different techniques are available.

4.2.2.2 Observations: Measurements

For simple counting and the recognition of a known chromosomal rearrangement it is not necessary that the cell be entirely flat and not overlying another cell. For detailed karyotype analysis, the requirements are much more stringent. An experienced cytologist will be able to make preparations with a sufficient number of complete but flat cells, where the chromosomes not only can be accurately counted, but also measured. Here a serious problem is encountered. When the cell is flattened, the chromosomes are stretched. They are more resistant to stretching than the cytoplasm, but they do stretch, and stretching is not homogeneous. It has been shown that some chromosome arms stretch relatively more than others, which will produce a bias (Sybenga 1959). After extended staining with orcein, chromosomes may even break. In addition, the contraction is different not only between cells, as is expected with slight differences in stage of (blocked) mitosis, but also within cells, even between homologues and chromatids of the same chromosome. It is necessary, therefore, to repeat the measurements in several cells and to analyse the results statistically.

For measuring chromosomes in a cell, several techniques have been developed. It is possible to measure directly in the microscopic preparation, using special measuring devices attached to the microscope. Static ocular micrometers are insufficiently accurate, but screw ocular micrometers are very precise. They are not convenient, however, for measuring large numbers of chromosomes in large numbers of cells. In addition, since the chromosomes can not be marked after having been measured, a drawing or other help to the memory must be made to indicate which chromosomes have had their turn. More convenient and equally accurate are systems using a camera lucida, provided these are equipped with satisfactory measuring devices. Simple pen or pencil drawings measured with a ruler, or rulers which are aligned directly along the chromosome in the microscopic image give insufficiently accurate results. Much more accurate are thin white lines in a black background that can be brought to the two ends of the chromosome or chromosome segment to be measured. A fixed system of two lines under a small angle (Fig. 4.2), between which the chromosome can be exactly fitted, was used by Sybenga (1964). Such direct measurements do not require the cell to be completely flat, provided the chromosomes are tilted only slightly. Partly automated variants of this method are available, but have not been published. Their accuracy is the same as measuring with the ocular screw micrometer, using the system in Fig. 4.2 or measuring in highly magnified photographs. Only straight chromosome segments can be measured, but one advantage, besides accuracy and convenience, is that the data can be processed directly when the measuring device is linked to a computer.

Enlarged photomicrographs made with maximal resolution, or their negatives projected on a screen, are quite useful, but the cells must be completely flat or several photographs of the same cell must be made to cover all chromosomes. With sufficent enlargment, the chromosomes can be measured with a common ruler. When the photograph has been taken with a lens with less than maximal resolution, replication of the measurement may, to a great extent, compensate for the loss of accurracy. The same is true for

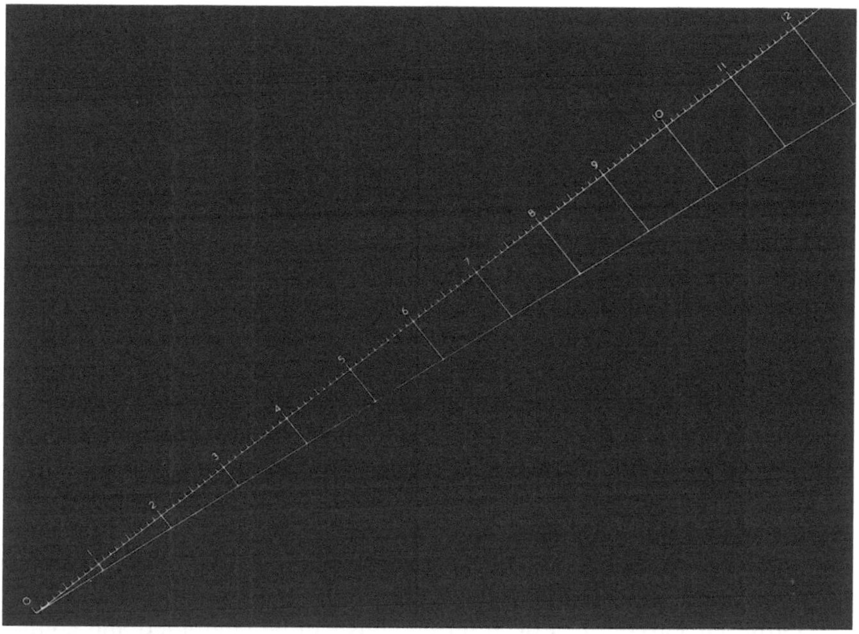

Fig. 4.2 Sets of two white lines on a black background for measuring chromosomes in the camera lucida. The chromosome segment to be measured is placed perpendicular to the calibrated line and where it fits exactly between the two lines, the value on the calibrated line is read. The units are arbitrary, but usually the optical system can be adjusted to make them equal 1 mM. (Sybenga 1964)

direct measurement in a microscope (Sybenga 1959). Measurements on a videoscreen connected to a videocamera on a microscope have been shown to be insufficiently accurate because of the thickness of the glass and distortion from electronic sources.

Most of the devices mentioned can measure straight objects only. A bend in a centromere or other clearly marked place presents no problem, but a bend, and even more a smooth curve, in a chromosome arm makes exact measurement almost impossible. In several cases, especially with small chromosomes, techniques are available to prepare sufficiently straight chromosomes. In other cases this is not possible and special solutions must be found. Direct measurement of curved chromosomes in photomicrographs is possible when marks can be made at the bends and the subdivisions can be measured separately, although the accurracy is disappointing. Another simple, but not very exact, solution is to overlay the chromosomes with pieces of wire and to measure these. With the use of a graphics (digitizing) tablet, curved objects can be measured directly through the camera lucida attached to the microscope. Here again, it is necessary for optimal resolution to have a black background and a very small white dot or light source at the measuring point in the "mouse". A simple pen is not sufficient. The advantage of a graphics tablet, besides being able to measure curved lines, is the possibility of feeding the results directly into a computing system. It should be noted, however, that measuring with a graphics tablet requires special corrections or special programs in order to prevent over- or underestimation of the length of the curved lines. The measurements can also be made indirectly from photomicrographs, but this again requires completely flat cells or a series of photographs. For C-banded preparations, this is usually no problem, but after gentle squashing without an air-drying step, which usually gives less distortion, flatness is a bottleneck.

There have been interesting developments in automated chromosome measurement and idiogram construction. The basis is a digitized picture of a cell produced by a video camera. By image processing, the chromosomes and their landmarks, such as primary and secondary constrictions but also different types of bands, are separated and individually processed. It is possible to make a karyogram (Sect. 4.2.3.1) of an individual cell with the chromosomes straightened out, aligned and ordered according to length and other characteristics. Different karyograms of different cells can be combined and compared. Data on length of chromosomes and location of well recognizable markers can be recorded numerically. The processed images can be stored on disk. Especially in human cytogenetics, considerable progress has been made (Piper and Lundsteen 1987; Lundsteen and Piper 1989). Yet full automation is usually not attained nor desired. Interaction between technician and machine appears to be more effective. The first steps in automation, automatic metaphase cell-finding and autofocusing, have been available for a long time (see, e.g. Green and Cameron 1972) and is applicable to plant cell preparations. Chromosome separation, ordering and straightening are possible for chromosomes lying free (see, e.g. Fukui 1986; Armstrong et al. 1987). The most advanced system presently available for plant chromosomes is probably that described by Houben et al. (1990). Especially when the chromosomes are bent and contain several markers, simple computer straightening and lining up is a great help. Without interaction with the operator, acceptable results have not yet been attained, even when chromosome overlapping does not occur. A gradual further development is expected. The bottleneck is not in the machine, but in the preparation and the software.

4.2.3 Presenting the Karyotype

4.2.3.1 Karyogram and Idiogram

The karyotype is a characteristic of the genome: the number and morphology of the chromosomes. It is not a figure. There are two main ways to present a karyotype graphically. The first is the *karyogram* (Fig. 4.3B), in which a representative cell is photographed (Fig. 4.3A), the chromosomes cut out from the photograph and collected in pairs of (presumed) homologues, and the pairs lined up in descending order of length. It is a very demonstrative way, but when the chromosomes do not have specific morphological characteristics that make individual identification possible, it may be misleading. Variation in the length of the chromosomes, not only the total length of the entire complement as a result in differences of contraction but, more important, the relative length of each chromosome in the complement, may be considerable. This variation is not expressed in the karyogram. Consequently, the incorrect impression is given that the differences between the chromosomes as shown in the karyogram are real and that the chromosomes can be distinguished and identified even when the actual differences between non-homologous chromosomes are small compared with the variation in the observations. Presenting the karyotype as a simple karyogram with uniformly stained chromosomes used to be the standard procedure for human chromosomes before the advent of Giemsa-banding techniques. At present all human chromosomes can be identified individually by their banding pattern and even small hetermorphisms between homologues can be recognized. The order according to length is maintained. Before G-banding was possible, large groups of chromosomes had to be taken together, as they differed too little in size and location of the centromere to be identified. For most plant species this is still the case, as either banding is not possible with sufficient detail or the technical procedure required is considered too tedious to be acceptable for large numbers of cells. In such cases the chance of misclassification is considerable.

The alternative representation of the karyotype is the *idiogram*, an ordered set of idealized chromosome diagrams, the length representing the length of each chromosome, and the place of the primary and secondary constrictions, as well as other recognizable markers like C-bands, drawn at the proper locations. Figure 4.3C gives four (apparent) idiograms in addition to the karyograms, but the lengths of the arms are not based on repeated measurements. The measures given should be the averages for the same chromosomes in a number of cells and should not represent a single example as in the karyogram and in Fig. 4.3C. The chromosome length, and the length of the arms or other recognizable segments as given in the idiogram, can be the relative length (percentage of total genome length) as in Fig. 4.5D, or the average length in μm at mitotic metaphase after a standard pretreatment. Because of variation between cells in chromosome contraction, the proportional length is usually preferred, with an average of the total complement length in μm given as additional information. If desired, the actual average chromosome length can then be derived.

The problem of non-identifiable chromosomes is even greater in an idiogram than in a karyogram because, in an idiogram, the sizes are by definition averages for particular chromosomes, which must first have been identified. When this is not possible, the relative chromosome length or location of the primary constriction for a particular chromosome in one cell need not at all correspond to the relative length or the location of the primary constriction of the same chromosome in another cell. Several serious mistakes have been made by simply ordering the chromosomes in each cell according to length and centromere position and assuming that this is indeed always the correct order. It is quite possible that, instead of homologues, merely the most similar chromosomes are pooled to derive average length and centromere

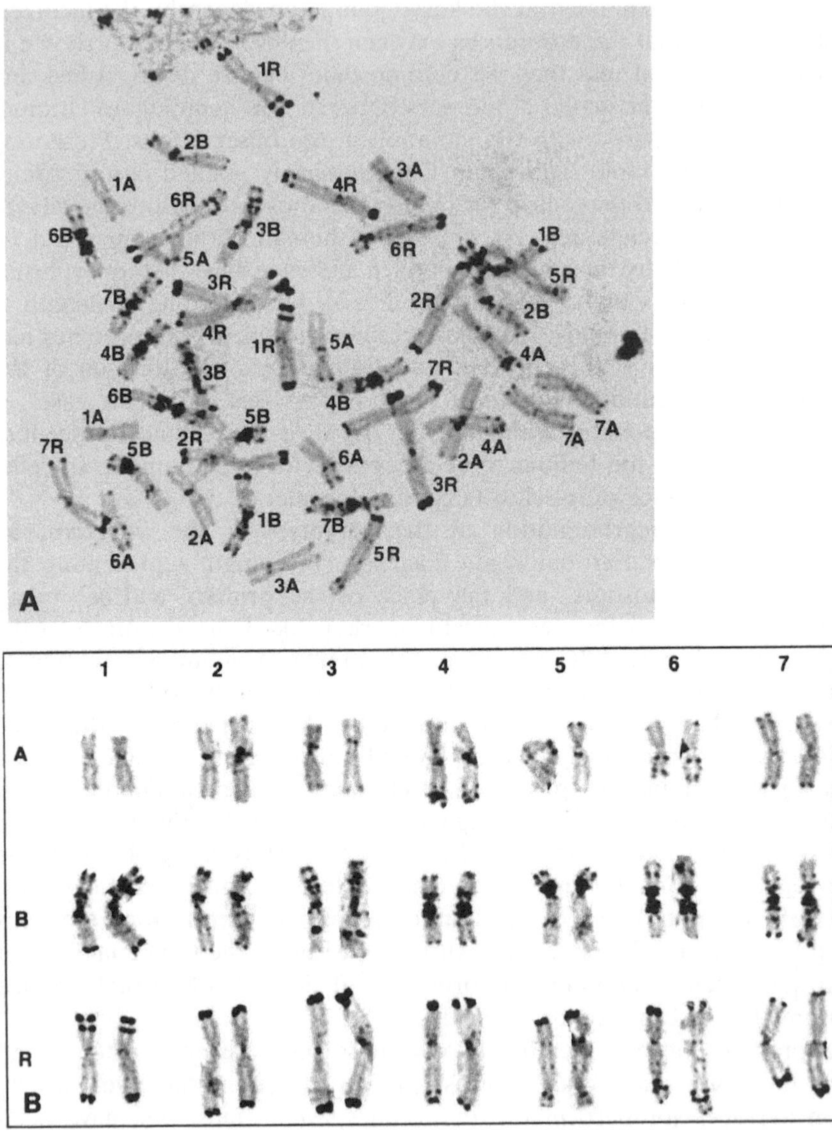

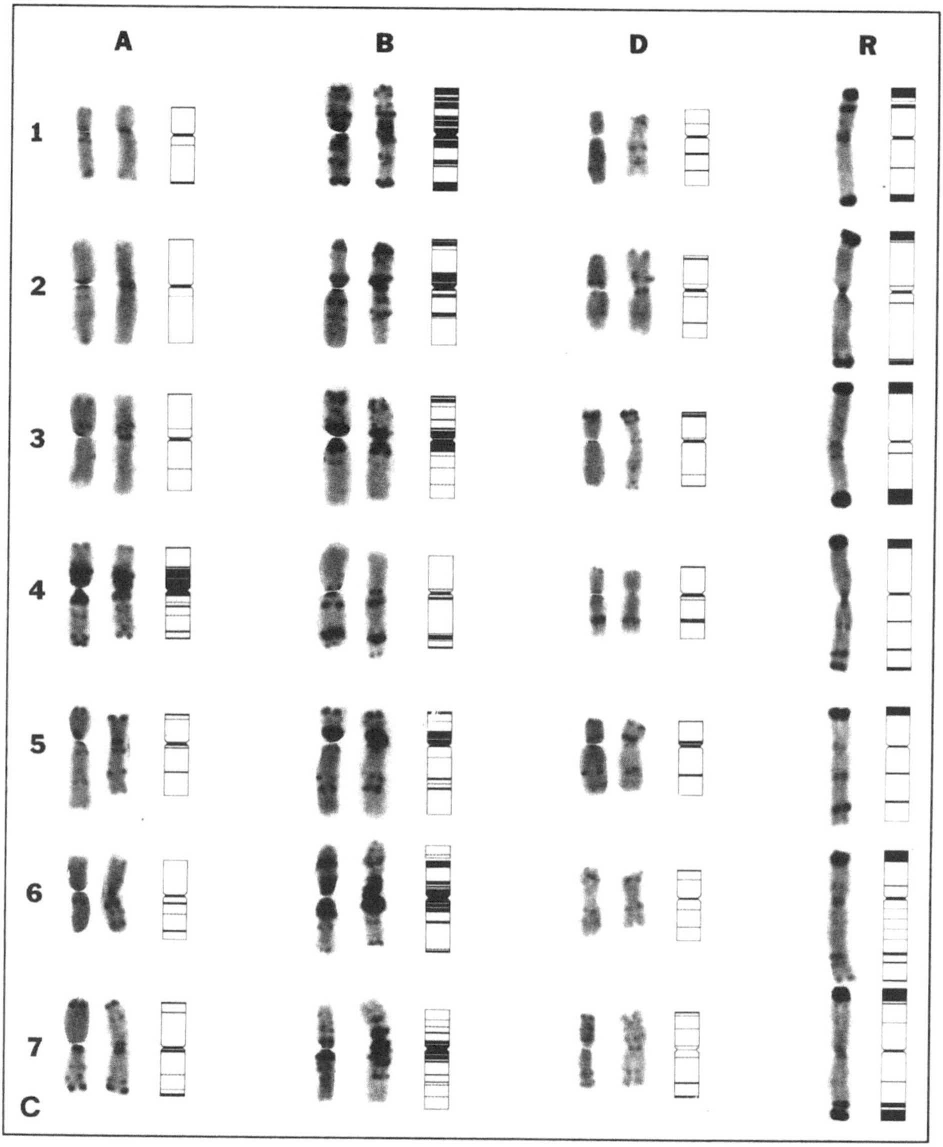

Fig. 4.3 A. C-banded squashed root tip cell of 6x triticale with 14 chromosomes of rye (*1R–7R*), having pronounced telomeric heterochromatic bands in most of the arms, and 28 chromosomes of wheat. The B-genome chromosomes (*1B–7B*) of wheat have pronounced banding around the centromeres. The A-genome (*1A–7A*) has a much less characteristic banding pattern. **B**. The karyogram of **A**. **C** The karyograms of wheat (Chinese Spring: *1A–7A*, *1B–7B*, *1D–7D*) and rye (Danskowskie Zlote: *1R–7R*). Diagrams of the banding pattern are given, resembling an idiogram, but not based on measurements (cf. Fig. 4.5D). In **C**, the *left* chromosomes of the pairs are composed of telocentrics. The nomenclature has not yet been adjusted to the generally accepted view that 4A and 4B should be interchanged. (Courtesy of A. Lukaszewski)

position. The result is an overestimation of the differences between chromosomes and a greatly reduced apparent variation in the sizes of the individual chromosomes. Approaches to karyotype analysis when the chromosomes cannot be individually recognized, and some of the statistics involved, will be briefly discussed below.

4.2.3.2 Idiogram Construction: Coping with Variation

In simple cases, for instance for cytotaxonomic purposes, it suffices to give approximate length and general location of the centromere, using generally accepted codes as given by Levan et al. (1964). When the centromere is located in or near the middle of the chromosome (a *median* centromere), the chromosome gets the indication *M* (*metacentric*); when the centromere is in the median region but not exactly in the middle, the chromosome is *m* and still considered metacentric; a submetacentric chromosome is *sm*; a subacrocentric chromosome with the centromere in the subterminal region is *st*; an acrocentric chromosome with the centromere in the terminal region is *t* and a really telocentric chromosome is *T*. Naranjo et al. (1983) present a simple template, three diverging lines along which photographs of the chromosomes can be aligned to see which category they belong to. The chromosome is thought to be divided in eight equal segments and the transition arm ratios are considered to be 5:3, 6:2 and 7:1. The karyotype can be simply described by giving the number of chromosomes belonging to each category. Satellited chromosomes (having a secondary constriction) are a separate category.

When more details are required, simple classification in categories is insufficient, and accurate measurements and statistical methods for dealing with variation are necessary.

There are seven sources of variation in chromosome and chromosome segment size that are important for karyotype statistics. The following five can be considered "random" variation: (1) errors in measurement; (2) variation due to techniques (pretreatment, fixing, squashing, spreading); (3) non-induced, non-intrinsic variation between chromosomes within cells, possibly due to variations in contraction; (4) variation between cells due to differences in cellular stage of contraction; and (5) variation between cells not due to differences in cellular stage of contraction. Two sources of variation are essential components of the idiogram itself: (6) intrinsic differences between non-homologous chromosomes, and (7) intrinsic differences between homologous chromosomes (heteromorphism). The first five sources of variation must be separated from the last two before a reliable idiogram can be constructed.

A common and serious problem in karyotype analysis arises when the non-intrinsic variation is of the order of magnitude of the differences between non-homologues. Then non-homologues can not be reliably distinguished from homologues. At the same time, chromosomes may look treacherously different and still fall statistically within the same confidence limits of size and centromere location and, in fact, may be homologues. To aid in chromosome definition in situations of unreliable recognition, different methods have been applied. A first step is visualizing the distribution of the measurements of chromosome segments, for instance by plotting pairs of different chromosome characteristics against each other. The length of the short arm against the length of the long arm, or the arm ratio against total chromosome length are common combinations (Fig. 4.5). Such plots, however, are not sufficient in themselves; they can even be misleading. And when they permit a direct, unambiguous classification of the chromosomes, this would have been possible without plotting. In order to construct the idiogram that best represents the real situation in cases where direct classification is impossible or suspect, the non-intrinsic or random sources of variation should be analyzed first.

1. Errors of Measurement

a) Gross mistakes in measurement or in recording, even though seemingly improbable, appear to be inevitable. A simple check on the probability of each measurement, although a boring excercise, is necessary to prevent such errors. Similarly often overlooked, but of a quite different nature, is statistical variation in the results of the measurements. This is mainly due to the difficulty of exactly determining where the chromosome segment considered starts and where it ends. The use of a graphics tablet or other (usually less accurate) devices introduce special errors, some of which can be corrected by the proper data processing. Other technical problems, for instance the difficulty of following the core or the outline of the chromosome with the measuring point, require replication of the measurements. It should be clear that, with small chromosomes, the errors of start and finish of each individual measurement are more important than with large chromosomes. With curved chromosomes, the difficulty of following the core or contours is important, and large chromosomes tend to be more susceptible to bending than small ones.

b) The resolution of the optical system used is a factor of importance. Sybenga (1959) compared the accuracy of measurement using a 40x, 0.65 n.a. lens with the accuracy using a 100x, 1.3 n.a. lens on orcein-stained squash preparations of rye root tips (*Secale cereale*). Measurements were made both directly with an ocular screw micrometer and indirectly in the projection of a photograph at large magnification. It appeared that the same reliability could be obtained using several times the number of replications of measurement and the 40x lens as compared to the 100x lens. The first was more convenient, because the chromosomes did not have to be completely flattened. Flattening does have an effect, as shown below. The effect of tilt of only slightly flattened cells appeared to be smalll.

c) Automated karyotyping using a digitized video system has its own requirements. The accuracy depends, in addition to the factors operating for all other methods of measurement, on the electronic systems, especially the number of lines used and the resolution within lines. It is possible to increase image definition by repeated registration of the image (image enhancement). Special high-resolution cameras are useful only when the microscopic image contains information that is not observed with standard cameras. With high microscopic magnification this is not necessarily the case, because the resolution is limited by the optical system, including the wavelength of the light used. With laser scanning microscopy, better resolution with relatively low magnification can be obtained than with normal microscopy and then the characteristics of the electronic system may become important. All these refinements make sense only as long as other sources of error are not considerably larger.

2. Variation due to Preparation

a) The clearness of chromosome outlines, as shown above, is an important character for maximizing accuracy of measurement. Pretreatment, fixation and staining affect not only the contours but also absolute and even relative chromosome size. Exactly how serious the effects of the pretreatment are is not clear. In any case, when presenting an idiogram with quantitative information, it is necessary to mention the pretreatment used for conditioning the chromosomes and the methods of fixation and staining.

b) Techniques of preparation, especially squashing and spreading, have pronounced effects on chromosomes size. This is not detrimental to karyotype analysis as long as all chromosome segments are equally affected. Mouras et al. (1986) compared chromosome characteristics of haploid cells derived from anther culture of *Nicotiana plumbaginifolia* after different preparative treatments. They concluded that hypotonic treatment of protoplasts followed by spreading and air drying did not result in chromosome sizes different from those after squashing. Flame drying instead of air drying resulted in a triple increase in chromosome volume. Variation in response between chromosomes was not studied. Squashing itself can have considerable and non-homogeneous effects. By comparing different intensities of squashing, Sybenga (1959) concluded that, in rye, the "better" the squash, i.e. the more flattened the chromosomes after the most intensive squashing, the more the chromosomes were stretched. More important was the observation that the long arms increased in length more than proportionally compared to the short arms. It is possible that differences in heterochromatin, which may be more compact and consequently slightly more resistant to squashing, may play a role, but this has been insufficiently documented. Several of the short arms of the non-metacentric chromosomes of rye usually have a larger heterochromatic segment than the long arms, which may be practically free of visible heterochromatin.

3. Random Variation Within Cells

Random variation in chromosome measurements within cells is best analyzed by comparing homologues, provided these can be assumed to be identical (which is not necessarily so) and can be distinguished from other chromosomes. Random variation between identical homologues gives an impression of the variation level in general. If polymorphisms in chromosome length are analyzed, this type of variation plays a role, but cannot be directly determined. This dilemma will be further considered in point 5.

4. Variation Between Cells due to Differences in Stage

a) Differences in contraction due to differences in mitotic stage result in large size differences, because in many organisms the chromosomes continue to contract during prometaphase-metaphase. The accumulation of cells in metaphase after treatments with spindle suppressors makes it possible to select cells with maximally contracted, and therefore straight, chromosomes, but the variation in average chromosome length per cell cannot be completely eliminated. The most obvious solution is not to express the length of chromosome segments in absolute length but in relative length, for instance as a percentage of the total chromosome complement. Occasionally, one readily recognizable chromosome is used as the standard and the other chromosomes or their component segments are given as a percentage of that standard. This is not a very attractive method from a statistical point of view. When comparing different species, different genotypes or different (pre)treatments, some information on absolute chromosome length may be desired. This may be the length in micrometers of the total chromosome complement or of the average chromosome. It should be noted that, because of the great change in chromosome length during mitosis, absolute chromosome length is not a very consistent parameter. What is given as chromosome length is usually the length after a certain pretreatment and a certain method of making the preparation. For idiograms, it is customary to give relative chromosome segment length when comparing different chromosomes and changes therein. The absolute length can better be given in a separate scale.

b) In addition to overall differences in length as a result of differences in contraction, there may be relative differences. Heterochromatin usually contracts earlier in the course of prophase, and consequently less at later stages than does euchromatin, and there may be more, less easily detected differences between chromosome segments. As a result, differences between chromosomes and even variation within chromosomes (between homologues and between cells) may result from contraction differences. These are generally difficult to trace and then are confounded with "random" variation between cells.

5. Random Variation Between Cells in (Relative) Chromosome Segment Length

The variation between cells within individuals is the most readily handled form of variation. The cells may be considered genotypically identical (except for rare systematic variation and incidental errors) and the number of replications of the measurements can be chosen to meet the requirements of statistical analysis. The between-cells variation can be estimated by measuring specific recognizable chromosomes. When two homologues of each chromosome are present in each cell, both between-cells and within-cells variation can be estimated, provided the homologues are a priori known to be identical. When this is not known, the analysis becomes much more difficult. In principle it is possible to detect a difference between two homologues by analyzing the length distribution of the chromosome or of one or more of the chromosome segments (usually two arms, occasionally one, with or without a satellite). When this distribution can be statistically shown to represent two distinct populations, the homologues may be assumed to be different. Such differences can be the result of rearrangements, but also of polymorphism for heterochromatic segments or of differences between homologues in facultative heterochromatinization. When there is apparently only one population, this does not exclude the possibility of slight differences, but large differences are improbable.

Between–cell variation can be very disturbing in distinguishing between different chromosomes with similar morphology. It may cause so much overlap in size or location of the centromere, that mere classification on the basis of direct measurement leads to misclassification in a large number of cases. Even when only two pairs are considered, the four chromosomes together will not usually permit a distinction of two different distributions even when the two pairs are not identical. The best reported approaches for testing whether closely similar pairs can be distinguished statistically are based on a preliminary classification of length or arm-length ratio. Preclassification on such an unreliable basis tends to increase the apparent difference between chromosomes, but is the simplest way in situations where differences between chromosomes are small.

Kessous et al. (1968) ordered the chromosomes of *Salamandra* spp. ($2n = 24$) per cell in sequence of decreasing length, as is customary in karyotype analysis. It appeared that three groups, each with four chromosome pairs, could be distinguished: one group with long metacentric chromosomes; a second group of metacentric to subacrocentric medium-sized chromosomes and a third group of small submetacentric chromosomes. The presumed homologues were pooled. Each set of two consecutive chromosomes in the ordered series of 12 chromosomes was tested with respect to total relative length and centromere index (length of short arm divided by total chromosome length) using the F-statistic. It appeared that, even after preclassification, no statistically significant difference in chromosome length or in centromere index between any of the chromosomes of the first and third groups (of the four groups distinguished) could be demonstrated. Between the groups and within the second group, length or centromere index differed, but not enough to exclude at least some overlap.

More sophisticated, but based on a similar preclassification, is the stepwise discriminant function analysis of Harris et al. (1973), applied to groups of similar human chromosomes. The analysis was designed before the general application of G-banding, which presently allows the distinction of all human chromosomes, but the approach remains interesting for cases where such

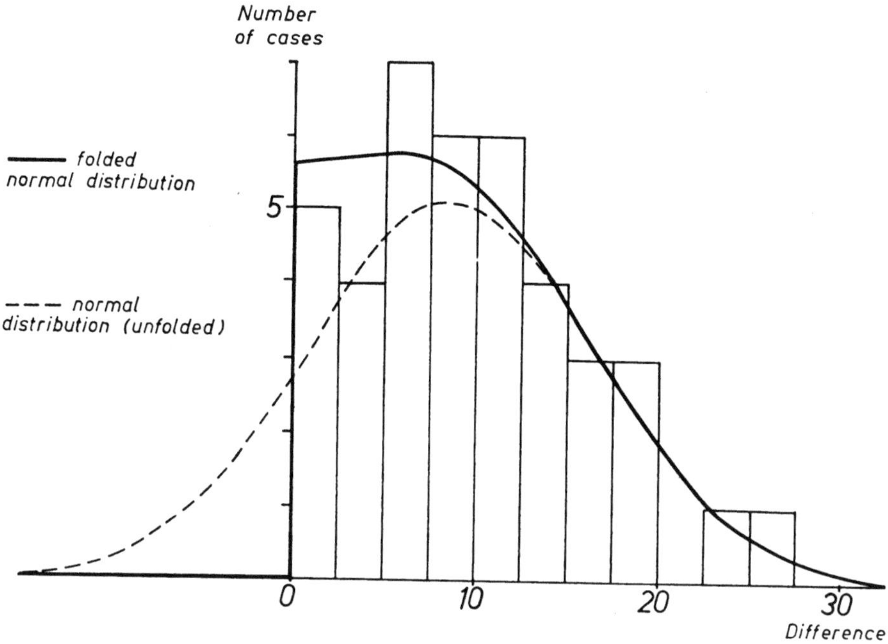

Fig. 4.4 Histogram of the distribution of 40 differences between the lengths of the longer and the shorter arm of chromosome II of *Larix decidua*. The histogram shows a folded distribution. The corresponding normal distribution is shown as a *dashed line*. (Matern and Simak 1968)

distinction is not possible. In their analysis, Harris et al. (1973) only considered the lengths of the short and the long arms of each chromosome. The frequency of misclassification could be estimated and gave an excellent impression of the errors made in classical karyotype construction. For details, the reader is referred to the original publication.

(Pre)classification on the basis of chromosome length, without strict identification, results in reduced variation between chromosomes and in a spurious increase in the difference, or in an unrealistic reinforcement of the conclusion that the difference is real (Essad et al. 1966). Variation in the difference between arms can also be spuriously reduced when the arm measured as the longer arm in (sub)metacentric chromosomes is systematically considered to be always the same. This may lead to the erroneous conclusion that the arms are different or that the difference is larger than it is in reality. The effect was analyzed by Essad et al. (1966) for the ratio between the short and long arms of specific chromosomes of diploid *Lolium perenne*, and by Matérn and Simak (1968) for the difference between the two arms of a representative chromosome in the haploid complement of *Larix decidua* (Fig. 4.4). The frequency distribution of the transformed arm-length ratio and of the arm-length difference showed truncation at ratio 1 (Essad et al. 1966) and difference 0 (Matérn and Simak 1968). Both can be considered a folded normal distribution (if necessary after transformation of the original observations), from which the actual distribution can be reconstructed. Matérn and Simak (1968) refer to Johnson (1962) for a method in applying this correction, for which Essad et al. (1966) give their own solution. The mean of the unfolded distribution corresponds to the actual ratio or difference. Matérn and Simak (1968) also present a method for estimating the risk of arm reversal when the arm-length ratio is close to 1.

The use of the haploid complement in karyotype analysis simplifies the analysis considerably. In gymnosperms, haploid cells are readily obtainable from endosperm tissue (Matérn and Simak 1968). In angiosperms, haploid mitoses can be obtained from haploid plants (Mouras et al. 1986) developed after in vitro embryogenesis in cultures of anthers or free spores, or arisen parthenogenetically from unfertilized egg cells. Haploid cells are also found in the spore divisions, especially in the readily available microspores either in the anther or in the germinating pollen tube in the style. Several methods for studying microspore mitosis have been available for many years (Conger 1953; Kwack and Kim 1967).

There may be tissue-specific differences in chromosome characteristics. Larsen and Kimber (1973), for instance, found that chromosome 5B of bread wheat, when measured in meiotic telophase II, had a much larger arm ratio (about 2.6) than the most heterobrachial chromosome in the somatic karyotype in the same plants (about 2.1). When two telos of 5B were present instead of the metacentric chromosome, it could be recognized without doubt and again this ratio was found. The phenomenon may be of scientific interest; for karyotype analysis it contains a warning.

4.2.3.3 Idiogram Construction: Plotting the Observations

The discussion in the preceding section of the different aspects of variation in chromosome measurements and some suggestions for coping with it do not yet

→

Fig. 4.5 Plots of the lengths of the chromosome arms of a homozygote of translocation T662W (3R/6R) of rye, where the two translocation chromosomes can be recognized in addition to the satellite chromosome. The short arm of this chromosome is measured including the satellite. No C-banding is applied; five diploid cells: each chromosome represented ten times.

A All chromosomes together. *Ordinate* Long arm; *abscissa* short arm, in % of total genome length. The distribution of some chromosomes is apparently folded around the line where the arms are equal. The recognizable chromosomes have not been marked.

B The three recognizable chromosomes: the small and large translocation chromosomes (*large dots*) and the satellite chromosome (*small dots*). *Ordinate* Long arm; *abscissa* short arm, both in % of total genome length. The two chromosomes from the same diploid cell are connected by a line. The three chromosomes are readily distinguished and the averges and standard error for the arm lengths can be determined. The distribution of the small translocation chromosome is clearly folded around the line where the arms are equal

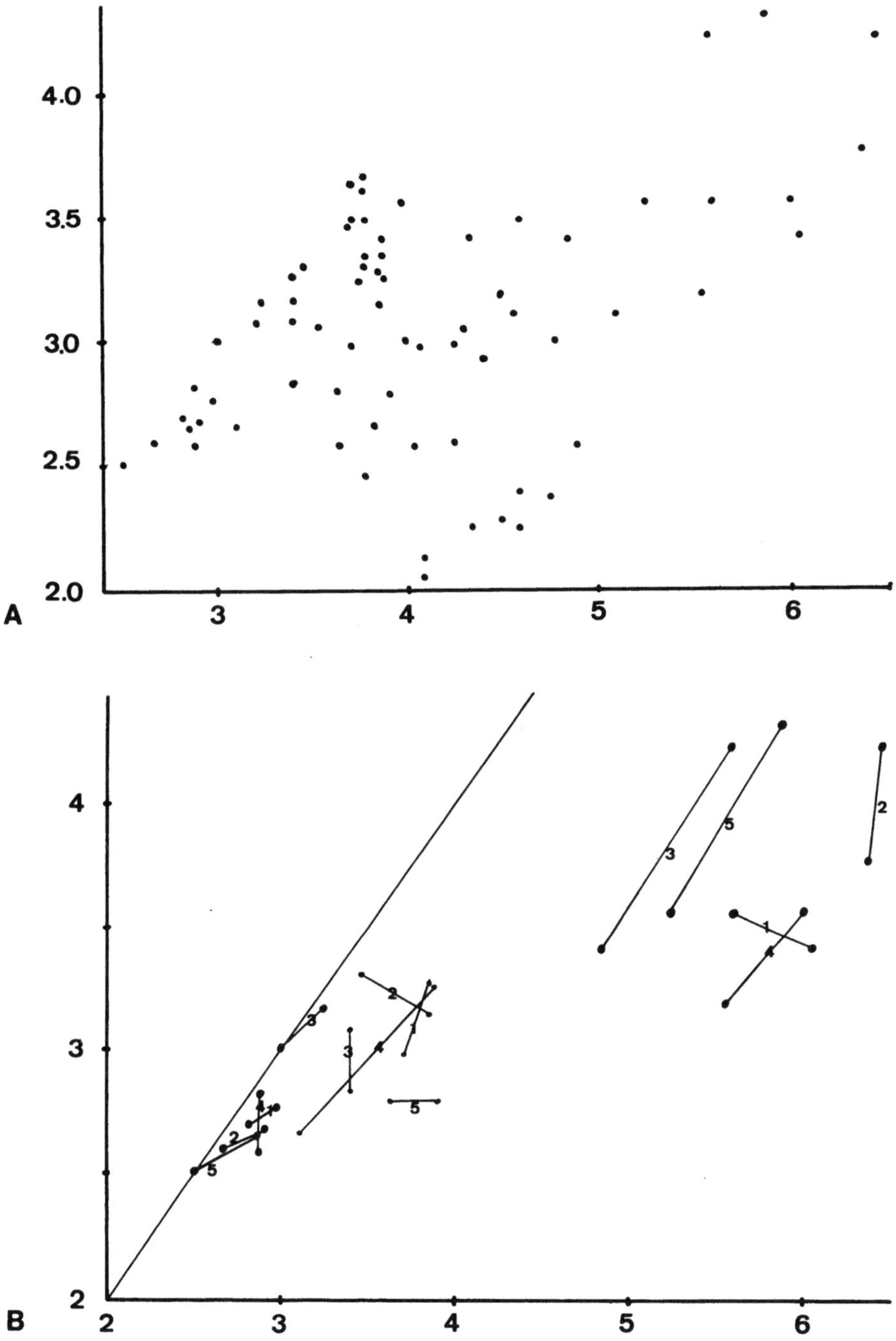

Fig. 4.5A,B

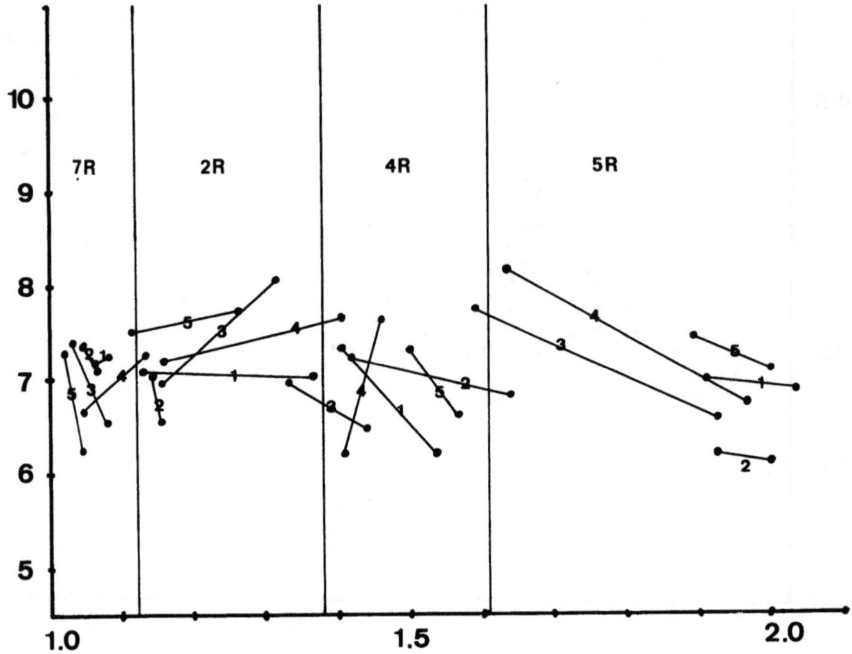

C

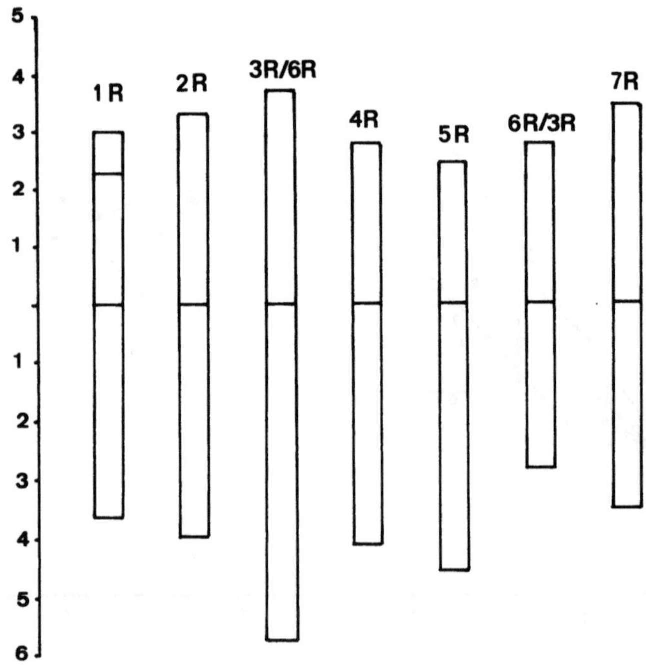

D

Fig. 4.5C,D

provide a direct method for idiogram construction when, due to non-intrinsic variation in the observations, unambiguous chromosome identification is not possible. A common intermediate step is the construction of a graphic representation of all measurements made. Because only two measurements are available for most chromosomes that cannot be identified with certainty (the two arms), a two-dimensional plot is usually sufficient. There are two basic ways to plot the two chromosome characteristics. One is based on the usual description of chromosome size and shape: total length and location of primary constriction, the latter either in the form of centromere index (short-arm length divided by total chromosome length, always smaller than 0.5) or arm-length ratio (long-arm length divided by short-arm length, always larger than 1), occasionally short-arm length divided by long-arm length (always smaller than 1). The length is usually expressed as a percentage of total karyotype length, which is independent of the degree of contraction. An alternative is to plot the two arms against each other. The advantage of this presentation is that any change involving a single arm can be traced to that arm directly, which is not possible when arm ratios and total length are plotted. A second advantage of direct plotting is that the distribution of the length variation approaches a normal distribution better than does a ratio. In the latter case, angular or other transformations are necessary for statistical

Fig. 4.5 C The four chromosomes that cannot be individually recognized in each cell (unlike the chromosomes of **A** and **B**). *Ordinate* Ratio long/short arm; *abscissa* total length.

There is considerable overlap. The length differences are quite small and the ratios provide the most critical information for the distinction between the chromosomes. When groups of ten (from 5 cells, 2 in each cell) are taken separately (*vertical lines*), the average ratios of these groups are reasonably representative for the actual chromosomes.

The rightmost group (*5R*) must have two chromosomes overlapping with the next group (*4R*), but which chromosomes cannot be deicided. The two chromosomes derived from each cell are connected by lines, with the cell number (*1–5*) indicated. This excludes certain combinations of 10 chromosomes. In a few more cases it cannot be decided which chromosomes to choose, and a probable pair has been chosen. In cells *3* and *4* the chromosomes attributed to *5R* were chosen because of their position at the right side of the group around ratio 1.5. It could also have been the two more to the left. This choice implies a potential error, reducing the variation actually present. This error cannot be large when the positions in the scattergram are clear, and when it is assumed that the difference between the two chromosomes within a cell is not excessive. On this basis, the average chromosome length and arm ratio, and their standard error can be reasonably well approximated.

With other recognizable translocations other sets of chromosomes can be characterized.

D The idiogram based on the averages of the chromosome measures as identified in the scattergrams. The standard errors have not been included, but can be readily supplied. The satellite in *1R* is shown. Here, *1R* is the smallest chromosome, as in many other genotypes. In others, it is *7R*. This variation is mainly due to heteromorphism for telomeric heterochromatin, as stained by C-banding analysis. Moore and Gregory (1963), naming the long arm x and short arm y, used artan y/x (in radians) as the parameter plotted against total length. Because the variation in length tends to increase with increasing length (for separate arms also), these authors used log(x + y) instead of total length. Chromosome length again was given as a percentage of the total genome.

Another advantage of the arm/arm plot is the much lower increase in the variance of the short arm itself than in the arm-length ratio with increasing difference between the arms (Fig. 4.5).

From the plots of Fig. 4.5, where the individual pairs of chromosome parameters form "clouds", each representing a chromosome or pair of homologues, a number of aspects of the variation described in the previous section appear. One is the compression of the clouds near the 45° line when the arms are plotted individually and near the ordinate when total length and arm ratio are used. Here the "average" (arm lengths or total length and ratio) for the chromosome is not the centre of gravity of the cloud, but a value closer to the line where the distribution folds back. The overlap between clouds is also clear when no preselection has taken place. A classification of the chromosomes in the plot is facilitated when the cells from which the chromosomes originate are marked per point in the plot.

The step from a plot to an idiogram when the distributions of the chromosome characteristics are not well separated remains complex and is essentially imperfect. The approach of Harris et al. (1973) developed for human chromosomes using discriminant function analysis (see above), although not as reliable as suggested because of artificial reduction of variation by preclassification, is interesting. It results in the delimitation of idealized areas of distribution of separate chromosomes from which length and arm ratio can be derived. With considerable overlap, exaggeration of the differences between chromosomes by this artificial separation is still inevitable and the real averages are closer together than suggested. In such analyses, deformation of the parameter distributions because of foldback of the distribution around the ordinate can be corrected just as well, as shown by Essad et al. (1966) or Matérn and Simak (1968) (discussed above). No applications of the model of Harris et al. (1973) to plants have been reported.

To simplify the statistical approach, an attempt can be made to separate fused distributions visually. Although not very exact, it may give a reasonable approach and, with the proper precautions, acceptable results can be obtained, as is shown by the example of Fig. 4.5. It is not relevant at present for rye itself, because C-banding techniques make most chromosomes recognizable at mitosis. The approach, however, is simple and can be applied to organisms where neither the C-banding pattern nor chromosome gross morphology is sufficiently discriminating to reliably identify the chromosomes. As seen in Fig. 4.5, after removal of the three recognizable chromosomes, the four remaining chromosomes vary in arm ratio but hardly in length. As in Sybenga and Wolters (1972), only the arm ratio contains sufficient information for distinction. The distribution appears to be continuous, so there must be considerable overlap, and the range covered is large: from 1 to more than 2. When each chromosome is present twice, for five cells analyzed, groups of ten chromosomes can be taken together and their limits indicated in the graph, disregarding for the moment overlap between neighbouring groups. The middle of each group of ten represents the average arm ratio for one particular chromosome. This introduces inevitable errors that can only be approached in a general way. The clearest is the error for the ten chromosomes on the right, whereas the true average must be more to the left. Yet separation of the averages is much better than visual inspection of the chromosomes in cells would permit. This gives the average arm ratio per chromosome. The average length can be derived from the original data.

In normal rye without marked translocation chromosomes, there are two more chromosomes in the plot. One has an arm ratio close to 1 and here two chromosomes coincide. In this case, the original two-dimensional plot is necessary where total length can be analyzed also. Both chromosomes show a foldback around ratio 1.

In rye, several translocations are available in which the shape of the chromosomes has changed sufficiently to make them recognizable. One was used in the example of Fig. 4.5 to simplify the analysis by eliminating the two recognizable translocation chromosomes from the plot. At the same time, the characteristics of both the normal and the translocated chromosomes involved in the translocations can be identified by comparison with the normal karyotype.

Using a Principal Components Analysis, developed originally for numerical taxonomy, Fillion and Walden (1973) tested the detectability of changes in chromosome morphology caused by six different interchanges in maize, all involving chromosome 9. The effectiviness of this approach appeared to depend on the stock used and on the chromosome involved besides 9, partly because of the original shape (metacentric, sub-metacentric, acrocentric) and partly because of other, not readily traceable, causes such as variation in contraction. As a check, relative size of the chromosomes and breakpoints of the translocations where known from pachytene analysis, where knobs and chromomeres, but also the approximately 13 times greater length, made analysis much easier than at mitotic metaphase. Decrements in size by translocations were generally more easily detected in the somatic karyotype than increments. A 50% decrease in pachytene length of the short arm of chromosome 9 was detected in mitosis, but a 40% increase was not. Surprisingly, an increment of only 10% of 5L at pachytene could be recognized in mitosis. This again demonstrates the difference in contraction of specific chromosome segments in different tissues.

The identification of chromosomes involved in trisomy is usually more complicated than that of chromosomes altered in shape. When the distributions of the chromosome parameters overlap, it is difficult to detect exactly where the extra chromosome fits in. Then it is usually better to rely on meiotic analysis. Similarly, for many translocations and other chromosome structural rearrangements that give only minor changes in chromosome shape, meiotic analysis is more suitable. In several cases, however, karyotype analysis will serve the purpose just as well, and does not require the specifically marked material often necessary for meiotic analysis (cf. Chap. 7).

Although several computer programs are available for aid in idiogram construction on the basis of measurements when the chromosomes can be visually distinguished, there is a clear need for more sophisticated systems that make use of the different approaches discussed above. In most cases the publications available do not contain sufficient technical details to apply the analytical methods used. The combination of techniques and computer programs necessary for a completely automated idiogram construction, when overlap in size of chromosomes and of chromosome arms within chromosomes occurs, has not been published.

Bennett (1984) suggests that the idiogram should contain information on the position of the chromosomes relative to each other in the mitotic nucleus. This has not become common practice, not only because of the difficulty in obtaining the data required, but also because it is not generally accepted that the chromosomes do consistently have a fixed position in the nucleus.

4.2.4 Markers Within Chromosome Arms

4.2.4.1 Chromosome Banding

The use of chromosomal markers other than primary and secondary constrictions has long been known for material where such bands can be observed without special treatment. The most obvious example is the dipterous polytene chromosome, where Painter (1934) first described the banding pattern for *Drosophila* in detail. There are a few examples where bands in polytene chromosomes in plants have been used for chromosome identification after standard chromosome staining. Bennett and Smith (1975), for instance, identified complete and translocated rye chromosomes as introgressive substitutions for wheat chromosomes in Feulgen-stained polytene antipodal cells of different wheat varieties. Bands that identified the rye chromosome segments appeared at the same places at which C-bands are found in mitosis.

In a limited number of plant species, heterochromatic segments are sentitive to low temperatures that induce them to be undercontracted at metaphase, where they appear as lightly stained segments or bands, or "tertiary" constrictions. Such less stainable segments were formerly supposed to suffer "nucleic acid starvation". Striking examples are different species of *Trillium*, where variation in the pattern of cold-sensitive segments permits the distinction between species and even populations within species. See, for instance, Fukada (1984), who also cites several other examples. For commercial plant species, this type of chromosome structural differentiation has not been reported to be superior to C-banding.

Differential staining of segments that are generally considered heterochromatic, because they tend to correspond with condensed interphase chromatin, was first reported to be possible with the fluorochrome quinacrine dihydrochloride, an acridine derivative, which is related to the non-fluorescing actinomycin D, which reacts with specific DNA bases (Caspersson et al. 1968). Shortly later, it appeared that quinacrine mustard gave better resolution than quinacrine alone (Caspersson and Zech 1970). It is especially suitable for high resolution microfluorometry along the length of the chromosome which permits the quantitative analysis of even small differences between homologues. Because the technique is not as convenient as the later developed banding techniques (since it requires special microscope adaptations and the stain fades rapidly) and the pattern is usually very similar to that of G- or C-banding, especially in plants (Vosa and Marchi 1972, see below), it is presently only used for special purposes.

Shortly after 1970, techniques were developed that involved partial denaturation of the chromatin in plants using Barium hybroxide denaturation, followed by renaturation in 2x SSC (Standard Sodium Citrate) and staining with Giemsa (BSG technique). This technique differentiates between segments with tighter and segments with looser packing, which is one major difference between euchromatin and heterochromatin. The bands obtained by such methods tended to stain primarily centromeric heterochromatin and were therefore called C-bands. Because such bands appeared also at other locations in several species, C-banding was later understood to mean constitutive heterochromatin banding.

More refined chromosome banding techniques involving different types of denaturations and trypsin treatments were first developed for human chromosomes and appeared to give a pattern (G-banding pattern = Giemsa-banding pattern) that is practically identical with that produced by quinacrine fluorescense (Evans et al. 1971; Seabright 1972). The major G-bands coincide with late replicating segments. This can be demonstrated by tracing the moment of tritiated thymidine incorporation. The late replicating segments can be seen in micro-autoradiographs of metaphase chromosomes fixed after a minimum interval after incorporation (Ganner and Evans 1971). Later developments gave even better resolution and involved pulse labelling with BUdR and the use of prophase instead of metaphase chromosomes.

It is assumed that typical G-banding is not possible in plants, because treatments with trypsin or other proteolytic enzymes have no effect. Occasionally, incompletely condensed chromosomes, as occur in prophase, may show a beaded appearance after trypsin treatment followed by the normal procedures for G-banding (Wang and Kao 1988; Yang and Zhang 1988). This may not be equivalent to the G-banding obtained in many animal species. After extended trypsin treatment, the major chromosome coil tends to become visible (*Secale*, *Hordeum*, *Vicia*: Yang and Zhang 1988). This suggests that trypsin treatment in plants reveals variations in the last period of chromosome condensation, possibly involving acidic or other non-histone proteins in the major chromosome scaffold. C-banding, on the other hand, clearly involves an earlier stage in chromosome condensation, because small bands in the contracted chromosomes can be found, not as short segments across the entire chromatid, but as small dark spots as the sides or in the middle of the chromatids (Fig 4.6). This implies that they are inside the major coil. In the two chromatids of the same chromosome and usually even in the two homologues, such dots are found on equivalent sides of the chromatid, demonstrating the existence of a relatively rigid system in coiling. Such spots are not normally found in mammals, where even the smallest bands are seen as thin bands across the entire chromatid.

The rather detailed banding found in some species of plants (*Anemone* spp., for instance; see Marks and Schweizer 1974) may look like G-banding, but they are obtained by regular BSG-banding techniques. The reason that genuine G-bands are not produced in plants has been attributed to the more compact condensation of plant chromosomes, but this is not of major importance because DNA density is not always higher in plants than in animals (Schubert et al. 1984). There must be other reasons.

A great number of variants of the C-banding technique are in use and there is considerable variation in the personal preference of different cytologists for specific variants. It is clear that several work quite well, even when they differ in the steps considered crucial.

Several other techniques have been developed for making specific chromosome segments visible, both for plants and animals: R-banding, which gives the reversed pattern of Q-, G-, or C-banding (Schweizer 1976); Hy-banding, involving special, usually rather excessive HCl hydrolysis and Feulgen or carmine staining (Greilhuber 1974, 1975); N-banding, involving a drastic modification of the C-banding technique (Funaki et al. 1975, later modified; see, e.g. Endo and Gill 1983). N-banding was formerly believed to preferentially stain Nucleolar organizers, from which the name was derived. Later it appeared that specific other segments with an unusual chromatin composition would also stain with the same N-banding technique, sometimes even when the NORs were not stained. The technique is relatively simple and marks segments that are not differentiated with C-banding. It is, incidentally, used in plants in addition to C-banding. Other banding systems have not found general application. Specific NOR-staining with silver, which is not a banding technique (Goodpasture and Bloom 1975, later adapted by others for many different types of material), has become a useful addition to the arsenal of techniques for marking specific chromosome segments.

There is considerable variation in banding patterns between species and there often is polymorphism within species (Sybenga 1983b; Pilch and Hesemann 1986; and several others). If a "standard" banded karyotype is presented, this is only the most frequent type, or it is based on the most frequently observed bands. C-band polymorphism is a great disadvantage for identifying specific chromosomes, especially when they occur as additions or substitutions in alien species. Polymorphisms are, however, interesting as markers in cytogenetic studies.

A further refinement of the banding pattern can be obtained by using the characteristic of BUdR (5-bromodeoxyuridine) to affect chromosome-staining properties when it has been incorporated in the DNA. This was first applied by Stubblefield (1975) to Chinese hamster chromosomes and adapted to plant chromosomes (*Allium*) by Cortés et al. (1980), and Cortés and Escalza (1986). BUdR is applied for a short time, washed out and replaced by thymidine, in which the root tips are kept for several hours. In the subsequent mitosis, the material is fixed and squashed, treated with RNAse, 1/2 SSC and the fluorescent stain Hoechst 33258 for 0.5 h. An extended period of intensive irradiation with UV, followed by Giemsa staining, results in decreased staining of the segments with BUdR. If BUdR had been given during the last stages of DNA synthesis, the late replicating segments would appear as distinct light bands. Given in intermediate stages, followed by a period of normal DNA synthesis, the last segments replicated stain dark again, but the segments replicated earlier contain BUdR and stain light. The pattern is more detailed than after normal C-banding, especially in late prophase chromosomes that are not fully contracted. The method has two disadvantages: the technique is very critical and the banding pattern depends on the moment of BUdR incorporation during S-phase.

Still another method of producing bands that has received attention in mammalian cytogenetics is restriction-enzyme banding (Lima-de-Faria et al. 1980; Bianchi et al. 1985). After the necessary pretreatment, the chromosomes are treated with one or more restriction enzymes. Segments where DNA sequences that are specific for the enzyme used are abundant are partly digested. The banding pattern, therefore, depends on the type of enzyme and the location of specific highly repetitive DNA families. Attempts to make the method applicable for plants have not yet been fully successful, but may become so in the near future.

With a detailed banding pattern, the necessity of making an idiogram based on repeated and statistically processed measurements of chromosomes becomes much less urgent, and we see that, for instance in human and mouse cytogenetics, karyograms are practically the only way karyotype morphology is presented. Only when size as such is of interest, apart from being the major characteristic for identification, is size given. For plants whose C-banding pattern is often not sufficiently detailed to permit unequivocal recognition of chromosomes and chromosome segments, especially for species with small chromosomes, the idiogram remains of considerable interest. Whenever possible, it should be augmented with any other characteristic available. In this case, too, it appears that the analysis of size does not get the attention the details of the data tend to suggest.

In the karyograms of humans, mice and a few other species, special code systems for different chromosome segments have been adopted similar to those used for *Drosophila* salivary gland polytene chromosomes. The sections carry a letter and the sub-sections, a number. The sections should be bordered by constant and readily identifiable bands. Occasionally this is possible and has been attempted in plants (Schubert et al. 1987), although it often remains difficult because of limited reproducibility of some bands and the necessity first to decide which are the consistent bands and which not. For plant chromosomes where insufficient bands are available for distinguishing more than one, two and occasionally three segments, such detailed systems are not applicable. A border between two segments, placed somewhere in an unmarked homogeneous region of a chromosome, cannot fulfil its function. Yet it is important to be able to identify chromosome segments on the basis of universal coding systems. It is regrettable, therefore, that, even for the two arms of a chromosome, no consistently used code is available. When long and short arms can be distinguished, it makes sense to call the long arm L and the short arm S. In wheat, for instance, this is the system used by most wheat cytogeneticists. Some, however, use p for the short arm and q for the long arm, as in human cytogenetics. Others again use a and b, or A and B, or I and II, etc. Especially when the chromosome homoeology relations between species are the subject of study, it is useful to use standard terminology. For rye, for instance, it has been agreed (Sybenga 1983b) to follow the nomenclature of the *Triticinae*, based on the nomenclature of wheat. As long as it has not been decided which system the wheat geneticists will ultimately agree to follow (L and S as previously, or p and q as in human cytogenetics), it is premature to decide definitely on the system to be used for other *Triticinae* (see: several contributions to the Proc. 7th Int. Wheat Genetics Symp. edited by T.E. Miller and R.M.D. Koebner. IPSR, Cambridge 1988).

The interesting question of how one knows which chromosomes in different species are homoeologous when their morphology is too different for identification will be considered in Chapter 9. In general, when the species are not very closely related, there is little correspondence between morphological and banding markers, partly because of large differences in the pattern of repetitive DNA and heterochromatin and partly because of the occurrence of translocations during speciation.

4.2.4.2 Molecular Markers

The use of molecular markers of specific chromosome segments preceded the introduction of banding techniques. Applying tritiated thymidine at the end of S-phase and subsequently determining the location of late replicating segments in the chromosomes by microautoradiography has been in use since the 1960s (Lima-de-Faria and Jaworska 1968). It is an effective way of marking late replicating heterochromatin. It is molecular only in a marginal sense. Hybridizing specific DNA or RNA with slightly denatured (single-strand) DNA is the most direct molecular way of marking specific chromosome segments. Gall and Pardue (1969; see also Pardue and Gall 1969) were the first to succeed in hybridizing external RNA and DNA in situ (in the preparation where the morphology of the chromosome could be studied) with the homologous DNA at the original chromosomal location. The DNA in the mouse leucocyte preparations made for light microscope chromosome studies was "molten" (thermally denaturated, i.e. made single-stranded) by moderate heating, and free single-strand nucleolar RNA (and later DNA) were applied to the preparation. This RNA was labeled with tritiated uridine and the DNA with tritiated thymidine, and after permitting time for hybridization and preparing for micro-autoradiography, the sites where the added nucleic acid had been bound could be detected by microautoradiography. As expected, with the use of nucleolar (ribosomal) RNA, these sites were the rDNA at the NORs. Later, special satellite DNAs were used in the same way and these hybridized with pericentromeric heterochromatin. The chromosomes were stained with Giemsa, and soon it became clear that the same process of denaturation used for in situ hybridization also caused differential staining ability for Giemsa stains. This led to the discovery of C- and later of G-banding.

In situ hybridization with known labeled DNA probes has become a very effective way of locating specific DNA sequences in chromosomes. A certain level of repetitiveness in the chromosome of the DNA involved is required because the resolution of the technique is limited. The mouse rRNA locus of the example given above represents a moderately repetitive gene. Highly repetitive DNA as occurs in many plant species in segments that stain heavily with C-banding techniques is especially favourable for in situ hybridization. Originally, DNA was taken from the organism and repetitive DNA was isolated as satellite DNA from Cesium chloride gradient preparations or from preparations used for reassociation analyses.

By applying purification techniques and later also by cloning in bacterial plasmids, different families of repetitive DNA with specific physical properties could be separated and identified in rye and later also in other Triticinae (Appels et al. 1978; Bedbrook et al. 1980; Flavell 1981; Appels and McIntyre 1985). When these different families were hybridized in situ with denatured rye chromosomes in cytological preparations, it appeared that most of the terminal heterochromatin blocks contained most of the families. It was also shown that removal of small segments from the blocks of heterochromatin tended to remove specific DNA families. This showed that these families did not occur in a mixture, but were physically separated. A few highly repetitive DNAs were restricted to specific interstitial segments and another group was distributed evenly over the chromosomes. Hybridization with related rye species and with wheat showed that even closely related species showed very different patterns of repetitive DNA. Several families are species-specific, but others are shared by species that are not especially closely related, such as rye and wheat.

To locate specific repetitive rye sequences on wheat chromosomes, Rayburn and Gill (1985) used an alternative in situ hybridization technique. It makes use of horseradish peroxydase, linked by avidin to biotin, which in turn is associated with the DNA probe. The peroxidase converts colourless DAB (diaminobenzidine tetrahydrochloride) into a brown precipitate. Where the probe hybridizes with the homologous DNA in the chromosome, a brown color appears that contrasts with the blue Giemsa stain of the rest of the chromosome. On several localized sites in the wheat chromosomes, segments homologous with the rye probe pSc 119 were present. By using a probe of rye repetitive DNA that is evenly distributed over the rye chromosomes but is present in only a few wheat loci, Lapitan et al. (1986) could readily trace rye chromosome segments introduced into wheat chromosomes. (Fig. 4.6A). With Tritium as the label, small segments could not be recovered unequivocally.

Most initial experiments locating specific molecular markers on plant chromosomes were done on *Triticinae*. Later other species followed. In the polytene chromosomes in cultured cotyledon cells of *Pisum*, in situ hybridization could be carried out effectively (Davies and Cullis 1982). Especially the technique of making "protoplasts" of fixed material by enzyme- (cellulase- and pectinase-) digestion of cell walls (Mouras et al. 1978, 1986) appeared to be very useful in preparing plant material for in situ hybridization. A 17 kb T-DNA segment introduced in *Crepis* by *Agrobacterium* transformation could be located on the chromosomes by in situ hybridization with Tritium as well as biotin labeling (Ambros et al. 1986). The position in its chromosome of a low-copy-number kanamycin resistance gene, with promotor, introduced into *Nicotiana* by *Agrobacterium* transformation, was demonstrated by in situ hybridization (Mouras et al. 1987). Single genes have not yet been located by this method in plant cells. When the gene is extended with a segment of adjacent spacer DNA. or, possibly, other unique non-gene DNA, it may become large enough. At present, segments that are small enough to be cloned by standard recombinant DNA cloning systems are on the border of the size requirement for making hybridization sufficiently frequent and for making the marked locus detectable by microautoradiography or by biochemical methods. Gustafson et al. (1990) located the seed storage protein loci (*Sec* genes) in rye by in situ hybridization. These too are multicopy genes, but with a low copy mumber (Fig. 4.6B).

Fig. 4.6 A Rye chromosome segment (1RS) translocated to Amigo wheat chromosome arm 1AL (*arrowheads*), break point in the centromere. The rye segment has been marked by hybridization with the rye probe pSc119 (Lapitan et al. 1986). It has been derived from a repetitive DNA family of rye, where it is distributed relatively evenly over the genome. It hybridizes with several short, distinct segments in the wheat genomes. The difference in hybridization pattern makes a distinction between rye and wheat segments possible (Courtesy of B.S. Gill). **B** The multicopy storage protein gene *Sec-1* of rye in 6x triticale (incomplete root tip cell), marked by in situ hybridization with the pSc503 c-DNA clone from P. Shewry. The pair of strong signals is the hybridization site of the *Sec-1* locus on the satellite of 1RS, containing at least 10 copies of the gene. The weaker signals are on 6RS where an inactive locus with fewer copies is positioned. In other cells cross-hybridization with other storage protein genes is occasionally observed (Gustafson et al. 1990; courtesy of J.P. Gustafson)

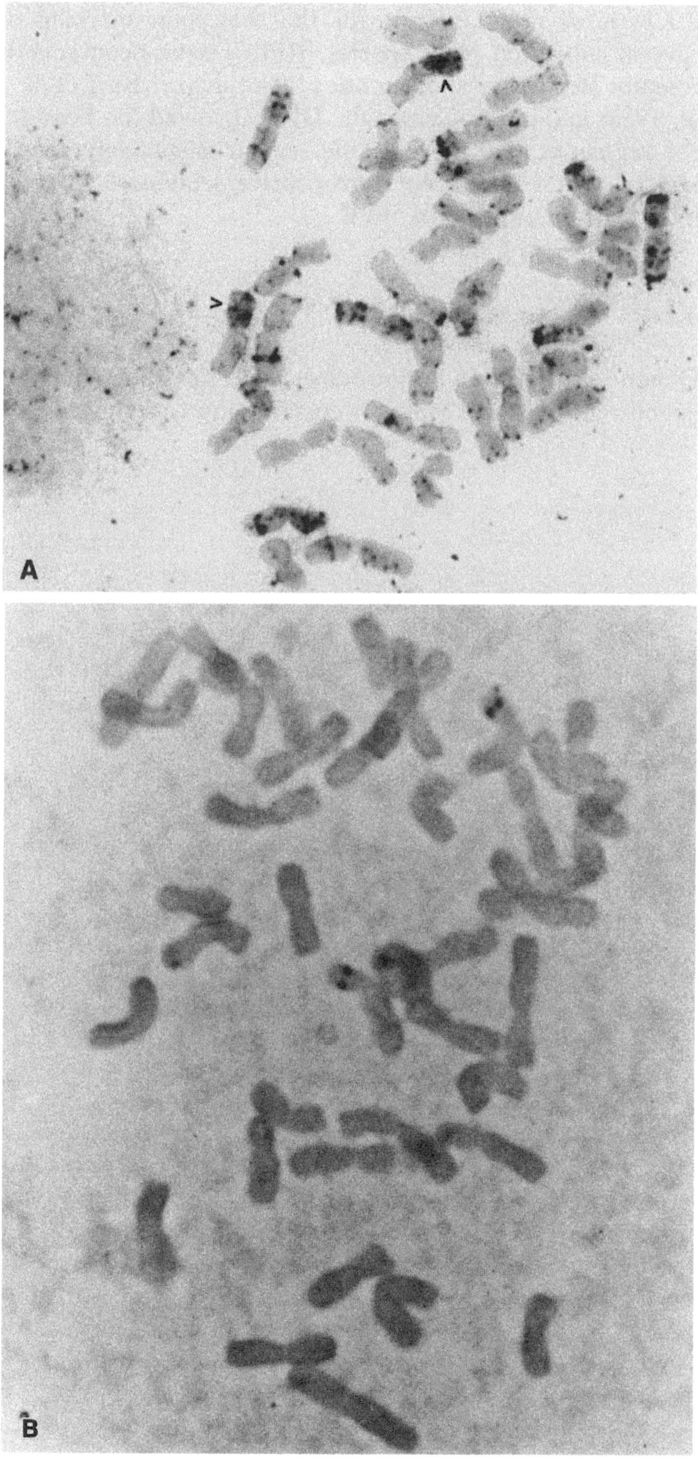
A
B

The use of restriction fragment length polymorphism (RFLP, see Sect. 2.1.2) is of great interest for the molecular marking of chromosomes. In several cultivated plant species, RFLPs have been genetically analyzed (e.g. tomato: Bernatzky and Tanksley 1986; maize: Burr et al. 1988; Helentjaris et al. 1988) and in some cases the DNA involved has been sequenced. However, the segments, like those of the regular single-copy genes, are of insufficient length. When they occur in repetitive sequences, they are no improvement over the repetitive DNA itself.

4.2.4.3 Genetic Markers

Genetic markers can be introduced into the idiogram, which then turns into a genetic chromosome map. Its construction is discussed in Section 8.3.

Chapter 5

Karyotype Variants A: Chromosome Structural Variants

5.1 Deficiencies

5.1.1 Types

A deficiency is the absence of a chromosome segment (of any size) that is present in the normal karyotype. An alternative term is *deletion*, which by some authors is used preferentially for the deficiency of a terminal segment of a chromosome. Deletion of entire chromosome arms from metacentric chromosomes results in telocentric chromosomes, also referred to as telosomes. These will be discussed in Chapter 6.

Deficiencies of genetically important segments are infrequently viable in diploids, even as heterozygotes. Homozygous deficiencies of genetic importance are almost never viable. Absence of heterochromatin segments or other forms of repetitive DNA is usually not considered a deficiency in the strict sense, but is one of the manifestations of *polymorphism*, especially when it is relatively frequent in a population. It is often too small to produce typical diagnostic meiotic characteristics. Such "deficiencies" are normally homozygous-viable.

5.1.2 Origin

One way of somatic formation of a deficiency from mutagenic treatment, or spontaneously, is shown in Figure 5.1. The meiotic origin from the special segregational behaviour of translocations and inversions will be discussed with these rearrangements in Sections 5.3 and 5.4. The combination of specific translocations may produce specific deficiencies in the progeny. Simple mitotic detachment of heterochromatic segments is occasionally assumed to be possible and has been reported by Gustafson et al. (1983)

5.1.3 Relevance

Deficiencies are infrequently encountered in plant breeding programs, even after mutagenic treatments applied for the induction of mutations. When

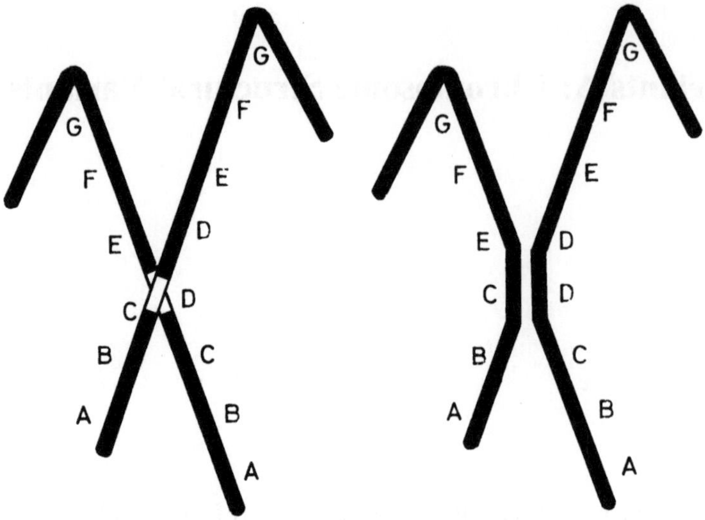

Fig. 5.1 The origin of a duplication and a deficiency from the interaction between lesions in two homologous chromosomes. The interstitial translocation of Fig. 5.16 can also result in a comparable duplication and a deficiency in later generations, but then due to special meiotic segregations of the chromosomes involved. (After Sybenga 1972)

specifically selected, they may be found (McClintock 1931). Deficiencies, usually accompanied by duplications, segregate in the progeny of most inversion and translocation heterozygotes (Sects. 5.3 and 5.4) and may then be undesired.

There are few reasons to be interested in the application of deficiencies. With molecular transformation, or when a gene is transferred by random translocation from an alien addition chromosome into a chromosome of a cultivated species (Sect. 10.4.4.2), the original gene is usually still present. With low-frequency molecular transformation, the frequency of homologous recombinational insertion (which removes the original allele) may be reasonably high, but the event itself is rare. With high frequency transformation, using large amounts of transforming DNA, the frequency of transformation may be higher, including multiple transformation, but it tends to be mostly random insertion. Then the chance that the original allele is replaced is small. If the original allele is dominant or epistatic and its expression not desired, it may have to be removed before an introduced gene can be expressed. This may be attempted by mutation or directed induced deficiency (Sect. 10.1). Because of the usually deleterious accompanying effects, this is expected to be feasible only in allopolyploids or highly duplicated diploids. The method is not simple and mutation is often a better solution. When the mutation is a deficiency, it may be difficult to find genotypes that compensate for a possibly accompanying deleterious effect.

The possibilities for the use of deficiences in differentiating homologous chromosomes in programs of allopolyploidization of autopolyploids are limited (Sect. 12.2.2).

5.1.4 Characteristics and Identification

Some larger deficiencies can be recognized in *somatic* chromosomes, especially when clearly marked segments are involved and especially when entire arms are involved (Melz and Winkel 1986). Large deficiencies may escape detection and smaller deficiencies fall within the range of random somatic chromosome size variation. Occasionally they include C-bands or other markers that make them scorable. This requires special cytological techniques, not always readily applicable on a scale large enough to identify deficiencies when their frequency is low. Also, it must be certain that natural polymorphism for the bands does not mimic the deficiency.

Deficiencies can be readily recognized in dipteran salivary gland or other polytene chromosomes, even when small and including only a single band (Sect. 8.3.2.2.1). The homologues are closely and very accurately paired. At the place of the deficiency, the normal chromosome forms a small buckle or loop opposite the place where the homologue has the deficiency. If the banding pattern is thoroughly known, homozygous deficiencies, which do not form a buckle or loop, can also be detected. Meiotic pairing is not as exact as polytene chromosome pairing and small deficiencies are fully absorbed by the pairing structures. Larger deficiencies, if heterozygous, form a small loop or "buckle" at pachytene in light microscope preparations. In electron microscope SC preparations the same may be found. Because the SC does not contain the bulk of the chromatin, whereas in the pachytene bivalent in regular light microscope preparations all components of the chromosome are present, the probability of the appearance of a buckle is not necessarily the same for both types of preparation. The buckle, if formed, is in the normal chromosome opposite the deficiency, but pairing is often so inaccurate that the size and the position of the buckle may vary, or the buckle may even be invisible. In this respect they closely resemble duplications, which will be discussed below (Sect. 5.2.4). A disadvantage of most SC analyses is that few morphological chromosome markers are available for specifying the location of the deficiency and the variation therein.

For checking if a mutation is a change at the DNA level or a (small) deficiency at the chromosomal level, meiotic pairing normally has insufficient resolution. Yet, if a deficiency is suspected because of an unexpected pattern of mutation or a reduction of recombination in a specific chromosome segment, it may occasionally be worthwhile to analyze the pairing structures, either at pachytene by light microscopy or in SCs (Fig. 5.2A). Whereas absence of a detectable structural deviation does not justify a conclusion, a positive result without further information must also be considered with

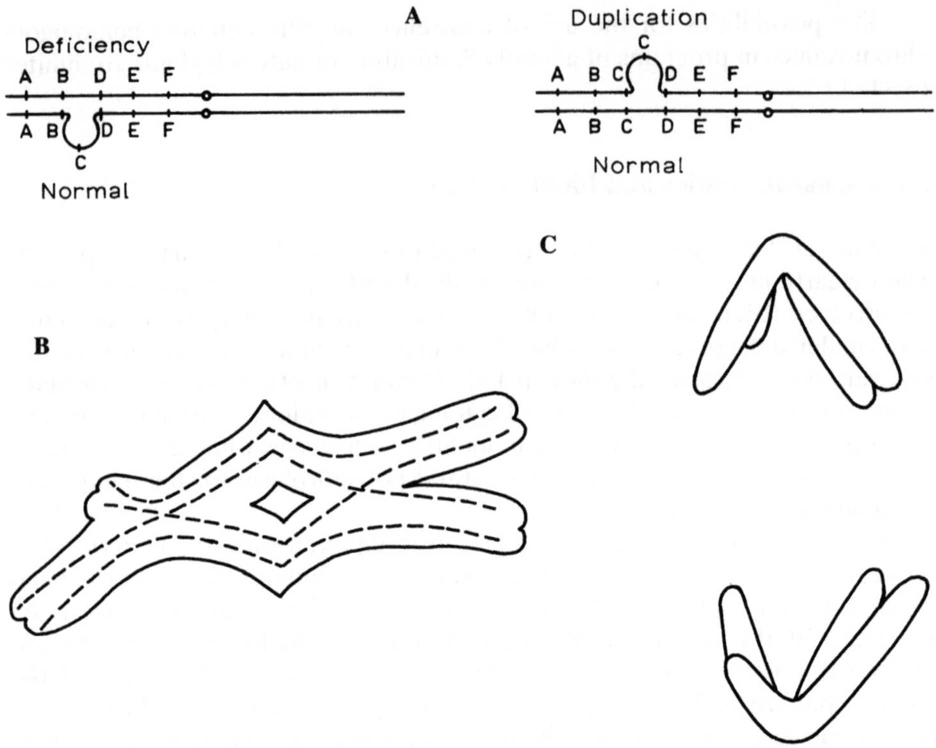

Fig. 5.2 A Diagram of pairing in heterozygotes for an interstitial deficiency and a duplication. The buckle would not be visible at pachytene in light microscope or SC preparations when pairing "correction" results in a segment of non-homologous but complete pairing in and around the segment involved. **B** Metaphase I bivalent of a heterozygote for a large terminal deficiency. A terminal duplication would give a similar, visibly heteromorphic bivalent. An interstitial chiasma makes the sister chromatids unequal at anaphase I (**C**). (After Sybenga 1975)

caution. Occasional presence of a pairing loop or buckle, for instance, may be a simple pairing irregularity, or due to a chance duplication, which has the same pairing morphology. When the pairing pattern does not give sufficient information, other approaches may help. Biochemical methods (linked isozymes) or, more effective but also more laborious, molecular methods (linked restriction fragment length polymorphisms, RFLPs, or specific DNA probes) may be necessary to distinguish gene mutations from deficiencies and deficiencies from duplications.

Large deficiencies can be seen even at meiotic metaphase I as heteromorphic bivalents (Fig. 5.2B). Heteromorphic bivalents may have other origins (heterozygous duplications, interstitial translocations), but when large enough to be recognized, such rearrangements will show additional types of configurations in other meiocytes, usually multivalents at metaphase I, which cannot result from deficiencies. These will be discussed later in this chapter and in Chapter 7.

At anaphase, heteromorphic bivalents formed by deficiency heterozygotes may have differently sized chromatids on the same chromosome as a result of an exchange proximal to the rearrangement (Fig. 5.2C). The frequency of heteromorphic chromatids at anaphase I and metaphase II is a good measure of the frequency of genetic exchange in the segment between centromere and deficiency and, as expected, appears to be correlated with the frequency of chiasmata observed at metaphase I (*Lilium*: Brown and Zohari 1955; *Allium*: Zen 1961; see also Sybenga 1975). The relation is not simple because two exchanges may cancel each other (Sect. 8.2.1.4).

5.1.5 Consequences

If the gametes and the heterozygous (and possibly homozygous) diploid progeny are viable, a deficiency will segregate like a gene. Small deficiencies of mainly heterochromatic material will hardly affect segregation and will be transmitted to the progeny through the male and female line. Larger deficiencies, some small enough not to be cytologically recognizable, will not be transmitted through the pollen because of reduction of the competitiveness in the haplophase. Deficiencies, especially interstitial deficiencies, reduce the effectiveness of pairing, usually in the close neighbourhood, but occasionally also some distance away (Rhoades 1968).

It should be noted that the same local effects will occur with interstitial translocations, where in one chromosome arm a segment is absent, but is present elsewhere. These are not accompanied by similar drastic phenotypic effects as deficiencies. In all such cases, disturbed pairing reduces recombination.

5.2 Duplications

5.2.1 Types

Duplications are segments of chromosomes present more than once in a genome. When entire chromosomes are involved, either in their original form or modified, it is polysomy (hyperploidy, cf. Sect. 6.2.2) and is not considered duplication. The presence of entire extra genomes is polyploidy (Sect. 6.1.2).

Different types of duplications can be distinguished (Fig. 5.3), mainly on the basis of their position relative to the original segment. In a few cases the extra segment seems to be attached simply to the end of a chromosome. The location may be in the same chromosome, in the same arm, in the other arm, or in another chromosome. Interstitial duplications may have different locations

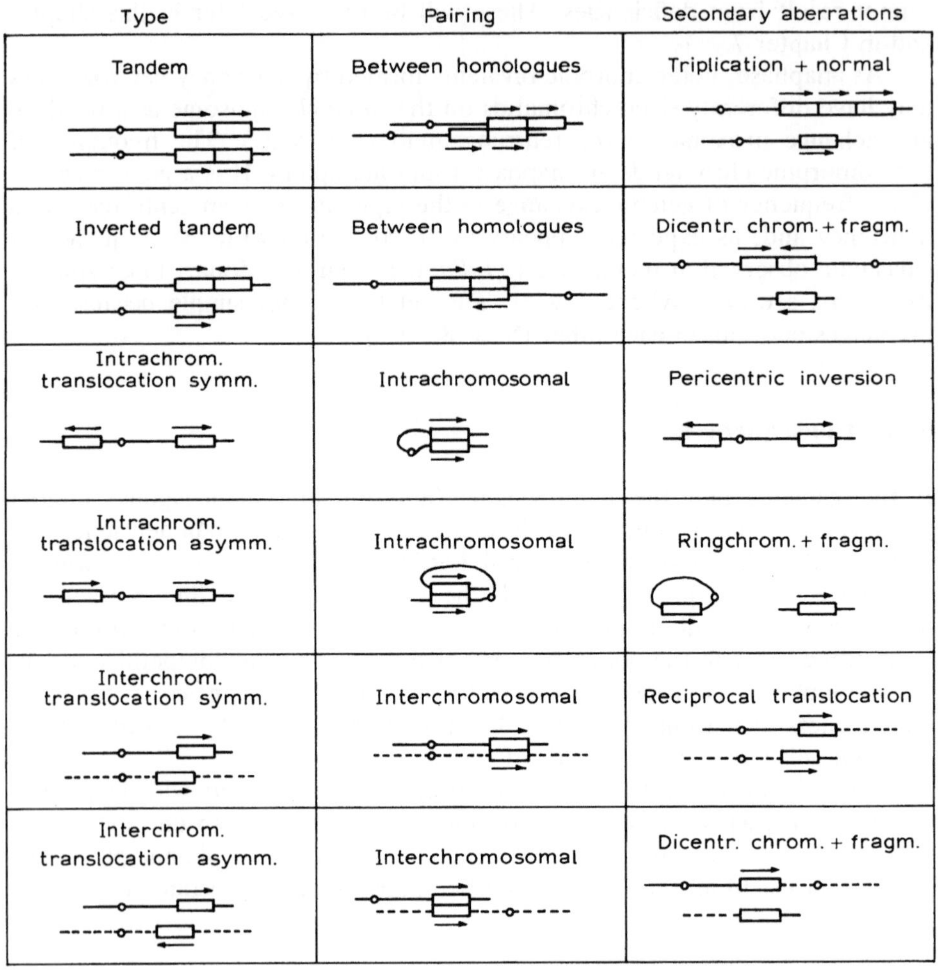

Fig. 5.3 Diagrams of some of the possible duplication types, with a number of possibilities for pairing, and the resulting secondary rearrangements after exchange

with respect to the original segment (Fig. 5.3). The orientation with respect to the centromere may be the same as the original segment or inverted. Location and orientation may have some effect on the functioning of the genes in the segment involved, but the most pronounced effect of location is on the meiotic behaviour.

In plants, duplications as incidental aberrations are not common. More frequent are, as with deficiencies, polymorphisms for heterochromatic chromosome segments or segments that contain only non-coding DNA of other types. These will not be considered here. Homozygous duplications are not uncommon as a stable part of the normal genome, but are often not readily recognized.

5.2.2 Origin

Duplications can be induced by ionizing radiation, other mutagenic treatments, or they can arise spontaneously. Tandem duplications are primarily a result of symmetric exchange between homologous chromosome arms or chromatids with breaks at different positions (Fig. 5.1). At the same time, a deficiency is produced. The production of non-tandem duplications is more complicated. It usually involves chromatid rather than chromosome rearrangement, followed by special segregation of the chromatids. These will not be discussed further here. The extremely low frequency of somatic induction of duplication of specific chromosome segments makes it unsuitable for practical purposes.

There are different ways for duplications to result from the specific meiotic behaviour of other chromosomal rearrangements. Here also, at the same time, a deficiency is produced, but occasionally the deficiency can be separated from the duplication. This origin will be discussed with the rearrangements causing them: inversions (especially pericentric inversions, Sect. 5.3.4), and translocations (reciprocal and interstitial, including A-B chromosome translocations, Sect. 5.4.2.4).

5.2.3 Relevance

There are different reasons for a plant breeder to be interested in duplications and to wish to know where they may be expected, how they are recognized, what their characteristics and effects are and possibly how they are induced.

Whereas small deficiencies may mimic recessive mutations when they involve chromosome segments with readily recognized marker genes, small duplications have only infrequently a readily recognized effect. Occasionally, when they involve a gene with a pronounced dose effect, an apparent dominant mutation may result, as in the case of the well-known *Bar* duplication of *Drosophila melanogaster*. Usually duplications have more general effects, which nevertheless are rather specific for the segment involved. Studies of gene dose effects in plants are rare and have been reported mainly for polysomics and polyploids where entire chromosomes and genomes are duplicated. There has been a certain interest in duplicating specific chromosome segments that carry commercially interesting genes with clear dose effects and special programs to induce these duplications have been developed. These will be considered in Section 11.2. Translocations are the main source. Such duplications have their disadvantages and homozygosity may be difficult to realize. Multiple transformation or gene amplification would be more promising if they were generally available (Sect. 11.2.1).

A different type of duplication that is of interest in plant breeding is the induced transfer of an alien chromosome segment carrying a desired gene from another species. Since it is most probably a related species, the segment is in principle a duplication when introduced by translocation and not by homologous recombination. The induction, consequences and application will be discussed in Section 10.4.4.2.

5.2.4 Characteristics and Identification

Like deficiencies, and with the same limitations, duplications can be detected at mitosis when they are large or involve clearly recognizable segments. When specific segments are duplicated with a reasonable frequency by the meiotic consequences of specific rearrangements (see below), cytological, biochemical or molecular markers may be effectively used to detect their presence.

Heterozygotes for interstitial duplications are characterized at meiosis by forming buckles or loops at pachytene, with the same limitations as deficiencies (Sect. 5.1.4).

A special form of duplication has been reported by Brandham (1990) in *Aloeacea*, where size differences between homoeologous chromosomes of different species can at least in part be explained by pericentric inverted duplications. In the hybrid between species with chromosomes of different size, one of the duplicated segments in the larger chromosomes may pair with the homoeologous segment in the corresponding smaller chromosome and form a chiasma. If the inverted segment is involved, an E-type or L-S bridge is formed.

Like deficiencies, duplications tend to disturb the normal pattern of genetic exchange, but unlike deficiencies, duplications can undergo exchange themselves, resulting in special diplotene-metaphase I configurations. These can be analyzed in addition to the pairing pattern. The position of the duplication in relation to the original segment (Fig. 5.3) has considerable effect on the characteristics of the resulting pairing and diakinesis-MI configurations when the duplication pairs with the original segment. This is possible only when it is relatively large. Figure 5.3 gives a few examples of the pairing patterns and the result of exchange between the duplication and one of its homologues in the original position. When the duplication is homozygous, more possibilities arise than when it is heterozygous, but in that case it is more probable that the duplication pairs with its similarly displaced homologue than with the original segment, as in naturally occurring, homozygous older duplications.

The morphology of the resulting metaphase I configurations is not always sufficiently specific to justify the definite conclusion that a duplication is involved. Meiosis in heterozygotes for terminal duplications, derived from translocations with a terminal break point, may closely resemble the translocation from which they have been derived, and the distinction is often possible only on the basis of quantitative characteristics. Large duplications may form heteromorphic bivalents at metaphase I, and unequal chromatids at anaphase I when exchange has taken place between the centromere and the duplication. This is very similar again to that seen in the deficiency bivalents in deficiency heterozygotes.

When no normal partner is available, as in haploids, old displaced duplications of several kinds may be found to pair and even form chiasmata. The often quite complete pairing of SCs in haploids should not be interpreted as hom(e)ologous pairing between duplicated segments. Most of it is non-homologous, extended late zygotene pairing which has special opportunities in haploids where homologous partners are lacking. The initiation of such pairing, however, may well be in old, duplicated segments. Already in 1934, Lammerts reported pachytene pairing and metaphase bivalent frequency in a haploid of tobacco that carried a known duplication in addition to possible ancient duplications (cf. Sybenga 1975). The duplication was derived from a translocation and was of considerable length. Since the configurations were more complex than expected on the basis of the duplication alone, it may be supposed that some homoeologous pairing between the two genomes of the allotetraploid tobacco had also taken place. The duplication was displaced and inverted, resulting in a bivalent and a fragment after exchange in the duplicated segment. A detailed analysis of metaphase I bivalents and trivalents involving C-banded chromosomes of a rye haploid by Neijzing (1982, 1985) suggested several duplications in different chromosomes, but apparently not in all arms. Several analyses of metaphase association of haploids, including other rye haploids, and using SC analyses have been published. In polyhaploids of allopolyploids, homoeologous pairing and metaphase I association can be quite extensive, but this has a different character than pairing of single duplications.

5.2.5 Consequences

Crossing-over in displaced duplications results in translocations between non-homologous chromosomes, in haploids as well as diploids. In haploids the probability of production of a translocation is considerable, but the probability of recovery is small. Only when restitution nuclei are formed instead of an (irregular) anaphase segregation will a functional gamete have a chance to be formed. Since exchange in a duplication produces a bivalent that has the possibility to maintain a functional spindle, the probability of anaphase segregation is enhanced by such an exchange, and the probability of restitution decreases. Nevertheless, translocations are a real possibility after recovery of progeny from haploids.

Duplications in diploids usually have a very limited possibility for exchange, but if it occurs, recovery of the resulting translocation is quite possible. If there is only one duplication, large enough to undergo repeated exchange, there will always be only one type of translocation. There is insufficient information on spontaneous translocations to conclude whether they

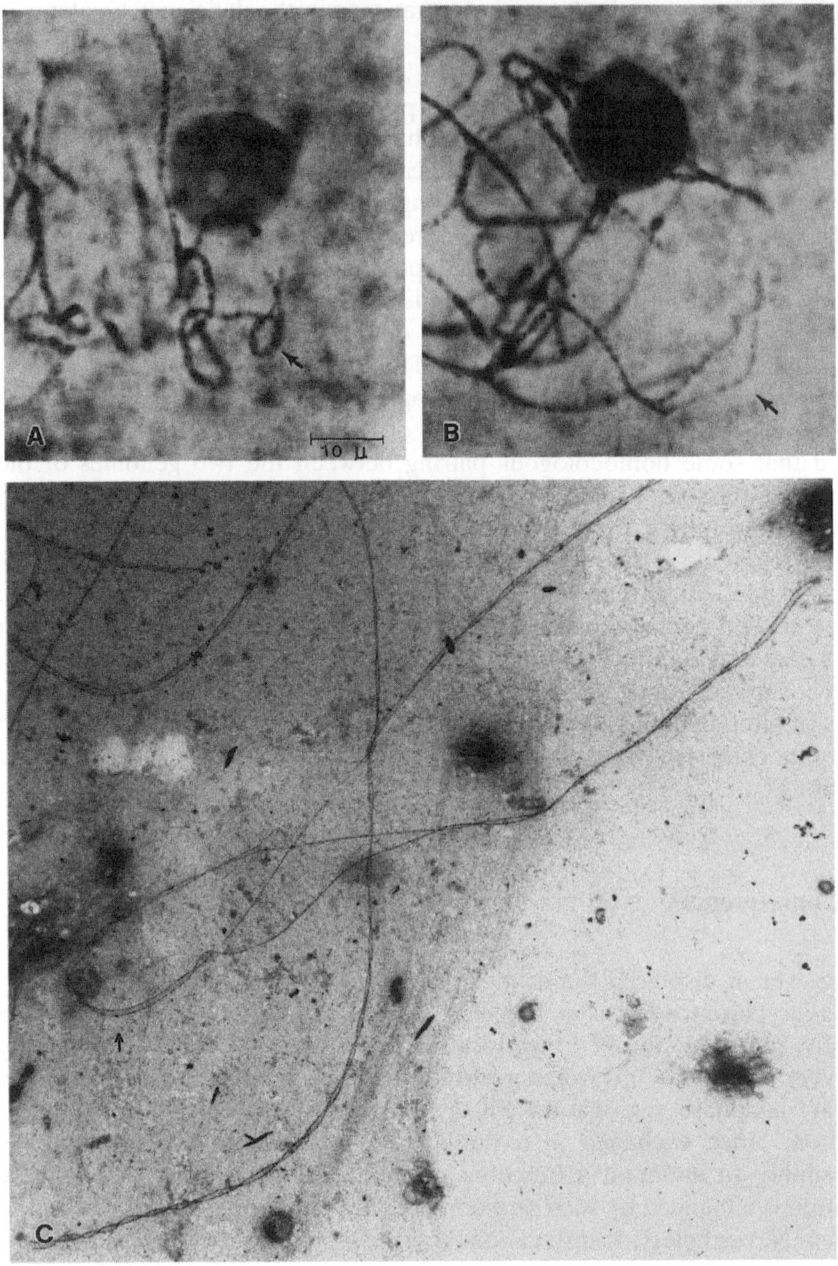

Fig. 5.4 Inversion heterozygote pairing at pachytene. For the structure of the loop, see Figs. 5.5 and 5.6. **A** Small paracentric inversion in maize, the inverted segment is not paired. **B** The same inversion, loop pairing (**A** and **B** courtesy of M.P. Maguire). **C** *Allium*: hybrid between *A. cepa* and *A. roylei*; synaptonemal complex spread of a nucleus containing a large pericentric inversion (courtesy of J.N. de Vries). Large unpaired segments; one arm twisted around another SC

are possibly the result of exchange in duplications or due to errors in replication or other relatively random effects.

5.3 Inversions

5.3.1 Types (Paracentric and Pericentric) and Origin

Inversions, as obvious from the name, are rearrangements in which a segment within a chromosome is inverted. The inversion either involves a segment in one arm, i.e. the inversion is at one side of the centromere (*paracentric inversion*), or the centromere is in the inverted segment (*pericentric inversion*). The paracentric inversion is the only type possible in telocentric chromosomes. A pericentric inversion is equivalent to an interchange between the two arms of the same chromosome. The most common origin is interaction between two lesions in one chromosome, occurring spontaneously or caused by ionizing radiation or radiomimetic substances (Fig. 2.4). Inversions can result from meiotic exchange in some very special types of duplication (Fig. 5.3) but this is not much more than a theoretical possibility.

5.3.2 Relevance

For plant breeding the main importance of inversions is negative: when occurring as heterozygotes in hybrids between two cultivars or between a cultivar and a wild relative, fertility is reduced. When a gene is to be transferred from the wild chromosome with an inversion to the normal homologue of the cultivar, recombination is reduced or even prevented. Relatively large inversions may still (infrequently) show recombination (Sect. 10.4.1), and then it is possible by meiotic analysis to estimate the probability of genetic exchange and to determine the size of the population required for a sufficient number of recombinants to be obtained. Inversions may be carried over to the cultivar from a wild relative or another cultivar without being detected. They may be carried over with a transferred gene or by accident. In later hybridization programs this may have undesired consequences. When an inversion seriously reduces the fertility of a hybrid, even transfer of genes from other chromosomes besides the inversion chromosome becomes difficult.

Inversions have been used relatively effectively in allopolyploididization of autopolyploids (Sect. 12.2.2; Doyle 1963).

5.3.3 Characteristics and Identification

Changes in the *somatic* karyotype, merely in the length of chromosome arms, are not observed with paracentric inversions, because the arm involved does

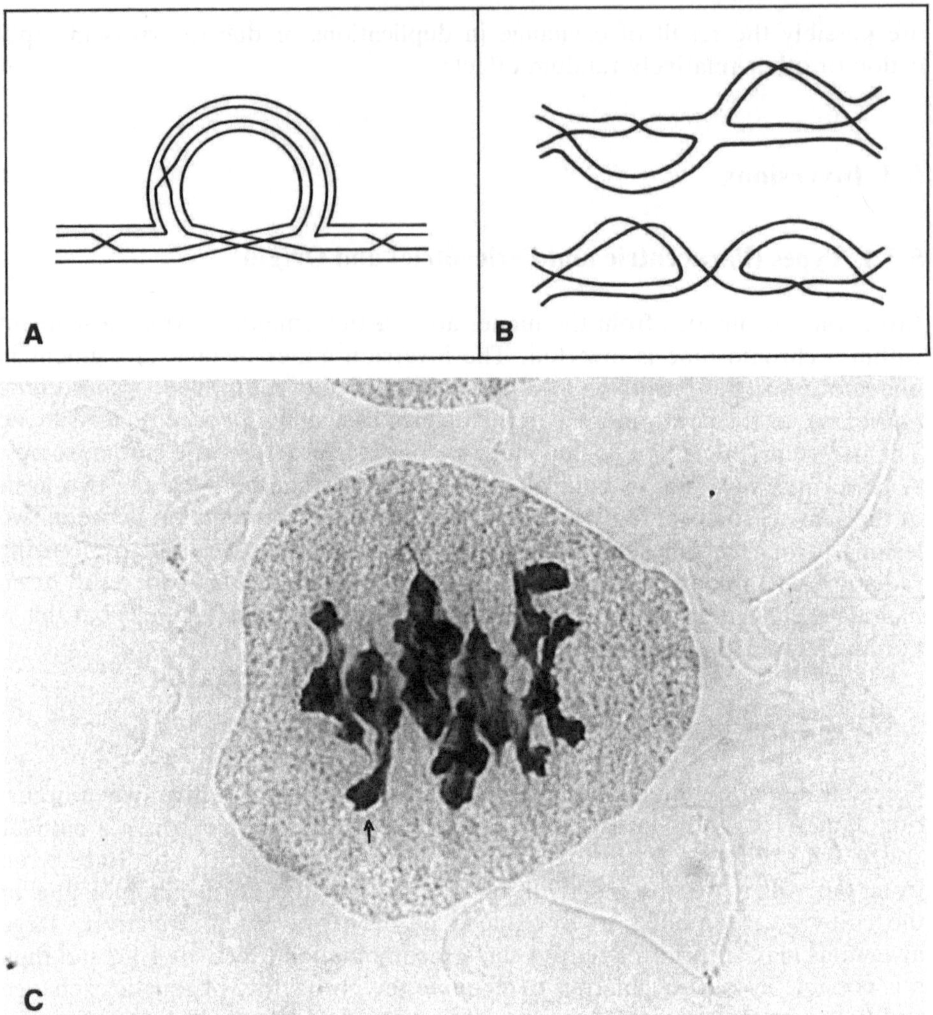

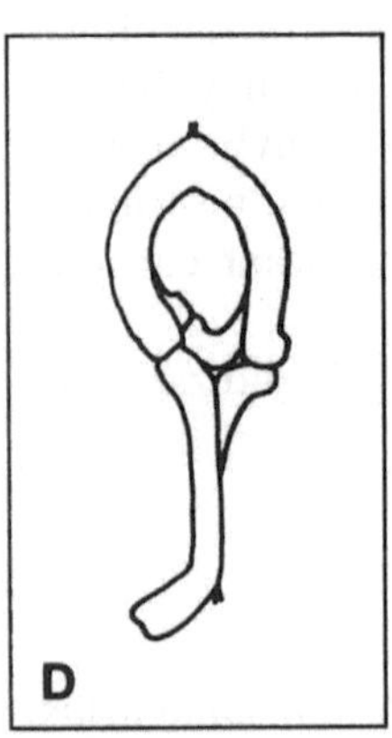

Fig. 5.5 **A** Pairing loop of inversion heterozygote with three chiasmata, one inside the loop and one on each side. **B** Diakinesis "pretzel" resulting from the combination of chiasmata of **A** (After Sybenga 1975). **C** Metaphase I "frying pan" or "spoon" with one chiasma in the loop and one in a terminal segment in the pericentric inversion heterozygote in the *Allium* hybrid of Fig. 5.4C. With a chiasma in both terminal segments a small "figure 8" is formed, as in **B** (courtesy of J.N. de Vries). **D** Diagram of **C**.

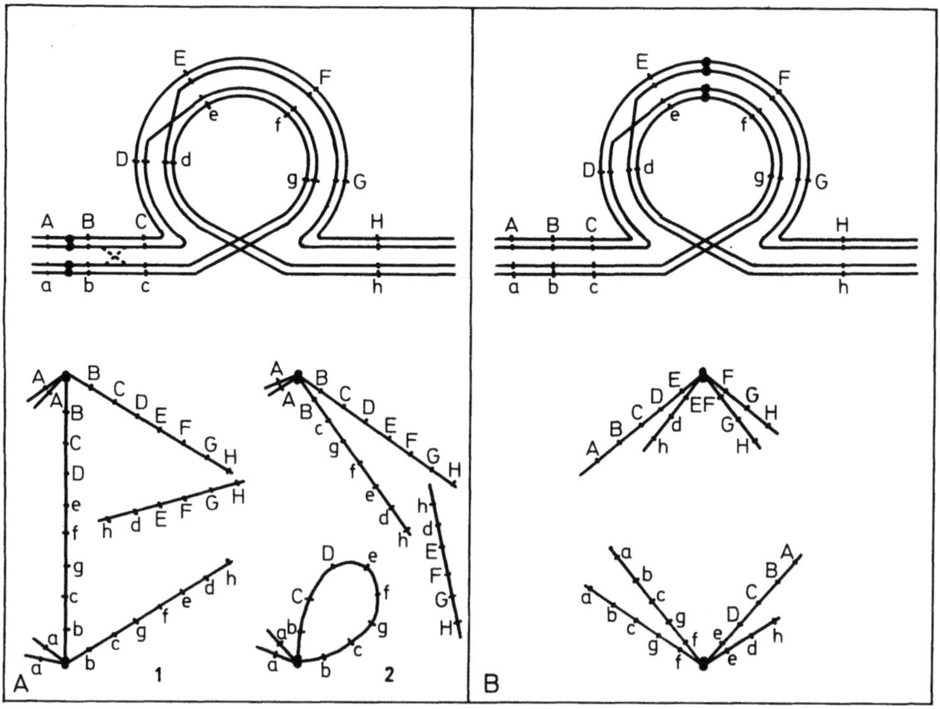

Fig. 5.6 A Paracentric inversion heterozygote, one chiasma in the loop; pairing and anaphase I configuration. A bridge and an acentric fragment result. With an additional chiasma in the interstitial segment (involving two specific chromatids as indicated) the bridge is converted into an anaphase I loop, forming a bridge at anaphase II. **B** Pericentric inversion heterozygote with one chiasma in the loop; pairing and anaphase I configuration. The separating chromosomes each have one normal and one duplication-deficiency chromatid. (After Sybenga 1975)

not gain or lose material. Detailed banding may reveal a change in pattern compared to the normal chromosome, but for cultivated plants this has not been reported.

Pericentric inversions may alter the length relations between the arms when the centromere is not in the middle of the inverted segment. When it is, the arms, although changed in genetic composition, do not change in length. When the inversion is asymmetric with respect of the centromere, it is still not detected merely on the basis of the length of the arms, when it changes the length of one arm into that of the other and vice versa. Only quite asymmetric pericentric inversions that cause the arms to change their length relations will be recognized in the somatic karyotype without detailed banding.

The *meiotic* consequences of inversions are considerable and quite characteristic (Figs. 5.4, 5.5, 5.6). Complete pairing requires that a loop be formed (Fig 5.4B,C). When pairing initiation is ineffective in the inverted segment, no loop will appear. This is common when the inverted segment is small, even when in principle pairing initiation would be possible. In electron microscope SC preparations and even in light microscope preparations of pachytene in favourable material, a short segment of unpaired chromosomes is then visible (Fig. 5.4A), but often complete non-homologous pairing in the inverted segment makes its detection impossible. Most larger inversions may show all three types of pairing, in different cells. When the centromere is not visible, the distinction between the two types of inversion is not possible except on the basis of general location in the chromosome in combination with knowledge of the centromere position.

The consequences of inversion heterozygosity at later stages depend on the formation of chiasmata in the loop. Due to pairing difficulties around the inverted segment (Fig. 5.4C), chiasma formation is usually reduced. In organisms with distal chiasma localization the typical, later meiotic consequences of pericentric inversions are often undetected because of the necessarily proximal location of the inverted segment, where chiasmata are rarely formed. Paracentric inversions also often remain without consequences at diakinesis-anaphase I or II because of interference with pairing and chiasma formation. Because the somatic characteristics of inversions are usually not very clear, and pairing stages not well accessible, inversions are much less frequently reported in plants than reciprocal translocations (Sect. 5.4).

At diplotene, chiamata in the inverted segment result in typical pretzel-shaped configurations. The details depend on the location of the chiasma(ta) in the loop and those outside the loop. The position of the centromere is not of importance at this stage (Fig. 5.5A,B; cf. Brown and Zohary 1955; Darlington 1965). At metaphase I, after the centromeres have become activated, the diplotene shape is modified but still recognizable. With pericentric inversions a "frying pan" or "spoon" bivalent, or an E- or 8-shaped bivalent appears (Levan 1941; de Vries 1989), depending on the combinations of chromatids participating in the chiasmata in the inverted segment and distal segments (Fig. 5.5.C,D).

A chiasma in the inverted segment of a *paracentric* inversion connects the two centromeres of the exchange chromatids. This results in a *bridge* (Fig. 5.6A), which can be drawn out and ultimately break at anaphase I. The terminal segments together form an *acentric fragment* that lags at the equator, but may also remain loosely attached to the bridge at or near the point of exchange. The fragment may occasionally be so close to the groups of segregating chromosomes that it is not immediately seen. The bridge and the fragment at anaphase I are the most reliable criteria for the presence of a paracentric inversion. However, the absence of a bridge with a fragment is not proof for the absence of a paracentric inversion: insufficient chiasma formation in the chromosome region concerned may conceal its presence. On the other hand, there are other possibilities for bridges to be formed at meiotic metaphase-anaphase I: spontaneous breakage resulting from premeiotic imbalance, for instance in unstable genotypes, as well as errors in chiasma formation (U-type exchanges, Jones 1969). In those cases the size of the fragment is variable, depending on the locations of the breaks or the chiasmata. Paracentric inversions always have the same size fragment: that of the loop plus the size of the two end segments. In favourable situations this can be a sufficient criterion to distinguish between a paracentric inversion and other causes of bridges in meiosis.

If, in addition to a chiasma in the paracentric inversion loop, one occurs in the interstitial segment between the centromere and the inversion, the anaphase bridge is converted into a loop in one of the chromosomes of the original bivalent (Fig. 5.6A). The fragment remains. At anaphase II the loop becomes a bridge that is not formed when a bridge is present at first anaphase. Two chiasmata in the inversion loop and none in the interstitial segment, usually restricted to large inversions, can either cancel each other's effect (reciprocal chiasmata) or lead to two bridges (complementary chiasmata) or to a single bridge (the two types of disparate chiasmata). With two chiasmata in the loop and a chiasma in the interstitial segment one or both bridges can be converted to loops and anaphase II bridges. The frequency of anaphase I and II bridges provides an indirect estimate of recombination in the inverted segment (Sybenga 1975).

Chiasmata in the inverted segment of *pericentric* inversions have quite different effects at anaphase I (Fig. 5.6B). Bridges are not formed, but the exchange leads to the interchange of end segments such that one of the two homologues has two identical end segments of one type and the other has two end segments of the other type. The resulting chromosomes are called "pseudo-isochromosomes": the arms of one chromosome are homologous except for the segment around the centromere. Such a chromosome is not functional. It carries a large duplication for one arm and a comparable deficiency for the other arm. When the pericentric inversion is not symmetric around the centromere, the effect of an exchange in the loop can also be seen at anaphase I when the two chromatids of the separating chromosomes are different lengths (Fig. 5.6B). This is usually not readily observed, but if it is, it is a good indication of a pericentric inversion. The frequency of unequal chromatids at anaphase I provides an estimate of the frequency of recombination in the inversion loop, as the frequency of anaphase I bridges does for paracentric inversions. However, a difference between the sister chromatids of an anaphase chromosome can also be observed in heterozygotes of other chromosome structural variants: deficiencies, duplications, different types of translocations (Sects. 5.1, 5.2, 5.4).

5.3.4 Consequences

The genetic and segregational consequences of inversions are derived directly from the meiotic pairing and chiasma formation patterns. A major effect is suppression of *crossing-over*, by interference with pairing as well as by the elimination of the imbalanced cross-over gametes. In *Drosophila*, they were known as "cross-over reducers" before they were identified as inversions.

Inversions can be classified like genes when they are recognized by the changes they cause in the karyotype, or, in heterozygotes, because of reduced fertility, when the inverted segment is large enough to contain sufficiently frequent cross-overs. Then they can be used as chromosome markers in a linkage analysis in segregating progenies of heterozygotes for the inversion and marker genes. If a gene is linked to an inversion, it is in the same chromosome; where in the chromosome is not immediately clear.

Fertility is reduced as a consequence of the formation of deficiencies in both types of inversions, although by somewhat different mechanisms. In paracentric inversions the bridges tend to break and result in deficiencies, while the acentric fragments will be lost or are occasionally randomly included in one of the gametes. In pericentric inversion heterozygotes, the deficiencies are always accompanied by duplications. In a number of organisms, of which several *Drosophila* species are good examples, the bridge of the paracentric inversion does not break in the female and keeps the recombinant chromatids together. The result is that in the second meiotic division the recombinant chromatids remain associated with the new cell membrane and are included in the secondary polar body. The non-recombinant chromatid is included in the functional cell which becomes the egg. The result is complete fertility, but no observed recombination in the inverted segment. In the male no recombination takes place anyway, so that with complete maintenance of fertility a block of genes remains intact. This has population-genetic consequences that need not be discussed in the present context (Swanson 1957). In maize, paracentric inversions often, but not always (Burnham 1962), behave similarly in the female. In pollen mother cells, in contrast to male *Drosophila*, crossing-over in the loop is as frequent as in the female embryo sac mother cells, but the crossing-over products are not eliminated. Consequently, pollen fertility is reduced. Because of the excess of pollen normally present, this does not seriously affect fertility.

5.4 Translocations

A translocation is the transfer (as well as the result of the transfer) of a chromosome segment away from its original position to another position in the genome. There are two basic types. The most common is the *reciprocal translocation* or *interchange* and the other is the *simple translocation*. In the simplest form of reciprocal translocation (Sect. 5.4.1), two chromosomes have exchanged terminal segments (of any size), including the telomeres. There are, basically, two types of "simple" translocation (Sect. 5.4.3): the interstitial translocation with different subcategories and the simple terminal translocation, where a terminal chromosome segment appears to be attached to the end of another chromosomes. This resembles the interchange in many respects and will be discussed together with the interchange (Sect. 5.4.1).

5.4.1 Reciprocal Translocation or Interchange and Simple Terminal Translocation

5.4.1.1 Types and Origin

There is only one basic type of reciprocal translocation, the interchange of two terminal segments between two non-homologous chromosomes. The reciprocal transfer of interstitial segments requires many more interactions between chromosomal lesions and is so rare that it need not be considered here. Occasionally, one of the interchanged segments is very small and can, for all practical purposes, be neglected. Then, in practice the translocation is a simple terminal translocation, from which it would be hard to distinguish. Possibly all simple terminal translocations are just variants of the interchange.

It is not uncommon that a chromosome is involved in more than one interchange. This introduces special complications that will be discussed separately (Sect. 5.4.2).

Interchanges, and translocations in general, can be formed in somatic or meiotic cells spontaneously or induced by ionizing radiation or other mutagens. In principle, for reciprocal translocations, and also for simple terminal translocations the interaction between two lesions is necessary (Fig. 2.4.). Present-day understanding of telomere structure does not exclude true terminal, simple translocation which requires a new telomere to be formed where the translocated segment is removed. A reciprocal translocation between the two arms of the same chromosome results in a pericentric inversion (Sect. 5.3.1, Fig. 2.4).

The meiotic origin from duplications has been mentioned above (Fig. 5.3). It is a rare event, but when a duplication, large enough to undergo regular exchange with one of the original segments, happens to be present in an individual or population, translocations can be relatively frequent in the progeny. They are all identical.

5.4.1.2 Relevance

Interchanges are the most common type of spontaneous and induced chromosomal rearrangement. Different species and less frequently different cultivars of the same species (wheat, for instance) may differ in one or more interchanges and these appear in a heterozygous state in their hybrids. Translocations are often carried along unknowingly in breeding programs into new varieties. When they occur in chromosomes that carry a specific gene, which is to be transferred from one cultivar or species into another, the level of *recombination* by exchange (crossing-over) in specific chromosome segments can be considerably reduced in the heterozygote, even to zero. In addition to being reduced, recombination is affected in another way: the linkage relations are altered, and all genes of the two or more chromosomes involved are linked in the heterozygote (Sect. 5.4.1.4). *Fertility* is usually reduced (Sect. 5.4.1.4), but not always. Homozygotes tend to be normal, but may occasionally show slight irregularities in their phenotype and reproduction, probably due to homozygous damage at the break points. In homozygotes the linkage groups have changed compared to the original homozygote.

Translocations are used to *transfer* chromosome segments with specific genes from an added alien chromosome to a chromosome of a cultivar (Sect. 10.4.4.2.2), but for this purpose interstitial, simple translocations are more suitable. Translocation heterozygotes can be used for making specific *duplications* (Sect. 11.2.2) and they are a source of *aneuploids*, including tertiary and compensating trisomics to be used in developing *balanced trisomics* for *hybrid breeding* programs (Sect. 12.4.2.2). Their effect on linkage, which is negative when specific genes must be transferred by exchange-recombination, can be used to keep large groups of allelic combinations together (*permanent translocation heterozygosity*: Sect. 12.3). Translocations, as well as other rearrangements, may play a role in the artificial differentiation of genomes in attempts to convert an autopolyploid into an allopolyploid: *allopolyploidization* (Sect. 12.2.2.1). For genetic research, translocations and their derivatives are used for gene localization (Sect. 8.3.2.2.2).

5.4.1.3 Characteristics and Identification

Translocations in plants have been extensively described by Burnham (1956). Since then much new information has become available.

Karyotype. When the segments interchanged in a reciprocal translocation differ in size, the morphology of the two chromosomes involved can change sufficiently to be recognizable in the karyotype. With clearly different banding patterns, recognition can be relatively easy. Even on the basis of chromosome size, several reciprocal translocations can be recognized, especially with careful karyotype analysis (Fig. 4.5). Drastic size changes make recognition sufficiently simple for use in somatic classification of the translocation. However, in spite of considerable size differences in the interchanged segments, recognition in the karyotype may still be difficult, for instance when the reconstructed chromosomes resemble other, unchanged chromosomes in the karyotype, or when the chromosomes have "exchanged morphology", etc. (Fig. 5.7).

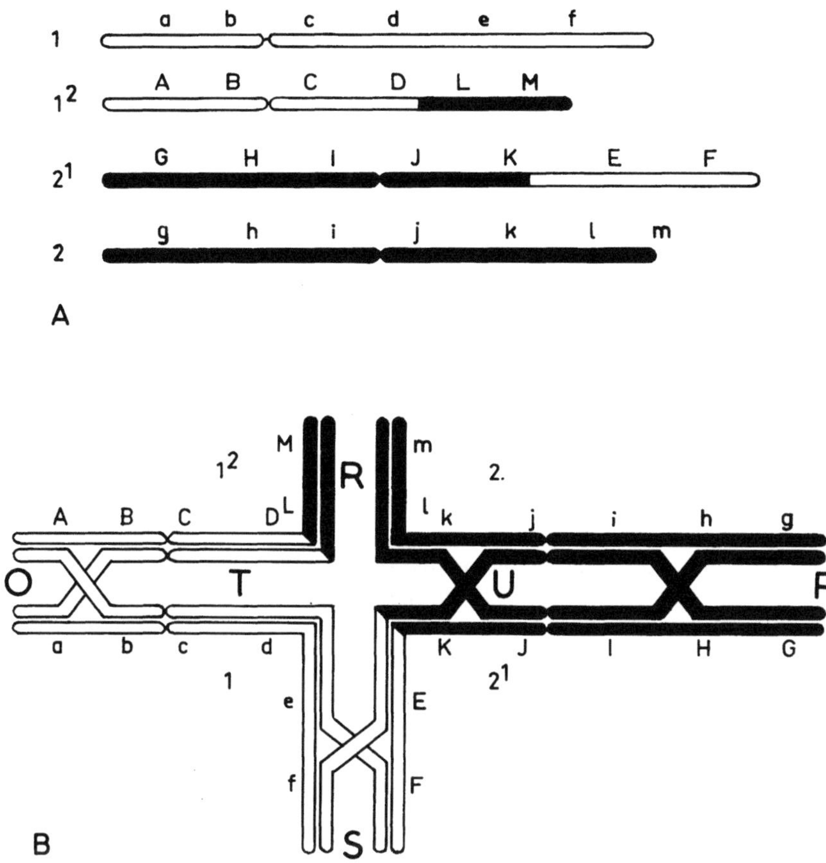

Fig. 5.7 A The chromosomes involved in a heterozygous reciprocal translocation (interchange): *1* and *2* are the normal chromosomes, 1^2 and 2^1 the translocated chromosomes. **B** The cross-shaped pairing configuration of **A**. There are six segments: *O* and *P* (arms not involved); *R* and *S* (the interchanged segments); *S* and *T* (the interstitial segments) of the translocated arms. Chiasmata have been drawn in four arbitrary segments. (After Sybenga 1975)

Meiosis. Homozygous interchanges do not behave essentially different from normal individuals in meiosis. Only when one of the chromosomes has become exceptionally small, can an increased frequency of open bivalents and even univalents for this chromosome be observed (Sybenga 1975). Heterozygotes, however, show specific meiotic abnormalities. The characteristic *meiotic pairing* behaviour of a reciprocal translocation (interchange) heterozygote is shown in Fig. 5.7B. When all homologous segments pair, a cross is formed. This can be seen in the synaptonemal complex and in favourable material at pachytene in the light microscope (Fig. 5.8).

In the pairing cross the point where the pairing chromosomes exchange partners is fixed at the place of the breaks in the chromosomes. Some variation

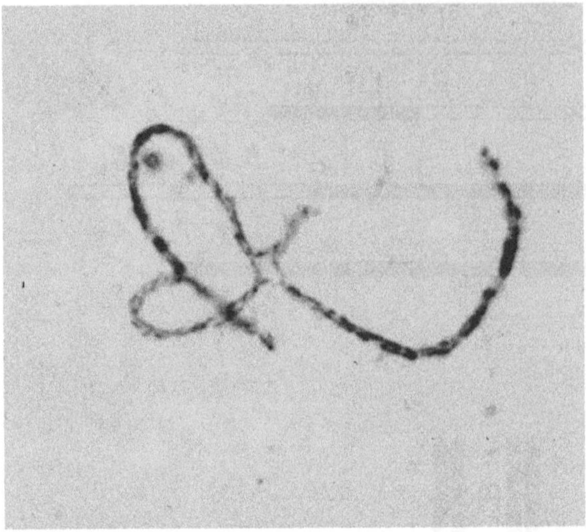

Fig. 5.8 Light microscope photomicrograph of the translocation cross at pachytene of interchange T8–10/N of maize. (Courtesy of M.M. Rhoades)

in pairing around the break point of a translocation heterozygote, even involving considerable stretches of non-homologous pairing (Burnham 1962; de Jong et al. 1989), is quite common. When the banding or chromomere pattern of the chromosomes is sufficiently specific to recognize individual chromosomes or even segments, the location of the point of partner exchange can be seen, as well as its variation. Chromomeres are clear at pachytene in a few species and sometimes in specific cultivars. They are not visible in electron micrographs of SCs, where the chromatin is either mostly removed or highly dispersed (basic formaldehyde fixation). When present in the chromosomes involved in the quadrivalent, nucleoli can be useful markers in the SCs of nucleolar chromosomes in EM preparations. Centromeres can be made visible more readily in some materials than in others and if they are, they are good markers for the location of the point of partner exchange, provided the distance is not too large. In general, the identification of the break point of the interchange is not very reliable in pachytene preparations. The difference in size relations between mitotic and meiotic chromosomes make the interpretation of the observations difficult.

When pachytene is not accessible, later stages must be used for the recognition or identification of translocations and for studying their meiotic behaviour. Chiasmata follow a genotype-specific system of frequency and localization, which is affected by pairing problems encountered around the pairing cross. The chiasmata keep the four chromosomes together, but the original cross shape is not maintained when the homologues separate at the end of pachytene. At diplotene, some remnant of the cross may be visible,

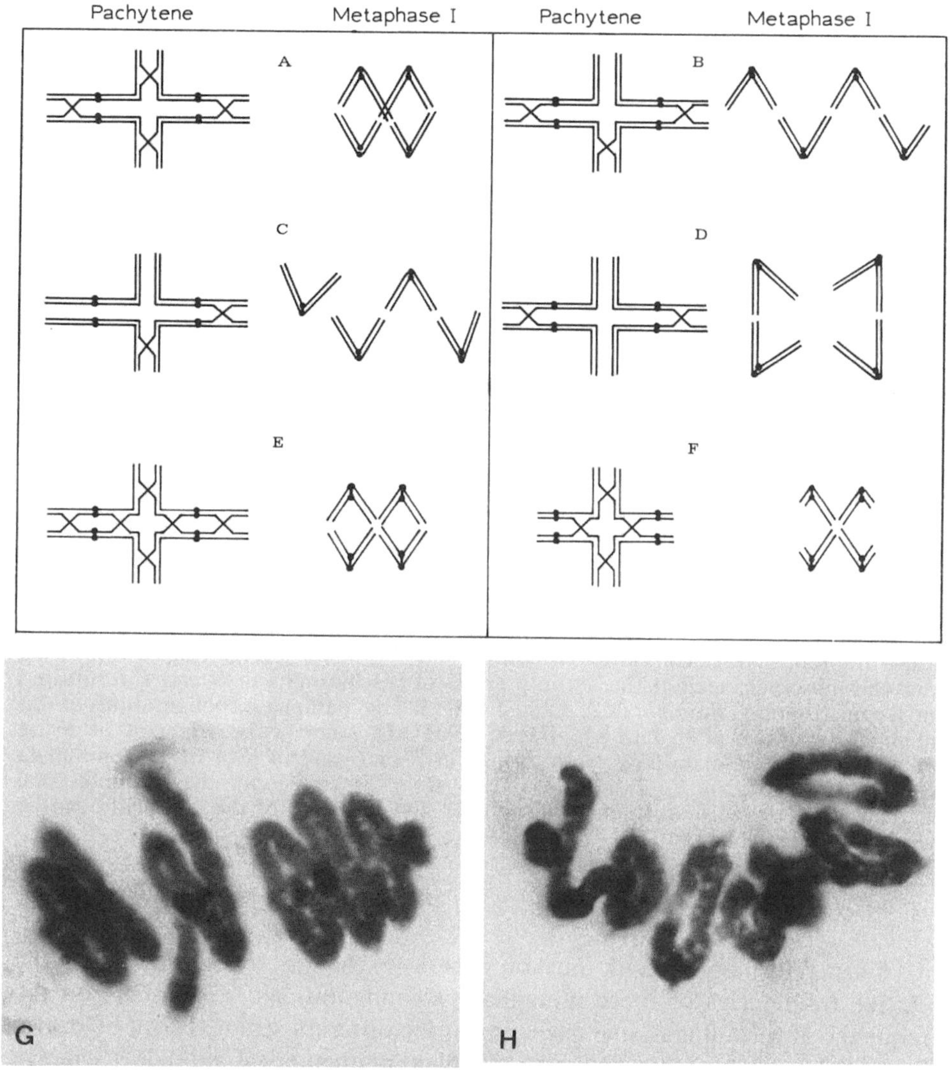

Fig. 5.9 A–F Different combinations of chiasmata at pachytene and the corresponding metaphase I configurations of interchange quadrivalents. **G** Chain quadrivalent in rye, corresponding to **B**. **H** "Frying pan" quadrivalent in rye, the configuration formed with the combination of chiasmata in Fig. 5.7. (After Sybenga 1975)

but at diakinesis this has disappeared. Most observations on meiotic configurations are made in diakinesis or first metaphase, when the centromeres of the four chromosomes in the quadrivalent stabilize their orientation on the spindle poles. Depending on the presence or absence of chiasmata in the different segments of the quadrivalent, different and specific metaphase I configurations result (Figs. 5.9, 5.10). These contain information on the system

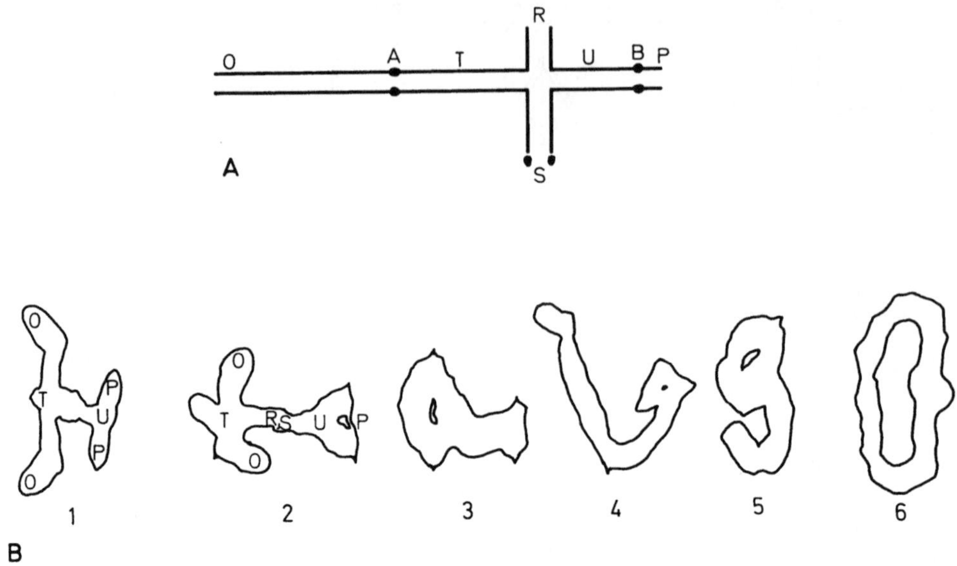

Fig. 5.10 A The pairing cross of the interchange heterozygote of *Delphinium* of Jain and Basak (1963). The interstitial segments are large and, like the non-translocated arms, different in size. Chiasmata in both *O* and *T* produce a large ring of two of the four chromosomes, seen at the *left* in *3* and *5* of **B**; chiasmata in *U* and *P* result in a small ring as seen at the *right* in *2*, *3* and *4* of **B**. **B** The metaphase configurations of this interchange, drawn after Jain and Basak (1963). In *1* and *2* the segments of **A** are indicated. In *1* chiasmata have been formed in *T* or *U* and in *R* or *S*; in *2* chiasmata have been formed in *P*, *T* and *U* and in *R* or *S*; etc. From the morphological characteristics of the multivalents the number and locations of the chiasmata can be derived. (After Sybenga 1975)

of chiasma formation, and thus of genetic exchange. Therefore, from the relative frequencies of these metaphase I configurations, conclusions on the frequency of recombinational exchange in the different segments can be drawn (Sect. 8.2.1.4), but this requires complex mathematical models (Sybenga 1975). Although occasionally important for estimating the probability of recombination between a gene and an interchange or between two genes close to an interchange, this very specialized analysis will not be discussed further here. A number of different metaphase I configurations and their origin from different combinations of chiasmata in the paired segments are shown in Fig. 5.10. Many of them are characteristic for translocations, but also for tetrasomic quadrivalents (Chap. 7).

At meiotic metaphase I, the most common interchange configurations are the ring and the chain quadrivalents. With metacentric chromosomes, rings result when chiasmata are formed in all end segments, including the two interchanged segments (R and S) and the two non-translocated arms (O and P, Fig. 5.7). Chains are formed when a chiasma is absent in any one of the four end segments, but with chiasmata in the other three. There are two more segments of interest for the resulting metaphase I configuration: the interstitial segments T and U (Fig. 5.7). A ring quadrivalent (Fig. 5.9A; Fig. 5.10, *6*) is converted into a real "figure 8" (Fig. 5.9F; Fig. 5.10, *3*) when a chiasma is formed in either one or both of the interstitial segments in addition to the chiasmata in the four end segments. Depending on the length of the segments and the degree of chromosome condensation, the locations of the interstitial chiasmata may be visible in the metaphase I configuration. In the example of *Delphinium* (Fig. 5.10) the differences between the segments permit recognition of the location of the chiasmata in all segments even without C-banding. When the chromosomes are more symmetric and the chiasmata tend to be localized distally, as in rye (Fig. 5.9G,H), without C-banding this is only possible with extreme differences between segment length. Meiotic C-banding can be very helpful in locating chiasma positions. *Delphinium* is a typical example of high frequency of chiasmata in the interstitial segments. An alternate ring (Fig. 5.9A) and "figure 8" are easily mistaken for two partly overlapping ring bivalents (to the left in Fig. 5.9G) when not observed very critically.

Quadrivalents are not always formed in interchange heterozygotes. Small interchanged segments may fail to have chiasmata in some or even many of the meiocytes and when chiasmata are present in the non-interchanged arms, two bivalents are formed. These may be heteromorphic, depending on the relative sizes of the different chromosomes and chromosome segments. With low overall levels of chiasma formation, bivalent pairs and trivalents with a univalent or even sets of four univalents or two univalents with one bivalent may result. The frequency of multivalents may then become so low that the interchange is not detected when a small number of cells is analyzed. Examples are hybrids between species that are not closely related or where the regulation of chiasma formation is disturbed by hybrid dysgenesis. Desynapsis will have similar effects within species.

Interchanges between pronounced acrocentric or telocentric chromosomes, where one arm is too short to form chiasmata, or even absent, have fewer segments available for chiasmata and consequently fewer types of configurations in the heterozygotes. In fact, although a pairing quadrivalent is formed at pachytene, this is maintained at metaphase I only when there are chiasmata in the interstitial segments in addition to at least one of the interchanged segments. Few plant species have karyotypes with predominantly pronounced acrocentric chromosomes, but in several plant species the occurrence of a few (sub)acrocentric chromosomes is not uncommon. Heterozygotes for interchanges between acrocentric chromosomes with chiasmata predominantly in the distal (or proximal) segments will not or very seldom form multivalents at metaphase I and the interchange may go undetected. The segregational irregularities inherent to interchanges will still appear, but it is difficult to recognize the interchange as the cause of these irregularities. It has been observed, however, that in such cases additional chiasmata may be formed

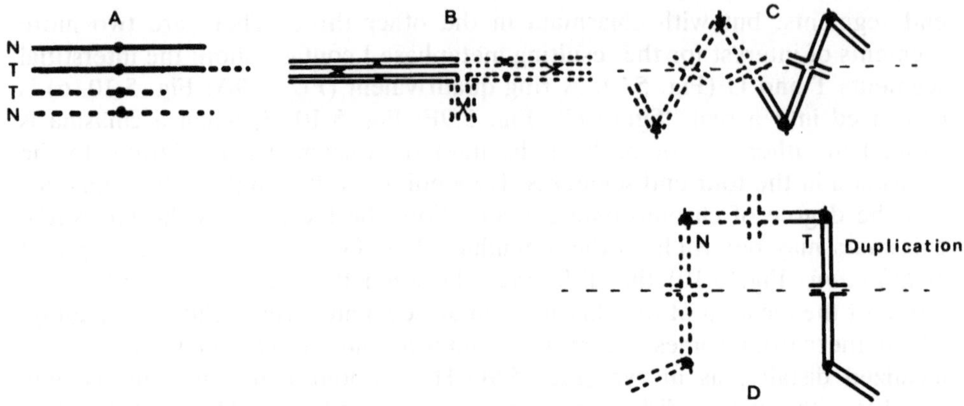

Fig. 5.11 The origin of a terminal duplication by adjacent orientation of a terminal translocation. **A** The chromosomes: *N* normal, *T* translocated. **B** Diagram of pairing configuration. **C** Alternate orientation, leading to segregation of one complete normal complement (*top*) and one complete translocation complement (*bottom*). **D** Adjacent segregation: terminal duplication (*top*) and deficiency (*bottom*)

in segments usually not forming chiasmata. Then quadrivalents will appear nevertheless (Parker 1987).

The *orientation* of the configurations of interchange and simple terminal translocation heterozygotes is of considerable importance, because only special combinations of the chromosomes of the complex result in balanced gametes. In addition, the process of orientation is complex because of the large number of co-orienting centromeres. Orientation and segregation will be discussed in Section 5.4.1.4.

The meiotic behaviour of *simple, terminal translocations* differs from that of interchanges only by not having a four-armed pairing cross but a three-armed configuration (Fig. 5.11). At metaphase I a ring quadrivalent is not formed, but a real "figure 8" is possible, provided both interstitial segments are large enough to have a chiasma simultaneously. The configurations formed by simple, terminal translocation heterozygotes resemble those of the duplication that is readily formed by special segregation of this translocation (Fig. 5.11). There are only few reports of meiosis of both the simple, terminal translocation and the duplication derived from it, one is T242W in rye (*Secale cereale*, Sybenga and Verhaar 1980).

Fig. 5.12 Different orientations of an interchange quadrivalent and the resulting segregation. **A** With alternate orientation of a ring quadrivalent, four balanced but two by two structurally different tetrad cells are formed **B, C** With the two adjacent orientations, imbalanced spores are formed. This results in linkage between all genes in the two pairs of chromosomes, but with the possibility of exchange recombination. **D** By an interstitial chiasma (cf. Fig. 5.7B) two balanced, but structurally different spores and two imbalanced spores and formed. (After Sybenga 1975)

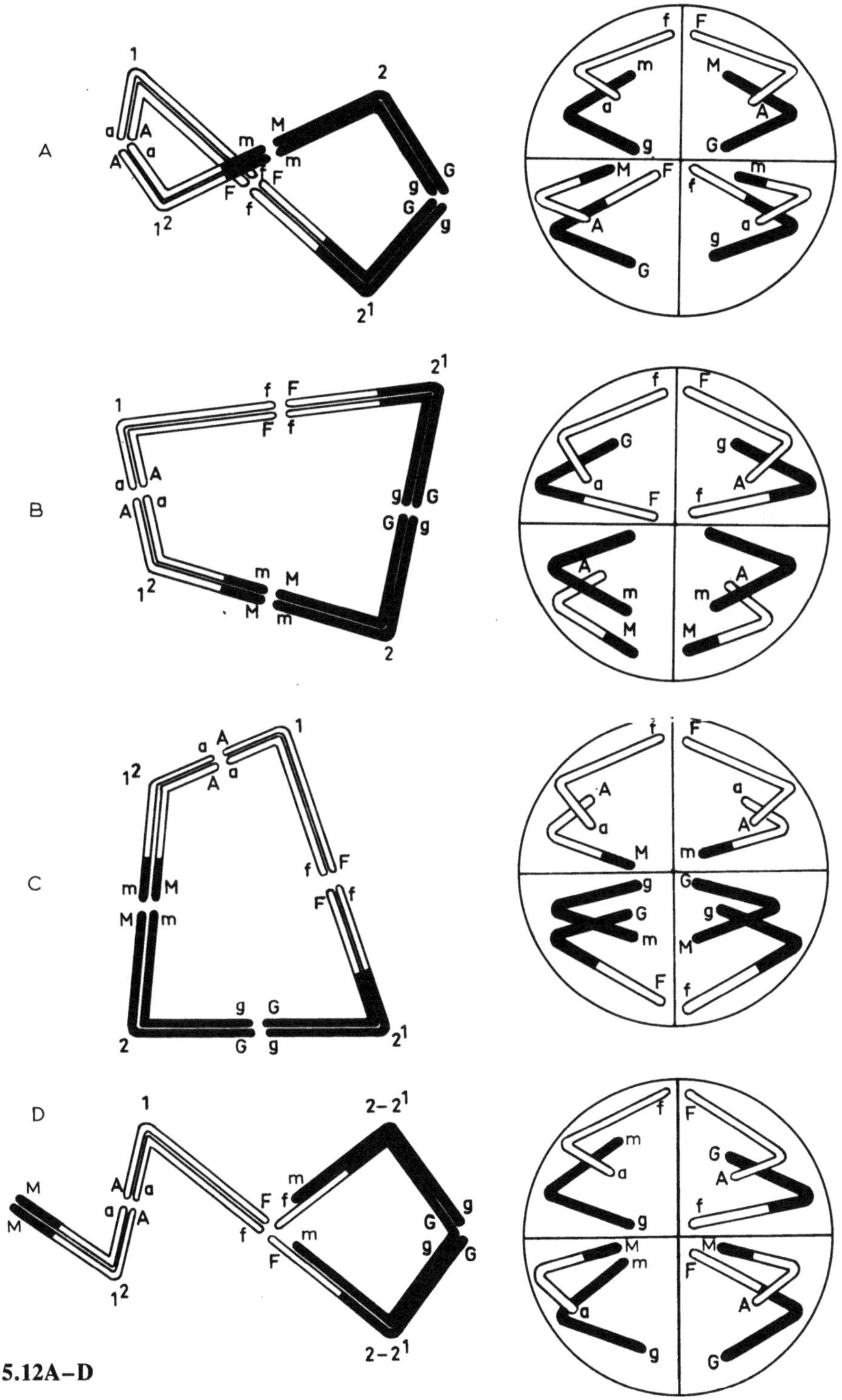

Fig. 5.12A–D

5.4.1.4 Consequences: Orientation and Segregation

The most important consequences of the meiotic behaviour of interchange heterozygotes result from the metaphase I orientation and the subsequent anaphase I segregation of the configuration. The final orientation is not established immediately at the beginning of centromere activation and attachment to the spindle microtubules in prometaphase. However, the initial position of the chromosomes and their centromeres in the nucleus have an effect on the final orientation. Both initial orientation and subsequent reorientation depend on various cellular and chromosomal factors. The subject is extremely complicated and will be discussed here only briefly, in spite of its importance, not only for interchanges but also for all situations in which multivalents are formed.

The orientation and subsequent segregation of the two chromosomes of a normal bivalent are relatively simple: as long as one of the chromosomes moves to one pole and the other to the other pole, meiosis can continue its normal course without undesired consequences. Two chromosome pairs in two bivalents will normally segregate independently, which implies that in the daughter cells the genes of the two pairs will occur in different combinations: chromosome recombination (Fig. 3.12). Because the two chromosomes of each bivalent are equivalent, no daughter cells with too few or too many chromosomes or chromosome segments (deficiencies and duplications) will result. In an interchange heterozygote this is different, because all four chromosomes of the complex are different. Even when bivalents are formed, not all combinations of two chromosomes will form a balanced set: chromosome recombination is mechanically still possible but in 50% of the cases it leads to imbalanced and thus non-functional duplication-deficiency gametes. When the exchanged segments or the original chromosomes differ considerably in size, the bivalents can be seen to be "heteromorphic". At metaphase, their orientation with respect to each other can then be scored directly in the preparation.

The same imbalanced segregation may result from a quadrivalent. Only the chromosomes in alternate positions in the quadrivalent form a balanced combination, either the two normal chromosomes together, or the two interchange chromosomes together (Fig. 5.7). "Alternate segregation" after *alternate orientation* (centromeres co-orient, which have alternating positions in the quadrivalent) results in balanced gametes. "Adjacent segregation" after *adjacent orientation* (centromeres adjacent in the quadrivalent co-orient) results in imbalanced gametes. Such imbalanced gametes will not normally function in plants. In animals they may do so because the haplophase is genetically not as demanding as in plants. In animals, however, the zygotes or embryos are more sensitive to genetic imbalance and may fail to develop properly. Occasionally, when the imbalance is minor, duplication and deficiency progeny may arise from interchange heterozygotes, mainly through the mother, where there is no serious competition between the spores. Selection against deviants through the male is usually strong. In exceptional cases, two imbalanced gametes of complementary types will be combined by fertilization. This results in a balanced, apparently normal heterozygote. In animals this is not especially rare, but in plants it is quite unusual, because of the failure of an imbalanced male spore to function.

Because the products of chromosome recombination are not viable, there is *no chromosome recombination* between the interchange chromosomes (Fig. 5.12), and all genes in both chromosomes appear to be linked as if in one linkage group. This is an important consequence of interchanges and translocations in general.

The least complicated multivalents are chain and ring quadrivalents, and these show the straightforward behaviour of linkage of all genes in the two chromosomes, only broken by exchange recombination, and further a reduction in fertility due to the occurrence of non-functional gametes. The frequency of these non-functional gametes depends on the frequency of adjacent orientation. Some factors determining this frequency will be briefly discussed below. Other configurations show additional complications. The trivalent with univalent, for instance, will show similar consequences with respect to linkage, but lower fertility because of the erratic behaviour of the univalent, which in plants is often lost.

Configurations with interstitial chiasmata show a somewhat different segregational pattern. As seen in Figs. 5.7 and 5.12, an interstitial chiasma results in a structural difference between the two chromatids of two of the four chromosomes. These chromosomes, originally one translocation chromosome and one normal chromosome, have now become identical, but each with two different chromatids. The other two chromosomes, originally also different, without interstitial chiasma remain different. The example is a typical "frying pan" quadrivalent at diakinesis/metaphase. Assuming disjunctional segregation of the two chromosomes in the "handle" of the pan, a 2:2 segregation results. In the second meiotic division, where the chromosomes are no longer connected, the orientation of the chromatids in different chromosomes in a cell is (assumed to be) independent. As seen in Fig. 5.12, two different combinations of chromatids are possible here, each with a 50% probability. Two combinations are imbalanced and the other two are balanced. Of these, one has the two normal chromosomes and one has the two interchanged chromosomes. There has been exchange recombination between these in 50% of the cases, between which genes depends on the location of the chiasma. As a consequence of this segregation, there are 50% non-functional gametes, systematically resulting in 50% sterility. In absence of interstitial chiasmata, the orientation of the quadrivalent determines fertility, and this can vary considerably.

In all these cases the interchange and the normal complement segregate 1:1 in the gametes, and 1:2:1 in the F2, exactly like a monofactorial marker.

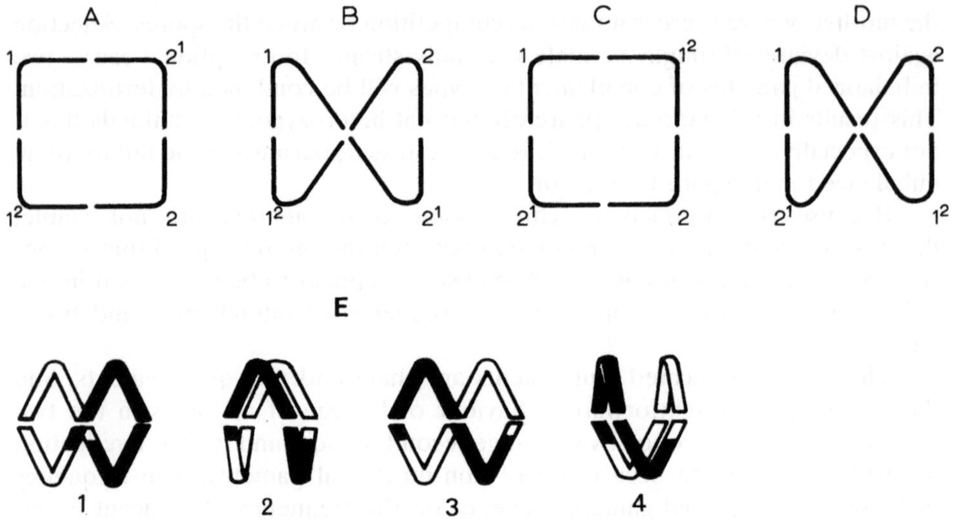

Fig. 5.13 The four types (**A–D**) of orientation of an interchange ring quadrivalent. In a three-dimensional space **B** and **D** are equivalent, as seen from the four views of an alternate ring shown from different angles (**E**). Centromeres *numbered* as in Fig. 5.7. (Sybenga 1984b)

An interchange, and in fact any other translocation with similar behaviour, can be considered as a monofactorial marker in any genetic experiment (Sect. 8.3.3.2.2).

The relative frequencies of the different orientation types depend on a number of factors. Only considering ring and chain quadrivalents, in addition to the adjacent and alternate orientations, a few more types are possible and for the adjacent orientation two distinct types can be distinguished. The determining factors in orientation and subsequent segregation have been discussed on several occasions (see, for example Sybenga 1975; Rickards 1983; Sybenga and Rickards 1987). Although it has been done in the past, it is too simplistic to consider merely the different segregational possibilities, assuming that all have an equal probability, and deriving expectations on fertility and frequency of abnormal segregants from these assumed probabilities. The different processes taking place in preparation of and during orientation are too many and too complex for simple model hypotheses. Nevertheless, there are a few simplified model situations that can lead to better insight in the factors involved. In Fig. 5.13 the simplified situation is assumed of a 2:2 orientation of a ring quadrivalent (Sybenga 1972, 1975; cf. Fig. 5.12). One of the four chromosomes is given a constant position: i.e. chromosome 1 (top left). It can co-orient with chromosome 1–2, and the other two chromosomes have a choice of two poles. One results in situation A (adjacent) and the other in B (alternate). However, chromosome 1 can also choose its other partner in the ring, viz. 2–1, with which to co-orient. Then chromosomes 2 and 1–2 have two possibilities for choosing their respective poles. One results in adjacent orientation again, the other in alternate. As a consequence, there are two adjacent and two alternate types. The two adjacent types are quite different. In one, the homologous centromeres (1 with 1–2, and 2 with 2–1, see Fig. 5.13) co-orient. This is called *adjacent 1*. In the other case the non-homologous centromeres co-orient (1 with 2–1 and 2 with 1–2), and this is called *adjacent 2*. As can be seen in Fig. 5.12, the resulting segregational products are quite different, but both imbalanced. In general, adjacent 2 will result in much larger (and of course quite different) duplications and deficiencies than adjacent 1.

With interstitial chiasmata the situation is different. Adjacent-2 coorientation is rare and with an interstitial chiasma, the difference between adjacent 1 and alternate disappears: both chromosomes have one normal and one interchange chromatid (Fig. 5.12).

Interchanges between acrocentric chromosomes do not have such problems with adjacent-2 orientation (Sybenga 1975), but are not sufficiently common in plants to justify a detailed discussion. Since the formation of a multivalent requires interstitial chiasmata, adjacent-1 and alternate cannot be distinguished.

Chain quadrivalents would seem to have the same possibilities for orientation as rings, but in practice they have more limitations. Depending on whether the segment lacking a chiasma is a translocated segment or a non-translocated arm, the two centromeres in the middle will be non-homologous centromeres or homologous centromeres. Adjacent orientation of a chain quadrivalent, with the two central chromosomes co-orienting and the outer two pairs not, is much less stable than a chain with the two outer sets of centromeres co-orienting and the inner two centromeres not. As a consequence, the latter type is much more frequent than the first type. When the shortest segment is a translocated segment, the first type is adjacent 2 and the second adjacent 1 (Fig. 5.13). In many interchanges between metacentric chromosomes the shortest segment is a translocated segment. In chains, therefore, adjacent 2 is seemingly improbable, and this has, incorrectly, been stated as a rule. It depends, however, entirely on which segment fails to have a chiasma. With acrocentric chromosomes this dilemma does not exist.

There has been considerable discussion about the relevance of a distinction between the two types of alternate orientation, which have identical segregation. This will not be discussed here (see Sybenga 1984b).

There are more categories of orientation, and these are the types most frequently producing aneuploidy in the progeny. One is the orientation where three of the four chromosome orient on one pole, and only one on the other. Although this type is not excluded and is part of certain theoretical models, it is probably not stable enough to persist. Another type is *linear orientation* of ring or chain quadrivalents (Figs. 6.7C, 6.11F). At anaphase I the chiasmata in the quadrivalent are not released entirely synchronously, and when one of the polar-oriented chromosomes is free early, it will go to one pole and the other three to the other pole, thus resulting in an extra chromosome at one pole and one less in the other (non-disjunction). Adjacent segregations result when two chromosomes move to one and two to the other. Aneuploidy has still other origins. One is the random segregation of univalents when a trivalent and a univalent are formed. In plants, univalents are often lost, and hypoploidy is the most common consequence, although usually not viable in diploids.

In the simple model of Fig. 5.13, the expected frequencies of the four 2:2 orientation types are equal. In practice, this is almost never realized. In many cases, alternate orientation is more and sometimes even much more frequent than adjacent. Only occasionally is adjacent orientation more frequent. Detailed theories on orientation (Sybenga 1975; Rickards 1983) do not explain this variation. Several factors may play a role (Sybenga and Rickards 1987), e.g. stiffness and rigidity of the quadrivalent, position in the nucleus, pre-orientation, reorientation, delay in cell development, etc. With so many factors involved, it is practically impossible to make a theoretical prediction of "expected" frequencies of final orientation types.

5.4.2 Two or More Interchanges Combined: Translocation Tester Set; Balanced Complex Translocation Heterozygotes

5.4.2.1 Types and Origin

The combination of two interchanges can be of three different types: (1) they have no chromosomes in common; (2) they have one chromosome in common; (3) they have both chromosomes in common. For simplicity, it is assumed that the chromosomes are metacentric, which is the most common situation in plants. In the second type, it is assumed that the chromosome common in the two interchanges has one arm involved in one translocation, and the other arm in the other translocation. There are two subtypes: (1) the two interchanges have been formed in the same genome, and the other genome is of the standard type; (2) one interchange occurs in one genome and the other in the other genome. The latter situation may originate from a cross between two parents, each with a different interchange, but also spontaneously or induced in the same diploid cell, simultaneous or in succession, each in one of the genomes (Fig. 5.14).

5.4.2.2 Relevance

The possibility to distinguish between the three combinations of two interchanges (no chromosomes in common; one chromosome in common; both

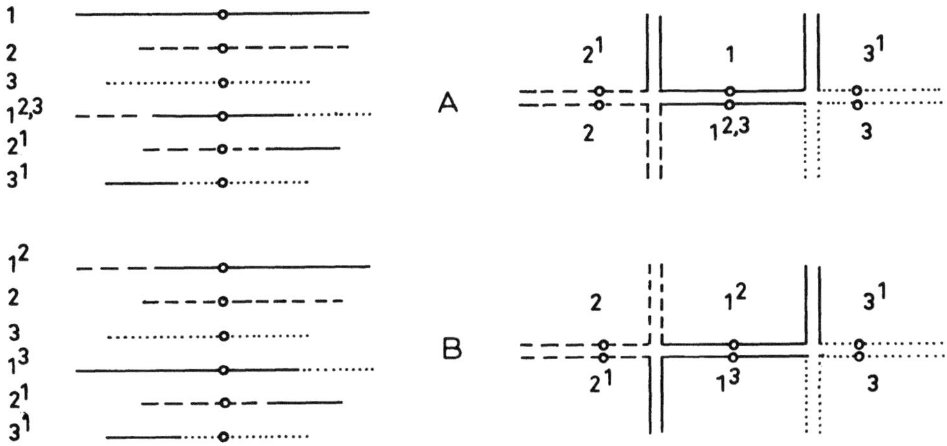

Fig. 5.14 Two interchanges involving three chromosomes. **A** Both genomes carry one translocation each; there is no normal homologue of the chromosome shared by the two interchanges. **B** One genome has both interchanges; there is one complete set of three normal chromosomes and one set of three translocation chromosomes in the diploid. Both types pair in a hexavalent, and with alternate segregation, the original types reappear unless there has been a chiasma in one of the two (differential) segments between the centromere and one of the pairing crosses

chromosomes in common) is used to identify unknown interchanges. A prerequisite is the availability of a set of known interchanges, together covering all chromosomes of the complement. Such a set is called a *tester set*. Tester sets are available for several cultivated species, including maize (Burnham 1954), barley (Ramage et al. 1961), pearl millet (Minocha 1991) and rye (Sybenga and Wolters 1972). Interchange (or translocation) tester sets can be used to identify simple translocations, trisomies and most other karyotype variants in ways very similar to that briefly discussed below. In allopolyploid cultivated species it is usually more convenient to make use of a series of monosomics (Sect. 6.2.1.3), which can be applied for several additional purposes, but cannot frequently be constructed in diploids.

Combinations of interchanges play an important role in nature as ways to conserve heterozygosity in populations exposed to inbreeding. Attempts have been made to apply two or more interchanges to create artificial, *permanent interchange heterozygotes* of cultivated species, with the same purpose. This will be discussed briefly below and in Section 12.3. Two translocations involving the same tow chromosomes can be used to produce specific *duplications* (Sects. 5.4.2.3 and 11.2.2).

Combinations of interchanges, like single interchanges and other chromosome structural rearrangements, have been proposed for use in differentiating pairs of genomes in autotetraploids: *allopolyploidization* (Sect. 12.2.2.1).

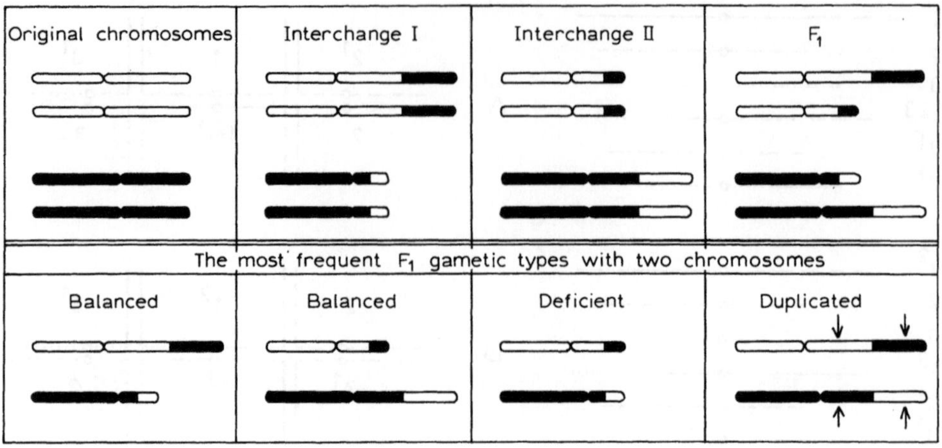

Fig. 5.15 The origin of a duplication from two heterozygous translocations involving the same two chromosomes. The location of the break must satisfy certain conditions, otherwise deficiencies occur

5.4.2.3 Characteristics and Identification

The effects of two interchanges on the *somatic karyotype* are sufficiently predictable to make further discussion unnecessary. The meiotic characteristics of the combination of two interchanges are much more interesting. They depend on how many chromosomes are involved and in which arms the breaks have occurred. When the two translocations have two chromosomes in common, at most one quadrivalent can be formed. Depending on which arms are involved in the two interchanges and where the breaks are located, the quadrivalent can have different shapes. When the breaks are in the same arms and not too far apart, the formation of a quadrivalent will be rare, and two (slightly) heteromorphic bivalents are formed (Fig. 5.15). These can orient independently, and thus a pair of balanced AI cells will be formed in 50% of the cells, one cell carrying one interchange and one with the other. In the remaining 50%, however, the combination is not balanced and results in daughter cells with a duplication and a deficiency. This is a method for producing "directed" *duplications* (see Fig. 5.15), and will be briefly taken up again in Section 11.2.2.

When the two interchanges have one chromosome in common, a total of three different chromosomes is involved. The two interchanges may have arisen due to one event in the same genome. Then the common chromosome must carry both breaks in the same homologue, and the other homologue (in a diploid) is free of breaks. The two interchanges may also have arisen in the two separate genomes; then no normal homologues are present of one chromosome: one carries the first, and the other the second translocation. In meiosis six chromosomes are united in one *hexavalent* configuration, independent of how the interchanges have arisen. The importance of the distinction between the two types becomes clear in the segregation at anaphase I and in the composition of the gametes (Fig. 5.14). When the two interchanges have been formed independently, assuming alternate orientation leading to balanced segregation, the two interchanges segregate to different poles. When the two breaks are in the same homologue, the two translocations will necessarily segregate together to one pole and the non-translocation genome will segregate to the other pole. There is one exception: an exchange in the segment between the centromere and one of the breaks in the common chromosome (the *differential* segment) by recombination transforms a double translocation chromosome and a normal chromosome into two translocation chromosomes, and vice versa (Fig. 5.14). With the proper orientation (which is not readily realized), one pole can now receive one interchange and the other pole the second, even when the two were combined originally. Similarly, the reversed may occur.

The possibility to distinguish between the three combinations of two interchanges (no chromosomes in common: two quadrivalents; one chromosome in common: hexavalent; both chromosomes in common: at most one quadrivalent, often only two bivalents) is applied to identify unknown interchanges by using a *tester set* of known interchanges, together covering all chromosomes. As noted above, a translocation tester set can also be used to identify trisomics and other chromosomal abnormalities, and to locate genes.

To identify an unknown interchange, it is crossed with each of the members of the tester set. If possible, the parents should both be translocation homozygotes. When heterozygotes are crossed, the double heterozygotes, of which the meiotic configurations must be analyzed, have to be selected in the F1 and this may produce undesired complications. Usually, metaphase I is analyzed. If two quadrivalents are found, the unknown and the known interchange have no chromosomes in common. If a hexavalent is formed, they have one chromosome in common. If never more than one quadrivalent is observed, often perhaps appearing somewhat irregular, or represented by two bivalents, the two interchanges have two chromosomes in common.

Not all possible combinations of chromosomes have to be present in a tester set, but it is not sufficient that each chromosome is represented only once. In a hypothetical example of a species with $n = 7$, a total of $(6 \times 7)/2 = 21$ different interchange types are possible. These can be identified by a combination of 1–2 (an interchange between chromosomes 1 and 2), 2–3 (between chromosomes 2 and 3), 3–4, 4–5, 5–6, 6–7, 1–7 (Table 5.1). In fact, one chromosome need not be represented because its presence in an unknown interchange (or other deviant) can be concluded by elimination. This makes two interchanges redundant and it does not matter which two. Absence of more than two interchanges from the set shown above still permits the identification of a number of unknown translocations, but not all.

Table 5.1. Testing an unknown interchange against a translocation tester set[a]

Configurations

Tester set											
1–2	*IV*	VI	VI	VI	VI	VI	VI	VI	VI	VI	VI
2–3	VI	VI	IV + IV	IV + IV	IV + IV	IV + IV	*IV*	VI	VI	VI	VI
3–4	IV + IV	VI	VI	IV + IV	IV + IV	IV + IV	VI	VI	IV + IV	IV + IV	IV + IV
4–5	IV + IV	IV + IV	VI	VI	IV + IV	IV + IV	IV + IV	VI	VI	IV + IV	IV + IV
5–6	IV + IV	IV + IV	IV + IV	VI	VI	IV + IV	IV + IV	IV + IV	VI	VI	IV + IV
6–7	IV + IV	IV + IV	IV + IV	IV + IV	VI	VI	IV + IV	IV + IV	IV + IV	VI	VI
1–7	VI	VI	VI	VI	VI	*IV*	IV + IV	IV + IV	IV + IV	IV + IV	VI
Unknown interchange:	1–2	1–3	1–4	1–5	1–6	1–7	2–3	2–4	2–5	2–6	2–7

[a] The species involved is supposed to have seven chromosomes, numbered 1 to 7. The unknown interchange is identified by the combination of diakinesis/metaphase I configurations in the double heterozygote. *IV*: single quadrivalent of possibly irregular shape, often not present in all cells. IV + IV: two quadrivalents. VI one hexavalent. Only unknown interchanges involving chromosomes 1 and 2 are shown. The tester set is redundant by at least two interchanges and may have interchanges with different chromosome combinations than shown here.

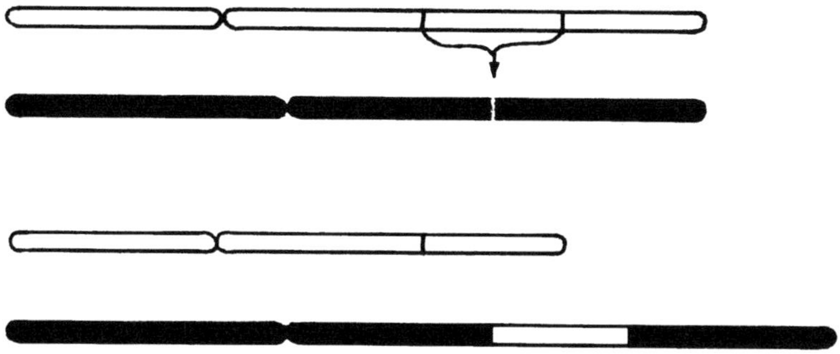

Fig. 5.16 The origin of a simple interstitial translocation, requiring three lesions

5.4.2.4 Consequences

Only alternate orientation will result in balanced daughter nuclei and consequently, all genes in the chromosomes involved in the combined interchanges will appear to be linked (Sect. 12.3). Fertility tends to be considerably reduced, disturbing *gene transfer* between species differing in more than one translocation.

Among the segregational products of the hybrid between two translocations involving the same two chromosomes specific duplications can be found (Sect. 11.2.7). Their origin is explained in Section 5.4.2.3.

5.4.3 Simple Interstitial Translocations

5.4.3.1 Types, Origin and Relevance

When the segment involved is translocated to a new location in the same chromosome, it is usually called a *shift*. It can also be transferred to any other chromosome. The orientation with respect to the centromere of the chromosome to which it is transferred can be the same as in the original position (symmetric) or reversed (asymmetric). This orientation has consequences for the resulting meiotic configuration and its behaviour in orientation and subsequent segregation of the component chromosomes.

The simple interstitial translocation requires three lesions (Fig. 5.16) and it is, therefore, much less common than the reciprocal translocation for which two lesions are sufficient. Interstitial translocations have been induced in somatic tissue in irradiation and other mutation experiments with the explicit aim to *transfer specific segments* from one chromosome to another (Sect. 10.4.4.2.2).

This is the main practical importance of interstitial translocations. Existing natural or induced interstitial translocations can *block gene transfer* by recombination from a donor species into a cultivated species because they have a considerably reduced opportunity to pair with the segment in the original position. Like any other chromosome structural rearrangement, interstitial translocations are potentially interesting for inducing pairing differentiation between the chromosomes in autopolyploids (*allopolyploidization*; Sect. 12.2).

5.4.3.2 Characteristics, Identification and Consequences

Somatically, interstitial translocations may be recognized when they are large or when readily identified segments are involved. Only when the karyotype has been described in detail will it be possible to definitely identify an interstitial translocation in somatic chromosomes.

The position and orientation of the translocated segment with respect to the centromere is important for the *meiotic configuration* that results. Their pairing in heterozygotes resembles that of deficiencies and duplications (Sects. 5.1.4 and 5.2.4). Examples are given by Darlington (1965) and Burnham (1962). When chiasmata are distally located, heterozygotes for interstitial translocations are not readily detected at metaphase I. In organisms with more randomly distributed chiasmata, detection at meiotic metaphase is easier. Because of the necessity to combine specific chromosomes in order to obtain viable combinations, heterozygotes for interstitial translocations, like those for interchanges, are characterized by linkage between genes in both chromosomes involved; reduced fertility because of the frequent production of imbalanced segregational products; increased probability to produce aneuploids and, if viable, other imbalanced progeny. Because the configurations are either more complex than in interchange heterozygotes or simply consist of two heteromorphic bivalents, the opportunities for balanced segregation are smaller, and fertility will be closer to 50% than observed in most interchange heterozygotes.

5.5 Other Rearrangements (Robertsonian Fission and Fusion; Isochromosome; Compound)

5.5.1 Types and Origin

Less common, but not negligible, rearrangements in plants are *centromere split* (fission) and *centromere fusion* (Fig. 5.17). In animal taxonomy and cytogenetics, and occasionally in plants, they are referred to as *Robertsonian translocations* (Robertson 1916). Since centromere splits in plants do not have their origin in translocation, it is not recommended to use the term Robertsonian translocation, but to refer to centromere split (or fission) or

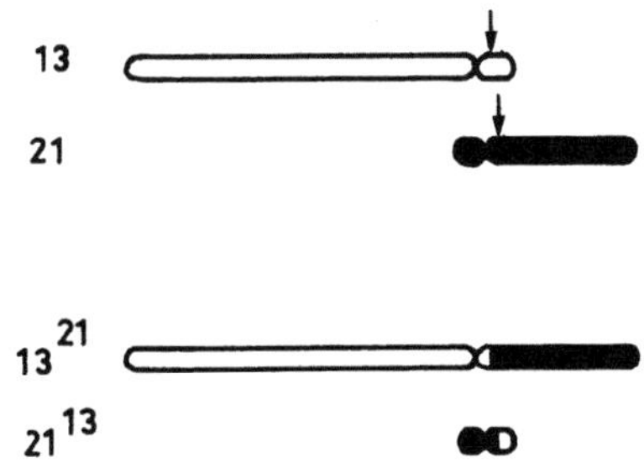

Fig. 5.17 Centric fusion. The short arms of the acrocentric chromosomes are heterochromatic and genetically inert. The new, small translocation chromosome may be lost without consequences. When homozygous, the fusion is stable

Robertsonian split (or fission). The fusions may be assumed to originate from translocations or translocation-like phenomena.

Related in origin to some forms of centromere split is the *centromere translocation*. The break is in or near the centromere, so that two (almost) entire arms are interchanged after simultaneous centromere split in two nonhomologous chromosomes.

Also related in origin are *isochromosomes*, consisting of two identical arms. When formed from an acrocentric chromosome with a genetically unimportant short arm, isochromosomes are equivalent to large duplications. When formed from a metacentric chromosome, one entire arm is lacking and it is equivalent to a duplication combined with a large deficiency. This is not uncommon in plants, but its viability is so low that it is observed mainly in allopolyploids and rarely in diploids as a deficiency/duplication. In diploids it may occur more readily as an extra chromosome. It will be discussed with the chromosome number variants in Section 6.2.2.2.

As a curiosity, the combination of two complementary isochromosomes (each representing one arm of a metacentric chromosome) can be mentioned, where no normal chromosome is present and no duplication and deficiency occur. It can be formed by centromere rearrangement at meiosis. A structure with a similar appearance can arise from the combination of the two different and complementing duplication-deficiency chromosomes from a pericentric inversion heterozygote (Sect. 5.3.3). For both cases the term *compound* is used. It will not be further discussed.

Fission of metacentric chromosomes into two telocentrics can occur spontaneously by centromere breakage (misdivision), especially in meiotic univalents (Figs. 6.8 and 6.9; cf. Sects. 6.2.1.4 and 6.2.2.1.4). The two telocentrics are infrequently recovered simultaneously, but more frequently one is found as an additional chromosome (telocentric trisomic: Sect. 6.2.2.2). In the telocentric trisomic the break is often not exactly in the centromere, but in the region of chromatid stickiness just outside the centromere, such that occasionally a very small segment remains of the other arm. Two independently formed telocentrics can be combined in the experiment (Sybenga 1975) and then made to replace one normal metacentric. True centromere fission, where both telocentrics are recovered together, has occurred for certain on several occasions in plants, mainly as a result of irregularity of meiotic segregation of univalents. The mechanism of origin will be discussed again in Sections 6.2.1.4 and 6.2.2.1.4 (Figs. 6.8 and 6.9). Recovery after induction by ionizing or UV irradiation or chemical mutagens has been reported in animals, but not with certainty in plants. *Fusions* are usually the result of translocations and are assumed to have accompanied speciation in several taxa (Stebbins 1971). Spontaneous and experimental fusions are rare in plants, but much more frequent in animals, where acrocentric chromosomes are more common. The origin of a fusion of two acrocentrics into one metacentric is shown in Fig. 5.17. *Isochromosomes* can be derived from acrocentric chromosomes or metacentric chromosomes by errors of centromere division of univalents at meiosis (centromere misdivision: Figs. 6.8 and 6.9). They share this origin with single telocentrics, centromere splits and centromere translocations. Their origin requires a fusion of two ruptures in the centromeres of the sister chromatids, whereas a centromere translocation requires centromere break and fusion in two non-homologous chromosomes. The isochromosome trisomic will be discussed further in Section 6.2.2.2 and centromere translocation in Sections 6.2.2.2 and 10.4.4.2.2.2.

5.5.2 Relevance

For the plant breeder these rearrangements are of little direct interest, except the centromere translocation, which is used to transfer chromosome segments with specific genes from one species to another (Sect. 10.4.4.2.2.2). Centric split and fusion heterozygotes sometimes cause a slight reduction in recombination, and at times this may be locally sufficient to make areas around the centromere practically free of recombination (Fu and Sears 1973; Sybenga et al. 1990). There are also a number of examples in which a centromere split has caused an increase in chiasma formation, apparently by reducing the effect of interference (Parker 1987). In principle, chromosome recombination can be altered by homozygous fusions and fissions, which could potentially be used for either maintaining (fusions) or breaking up (fissions) specific gene blocks (Sect. 8.4).

Centric splits and fusions are good chromosome morphological markers for the centromeres in genetic analyses, and as such very useful in gene localization studies. The effect of the centromere condition on recombination must be taken into account (Sybenga et al. 1990).

In many insects centric fusions or splits form polymorphisms that apparently play a role in the genetic system of the populations in which they occur. They have not been reported as stable polymorphisms in plants. Also, in speciation they have much less importance in plants than in animals, except for a few genera (e.g. *Cymbispatha*: Jones 1976). Infrequently, centric fission is found as a polymorphism, even homozygous in certain cultivars (e.g. tomato: Banks 1984).

5.5.3 Characteristics, Identification and Consequences

The recognition of centric fusion and centric split in the *somatic karyotype* is generally simple. The chromosome number is increased by one in centric split heterozygotes and by two in homozygotes. With fusions the number is reduced. The somatic recognition of compounds is not as simple. Centromere translocations can be recognized at mitosis, just like normal interchanges, from which they do not differ in principle.

At meiosis fissions and fusions behave essentially the same. When homozygous, the telocentrics or acrocentrics and the metacentrics form their specific bivalents with one and two arms respectively. Spontaneous or experimentally induced fissions of submetacentric or subacrocentric chromosomes may produce one small chromosome that may frequently form univalents.

When heterozygous, fissions and fusions form very characteristic trivalents with the metacentric in the middle and a telo- or acrocentric at each side. In many cases, and especially in insects with the condition as a stable polymorphism, the trivalent is formed consistently, and systematically segregates 2:1, with the metacentric moving to one pole and the two acrocentrics to the other. This results in balanced segregation (Fig. 5.17). The study of orientation, including preorientations and reorientation of the trivalents of centromere fusions and splits, has been very revealing regarding the processes involved in centromere co-orientation and the variation therein (Sybenga and Rickards 1987). When irregularities occur in meiosis (lack of sufficient chiasmata, non-alternate orientation), aberrant types, primarily trisomics, are found in the progeny.

The meiotic behaviour of *compounds* is also straightforward, but always consistently results in imbalanced gametes, with total sterility as a consequence. In animals the deficiencies and duplications can be transferred to the progeny and there they compensate each other, resulting in some fertility. In plants they are too infrequent to be really important. *Centromere translocations* behave practically the same as normal interchanges and in meiosis the two are difficult to distinguish. All consequences are fundamentally the same (Sect. 5.4.1.4).

Chapter 6

Karyotype Variants B: Chromosome Number Variants

6.1 Euploidy

6.1.1 Haploidy

6.1.1.1 Types and Terminology

The term haploid is used as an adjective (referring to a specific number of genomes or chromosomes of an individual) and as a noun (referring to the individual itself). In both cases it can have two meanings that often overlap, but that in some situations may cause confusion. Haploid is used: (1) for the chromosome number of the gamete (the haplont); (2) for the chromosome number of a single genome (in diploids) or of a set of combined genomes segregating together (in allopolyploids). In diploids and allopolyploids the two meanings coincide. In autopolyploids they are different: the gametic chromosome number is half the polyploid number and, consequently, a multiple of the basic genome number. For this reason the basic chromosome number is often referred to as *monohaploid* or *monoploid*. Linguistically, these terms are far from elegant, as is not unusual in biological nomenclature. In allopolyploids and, less appropriately in autopolyploids, the gametic chromosome number is also referred to as dihaploid, trihaploid, etc., in general: *polyhaploid*. If an allopolyploid is reconstructed by doubling the chromosome number of the hybrid between the parental species, this hybrid has the same chromosome number, but is not normally referred to as a (poly)haploid. Somatic reversion of a tetraploid number to the diploid number (Gottschalk 1976) would not be a reason to use the term haploid for the plant originating from it. Also, for all practical purposes, a plant derived from an unreduced gamete of an autotetraploid is a diploid, and this is the preferred term in this case. For an individual plant, the principal criterion for deciding whether the ploidy level is haploid or polyhaploid is the number of genomes rather than its origin.

6.1.1.2 Origin

Haploidy results from spontaneous or induced parthenogenetic development of normally reduced, unfertilized eggs or male gametes or gametophytes, rarely through a process of somatic reduction (Gottschalk 1976). Ways to select spontaneous haploids from large diploid populations, to increase their frequencies and to grow haploid plants directly from male and female spores by in vitro culture will be considered in Section 11.4.

6.1.1.3 Relevance

There is considerable interest in haploids in plant breeding. Very exceptionally, haploids are directly, economically attractive, for instance in floriculture (Sect. 11.4.1). Diploids produced by the same procedures from autopolyploids are of more direct interest. Rather than being applied as such they are used in breeding programs with the purpose to make polyploids from them again at a later stage. Breeding at the tetraploid level presents considerable problems which can be partly circumvented by preliminary breeding at the diploid level.

A more common application of (mono)haploids is to double their chromosome number in order to produce completely homozygous diploids. These are used in breeding programs to avoid laborious and time-consuming inbreeding, either as varieties of self-fertilizing species, or as lines to be used in hybrid varieties in self- or cross-breeders (Sect. 11.4).

6.1.1.4 Characteristics and Identification

Phenotypically, haploids tend to have a reduced stature, and are not as vital as diploids. The *somatic karyotype* is readily recognized as haploid when the chromosome number of the species is known.

Although the *meiotic* behaviour may be confusing, the expectation that no pairing and no chiasma formation would be possible is not realized. Light microscope analysis of pachytene pairing has been reported on several occasions (Kimber and Riley 1963; Magoon and Khanna 1963). It shows that pairing is very extensive, but also very variable. Electron microscope analysis of synaptonemal complexes is possible in many organisms where light microscopy of pachytene is not successful. However, the absence of a chromomere pattern, and sometimes even the impossibility to recognize the centromeres, make the results of limited interest. De Jong et al. (1991) report a non-random distribution of paired segments at a stage comparable to zygotene/pachytene in a rye haploid, similar to the non-random distribution of chiasmata at metaphase I (see also Neijzing 1982, 1985). Homologous pairing in a large duplication in a haploid of (allotetraploid) *Nicotiana tabacum* (Lammerts

1934) has been referred to in Sect. 5.2.4. As discussed in Section 5.2.4, duplications in haploids may result in bivalents. It is not always easy to decide whether the duplications have their origin in old allopolyploidy or in segmental duplications. In case of segmental duplication, a less random position of chiasmata in the genome (cf. Neijzing 1985) is expected than in the case of allopolyploidy, where entire genomes have at least some homology. There are no comparable reports of quantitative analyses of chiasma formation in polyhaploids.

In polyhaploids derived from allopolyploids (see Sect. 6.1.2.3) with genomes just different enough not to pair in the presence of complete homologues and without strong genetic pairing differentiation systems, or where this system has been inactivated, a considerable amount of pairing between the component genomes may be observed. An example is the dihaploid of allotetraploid cotton (*Gossypium hirsutum*). The behaviour is like that in a hybrid between related species and not representative for a haploid.

In absence of bivalents, meiosis cannot develop normally. A metaphase I spindle may not even be formed, especially when only univalents are present. When a few bivalents are present, an irregular anaphase usually follows, but only the chromosomes of the bivalents segregate normally. The univalents segregate more or less (sometimes less) randomly over the poles (Belling and Blakeslee 1927; Levan 1942; Sficas 1963; cf. Sybenga 1972, 1975), or lag in the equator, or split into chromatids. The latter is a common, although rather variable characteristic of univalents in meiosis. Cells with only univalents may pass to interkinesis without reduction (*restitution nucleus*), which may continue into an operationally normal, second meiotic division from which two (one dyad) balanced haploid spores develop. The univalents may also divide without interkinesis. When this occurs with sufficient frequency, the haploid may reproduce. This has been observed, for instance in the haploid "variety" Monosperma of *Coffea arabica*, arabica coffee, in which most berries contain only one seed, thus the name "monosperma" (cf. Sybenga 1972).

Bivalents, from which the component chromosomes segregate, result in genetic imbalance. Then, restitution to interphase (interkinesis) is a prerequisite for genetic balance, but translocations are produced as a result of chromatid exchange.

6.1.1.5 Consequences

Haploids are usually sterile, but may occasionally produce functional haploid gametes (Sect. 6.1.1.4). Exchange in duplications results in translocations between non-homologous chromosomes that carry the duplication, but bivalents may prevent restitution. In polyhaploids the translocations are predominantly between homoeologous chromosomes, and the resulting gamete carries translocations between homoeologues.

6.1.2 Polyploidy

6.1.2.1 Types

In the present discussion four types of polyploids will be distinguished:

a) Autopolyploids. The same basic genome is present more than twice, they are fully homologous and may even be identical but not necessarily so. Allelic variation, as found in any diploid species, is quite common. A representative genome formula is AAAA: an autotetraploid with four homologous genomes of type A.
b) Allopolyploids. Different genomes are combined, but each is present twice. The different genomes are genetically compatible, but differentiated with respect to meiotic pairing. There is no exchange between them. Any number of genomes may be present, always occurring in pairs. A representative genome formula is AABBDD, the common bread wheat, an allohexaploid with three pairs of different genomes: A, B and D.
c) Autoallopolyploids. Some genomes are present in more than a double dose, usually in even numbers (the autopolyploid segment of the genome), and addditional non-homologous but genetically compatible genomes are present in a double or possibly higher dose. A representative genome formula is AAAABB, where genome A is present in four doses, and genome B in two. An example is possibly timothy (*Phleum pratense*; Cai and Bullen 1991). Autoallopolyploids are rare.
d) Polyploid Hybrids. More than two genomes are present, which may have any relation except forming autopolyploid or allopolyploid combinations. Here, the terminological difficulty arises concerning the origin by hybridization or by reduction from a higher order allopolyploid. In the latter case the polyploid would be called a polyhaploid, a term appropriate for an interspecific hybrid also. An example is the hybrid between an allotetraploid wheat such as *Triticum turgidum* ($2n = 28$; genomes AABB) and a diploid ($2n = 14$) *Aegilops* species such as *Ae. squarrosa* with the D-genome, resulting in the triploid ABD ($3x = 21$). The same genomic combination (ABD) is found in the trihaploid derived from the allohexaploid *Triticum aestivum* ($2n = 6x = 42$).

6.1.2.2 Autopolyploidy: Types

Autopolyploids can have from three genomes (autotriploids), to as many genomes as are compatible with viability. The terms are based on Greek numerals, but beyond ten simply the arabic numerals are usually applied instead. With four genomes, the term (auto)tetraploid is used, with five: pentaploid, further: hexaploid (6 homologous genomes), heptaploid (7), octoploid (8), nonaploid (9), decaploid (10), and, occasionally, dodecaploid

(12). Triploids and tetraploids will be considered separately and in some detail, the higher polyploids only superficially.

6.1.2.2.1 Autotriploids

6.1.2.2.1.1 Origin

Autotriploids cannot be produced directly by doubling the chromosome number. Somatic origin by genome segregation or genome or chromosome loss starting from the tetraploid or higher level has not been reported. The most obvious origin for an autotriploid is hybridization between a diploid and the corresponding autotetraploid. Due to the unfavourable interactions between embryo, endosperm and plant tissue with abnormal chromosome number relations, the development of endosperm and embryo is often blocked at early stages: *triploid block*. Only with *embryo rescue* by in vitro culture can triploid progeny then be obtained. Using the tetraploid as the female parent, success is often greater than in the reciprocal cross: 3x (embryo), 5x (endosperm) and 4x (plant). With the diploid as the female, the relations are usually 3x (embryo), 4x (endosperm) and 2x (plant). Spontaneous triploid embryos on diploids, formed by fertilization involving one unreduced gamete (embryo 3x, endosperm 5x and plant 2x), often appear to be more viable, but are rare. Provided it is tenchnically possible, a *somatic hybrid* between a somatic (diploid) protoplast and a (haploid) tetrad protoplast (Pirrie and Power 1986) directly replaces the making of a sexual hybrid.

6.1.2.2.1.2 Relevance

The greatly reduced fertility of triploids (see below) makes them quite valuable for fruit crops where seeds are undesired (Sect. 11.3.1.1). The banana is the best-known natural example. For the same reason (pronounced sterility), artificial triploids have proven successful in water melon (*Citrullus*), where the seeds, although not as detrimental as in the banana, are still undesired. Among ornamentals, the triploid tulip is a good example. Here, sterility is an advantage because flowers that do not set fruit stand longer. The use of autotriploids in field crops is still limited, although occasionally important. An example is the sugar beet (*Beta vulgaris*) where the triploid has become very successful. Triploids would also be expected to be successful in field crops that, unlike the sugar beet where the hybrid has to be reproduced every time, can be reproduced vegetatively. For instance for potatoes (presently, an autotetraploid), triploidy might be interesting and in casava (presently, in practice, a diploid) artificial autotriploids are promising already. Such applications will be discussed in more detail in Section 11.3.1.1.

Triploids are an excellent source of primary trisomics, which are used in gene localization and as an intermediate in the production of other types of trisomics.

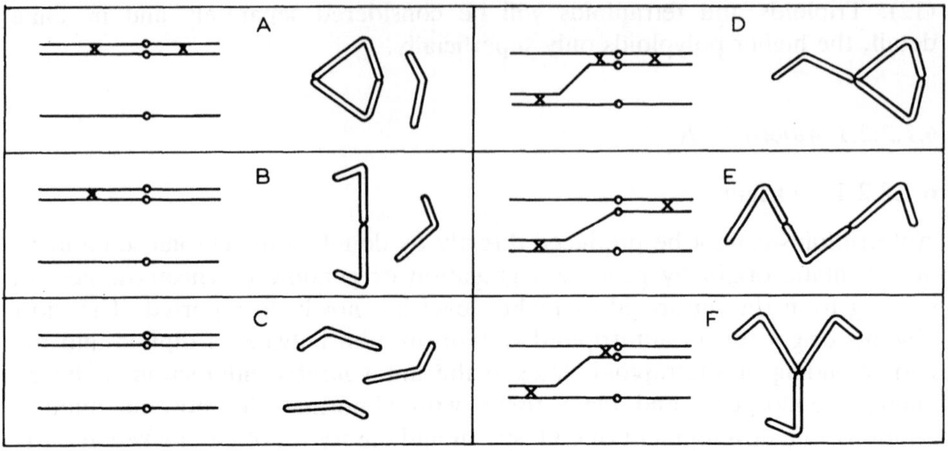

Fig. 6.1 Pairing and metaphase I configurations for one set of homologous chromosomes in an autotriploid. The principle is the same in a primary trisomic. **A**, **B**, **C** No partner exchange, pairing as bivalent and univalent, no other pairing configurations possible. **D**, **E**, **F** One point of partner exchange; there is an interstitial segment which can have a chiasma, and this affects the shape of the configuration at later stages. With only one chiasma, same metaphase configuration as in **B** is formed, without chiasmata there are three univalents as in **C**. (After Sybenga 1975)

Table 6.1. The nine modes of pairing of three identical chromosomes occurring in an autotriploid and in a primary trisomic

The chromosomes:			
	*a*1 —o— *b*1		
	*a*2 —o— *b*2		
	*a*3 —o— *b*3		
Arm *a* pairing:	*a*1/*a*2	*a*1/*a*3	*a*2/*a*3
Arm *b* pairing:			
*b*1/*b*2	II + I (*a*3 – *b*3)	III	III
*b*1/*b*3	III	II + I (*a*2 – *b*2)	III
*b*2/*b*3	III	III	II + I (*a*3 + *b*3)

Pairing is assumed to start at both ends exclusively. There can be one point where partners are exchanged. The chromosomes have arms *a* and *b*, and both arms occur three times; arm *a*1, arm *a*2, arm *a*3; arm *b*1, arm *b*2, arm *b*3. Arm 1 can pair with 2 and then 3 is free, etc. III is a trivalent, II a bivalent and I a univalent.

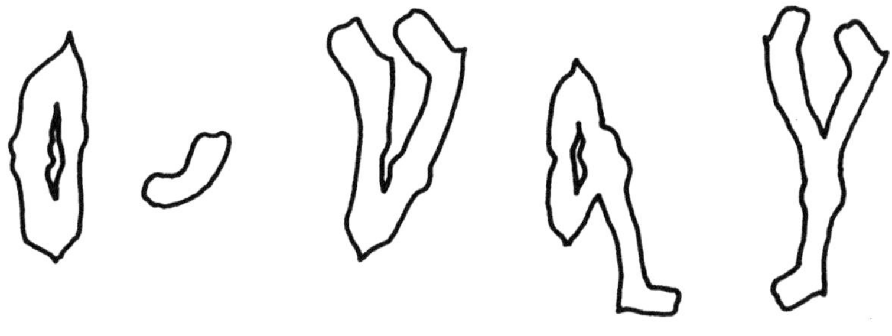

Fig. 6.2 Drawings of meiotic metaphase I configurations of rye involving three homologous chromosomes. **A**, **B**, **C** and **D** represent A, E, D and F respectively of Fig. 6.1

6.1.2.2.1.3 Characteristics and Identification

Phenotypically, triploids tend to be somewhere in between diploids and tetraploids, but they can be superior to both. Not much is known about the chemical composition etc., which has been studied much more widely in tetraploids (Sect. 6.1.2.2.2.3).

The *somatic karyotype* will readily identify the triploid level when the chromosome number of the species is known. Only with detailed C-banding patterns and clear differences between related species is it possible to distinguish an autotriploid from a triploid hybrid purely on the basis of somatic chromosome analysis.

The *meiotic behaviour* of autotriploids is characteristic. Three copies of each chromosome are available. These can pair in all combinations, in principle with the same frequency (Table 6.1). When the chromosomes of each set of three are numbered 1, 2 and 3, the three combinations 1–2, 1–3 and 2–3 are equally frequent for each point where pairing can start. When the two arms (say A and B) of each chromosome pair independently and as units, this results in $3 \times 3 = 9$ possibilities. Six result in a trivalent and two in a bivalent with a univalent. This is a simplified, but not very unrealistic starting model. On this pairing pattern, the system of chiasma formation is superimposed. In Fig. 6.1, different possibilities of chiasma formation are shown for each pairing combination for one chromosome. When one arm has few chiasmata, for instance when it is relatively short, a trivalent will not be found at diplotene-diakinesis-metaphase I, in spite of trivalent pairing. Drawings of examples of some of the metaphase I configurations are shown in Fig. 6.2.

When the trivalents segregate 2:1 and all bivalents segregate 1:1, the anaphase segregation of autotriploids is irregular even when all univalents are recovered, because the orientation of the different trivalents is not coordinated. From some trivalents two chromosomes will go to the pole where

Table 6.2. Anaphase I chromosome distribution in PMCs of triploid *Datura*, and the chromosome numbers in pollen grains and progeny (Satina and Blakeslee 1937). The extremes are more frequent than expected

AI	12–14	13–23	14–22	15–21	16–20	17–19	18–18	Total
% Observed	0.8	4.5	8.5	14.5	22.9	30.8	18.0	1000
Expected	0.05	0.6	3.2	10.7	24.2	30.7	22.6	
Pollen	**12–24**	**13–23**	**14–22**	**15–21**	**16–20**	**17–19**	**18–18**	**Total**
% Observed	2.6 1.2	4.0 2.6	7.2 3.8	11.0 5.0	16.4 9.2	16.0 10.8	11.2	500
Expected	0.3 0.3	0.3 0.3	1.6 1.6	5.4 5.4	12.1 12.1	19.3 19.3	22.6	

Progeny of the cross $3n \times 2n$ (seed set 20%; germination 70%):

$2n$	$2n + 1$	$2n + 1 + 1$	$2n + 1 + 1 + 1$	Total
58	138	79	10	285

only one goes from another trivalent. In addition, the univalents either segregate at random, or split into chromatids, followed by loss at the second division, or they are lost directly during the first division. Consequently, gametes with a wide range of chromosome numbers are produced, including deviant chromosome types resulting from centromere breakage of univalents. Anaphase segregation of autotriploids has been studied on a number of occasions. When the segregation products of bivalents (expected to be consistently 1:1) are subtracted from the anaphase I groups, a random distribution is often, but not always obtained (Table 6.2).

6.1.2.2.1.4 Consequences

The variable numbers of chromosomes in the gametes are the cause of a highly disturbed genetic balance in most gametes. This results in greatly reduced fertility, often even to zero, which makes autotriploids desirable in the cases mentioned in Section 6.1.2.2.1.2, but which prevents generative reproduction. Autotriploids must be made anew by crossing tetraploids and diploids when vegetative reproduction (natural or in vitro) or apomixis is not possible.

Triploid sterility is not only due to non-functional gametes, but also to the unfavourable combinations of numbers of genomes or chromosomes in the zygotes. Triploids of rye, when crossed with a diploid as male partner, produce few seeds, but those that do develop are mostly normal diploids, single primary trisomics and a few double primary trisomics. Other types are infrequent. When crossed with a tetraploid as the male parent, however, most chromosome numbers in the progeny are close to the tetraploid level (own unpublished results). Apparently, the triploid level is not favoured, although chromosome counts at anaphase I show that numbers around the haploid and the diploid level occur, and the two types of crosses demonstrate that, in principle, both can function. Especially in the cross between an autotriploid and an autotetraploid, a few progeny have numbers close to the triploid level. To what extent genomic imprinting (Lin 1984) plays a role here is uncertain.

6.1.2.2.2 Autotetraploids

6.1.2.2.2.1 Origin

Autotetraploids result from a simple doubling of the chromosome number of a diploid. This may take place spontaneously, although infrequently, in normal somatic reproductive tissues and more frequently in callus, either in vitro or on the plant. Regeneration of leaf mesophyll protoplasts may result in tetraploid plants when the protoplasts have originated from cells with an endomitotically or similarly doubled chromosome number (Sect. 3.1.4.1.2.1). Somatic chromosome doubling is common after treatments with suitable chemicals, among which colchicine is the most successful. Doubling results in two identical genomes, unless the process of induction is accompanied by mutation. This is not uncommon for in vitro callus culture (somaclonal variation) and after chemical induction.

The alternative way for chromosome doubling is meiotic induction, by development of a gamete from a restitution nucleus, or failure of reduction in other ways. The resulting diploid gamete has the same genomic constitution as the parent or slightly different, depending on the manner of origin (Sect. 11.3.1.2.1.2). When combined with a similar unreduced gamete or the diploid gamete of an existing tetraploid, the tetraploid progeny will have a higher level of heterozygosity than after somatic doubling. Different systems of induction of autopolyploidy and their specific merits and consequences will be discussed in more detail in Section 11.3.1.2.1.

6.1.2.2.2.2 Relevance

The practical importance of autopolyploidy, specifically of autotetraploidy is great. Several cultivated species are natural autotetraploids: potato (*Solanum tuberosum*), cocksfoot or orchard grass (*Dactylis glomerata*), alfalfa or lucerne (*Medicago sativa*); a few are autohexaploid: timothy (*Phleum pratense*), which is possibly autoallohexaploid. Many successful attempts have been made to induce autotetraploidy in normally diploid cultivars. As in nature, the most successful autopolyploids are expected to be cultivars where the vegetative parts yield the main product, because fertility tends to be disturbed. Practical application of autopolyploidy will be discussed in Section 11.3.1.

6.1.2.2.2.3 Characteristics and Identification

The *somatic karyotype* alteration due to genome doubling is straightforward, but when the origin is not known, an autopolyploid may not be readily distinguished from an allopolyploid.

Morphologically, autopolyploids tend to be more robust than their corresponding diploids (the "gigas characteristic"), and somewhat slower in development. This is due to the larger nuclei and consequently larger cells.

The larger cell volume is not fully reflected in larger organ size because the number of cells is somewhat reduced.

In established natural autotetraploids gigas characteristics and larger cell size are not necessarily expressed as strongly as in those made recently (van Dijk and van Delden 1990, where older literature can be found). In *Plantago media*, pollen grain volume was doubled in colchicine-induced autotetraploids, but was increased only 40% in natural autotetraploids. However, stomatal guard cells were longer in natural than in artificial tetraploids. Segregation for cell volume and DNA content in later generations derived from hybrids between the two types of tetraploids indicated that regulation was by additive genes rather than DNA content.

With higher ploidy levels the negative effects of the larger numbers of genomes more than compensate the positive effects and at a certain number of genomes growth tends to be reduced. There is apparently an optimum genome number, which is different for different species. In addition to changed physical relations, the change in gene dose and the different effects of dose for different genes can cause specific effects, especially in delicately balanced biological processes. Therefore, reproduction can be severely affected, in addition to the common negative effect of the errors of meiotic behaviour. On the other hand, secondary metabolic products may be formed in considerably larger amounts than in diploids (Sect. 11.3.1.2.2.1). Specific effects of the combination of different doses of different alleles have also been reported to have specific effects on the functioning of genes with potentially practical application (cf. Sybenga 1972). Heterosis, inbreeding and selection proceed quite different in autotetraploids compared with diploids.

In *meiosis*, the presence of four instead of two homologues has striking consequences. All four are capable of pairing and although at any position on the chromosome normally only two partners pair at a time, at different positions, these may be different partners: *pairing partner exchange*, linking all homologous chromosomes together (Figs. 6.3 and 6.4). The maximum number of chromosomes joined together in one configuration equals the number of homologous genomes in the autopolyploid. Translocations and aneuploids, which may occur in polyploids as well as at the diploid level, cause larger or smaller configurations, but these are exceptional and will be briefly discussed later. With random pairing between all homologues, the expected number of multivalents per cell can be predicted when certain assumptions are made (Table 6.3). The simplest assumption is that pairing starts at or near the chromosome ends and that pairing initiation in other chromosome segments is restricted. With four homologous ends, there are three possibilities for combinations of two in a pair. With two arm ends pairing independently, the total number of combinations is nine (Fig. 6.3; Table 6.3). Six of these involve partner exchange and produce a *quadrivalent*, and three form *two bivalents*: twice as many chromosomes are involved in quadrivalents as in bivalents and the number of quadrivalents equals the number of bivalents. There are numerous deviations from this model.

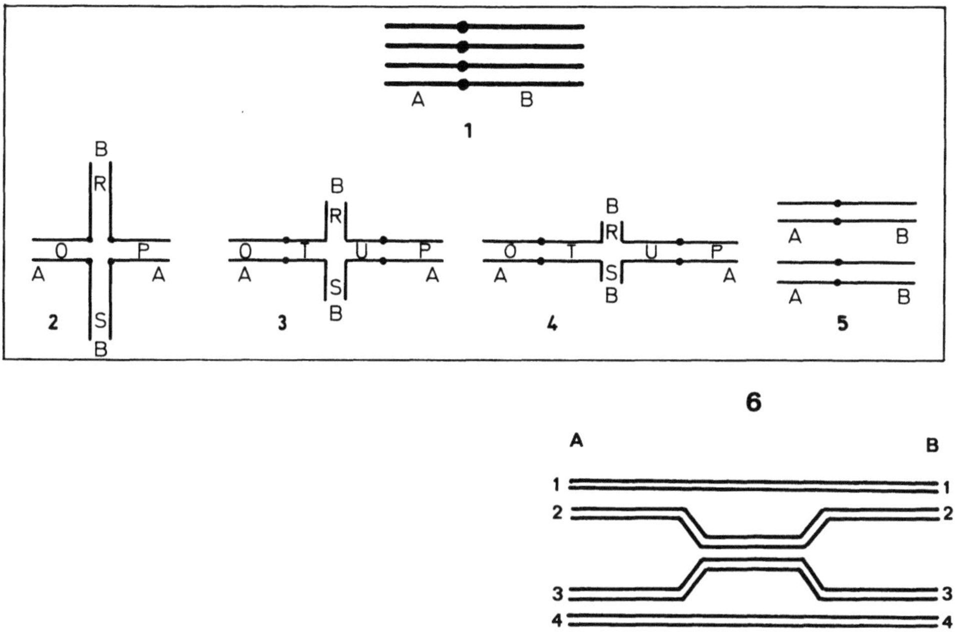

Fig. 6.3 **1** The four chromosomes of an autotetraploid or tetrasomic with arms *A* and *B*. **2–5** Different pairing configurations, including quadrivalent (**2–4**) and bivalent (**5**) pairing; cf. Table 6.3. In quadrivalents six segments (*O*, *P*, *R*, *S*, *T* and *U*) can be distinguished as in an interchange heterozygote (Fig. 5.7), but the point of partner exchange is not fixed. When its position is relatively distal, the two (equally long) interstitial segments *T* and *U* have a considerable chance of chiasma formation, resulting in branched configurations as discussed with the translocation heterozygotes. In **4**, the homologous segments *R* and *S* in arm *B* have become so small that they may fail to have a chiasma. Then two bivalents are formed, which may be rings when there are chiasmata in *T* and *U*.

With an extra point of partner exchange (**6**), the configuration that otherwise would have been two bivalents can now turn into a quadrivalent. At metaphase I this will be recovered as a quadrivalent only if there are chiasmata in the critical segment

Several instances are known where no or only few multivalents are observed, while on the basis of segregations (Sect. 6.1.2.2.2.4) or by studying the diploid derived from the polyploid, the autopolyploid character was beyond doubt. There are different reasons for an autotetraploid to fail to produce multivalents at meiosis. Because all genomes are fully homologous, pairing differentiation can hardly play a role. Segments that pair preferentially with other specific segments will be recombined into other chromosomes. Other segments of these chromosomes may pair preferentially with the corresponding segments of other homologues. Then, multivalents will be formed with increased frequency (Sybenga 1984b). In the population as a whole, therefore, the average frequency of multivalents should still be that expected with random pairing and chiasma formation between homologues.

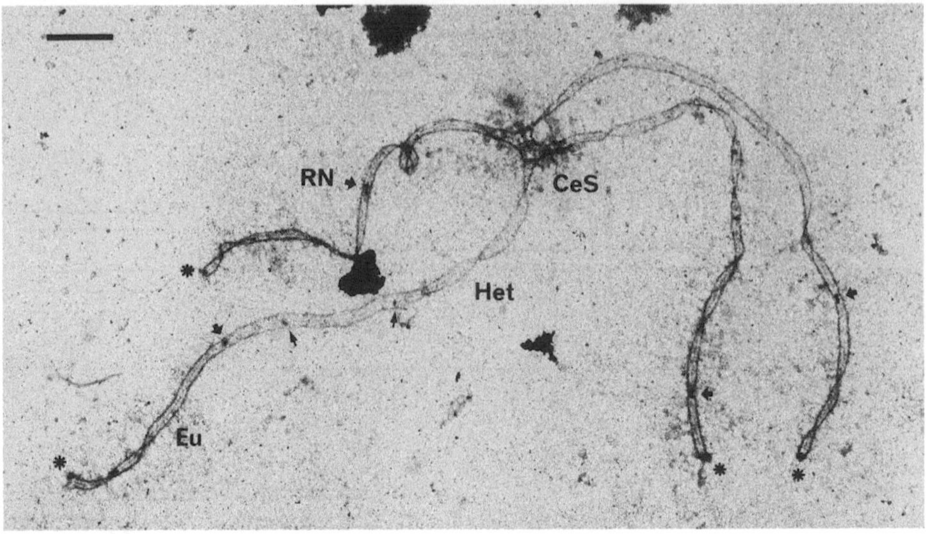

Fig. 6.4 Quadrivalent synaptonemal complex in the tetraploid hybrid (2n = 4x = 48) between *Lycopersicum esculentum* (tomato) and the related wild species *L. peruvianum*. Plant OH1B from J. Wijbrandi, electron micrograph by J. van Eden and E. Schabbing, Department of Genetics, Wageningen Agricultural University. Spread according to Stack and Anderson (1984). *Bar* = 1 µm.

There is sufficient homology for complete pairing. The point of partner exchange happens to be in the centromeric region. In most other quadrivalents it occurred elsewhere in the quadrivalent, and in several cases there were more points of pairing partner exchange. *RN* Recombination nodule; *large arrows*: examples of late RNs; *small arrows* early RNs. *Het* Proximal heterochromatic region staining dark in LM preparations, but with lightly staining SC. *Eu* Distal euchromatic region staining light in LM preparations, but with darker SC. All late RNs occur in euchromatic segments. *CeS* Centromere structure with heavy lateral elements. *Asterisks* at telomeric knobs; no relation to RNs. (Courtesy of J.H. de Jong)

There are several causes of predominant bivalent pairing, which operate even when all genomes are completely homologous. One is initiation of pairing in a single area in the chromosomes; partner exchange is not possible. The pairing in bivalents, however, is random and the inheritance is tetrasomic (Sect. 6.1.2.2.2.4). The localization of chiasmata in single chromosome regions, possibly affected by the pairing pattern, results in the failure to maintain pairing partner exchange. A well-known example is the natural tetraploid *Alium porrum* where at pachytene a large number of quadrivalents can be observed, but where at metaphase I only bivalents are present (Levan 1940). In this species, as in some other *Allium* species, chiasmata are concentrated around the centromere, at most one on each side. The point of partner exchange is only infrequently positioned exactly between two such chiasmata and, as a consequence, the pairing quadrivalents fall apart. Here, too, association in bivalents is random and the inheritance is tetrasomic.

Table 6.3. Pairing configurations in an autotetraploid. Both arms of each chromosome are present four times

The chromosomes:		
*a*1 ——o—— *b*1		
*a*2 ——o—— *b*2		
*a*3 ——o—— *b*3		
*a*4 ——o—— *b*4		

Pairing:	*a*1/*a*2 *a*3/*a*4	*a*1/*a*3 *a*2/*a*4	*a*1/*a*4 *a*2/*a*3
*b*1/*b*2 *b*3/*b*4	II + II	IV	IV
*b*1/*b*3 *b*2/*b*4	IV	II + II	IV
*b*1/*b*4 *b*2/*b*3	IV	IV	II + II

Pairing is assumed to start at the ends only. When arm *a*1 pairs with *a*2, *a*3 pairs with *a*4; when *a*1 pairs with *a*3, *a*2 pairs with *a*4; when *a*1 pairs with *a*4, *a*2 pairs with *a*3. The same for *b*. Partner exchange occurs when different arms pair with different chromosomes, and results in a quadrivalent. There are six quadrivalents against three bivalent pairs. With interstitial pairing initiation in addition to terminal pairing initiation, more points of partner exchange are possible and more probabilities for quadrivalent pairing. II is a bivalent, IV a quadrivalent.

Yet another reason for exclusive bivalent formation in an autopolyploid is again based on the pairing system and has been demonstrated by Rasmussen and Holm (1979) for the silk worm *Bombyx mori*. There is no reason why it would not operate in plants. It was first detected in three-dimensional EM reconstructions of zygotene and pachytene synaptonemal complexes. At early stages many more quadrivalents were present than at later stages, and it was concluded that there was a form of synaptic adjustment, leading to an extension of the SC between two partners at the expense of the SC between other partners in the quadrivalent. The point of partner exchange can apparently move to a chromosome end, and finally disappear. This is possible only up until the moment recombinational exchange fixes the SC, and perhaps is stopped already in the preparatory stages of exchange. The duration of the pairing process is apparently crucial for the effect of pairing adjustment. When the period is short, the frequency of quadrivalents will remain relatively high. Quantitative observations in SC spreads of maize suggest that synaptic adjustment, although present, is not strong enough to explain the observed shortage of quadrivalents (Gillies 1989).

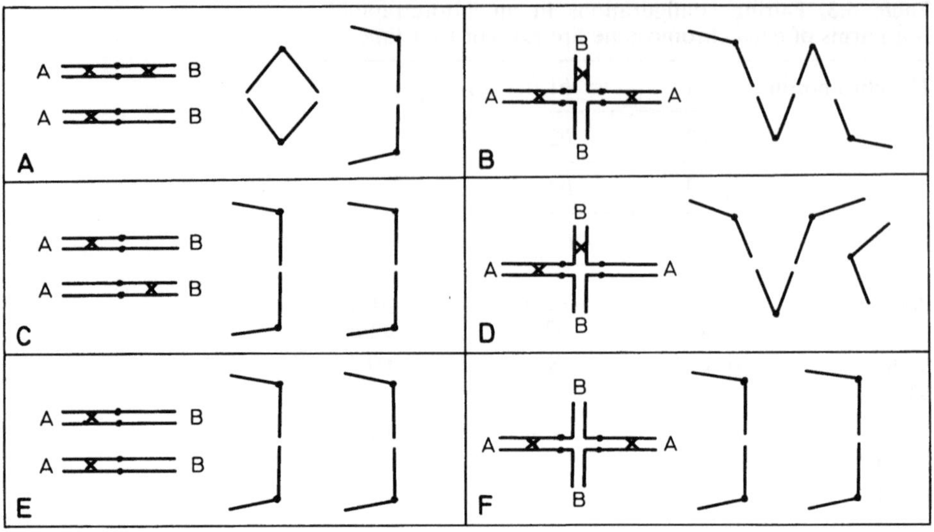

Bivalent pairing   Quadrivalent pairing

Fig. 6.5A–F Examples of metaphase I configurations after bivalent and quadrivalent pairing in an autotetraploid. The formation of chiasmata in the different segments or arms is critical for the resulting configuration. When two homologous arms fail to have a chiasma after quadrivalent pairing (**F**) two open bivalents appear as after the formation of one chiasma in each of the two bivalents after bivalent pairing. (After Sybenga 1975)

It is clear that purely on the basis of absence or low frequency of multivalents at metaphae I, one can not conclude that a polyploid is an allopolyploid. The segregation pattern is the most reliable, final criterion, provided the pitfalls inherent of tetrasomic inheritance and aberrant segregation ratios are avoided. More details will be given in Section 6.1.2.2.2.4.

Infrequently, polyploids are observed with higher than the expected frequency of multivalents, caused by a higher frequency of points of pairing partner exchange than two. The greater number of multivalents can only be realized if chiasmata are formed not only near the ends, but also between the points of partner exchange. As shown in Fig. 6.3, configurations with interstitial chiasmata and a complex shape are then observed.

Structural rearrangements in autopolyploids result in complex configurations. A low frequency of higher multivalents in autopolyploids, or in amphiploids between closely related species, will most probably be a good indication of a normal reciprocal translocation. The subject is discussed in Section 12.2.2.1 on manipulation of the genetic system because it plays a role in differentiating the genomes of an autopolyploid (allopolyploidization).

Different metaphase I configurations, bivalents as well as quadrivalents, and their origin, are shown in Fig. 6.5. The use of their relative frequencies in estimating pairing and chiasma parameters, and the effect of the variable point of partner exchange are discussed in Sections 8.2.1.4, 9.3.2.2 and 11.3.1.2.3.

Like the quadrivalent of the interchange heterozygote, the orientation of the centromeres in the quadrivalent or the smaller configuration of the autotetraploid are essential for chromosome segregation.

6.1.2.2.2.4 Consequences

The term *tetrasomic inheritance* has been used in the previous section. It refers to the fact that autotetraploids have a different system of inheritance than diploids, which have *disomic inheritance*. The basis is explained in Table 6.4A. The diploid has three different possibilities to combine the two different alleles *A* and *a* of a gene: *AA*, *Aa* and *aa*. The autotetraploid has five: *AAAA* (quadruplex, after the number of dominant alleles), *AAAa* (triplex), *AAaa* (duplex), Aaaa (simplex) and aaaa (nulliplex). Two are homozygous and three are heterozygous, but of different composition. These have different segregation ratios.

Numbering the four chromosomes of a homologous group 1, 2, 3 and 4, and assuming two-by-two segregation of the chromosomes (balanced segregation), from the four chromosomes six combinations of two chromosomes are possible: 1–2, 3–4; 1–3, 2–4; 1–4, 2–3 (Table 6.4). When in a *triplex* chromosome 1 has the recessive allele, in the gamete it will always be accompanied by a chromosome with the dominant allele and the *recessive phenotype* will *not* be *expressed* in the progeny. Upon selfing, or hybridization with another triplex, duplex types may be formed (in addition to triplex and quadruplex types) and these may, in the next generation, produce gametes and consequently progeny, without the dominant allele.

The duplex also forms six types of gametes, but one of these (Table 6.4) has a combination of two recessive alleles. Selfing (or hybridization with another duplex) results in $1/6 \times 1/6 = 1/36$ of the progeny being nulliplex, the same frequency being quadruplex, etc. This is a typical example of "tetrasomic inheritance", but it should be noted that the absence of segregation among the progeny of the triplex, and the simplex segregation identical to that in a diploid, are also tetrasomic inheritance.

Especially the seemingly disomic inheritance of the simplex (Table 6.4A) can be confusing. If a disomic segregation is found in a species with the tetraploid chromosome number, this does not imply that it is an allotetraploid, which by definition should have disomic inheritance. It may well be a simplex autotetraploid. Additional segregations or other characteristics must help to establish the type of polyploid. As an extra complication, genes may occasionally show aberrant segregation or interactions with other genes or even alleles. Therefore, the segregation of a single gene, by chance found to segregate in an autotetraploid progeny, should be interpreted with great caution.

Under exceptional circumstances it is possible that recessives appear in the progeny of a triplex where they are not expected, or that they are more

Table 6.4. A Gametic ratios in simplex, duplex and triplex autotetraploids, assuming 2:2 anaphase distribution. **B** Segregations with double reduction in triplex and duplex autotetraploid *Datura* (Blakeslee et al. 1923). For *a* the deviation is significant in the duplex: the locus will be far from the centromere

A.

Simplex	Duplex	Triplex
1 *A* ———o———	1 *A* ———o———	1 *A* ———o———
2 *a* ———o———	2 *A* ———o———	2 *A* ———o———
3 *a* ———o———	3 *a* ———o———	3 *A* ———o———
4 *a* ———o———	4 *a* ———o———	4 *a* ———o———

Chromosome combinations, each pole receiving one set of two

Pole a	Pole b			
1 and 2	3 and 4	*Aa* and *aa*	*AA* and *aa*	*AA* and *Aa*
1 and 3	2 and 4	*Aa* and *aa*	*Aa* and *Aa*	*AA* and *Aa*
1 and 4	2 and 3	*Aa* and *aa*	*Aa* and *Aa*	*Aa* and *AA*
Segregation *AA*:*Aa*:*aa*		Simplex 0:3:3	Duplex 1:4:1	Triplex 3:3:0

B.

Parents	Dominant	Recessive	% Recessive	Expected (chromosome segregation)
AAAa × *aaaa*	257	6	2.3	0.0
AAaa × *aaaa*	518	137	20.9	16.7
PPPp × *pppp*	160	1	0.6	0.0
PPpp × *pppp*	905	179	16.5	16.7

frequent than expected in the progeny of a duplex. This implies that more often than expected with the model described, two chromatids with recessive alleles are combined in the same gamete. These must have derived from the same original chromosome, from which normally the two chromatids would separate at anaphase II, and they would not be expected to be combined in one gamete. The deviant process is called *double reduction*. It is based on genetic exchange between the gene locus and the centromere, followed by adjacent segregation of the two chromosomes between which this ex-

Fig. 6.6 Double reduction in a tetrasomic or autotetraploid. **A** *Triplex*. Recessives are recovered in a single cell of the tetrad, where the two chromatids from the same original chromosome have come together. **B** *Duplex*. The system of pairing is important for the recovery of extra double recessives: only when *a* pairs with *A* will the combination *aa* and *AA* appear instead of *Aa* and *Aa*. This occurs simultaneously at two opposite places in the quadrivalent. In both *triplex* and *duplex* the location of the marker genes in the adjacent quadrivalent is important: only in the position drawn is double reduction possible, not when the markers are located at the sides

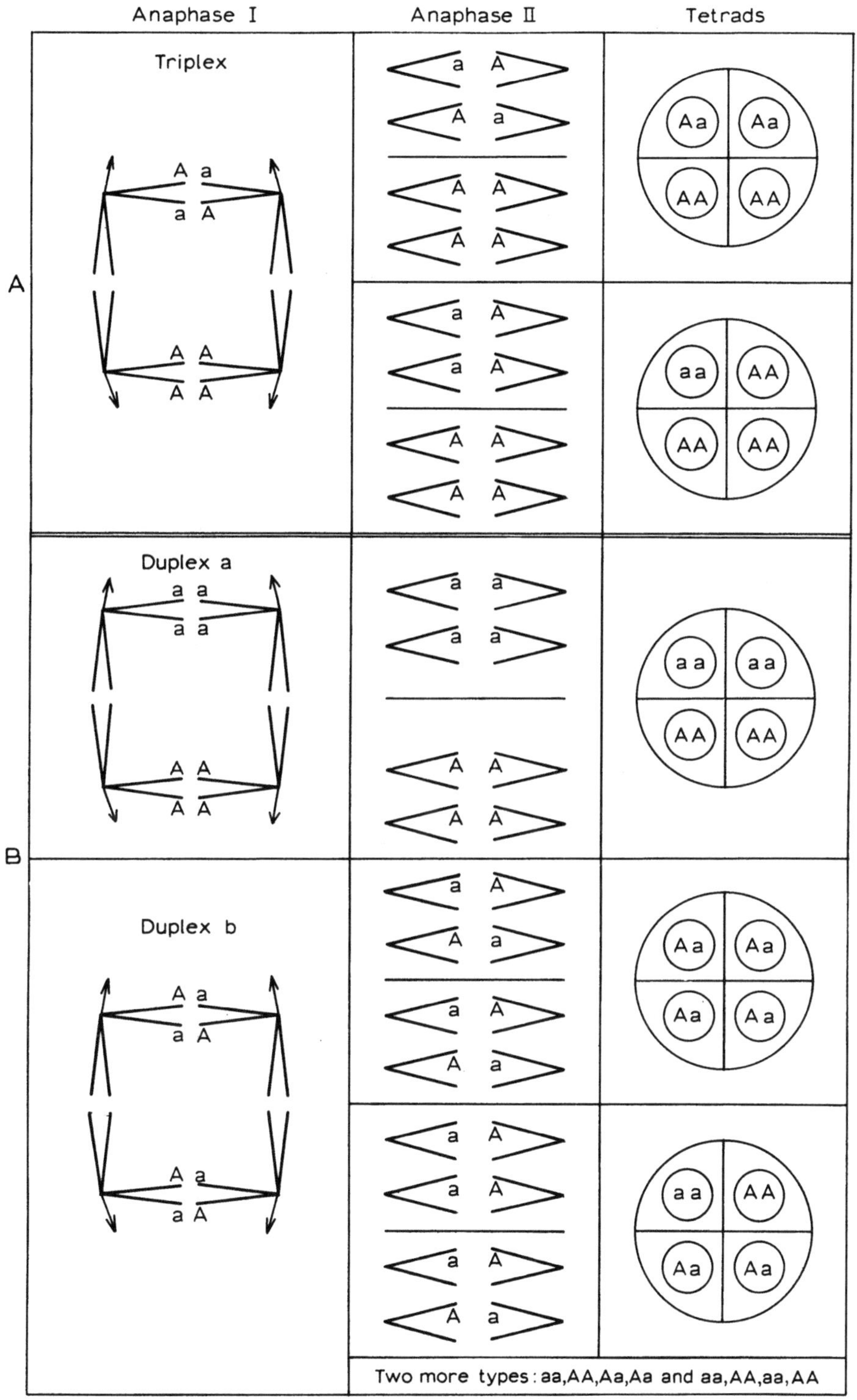
Anaphase I
Anaphase II
Tetrads
A
Triplex
A a
a A
A A
A A
a A
A a
A A
A A
Aa
Aa
AA
AA
a A
a A
A A
A A
aa
AA
AA
AA
B
Duplex a
a a
a a
A A
A A
aa
aa
AA
AA
Duplex b
A a
a A
A a
a A
Aa
Aa
Aa
Aa
aa
AA
Aa
Aa
Two more types: aa,AA,Aa,Aa and aa,AA,aa,AA

change had occurred (Fig. 6.6). Finally, the correct combination of the two chromatids of these two chromosomes at anaphase II must be realized.

Double reduction is clearly possible only when the locus is far enough from the centromere to permit exchange recombination between centromere and locus. The frequency of double reduction is apparently correlated with the frequency of genetic exchange between the locus and the centromere. This relation is proportional only as long as one chiasma is formed, because the requirement for double reduction is that both chromosomes moving adjacently to the pole, have two genetically unequal chromatids. This is always so with one chiasma. However, with a second chiasma, the two chromatids will become equal again with respect to the locus considered when these chiasmata are either compensating or complementary. Only two disparate chiasmata will result in unequal chromatids in the two adjacent chromosomes (Fig. 3.13). For three chiasmata the same result can be demonstrated. Maximum double reduction is found, therefore, consistently with one chiasma; with increasing numbers of chiasmata it levels off to one-half of this value.

The second factor determining the level of double reduction is orientation. In several species the frequency of alternate orientation is so high that double reduction is almost never observed. In theory, adjacent orientation might be expected to be found in 50% of the cases, when orientation is "random" (Fig. 5.13), as with interchange quadrivalents. Of this 50%, however, only half is the type required for double reduction, because it combines the two critical chromosomes in the same anaphase I daughter group. In the other type of adjacent orientation, the recombined chromosomes segregate. Finally, random anaphase II segregation reduces the combination of two original sister chromatids in the same cell again by 50%. In the triplex, again only half of these combinations is the combination of two recessives, the other has two dominant alleles. In fact, the outcome is no more than the replacement of two "heterozygous" allele combinations in the gamete by two homozygous combinations.

The frequency of double reduction, expressed as the frequency of combination of two sister chromatids, irrespective of the alleles involved (not, therefore, the frequency of double recessive gametes, and even less than that of recessive progeny) is usually called α. It is maximally 1 (always one chiasma) $\times$ 1/4 ("random" orientation) $\times$ 1/2 (anaphase II combination) = 1/8. It will usually be lower, either because the chiasma frequency is not optimal, or because the orientation is less frequently adjacent. It may be higher, when the orientation is more often than expected of the correct adjacent type. In autotetraploids of *Arabidopsis thaliana*, van der Veen and Blankestijn de Vries (1973) indeed observed double reduction as high as 1/8. Different theoretical maximum α values (1/6, 1/7, 1/12) have been proposed, but are based on less realistic assumptions.

Double reduction is of interest because it can cause segregation of recessives in triplex autotetraploids. It will, to a variable extent, affect segregation in duplexes (Table 6.4B). It should be noted that the loss of a chromosome, not

Table 6.5. The fraction of a population that is homozygous for one locus (recessive or dominant) after selfing for the indicated numbers of generations in a heterozygous diploid and in simplex and duplex autotetraploids

Generation of selfing	Diploid *Aa*	Tetraploid *Aaaa*	Tetraploid *AAaa*
1 (F_2)	0.5	0.25	0.05
2	0.75	0.38	0.194
3	0.875	0.493	0.326
4	0.938	0.558	0.438
5	0.968	0.648	0.531

infrequent in autotetraploids, may also increase the frequency of recessive phenotypes in the progeny, when the null phenotype resembles the recessive phenotype, which is quite common. The frequency for each specific chromosome to be lost is low. The phenomenon becomes more important when by accident the parent happens to be a trisomic (one chromosome lacking in the tetraploid) for the specific chromosome. Occurring for instance in a cross progeny of a nulliplex and a quadruplex, the aneuploid could be *Aaa* instead of *AAaa*, and could produce an entirely unexpected disomic ratio, as it would in a partly diploidized allotetraploid. It is necessary, therefore, to check the chromosome number when, on the basis of segregation, the distinction between autpolyploids and allopolyploids is made.

Inbreeding proceeds much slower in an autotetraploid than in the corresponding diploid, because much more frequently at least one of the four alleles of any gene will be of another type than the other three. Table 6.5 gives the fraction of the population homozygous for one locus (dominant or recessive) after selfing, for the diploid, the simplex tetraploid and the duplex tetraploid. Double reduction is not considered; it would give slightly higher frequencies. In a tetraploid, starting from a uniform duplex population, *random mating* results in the same level of heterozygosity as the F2. This is like in the diploid, but it represents a much higher level of heterozygosity. With continued random mating the 50% heterozygosity level will be reached only very gradually. In the diploid this level is reached after one generation.

Maximum heterozygosity is obtained in the diploid in an F1 between two homozygotes, irrespective of the number of alleles available for a specific locus in the population (Aa, or a1a2, or a1a3 etc.). In autotetraploids this is so only when no more than two alleles are available. With multiple allelism, for instance with the alleles a1, a2, a3, a4, maximum heterozygosity would be the combination of all four in the same plant. This cannot be realized in an F1 between two homozygotes, but in a cross between two different hybrids (Demarly 1963). For such genes a double cross hybrid is more heterozygotic than a single cross hybrid. Even without multiple allelism, selection in auto-

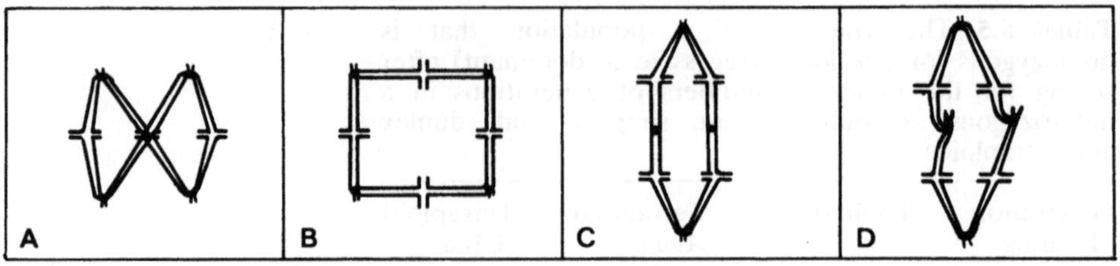

Fig. 6.7 Different modes of orientation of a ring quadrivalent in the meiosis of an autotetraploid, some of which lead to aneuploidy in the gametes (Sybenga 1987). **A** Alternate: segregation 2:2 (balanced). **B** Adjacent: segregation 2:2 (balanced). **C** Linear with centromeres amphitelic or inactive: segregation 1:1 or 1:2 with laggard(s), or 1:3 (all imbalanced), or 2:2 (balanced). **D** Linear: segregation 3:1 (imbalanced) when centromeres orientate as indicated

tetraploids proceeds slowly due to tetrasomic inheritance. With multiple alleles and erratic dose effects, it is even less effective, and stabilization of lines is slow. Selection in autotetraploids is discussed by Wricke and Weber (1986).

The *orientation* of the multivalent has important consequences for the outcome of meiosis. A trivalent with a univalent carries a considerable risk of imbalanced segregation. In many cases two chromosomes of the trivalent move to one pole, and one to the other. It is uncertain what the univalent will do. There is a chance that it lags at the equator and gets lost. It can also split into two chromatids which segregate to the poles, but do not function at the second division. The univalent can also break in the centromere (Sects. 5.5.1 and 6.2.1.5). Finally, it can move to the same pole to which the two chromatids of the trivalent move. In all these cases *aneuploid* progeny will result. The quadrivalent also will not always segregate 2:2. Depending on many factors intrinsic to the chromosomes, the genotype and the environment, certain orientations may prodominate. With linear and the less stable 3:1 orientation, aneuploidy results (Fig. 6.7). Aneuploidy is common in autotetraploids, but not favourable for the growth and reproduction of the plant. For practical applications, the level of aneuploidy in the population is of considerable importance. This subject will be considered again in Section 11.3.1.2.3.

6.1.2.2.3 Higher Autopolyploids

Of the autopolyploids with higher numbers of genomes than four, only the hexaploids play a role of any importance. The cultivated grass species *Phleum pratense* (timothy) has long been assumed to be a typical example. It forms exclusively bivalents at meiosis, but the genetic system was concluded to be hexasomic (Nordenskiöld 1953). The cause of the absence of multivalents is not clear in this case, but may be similar to that operating in autotetraploids with exclusive bivalent formation. Later, Cai and Bullen (1991) produced evidence from karyotype analysis that timothy might be better designated as an autoallohexaploid with genomes AAAABB rather than AAAAAA. Pentaploids and autopolyploids higher than hexaploids have been produced in the experiment, the polyploids with uneven genome number by hybridization between higher and lower genome number forms. The optimum number of genomes has generally been passed at the octoploid level. For practical breeding even the autohexaploids, except for the example mentioned, are of very limited interest. One important remark must be made: the higher the number of genomes, the more difficult it is to obtain results of selection, due to the conservative character of polysomic inheritance.

6.1.2.3 Allopolyploidy

6.1.2.3.1 Types

Allopolyploids are polyploids in which genomes of different origin are combined, usually derived from (not very) closely related species, while each genome is present twice as in a diploid. Like autopolyploids, allopolyploids are classified according to the number of genomes. Upland cotton, *Gossypium hirsutum* ($2n = 4x = 52$), is an allotetraploid composed of the genomes A and D, each occurring twice: AADD. A stands for a genome of *Gossypium herbaceum* and D for one of *G. raimondii*. The genome indication A happens to be used also for a *B. campestris*-related genome in allotetraploid rape seed, *Brassica napus* ($2n = 4x = 38$), where it is combined with a C genome from *B. oleracea*: AACC. Allohexaploid bread wheat has the genome composition AABBDD, and has an A genome derived from *Triticum monococcum*, *T. boeoticum* or another *Triticum* species with the A genome. The D genome is derived from *Aegilops squarrosa* (= *Triticum tauschii*) and the B genome probably from an *Aegilops* species related to *Ae. speltoides* and probably secondarily modified (Kimber 1983). Within a genus or group of related genera, it is attempted to give different symbols to different genomes, but between taxonomic groups the same symbols may be used. There are several more allopolyploid crop plants, especially among the Gramineae, including a number of grasses. The genomic composition of all important crop plants, including allopolyploids, is discussed by Simmonds (1976).

There are two types of intermediates between allopolyploids and autopolyploids: the *auto-allopolyploid* and the *segmental allopolyploid*. In the former, one set of more than two (an even number, usually four) genomes is identical and forms the autopolyploid part. There is at least one more set of two genomes that is different. There must be at least six genomes, and the genome composition can then be given, for instance, as AAAABB.

The segmental allopolyploid can be of two different types. In one type the differentiation between the genomes is insufficient for effective allopolyploidy (Stebbins 1947). This is not a stable construction, and will evolve into a true autopolyploid (Sybenga 1984b) or into the second type of segmental allopolyploid. Here, some chromosomes of a polyploid are completely homologous and others form pairs of homoeologous chromosomes. This implies that for the chromosome represented by a set of identical copies, the homoeologues of the other species are absent, and replaced by homologues. In the allotetraploid AABB, for instance, in the A genomes one chromosome would be present four times (chromosome 1A in the example), but the homoeologous chromosome of the B genome would then be absent: 1A1A1A1A; 2A2A2B2B; 3A3A3B3B, etc.: B1 is replaced by A1, which represents the autotetraploid segment of the allopolyploid. This is possible only when the genomes or at least the chromosomes involved are closely related and can readily replace each other. It has been artificially produced in allohexaploid bread wheat, where chromosomes from one genome can replace chromosomes from another genome: *nulli-tetra* substitution (Sears 1969). In wheat, not all chromosomes can be successfully replaced by their homoeologues. When translocations have occurred in the ancestry of the component species, the correspondence between the presumed homoeologues is insufficient for compensation.

Although on the basis of meiotic behaviour and sometimes genetic segregation the two forms of segmental allopolyploidy have been claimed to have been encountered in nature, the evidence is not very convincing. This is especially so for the first type of segmental allopolyploid (incomplete differentiation between genomes), which may be expected to be unstable.

6.1.2.3.2 Origin

Little is known of the origin of natural allopolyploids. They have probably been derived from spontaneous hybrids by meiotic doubling (Harlan and DeWet 1975). For self-fertilizers a single origin of the original hybrid is in principle sufficient, because selfing involving unreduced male and female gametes would result in the immediately fertile allopolyploid. However, there are no indications that the presently established allopolyploids have a pronounced tendency to form unreduced gametes, nor are they all perennials, which would permit a long period of unsuccessful attempts at doubling the original hybrid before it dies. Somatic doubling is not excluded, but it is very infrequent and would result in a minimum of heterozygosity. Clearly, the way most established allopolyploids have originated cannot be very faithfully reconstructed. The origin of artificial allopolyploids is discussed in Sect. 11.3.2.2.

6.1.2.3.3 Relevance

Allopolyploids have been very important in the evolution of many plant families (Simmonds 1976), and several important crop plants are allopolyploids.

The combination of specific genomes can apparently be quite successful, and it is easily understood that attempts to make artificial allopolyploids were made years ago and continue to be made, however, success is limited. On the one hand, there must be sufficient differentiation to prevent meiotic pairing between the genomes of the different species. On the other hand, the species to be combined must be close enough to have very similar control of the major biological processes, otherwise the interaction between the genomes can be deleterious. This balance appears difficult to find in experiments. The construction of allopolyploids, new or as a reconstruction of existing allopolyploids, to be used to introduce new genetic variation in the original species will be considered in Section 11.3.2.2.

6.1.2.3.4 Characteristics and Identification

The *somatic karyotype* should reveal the polyploid number, and when the chromosome structure of the component genomes is sufficiently different, the chromosomes of these genomes can be recognized. The genomes of bread wheat, for instance, can be distinguished on the basis of their C-banding pattern (Fig. 4.3C) and especially within the B genome the individual chromosomes can be recognized. This is somewhat more difficult for the A and D genomes. The genomes of allotetraploid cotton can be distinguished because of a considerable difference in the length of the chromosomes that has been maintained from the origin of the allopolyploid. Within the genomes, these relatively short chromosomes are not easy to identify. The identification of the component genomes is discussed in Section 9.2.1 and will not be considered further here.

The *meiotic behaviour* resembles that of a diploid, but the reason for this may be rather complex. In many allopolyploids the component genomes are too closely related for effective pairing differentiation at meiosis. Then genetic systems enhance the effect of the existing differentiation. Often such systems are not available in all genotypes of the parental species, and only specific, doubled hybrids can produce stable allopolyploids. The best-studied example is bread wheat, *Triticum aestivum*. In the long arm of chromosome 5 of the B genome a dominant allele of a gene (*Ph* = Pairing homoeologues) is present that inhibits pairing between homoeologous chromosomes, even when they have sufficient homology to pair in absence of this gene (Riley and Chapman 1958; Sears and Okamoto 1958). In most genotypes of the species that are potential contributors of the B genome to wheat, this allele is not present, but it is in a few (Dover and Riley 1972). This has been found by crossing genotypes of bread wheat lacking chromosome 5B, or that had an inactive mutation of *Ph*, with different genotypes of *Ae. speltoides*. In most combinations the effect of chromosome 5B or of the relevant gene was not compensated by the corresponding chromosome of the most common genotypes of *Ae. speltoides* and related species. *Ae. speltoides* normally even counteracts the action of *Ph*. However, in hybrids involving a few specific genotypes homoeologous pairing was restricted as in the presence of *Ph*. Surprisingly, a B-chromosome (the correspondence in nomenclature with the B genome is entirely coincidental), occurring in *Ae. speltoides* and in related *Aegilops* species, had the same effect: its presence prevented homoeologues to pair (Dover and Riley 1972). The relation between this B-chromosome and chromosome 5B is, however, unknown. Possibly in the evolution of these species a fragment of this B-chromosome was translocated to a chromosome 5 and was maintained in the population at a low frequency.

B-chromosomes of other species appear to have a similar effect. Evans and Macefield (1972) observed that the genomes of *Lolium perenne* and *L. temulentum*, both $2n = 14$, with considerable differences in chromosome size, pair without problem and form seven (heteromorphic) bivalents in meiotic metaphase I of the F1, and several quadrivalents in the amphidiploid (doubled hybrid). In the presence of a B-chromosome from *L. temulentum*, however, quadrivalents were practically absent from the tetraploid and in the diploid hybrid pairing was greatly reduced. On the other hand, B-chromosomes of rye (*Secale cereale*) were not found by Roothaan and Sybenga (1976) to compensate for chromosome 5B in wheat-rye hybrids.

The *Ph* gene in chromsome 5B of wheat is the strongest pairing regulating gene in wheat, but it is not the only one. There are several genes that restrict pairing and several others that promote pairing in general, both including homoeologous pairing. Different mutations of the *Ph* gene on chromosome 5B, with different degrees of inactivation of the gene, and mutations of the less active *Ph*2 gene on chromosome 3A (Sears 1976, 1984) have been induced.

In most other allopolyploids similar, although less pronounced, genetic pairing regulating systems, superimposed on an existing chromosomal differentiation, have been observed. In oats (*Avena sativa*) certain nullisomics have been shown to have multivalent pairing (Rajhathy 1983; 1991). Here, mutants with the same effect or B-chromosomes compensating for the absence of the critical chromosome have not yet been reported. In allotetraploid cotton, the differentiation between the A and C genomes is minimal and probably primarily based on a large difference in the amount of repetitive DNA. Yet, it is sufficient for effectively keeping the genomes genetically well separated (Endrizzi et al. 1985). In tall fescue (*Festuca arundinacea*) Jauhar (1975) found a similar system as in wheat, although not as pronounced, which was concluded to be haploid-inactive. Although in all these systems a major pairing regulating gene appears to be present, there are probably several more genes present with a similar action and probably also genes that act in an opposite direction. The required effectiveness of the genetic pairing regulatory system depends on the initial degree of differentiation between the genomes. Provided the system is in principle sufficient, there appears to be an array of possibilities for its organization, and judging from the presence of numerous reinforcing and counteracting genes, the efficiency of the system as a whole is not necessarily always great. Apparently, several genes are involved that have an effect on pairing but that could as well, or not, be present. They may have another main function, or they are merely redundant.

To introduce genes from alien species into a cultivated species, the possibility to manipulate homoeologous pairing is of great importance (Sect. 10.4.4.2.1).

6.1.2.3.5 Consequences

Normal, established allopolyploids reproduce normally, are fully fertile and have disomic segregation, like diploids. Newly synthesized allopolyploids (amphidiploids) may show reduced fertility as a result of unadjusted interaction between the genomes, and they may show insufficient pairing differentiation. This is comparable with the situation in established allopolyploids with a defective pairing regulatory system and it may lead to the segregation of novel aberrant types and to special segregation ratios (cf. Sects. 9.3 and 11.3.2). Even in balanced old and new allopolyploids the characters do not always simply add up or show dominance and epistatic relations. The interaction between the genomes may produce novel effects. An example is the amphidiploid between *Lolium multiflorum* and *Festuca drymeja* (Morgan and Thomas 1991) in which the chiasma localization patterns of the two parents deviate considerably from those in the diploid species.

6.1.2.4 Polyploid Hybrids

6.1.2.4.1 Types, Origin and Relevance

Polyploid hybrids (other than allopolyploids) are occasionally found in nature as spontaneous and transient events and, very rarely, as established apomicts. An occasional polyploid hybrid may be observed in a botanical garden or other collection. There is no particular reason to pay special attention to such cases, although in their identification and genetic behaviour, the fact that they are polyploid may play a role. Their importance for the plant breeder, especially the artificial polyploid hybrids, is primarily as experimental material, for the analysis of genome relationships and for the introduction of genes. All conceivable types can be constructed, provided they are viable. They are mentioned here as a special category for completeness, and to point out a few special characteristics. They will be discussed further in Sections 9.3.2 and 10.4.

6.1.2.4.2 Characteristics, Identification and Consequences

In the phenotype, the hybrid nature and usually also some characteristics of polyploidy are expected to be expressed. It should be noted that this is not always the case: the expression of the characteristics of one of the parents may appear to be more pronounced than those of the other parent and gigas characteristics are not always obvious. Unknown polyploid hybrids, which may occur spontaneously in collections, like many other spontaneous hybrids, are not always easily identified. The methods mentioned to identify the component genomes of an allopolyploid, as discussed in Section 9.2, are applicable also to polyploid hybrids.

Because of the imbalanced genomic composition of polyploid hybrids (other than allopolyploids), they will tend to be largely sterile for meiotic and developmental reasons, and the few progeny formed, unless of a parthenogenetic origin, will usually have an abnormal chromosome constitution. This presents a bottleneck in their use in plant breeding as intermediates in gene transfer or for constructing special genetic systems. On the other hand, the segregation of a variety of novel types presents precisely the opportunity to select progeny with a specific chromosomal composition.

6.2 Aneuploidy

The following aneuploids will be considered: *primary*, *secondary* (= *isochromosome*), *tertiary*, *translocation* and *telocentric* aneuploids, *monosomics* as well as *trisomics*, in diploids and allopolyploids; and trisomics and pentasomics in autotetraploids. The information provided here will be used in later chapters.

6.2.1 Hypoploidy, Monosomy

In hypoploidy one or more chromosomes are absent from the normal karyotype. When there is one less, it is monosomy. With two different chromosomes less, double monosomy, etc. When both copies of a normally disomic (diploid or allopolyploid) chromosome are absent, it is *nullisomy*. Emphasis in the following will be on monosomics; nullisomics will be mentioned only in passing. A chromosome less in an autotetraploid results in trisomy instead of tetrasomy for that particular chromosome, and will be considered later.

6.2.1.1 Types of Monosomy

If a normal chromosome is monosomic, the individual is a *primary monosomic*. If in a translocation heterozygote one of the two normal chromosomes involved in the translocation is lacking, the monosomic may be called a *trans-*

location monosomic, following the nomenclature used for trisomics (Sect. 6.2.2.1.1). If one of the translocation chromosomes is absent, the monosomic is a *tertiary monosomic*. Especially with translocations in which one of the translocated chromosomes is small, its absence can be relatively harmless. The remaining single chromosome in monosomics may be a normal or a translocation chromosome, but occasionally it may be a telocentric or an isochromosome, formed during the process from which the monosomy originated. This is an additional disturbing complication for the genome and is found less frequently in diploids than in allopolyploids where many forms of hypoploidy are tolerated.

For reasons of survival and transmission, monosomy is rare among *diploids*, and even when monosomics are formed, the chance of recovery is generally small. Species with a high level of duplication, possibly related to an ancient polyploid origin, like maize (2n = 20; some rather distant relatives have 2n = 10) and tomato (2n = 24) are known to tolerate surprisingly frequent monosomy. Khush (1973) lists monosomy in several diploid species: *Avena strigosa*, *Datura stramonium*, *Hyoscyamus niger*, *Lycopersicum esculentum* (tomato), *Nicotiana alata*, *Nicotiana langsdorffi*, *Petunia* sp. and *Pharbitis nil*, *Zea mays*. Most are primary monosomics, in which a normal chromosome is lacking, and usually only one was found. In *L. esculentum* (tomato) many tertiary monosomics were recovered, lacking a translocation chromosome (usually the smallest) from an interchange heterozygote or homozygote (Khush and Rick 1966; Khush 1973). In maize also, tertiary monosomy is easier to obtain than primary monosomy (Weber 1983). Most monosomics in diploids are weak and hardly viable, but in tomato and maize many are sufficiently vigourous to flower and even to reproduce; in maize, meiosis could be studied even in triple monosomics (Weber 1983).

6.2.1.2 Origin

In his review of monosomics in diploids, Weber (1983) reports several types of origin, but few are somatic: pollen irradiation and colchicine treatment of grown plants. In the latter case the disturbed spindle resulted in somatic chromosome segregation and aneuploid cell progeny. The most common origin is through deviant meiosis in abnormal karyotypes and in species hybrids; interaction between B-chromosomes and heterochromatic knobs in microspores of maize; in r-X1 deficiency heterozygotes of maize, which even produce double monosomics (Weber 1991). Translocation monosomics are due to 3:1 segregation of the translocation complex at meiosis. They are relatively frequent in tomato and in maize, but especially thoroughly studied in tomato (Khush 1973). Translocation monosomics have very infrequently been recovered in rye (*Secale cereale*; Sybenga unpublished), and occur in other species occasionally, too, when specifically looked for. The recovery of monosomics is simplified by using heterozygotes for chromosome-specific marker genes.

Monosomics are expected to arise relatively frequently in *tissue culture* during a callus phase, but monosomic diploid regenerants have not been reported. In polyploids, however, such as bread wheat (d'Amato 1977, for instance, and own unpublished results) monosomic regenerants are relatively common. Another possible somatic origin could be the addition to the medium of colchicine, benomyl, parafluorophenylalanine, chloramphenicol, amiprophos-methyl, CIPC. Plants have been recovered from such callus, but unequivocal reports on monosomy are not available.

Monosomics in allopolyploids occur spontaneously at a low frequency. In wheat there are two main sources, as reported by Sears (1954). One is haploids, which are rare, and which produce diploid progeny only at a very low rate. The second source is the partly desynaptic nullisomics for chromosome 3A, which had been isolated from the progeny of the selfed monosome. This again was isolated from the progeny of a haploid. In natural populations of wheat about 1% of all plants is monosomic, with considerable variation between varieties (Riley and Kimber 1961). These, too, can be an important source of monosomics. All monosomics of wheat have been recovered in the variety Chinese Spring and transferred to other varieties by backcrossing. In other allopolyploids (oats, for instance), the beginning of the construction of a monosomic series was usually a spontaneous monosomic.

6.2.1.3 Relevance

Monosomy in diploids is of considerable interest in locating genes in chromosomes. The combination of monosomy with a recessive gene in the remaining chromosome results in hemizygous expression of that gene. Conversely, when the location of the gene is known, it will enable the chromosome involved to be identified. Normally, however, this is not the case and special test-crossing programs with stocks carrying known genetic or chromosomal markers may be needed to identify the lacking chromosome. The phenomenon is sufficiently frequent in only very few diploids. For further details, the reader is referred to the literature (Khush 1973; Weber 1983, 1991).

Hypoploidy is much more common in allopolyploids and also of much more practical interest. It is used for locating genes on chromosomes, like in diploids, but with more possibilities (Sect. 8.3.2.1.1). It is also used as an important intermediate step in different forms of genetic manipulation, especially the transfer of specific genes (Sect. 10.4.3), and for special forms of quantitative genetic analysis (Law 1966). In semi-dwarf varieties of wheat carrying the *Rht2* gene, which has a dose-dependent expression, monosomy for chromosome 4D causes tall rogues to appear in the field, in some varieties with frequencies unacceptable for seed certification. For semi-dwarfs with Rht1, monosomy for chromosome 4A has the same effect (Worland and

Law 1985). Monosomy for 5A results in the appearance of the less striking "speltoid" character.

In allotetraploids the effect of the absence of entire chromosomes is more pronounced than at higher ploidy levels, and closer to that in the diploid. In allotetraploid *Nicotiana tabacum* (tobacco, 2n = 4x = 48) not all monosomics have been recovered, and those that have, are morphologically more distinct than in wheat. Nullisomics are practically inviable in tobacco, but most have been isolated in wheat. In another important allotetraploid crop plant, cotton (*Gossypium hirsutum*), the situation is comparable to that in tobacco (Endrizzi et al. 1985). Allohexaploid oat (*Avena sativa*, 2n = 6x = 42), is more comparable to wheat, but still the isolation of monosomics and nullisomics has been much less successful than in wheat, which is not solely due to the fact that wheat has received more attention from cytogeneticists than oats (Rajhathy 1991).

6.2.1.4 Characteristics and Identification

In the *somatic karyotype* monosomics are readily recognized, but, as usual, the chromosome involved can only be identified when it is readily distinguished on the basis of specific morphological characteristics (shape, size, presence of an NOR, detailed C-banding). Monosomy recovered among the progeny of known translocation heterozygotes can be assumed to involve one of the chromosomes involved in the translocation. Only when the translocation chromosomes have clear morphological characteristics can they be identified with certainty. In diploids, small translocation chromosomes tend to be favoured as monosomics. Usually, the identification of monosomics requires some experimentation, involving hybridization and meiotic analysis.

Monosomy, unless so disturbing that normal meiosis is impossible, simply results in the presence of a single *univalent* at *meiosis*. In translocation monosomics, instead of the normally expected quadrivalent a trivalent appears. Without further analysis it is not easy to determine which of the four chromosomes of the quadrivalent is absent.

The univalent often suffers centromere breakage at anaphase I, and may produce telocentrics or isochromosomes (Morrison 1954; Sears 1954). Especially in wheat, the monosomics of Chinese Spring have been an excellent source of these aberrant chromosomes types, and the entire set is available (Sears 1954).

In allopolyploids where the differentiation between the genomes is not as strong as in wheat, and where it is only sufficient to prevent homoeologous pairing in the presence of complete homologues, such as allotetraploid tobacco (*Nicotiana tabacum*, 4x = 48), a trivalent between the single remaining partner and its homoeologues may form in monosomics, in tobacco in up to 25% of the cells for some of the monosomics. This enhances regular segregation, but unless the single chromosome is always at the end, it may lead

Table 6.6. The average chromosome constitution in the progeny of selfed wheat monosomics (Sears 1954)

♀ \ ♂	$n = 21$ 96% (90 – 100)	$n - 1 = 20$ 4% (0 – 10)
$n = 21$ 25%	$2n = 42$ 24%	$2n - 1 = 41$ 1%
$n - 1 = 20$ 75%	$2n - 1 = 41$ 72%	$2n - 2 = 40$ 3%

to "non-disjunction" of the homoelogous pair with the result of an extra chromosome in the daughter group where the original monosome is absent, and a chromosome lacking in the group where the original monosome is present. The results are rather drastic: a change in the monosome in the progeny, and the origin of a substitution of one chromosome by a homoeologue.

6.2.1.5 Consequences

The monosomics of diploids tend to have greatly reduced fertility, but there are exceptions. Both in maize (Weber 1983) and tomato (Khush and Rick 1966) some, especially the translocation monosomics involving small chromosomes, reproduce sufficiently to permit a genetic analysis. If the viability of the monosomic is not a bottleneck, the majority of the progeny is monosomic again because the remaining homologue is univalent and tends to be lost. For translocation monosomics this risk is smaller because usually a trivalent is formed that can regulate segregation. When both gametes involved in fertilization lack the same chromosome, a nullisomic will result. These are only exceptionally viable in diploids and consequently very rare.

Monosomics in allopolyploids have been studied most extensively in wheat (*Triticum aestivum*, 6x = 42, Sears 1954), and they continue to be important in this species, especially the primary and telocentric monosomics. Because the univalent is included in considerably less than half of the anaphase I groups, the majority of the gametes (on average about 75%) is nullisomic (at the haploid level). Nullisomic gametes function on the female side, but on the male side the competition from the minority of normal pollen is strong. In some cases pollen lacking the particular chromosome does not function at all. On average, in wheat, over 95% of the fertilizing male gametes is normal and only 5% lacks the chromosome. On the female side this is 75%, close to the level of formation. As a consequence, in the selfed progeny of a wheat monosomic some 73% is monosomic again, only 3% is nullisomic and 24% normal disomic (Table 6.6). Individual monosomics may deviate considerably from these averages.

Table 6.7. The female gametic composition of a wheat plant monosomic for chromosome 5A, tested on 786 plants in the progeny of a cross mono 5A × normal (Sears 1954)

Chromosome constitution	Number of progeny	%
$n = 21$	129	16.4
$n - 1 = 20$	628	80.0
$n - 1 + \text{telo} = 21$	16	
$n - 1 + \text{iso} = 21$	11	3.6
$n + \text{telo} = 22$	1	
	785 = Total	

Gamete functionality is estimated most reliably in a test cross of a monosomic as either the male or the female with a normal tester. An example from the classical paper by Sears (1954) of the female gamete composition of one particular monosomic is given in Table 6.7.

For a number of allopolyploid cultivated species a standard set of monosomics (not always complete) is maintained and used to identify the chromosomes in chromosomal abnormalities. The maintenance of a set of monosomics is not particularly difficult (Table 6.6).

To identify an unknown monosomic, it is crossed with the series of standard monosomics and in the progeny the plants with two chromosomes less than normal are selected. The nullisomic has no obligate univalents, the double monosomic has two. In the first case the original monosomics were identical, in the second case they were not. It is clear that the reciprocal cross should be attempted when one of the parents has no or greatly reduced pollen transmission. As long as no standard set is available, the monosomics can be classified as identical or different and each type subsequently identified on the basis of chromosomal characteristics, morphology of the monosomic or the nullisomic or simply given an arbitrary identification number.

Person (1956) was the first to warn against an important problem regarding the maintenance of standard monosomic lines. Some monosomics, and even more so the corresponding nullisomics, have a slightly disturbed meiosis and may form univalents of the chromosomes that are not monosomic. As a result, new monosomies may arise in the progeny. When the originally monosomic chromosome is recovered in the gamete (an average of 25% of the cases) and a new chromosome is monosomic, monosomy is observed in the resulting progeny. However, it is not observed for the original chromosome. A univalent is observed in meiosis, but it is a different one. This phenomenon is called *univalent shift*. It makes it necessary to check the entire set regularly. Checks tend to be laborious and to be postponed too long.

The use of a standard monosomic series for analyzing chromosomal abnormalities in allopolyploids may meet with another difficulty. Between varieties translocations may occur, and these are carried over in breeding programs. Within the varieties they are homozygous and not readily detected. The variety Chinese Spring, the variety used most frequently as an experimental standard, differs from several common commercial varieties in one or even two translocations. When a chromosome is monosomic in such a variety and checked against the standard of Chinese Spring, it will not give a conclusive result if it is a translocation chromosome. There is an additional problem with such translocations. For use in chromosome manipulation programs, either for genetic analyses or for transferring genes between varieties or species, the monosomics have to be isolated in the variety with which the manipulation is to be carried out. The simplest way is to transfer the different monosomies from the standard (in wheat this is Chinese Spring) by backcrossing. The European Wheat Aneuploid Cooperative (EWAC), including wheat breeders and cytogeneticists, has been created to coordinate these programs for several different varieties. A translocation in the variety to which the monosomies are to be transferred causes meiotic irregularities and it is fundamentally impossible to introduce monosomy for the same chromosome, because this is not present in the variety with the translocation.

For further genetic consequences of monosomy and its application in estimating genetic parameters and in manipulation, see Sections 8.3.2.1.1, 9.2.2, and 10.4.3.

6.2.2 Hyperploidy

6.2.2.1 Primary Trisomy

6.2.2.1.1 Types

In primary trisomy one normal chromosome is extra in a diploid or allotetraploid: it is present three times instead of twice. In an autotetraploid, trisomy in contrast implies that one chromosome is absent: only three chromosomes of one type remain instead of the original four.

Basically, there is only one type of primary trisomic. A normal chromosome involved in a translocation and in excess in a translocation heterozygote does not result in the presence of three copies of one particular chromosome; this is not called primary trisomy, but translocation trisomy. It will be briefly discussed below.

6.2.2.1.2 Origin

Primary trisomics are almost never found spontaneously in *somatic tissues* of normal diploids. After treatment with colchicine or other spindle-disturbing chemicals, they may be formed and recovered incidentally and somewhat more frequently than monosomics. Paraphenylalanine, inducing haploidy in diploid fungi, can produce aneuploidy in somatic plant cells (cf. Sect. 11.2.3), including primary trisomy. In vitro culture, especially of callus, often disturbs regular nuclear and cell divisions and then results in chromosomal abnormalities, including primary trisomy. Plants have been recovered from such callus. It is a potential, but not the most efficient source of primary trisomics. A disadvantage is the simultaneous production of other chromosomal and genetic abnormalities.

Primary trisomics are infrequently the result of spontaneous errors in *meiosis* of normal diploids or allopolyploids. The most effective sources are partial *desynaptics* and *triploids*. Desynaptics produce trisomics in the same way as they produce monosomics (Sect. 6.2.1.1) but, at least in allopolyploids, with a somewhat lower frequency because the univalents tend to get lost instead of moving to the same pole together (non-disjunction). Autotriploids, if at all fertile, produce primary trisomics at a high frequency, either directly or as double trisomics from which in a later generation single trisomics can be isolated. For details of the formation of trisomics in the progeny of autotriploids, see the consequences of the meiotic behaviour of triploids (Sect. 6.1.2.2.1.4).

6.2.2.1.3 Relevance

Primary trisomics are a useful tool in genetic analysis, especially for locating genes on chromosomes in diploids, where monosomics, the preferred system for gene localization in allopolyploids, are not generally available. Primary trisomics are a source of secondary deviants such as telocentrics and isochromosomes, which are also used in gene localization, but in addition can serve as starting material for special constructions. These will be considered with the chromosomal types concerned. There is no obvious, direct practical application of primary trisomics.

6.2.2.1.4 Characteristics and Identification

As in the many cases discussed above, the *somatic karyotype* of the primary trisomic is rather typical, especially when the chromosome involved can be recognized by its somatic characteristics. The phenotype can be very characteristic of the chromosome involved, especially in diploids, and if a series of primary trisomics is available, comparison of the morphology of a new trisomic with the series may suffice to identify it. The oldest series of primary trisomics was produced by Blakeslee (Blakeslee and Avery 1938; see Fig. 11.1A). Complete or incomplete series of primary trisomics have been isolated for most crop plants, but have not always been maintained.

Table 6.8. Meiotic configurations in two rye plants trisomic for chromosome 1R. The frequency of trivalents at metaphase is dependent on the chiasma frequency, expressed in the number of associated chromosome arms

% Chromosome arms associated by chiasmata	No. of cells	Trivalents			
		Chain	"Frying pan"	"Y" shaped	Total
96.0	300	122	13	10	145
89.6	300	91	5	10	106

The *meiotic behaviour* of primary trisomics is in principle the same as that of the individual chromosomes of a triploid. There are three chromosomes of one type, and these can pair in all combinations. The situation is closely related to that in an autotetraploid. Again, there are three pairing combinations for each arm and for both arms combined this makes nine pairing types. Three of these are a bivalent with a univalent, six are a trivalent. Trivalent pairing can be less than 2/3 when the number of effective points of pairing initiation is less than two per chromosome. There may be more trivalents when there are more points of effective pairing initiation and when, in addition, these are sufficiently far apart to permit partner exchange between them, as in autotetraploids (Fig. 6.3). On the other hand, a reduction in the trivalent pairing frequency may be the consequence of pairing adjustment in the synaptonemal complex as discussed for the autotetraploid.

Chiasma formation determines the final metaphase I configuration frequencies. The types are shown in Fig. 6.1 and include the cases in which chiasmata are formed in the interstitial segments. Examples of the frequencies of different meiotic configurations found in primary trisomics in rye are given in Table 6.8. With lower chiasma frequencies the number of trivalents at metaphase I decreases.

The trivalent may orient in various ways in *metaphase I* of meiosis. The bivalent with univalent, has its own segregation pattern. The bivalent will segregate normally, but the univalent will often be lost in plants. In several cases it orients amphitelically and separates into chromatids at anaphase I (Fig. 6.8). The centromere may break instead of splitting (Fig. 6.9), and this has important consequences.

6.2.2.1.5 Consequences

Primary trisomy has several important consequences that make it interesting for the plant breeder, even when only indirectly. It is an efficient means of locating genes on chromosomes because marker segregation deviates considerably from that of the corresponding diploid. It does so, of course, only for the

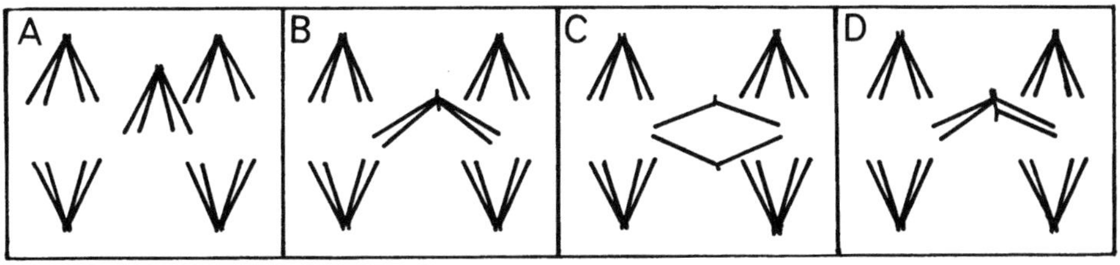

Fig. 6.8 Different modes of univalent behaviour (Sybenga 1987). **A** Centromere syntelic: the entire chromosome moves to one of the poles. **B** Centromere amphitelic without splitting; chromosome tends to lag at the equator. **C** Centromere amphitelic, splitting: single chromatids move to the poles, unable to divide at anaphase II. **D** Centromere misdivision of one of the chromatids (cf. Fig. 6.9)

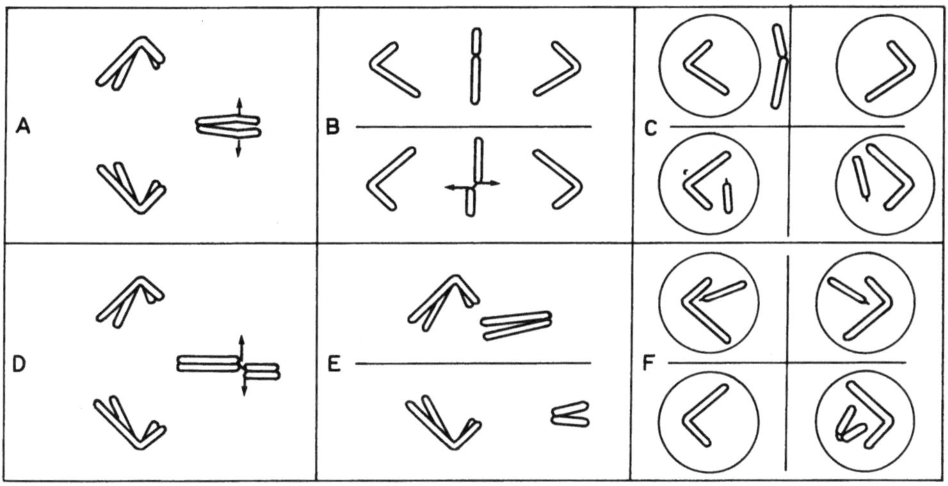

Fig. 6.9 Centromere misdivision of a univalent. A normal pair of chromosomes is shown to segregate normally. The univalent may separate into chromatids at first anaphase and break in the centromere at second anaphase (**A–C**). Two telocentrics are formed, but not necessarily in time to reach the poles to be included in the telophase nuclei. Alternatively, the centromere may break at first anaphase, sometimes leading to free telocentrics, sometimes followed by centromere fusion and resulting in isochromosomes (**D–F**). If a second, similarly behaving univalent is in the neighbourhood, fusion may take place between centromeres of the different chromosomes: centromere translocation

chromosome involved. In general, the fertility is not greatly disturbed unless the chromosome concerned has a strong negative dose effect.

With three homologous chromosomes, and two different alleles for a gene on that chromosome, there are two different homozygotes: *AAA* (triplex) and *aaa* (nulliplex), but two different heterozygotes which will show segregation in the self or test-cross progeny: *AAa* (duplex) and *Aaa* (simplex). This is one

Table 6.9. Trisomic segregations without double reduction. **A** *Simplex*, Aa1a2, gametic ratio 1:1. When disomic male gametes are not functional, the male gametic ratio is 1:2. The F1 segregates 3:1 when all male gametes function, 2:1 when disomic gametes do not function. This is a higher recessive frequency than in the diploid. **B** *Duplex*, A1A2a, gametic ratio 5:1. When disomic male gametes do not function, the male gametic ratio is 2:1. In the F1 the segregation is 35:1 when all gametes function, and 18:1 when disomic male gametes do not function. For several reasons the ratio is 12:1 to 10:1 in practice.

A	Aa_1	Aa_2	a_1a_2	A	a_1	a_2	B	A_1a	A_2a	A_1A_2	A_1	A_2	a
Aa_1	+	+	+	+	+	+	A_1a	+	+	+	+	+	+
Aa_2	+	+	+	+	+	+	A_2a	+	+	+	+	+	+
a_1a_2	+	+	−	+	−	−	A_1A_2	+	+	+	+	+	+
A	+	+	+	+	+	+	A_1	+	+	+	+	+	+
a_1	+	+	−	+	−	−	A_2	+	+	+	+	+	+
a_2	+	+	−	+	−	−	a	+	+	+	+	+	−

(+: at least one dominant allele).

less than in the autotetraploid. With full recovery of all chromosomes, and no other abnormalities, the expectation is that at anaphase I one pole receives two chromosomes and the other pole one: there will be as many gametes with two chromosomes as with one. In the duplex, when the two *A* alleles are called *A1* and *A2*, the combinations and single chromosomes with *A1A2*, *A1a*, *A2a*, *A1*, *A2* and *a* will be formed with equal frequency (Table 6.9). Only one of the six has no dominant allele. In a test cross with a double recessive, the segregation will be 5:1, when all assumptions are fulfilled. The F2 segregates 1 out of 36 or 35:1.

In practice this segregation is infrequently realized, for various reasons. One is that with selfing and ample pollination the pollen with the extra chromosome cannot compete with euhaploid pollen. With scarce pollination the transmission of the extra chromosome through the pollen is greatly increased (Janse 1987), but seed set is then reduced. Failure of male transmission of the extra chromosome reduces the segregation ratio up to about 18:1. Due to the loss of most of the univalents as well as the failure of some female disomic gametes and some trisomic embryos to function, the contribution of gametes with an extra chromosome is further reduced, until the rather common F2 segregation ratio of about 12:1 to 10:1 for primary trisomics is reached. This is sufficiently different from the disomic ratio to be distinguished, even in relatively small segregating populations.

The simplex (Aa1a2) also forms six types of gametes. Now the a alleles occur twice and are numbered 1 and 2: *Aa1*, *Aa2*, *a1a2*, *A*, *a1*, *a2*, all with equal frequency. When all chromosomes are recovered and all gametes function, the segregation will equal that of a disomic. However, as for the duplex, disomic gametes are less frequent because of univalent loss, and are often less

Table 6.10. A Trisomic segregations in maize and barley (cf. Sybenga 1972). **B** Progeny of three primary trisomics of *Datura* (Blakeslee and Avery 1938)

A.

Plant	Trisomic	Factor	Type	Dominant	Recessive	% Recessive
Maize[a]	10	*r*	*RRr* self	396	41	9.4
Maize	10	*r*	*RRr* × *rr*	819	213	20.6
Maize	10	*r*	*rr* × *RRr*	941	486	34.1
Maize	10	*r*	*Rrr* × *rr*	679	836	55.2
Maize	–	*r*	*Rr* × *rr*	1161	1196	50.7
Barley[b]	1	*br*	*Br Br br* self	370	25	6.3
Barley	3	*uz*	*Uz Uz uz* self	493	17	3.3
Barley	7	*ss*	*Sr Ss ss* self	145	11	7.1

B.

Chromosome	Total number	2*n*	2*n* + 1	% 2*n* + 1	Iso Same chromos.	Iso Different chromos.	2*n* + 1 Different chromos.	4*n*	*n*
1.2	2049	1780	213	10.40	6	0	27	23	0
3.4	2089	1634	452	21.64	0	0	1	1	1
19.20	4058	4498	141	2.96	7	4	100	3	1
Average all 12 trisomics				22.08					

[a] In maize, in the selfed duplex the segregation approaches 10:1, the extent to which this is due to double reduction is not clear. In the test crosses the excess of recessives (20.6 versus expected 16.6 in the duplex, and 55.2 versus 50% in the simplex) may well be due to double reduction.
[b] In barley, the frequency of recessives is low. This may be due to some functioning of disomic male gametes.

viable, as sometimes also trisomic zygotes. Consequently, the frequency of recessives can be considerably in excess compared to disomic segregation.

As with the autotetraploid, trisomic inheritance has the complication of *double reduction*. It has the same prerequisites: recombination between locus and centromere, adjacent segregation of the two chromosomes between which recombinational exchange has occurred (either on the basis of linear or adjacent orientation), and proper Anaphase II segregation. The effect is not as drastic as in autotetraploids, where recessives can be caused to segregate among the progeny of a triplex, which otherwise is not possible. The duplex in the primary trisomic segregates also without double reduction: double reduction can only increase the frequency of recessives (Table 6.10).

The transmission of the extra chromosome is variable. In inbred lines of normally outbred diploids, the tolerance for trisomy is low. Even in normal, outbred strains some trisomics are apparently inviable. In rye (*Secale cereale*), for instance, although trisomy for chromosome 7R has been claimed, it is probable that it has reduced viability and even in the best genotypes it can hardly be maintained. The telocentric trisomic for one of the arms is quite viable (own and other unpublished results). Because of serious inbreeding effects, the maintenance of trisomics of outbreeders by selfing is quite difficult. Vegetative reproduction here is the best solution, but requires special methods (Melz et al. 1988). In the inbreeding species *Hordeum vulgare* (barley), all seven primary trisomics have been isolated and can be maintained by selfing (Tsuchiya 1991). They can be recognized by their specific plant morphology. The same is true for the 12 primary trisomics of rice (Khush 1991). In outbreeders there is also a clear specific effect of each trisome on plant morphology, but the genetic variation makes it very difficult to distinguish adequately between the effects of genotype and trisomy.

In allopolyploids trisomics are readily recovered and maintained. The complete series has been available in wheat since the early 1950s (Sears 1954). Secondary deviants, like telocentrics and isochromosomes, but also other trisomics, are common. In general, primary trisomics are a good source of secondary variants, primarily telocentrics and isochromosomes. These result from centromere breakage of the univalent, and are discussed in Section 6.2.2.2 (see also Table 6.10B).

6.2.2.2 Secondary and Telocentric Trisomy

6.2.2.2.1 Types, Origin

In secondary and telocentric trisomy, an isochromosome and a telocentric chromosome respectively are present in excess of the normal disomic chromosome number. In principle, isochromosomes and telocentrics can be formed for each chromosome arm. When pseudo-isochromosomes are not taken into account, there is no variation as to type.

The origin can be in somatic tissues after irradiation or chemical mutagenesis, but this is rather unusual. Isochromosomes may be formed by induced or spontaneous asymmetric exchange at or near the centromere between homologues in somatic tissues, but this does not produce a trisomic. Similarly, telocentric chromosomes may be induced. They are somewhat more common in somatic cell or tissue cultures from which they can be recovered in differentiated plants. The most common origin is in meiosis of monosomics and trisomics, as discussed in the previous sections. Centromere breaks result in free chromosome arms with unstable ends, which may reach one of the poles, in half of the cases as an extra chromosome. Subsequent fusion of the breaks of two sister centromeres results in an isochromosome, and healing of the centromere break by production of a telomere results in a telocentric. Because centromere breaks are the result of centromere activity, pole-directed movement and subsequent recovery represent the rule rather than the exception.

6.2.2.2.2 Relevance

Secondary and telocentric trisomics can be used to locate genes on chromosome arms (Sect. 8.3.2.1.3), which is more specific than location with primary trisomics. Telocentric trisomics can also be applied in certain chromosomal contructions to modify the genetic system (Sect. 12.4.2), but this potential is rather limited. Both types of trisomic appear in the progeny of plants that have a high frequency of univalents, and may be disturbing. Recognition of their presence and nature, therefore, is important.

6.2.2.2.3 Characteristics and Identification

Telocentrics are readily distinguished from (sub)metacentric and subacrocentric chromosomes. Their further identification requires distinct cytological markers, or genetic or cytogenetic experimentation. When genetic markers are available, the specific segregations (Sect. 6.2.2.2.4) will identify the arm involved. Crossing with a standard translocation tester set and checking meiosis for association between the quadrivalent and the telocentric will permit the identification of the chromosome involved, but is not conclusive with respect to the arm. This requires specific chromosomal markers, which must discriminate only between the two arms of the chromosome involved.

Secondary trisomics (isochromosome trisomics) are not always readily recognized in the *somatic karyotype* because isochromosomes do not differ much in shape from normal metacentrics. When a special marker for instance a small or a large size, the presence of an NOR or a conspicuous C-banding pattern is present, recognition is easier. By using a tester set, the chromosome (arm) involved can be identified. Secondary trisomics have special *meiotic* characteristics, and are the only chromosomal construction (together with the very similar pseudo-isochromosome trisomics) that can form a ring trivalent in meiosis (Fig. 6.10), which usually identifies it. There are four copies of the arm involved in the isochromosome. With random pairing of these four arms, the ring trivalent is expected to be formed twice as frequently as the alternative configuration, a *ring univalent* (the univalent isochromosome) with a normal bivalent. The ring univalent is characteristic of an isochromosome as such.

The point of partner exchange in the trivalent is variable, and may be close to the chromosome end or at the centromere (Fig. 6.10). When there is a considerable interstitial segment, this may contain a chiasma, and a small "frying pan" or "spoon" is formed or even a double ring, somewhat resembling that of the pericentric inversion metaphase configuration. It is distinguished because of the presence of three chromosomes in the configuration. In a ring trivalent the isochromosome may orient towards one pole and the two normal chromosomes to the other. This results in primary trisomy in the progeny of a secondary trisomic. The ring univalent will often get lost or split and thus, occasionally, produce telocentrics.

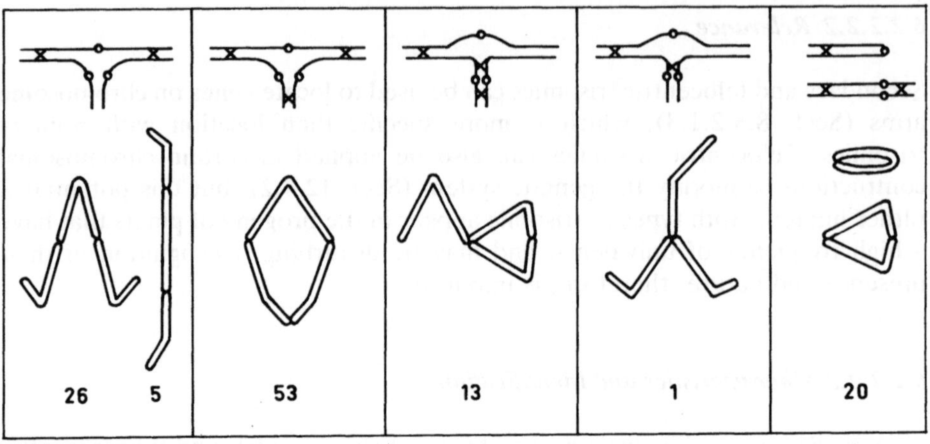

Fig. 6.10 The meiotic behaviour of a secondary trisomic (isochromosome trisomic). A ring trivalent may be formed, or a ring univalent. A "frying pan" with the small ring attached to a normal metacentric is also possible. This may resemble the pericentric inversion bivalent of Fig. 5.5C. The point of partner exchange is not fixed. The frequencies shown are for a secondary trisomic in *Datura* (Belling and Blakeslee 1924; total of 118 cells analyzed)

The behaviour of the *telocentric trisomic* at meiosis is simpler. There is again the necessity of choosing the pairing partner for the telocentric. In two-thirds of the cases it is expected to pair with a normal chromosome, in one-third it is univalent, a relation as in primary trisomics. The frequency with which this is realized depends on the same factors discussed above for autotetraploids and primary trisomics.

6.2.2.2.4 Consequences

The segregation of genes in telocentric and secondary trisomics does not follow the same rules as for primary trisomics. A telocentric and an isochromosome cannot normally replace a (sub)metacentric chromosome. In principle, therefore, the inheritance is disomic, with two complications.

1. When the extra chromosome has a dominant allele and both normal chromosomes a recessive allele, the plants with the extra chromosome in the progeny of a test cross or self will have the dominant phenotype, those without the extra chromosome will have the recessive phenotype. This can be used to locate genes in specific chromosome arms (Sect. 8.3.2.1.3).

2. There may be recombination between the locus of the gene and the centromere, transferring the allele of the extra chromosome to a normal chromosome and vice versa. The frequency of recombination is a measure of the genetic distance between the gene and the centromere (Sect. 8.3.2.1.3). The situation in secondary trisomy is similar, but slightly more complicated.

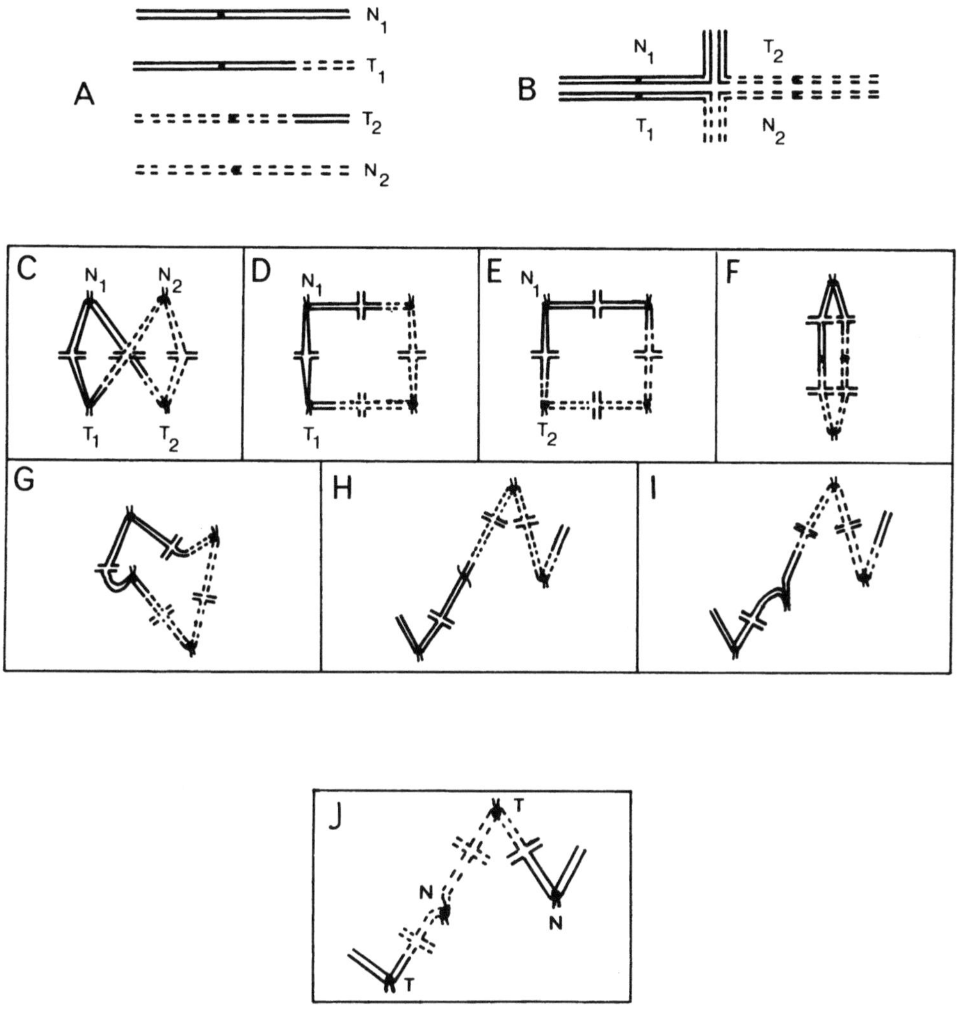

Fig. 6.11 The origin of translocation and tertiary trisomics from an interchange heterozygote. **A–G** Ring quadrivalents; **H–J** chain quadrivalents (Sybenga 1987). **A** The chromosomes of the interchange. **B** The pairing cross (cf. Fig. 5.7). **C** Alternate orientation leading to balanced spores. **D**, **E** Adjacent 1 and adjacent 2 orientation respectively, leading to imbalanced spores with the normal chromosome number. **F** Linear orientation with inactive or amphitelic centromeres, resulting in chromosome loss by lagging, or in irregular segregations. **G** Orientation resulting in 3:1 segregation with one normal and two translocation chromosomes moving to the same pole. The other pole is deficient. If the gamete with the extra chromosome is combined with a normal gamete, a translocation trisomic results. The same is true for **I**. **H** Part-linear orientation with one amphitelic centromere, resulting in irregular segregation, with the possibility of forming a translocation trisomic. **J** Part-linear segregation leading to the combination of two normal with one translocation chromosome. If the resulting gamete is combined with a normal gamete, a tertiary trisomic results

As long as the heteromorphic trivalent formed by a telocentric and two metacentric chromosomes orients alternately, the segregation will result in gametes with the normal complement and gametes with an extra telocentric. With linear orientation, however, a primary trisomic may result.

6.2.2.3 Tertiary and Translocation Trisomy

6.2.2.3.1 Types, Origin

In *tertiary trisomy*, a translocation chromosome is the extra chromosome. The term is from Blakeslee, who found this type as the third type after first finding primary trisomics in *Datura stramonium*, and subsequently secondary trisomics. This terminology is well established, but for new trisomic types terms are used that correspond more with the chromosomes and the further conditions involved. For each interchange, two tertiary trisomics are possible.

In *translocation trisomics* one of the four chromosomes involved in a translocation complex is the extra chromosome, in the background of a translocation heterozygote. With respect to total gene constitution, therefore, a translocation trisomic can be equivalent to one of two primary trisomics or one of two tertiary trisomics.

Tertiary trisomics almost exclusively originate from non-alternate segregation of a translocation complex in meiotic anaphase I. One of the translocated chromosomes moves to the same pole as the two normal chromosomes (Fig. 6.11). The *translocation trisomic* has a comparable origin: a normal chromosome moves to the same pole as the two translocation chromosomes (Fig. 6.11). Neither a primary trisomic nor a translocation trisomic with a translocation chromosome as the extra chromosome is directly produced by irregular segregation in a translocation heterozygote. A translocation trisomic with a translocated extra chromosome in the translocation background is produced for instance by crossing a tertiary trisomic with a translocation carrier.

6.2.2.3.2 Relevance

Tertiary trisomics can play a role in gene localization in specific chromosome segments (8.3.2.1.4.). In addition, they or their derivatives, the compensating trisomics, can be used to construct systems for the maintenance of male sterile lines used in hybrid varieties (12.4.2.2.). Tertiary and translocation trisomics can occasionally be frequent in the progeny of translocation heterozygotes and then disturb their behaviour in gene localization or other studies.

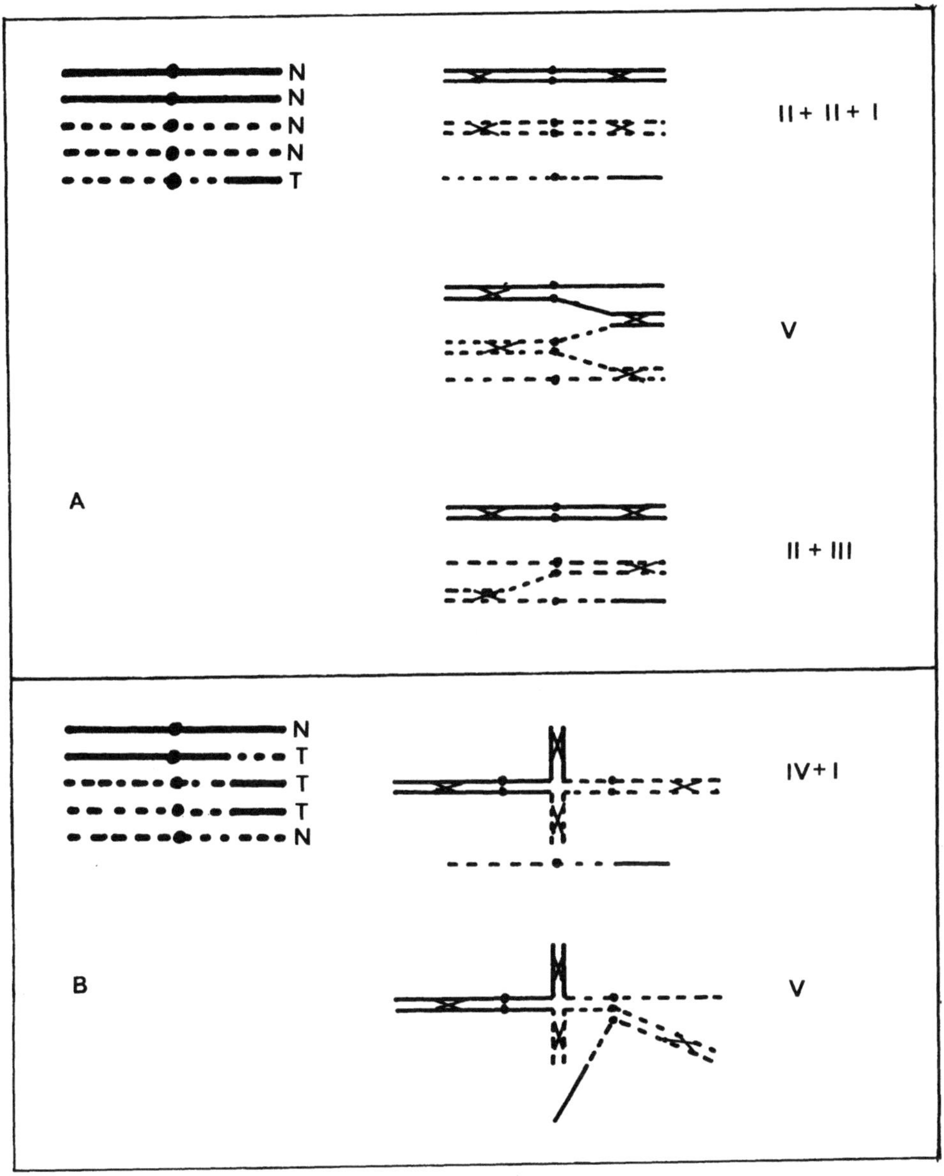

Fig. 6.12 Some of the pairing configurations of a tertiary (**A**) and a translocation trisomic (**B**). Both can form a quinquivalent, but only the translocation trisomic can form a quadrivalent that may become a ring at metaphase I

6.2.2.3.3 Characteristic and Identification

Tertiary and translocation trisomics can be recognized in the *somatic karyotype* as trisomics on the basis of chromosome number, but, as before, only detailed

C-banding or other typical chromosome characteristics can reveal their more exact nature.

The *meiotic behaviour* is rather characteristic, but only quantitative meiotic analysis can distinguish between the tertiary and the translocation trisomic. As in any trisomic, the extra chromosome must choose a partner. In the tertiary trisomic, however, the extra chromosome is composed of two parts, each from a different chromosome. Therefore, it can pair with one of the homologues of two pairs of (normal) chromosomes (Fig. 6.12). The other arms of these chromosomes will each pair with their specific homologues. The combination of the two pairs of normal chromosomes with the extra chromosome between them results in a quinquivalent. The extra chromosome can also be excluded from pairing, when the two normal pairs form bivalents. The third and fourth possibilities are pairing of the translocation chromosome with either one of the normal pairs, the other pair forming a bivalent. This results in a heteromorphic trivalent and a bivalent. Of the nine choices, one results in a univalent with two bivalents, four in a quinquivalent, and four in a trivalent with a bivalent (two types).

With a short translocated segment, the frequency of quinquivalents and one type of trivalent at Metaphase I will be small and the number of univalents increases. In general, however, the univalent frequency is not high and in principle a high recovery rate of the trisomic in the progeny is expected. When there is strong preferential pairing between the fully homologous chromosomes of the two normal chromosome pairs over pairing with the smaller homologous segment of the translocation chromosome, the frequency of formation of two bivalents and a univalent is larger than with random pairing, and the probability of the loss of the extra chromosome is also larger. The quinquivalent has a higher probability of irregular segregation than the smaller configurations. This may result in novel types of trisomics or imbalanced disomics.

Because of their role in balanced cytogenetic systems for maintaining male sterile lines for hybrid varieties, considerable attention has been given to the meiotic behaviour of tertiary trisomics (Ramage 1965; de Vries 1984).

Translocation trisomics have not been studied as extensively (Sybenga 1972). Whereas tertiary trisomics will not give a ring quadrivalent at meiosis, translocation trisomics, especially interchange trisomics, can form one, accompanied by a univalent (the extra chromosome). The translocation trisomic can also form quinquivalents, not only with the extra chromosome in the middle, as in tertiary trisomics, but also with the extra chromosome in the end position. There are six possibilities for this configuration. For the formation of a quadrivalent with a univalent there are two possibilities, and there is one for a trivalent with a homomorphic bivalent. Again, the formation of chiasmata determines which configurations at which frequencies will be observed at metaphase I. The orientation of these configurations determines the composition of the gametes and consequently of the progeny.

6.2.2.3.4 Consequences

When the maximum association is realized at metaphase, all multivalents orient alternately, and the univalent is lost, then the tertiary trisomic will be transmitted to 4/9 of the progeny. Breakdown of a quinquivalent into a trivalent and a bivalent will increase regular segregation rather than decrease it, because trivalents and bivalents tend to have a more frequent alternate orientation than quinquivalents. Tertiary trisomics sometimes have primary trisomics in their progeny.

Translocation trisomics have similar consequences. In addition, alternate orientation (common in certain species) results in a special form of preferential segregation, enhanced by preferential pairing between the completely homologous chromosomes over pairing between chromosomes that differ with respect to the translocation (Sybenga 1972).

Segregation of genes is, in principle, disomic with the complication that a dominant allele in the extra chromosome gives the dominant phenotype to the plant with the extra chromosome, irrespective of the genetic composition of the normal chromosomes.

6.2.2.4 Other Trisomies and Higher Polysomies

The number of different trisomies is theoretically almost unlimited. An inversion chromosome as the extra chromosome makes an inversion trisomic, and it can be a paracentric or a pericentric inversion. These will not be discussed. *Compensating trisomics* (Sects. 6.2.2.4.1 and 12.4.2.2) will be briefly introduced here, as well as the *alien addition* (Sect. 6.2.2.4.2).

Higher polysomies, tetrasomy for instance, are so similar to higher polyploids in chromosome behaviour that a separate discussion is not necessary. They may originate from selfing of a primary trisomic, and tend to be viable in diploids only when the extra chromosome does not have a very pronounced effect. In polyploids, they are more common and in wheat the whole series is available. They are used to identify the members of homoeologous groups of chromosomes (Sect. 9.2.2).

6.2.2.4.1 Compensating Trisomics

In a compensating trisomic one normal chromosome is replaced by two structurally altered chromosomes. Neither one carries the complete genetic material of the chromosome they replace but together, they contain all its genetic material, and more (Khush 1973). There are two main types: (1) the compensating pair consists of two translocation chromosomes of different origin (Fig. 6.13A); (2) the compensating pair consists of a translocation chromosome

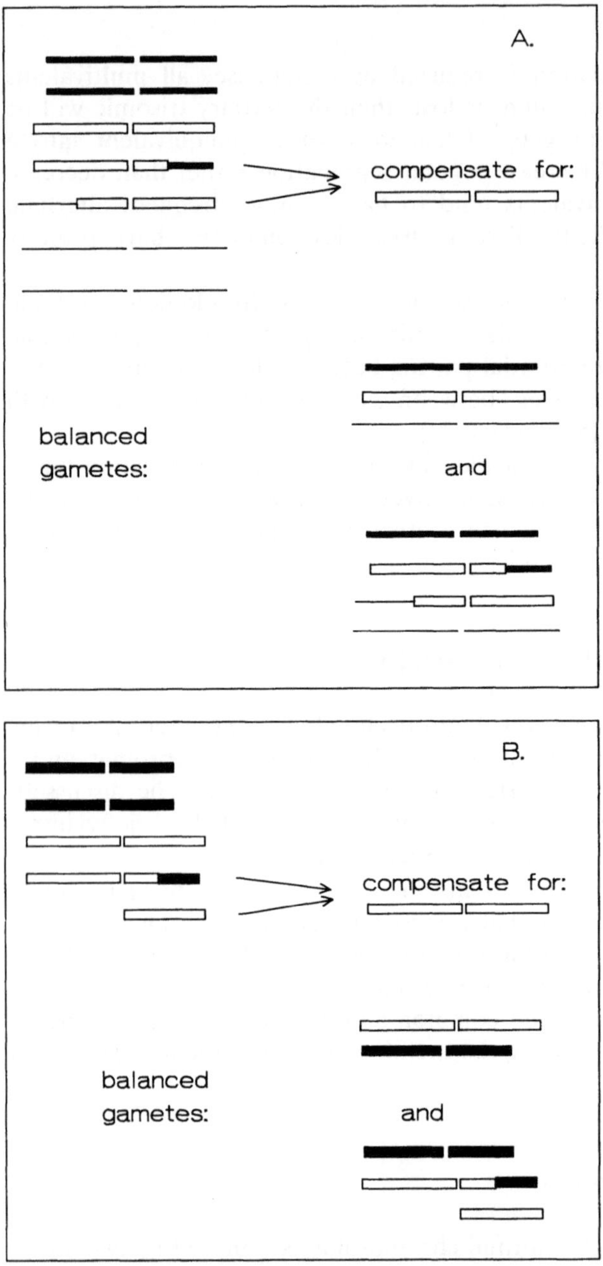

Fig. 6.13 Compensating trisomics. **A** Two complete translocation chromosomes derived from two different translocations compensate for one normal chromosome: both are necessary for a complete set of genes, but there is an excess of chromosome material. Balanced gametes are formed regularly, but their frequency depends on the meiotic behaviour of the rather complex configuration. There are two types of gamete: a normal set of chromosomes and the compensating set. **B** The absence of one normal chromosome is compensated by a translocation chromosome and a telocentric: both are necessary and again there is an excess of chromosome material. This is meiotically stabler than A

and a telocentric (Fig. 6.13B). The latter is meiotically the more stable of the two (de Vries 1985).

A compensating trisomic is constructed from the combination of two translocations or a translocation and a tertiary trisomic, or a translocation and a telocentric for the correct arm. In the selfed or cross progeny the compensating trisomics must be selected. If a marker is present, part of the population can be discarded after a simple classification, but among the rest a cytological test is usually required. Chromosome number and, if possible, the type of chromosomes (translocation chromosomes, telocentrics) are first checked in the karyotype. The characteristics of the somatic karyotype are shown in Fig. 6.13. Subsequently, the presence of the compensating trisomic is tested by studying the meiotic configurations. The most critical meiotic configurations can differentiate between compensating and tertiary or other trisomics. Usually, however, the parentage is known, and identification can start from certain presumptions. Nevertheless, a final meiotic test is indispensable (de Vries 1985).

Compensating trisomics have no other obvious use than application in balanced trisomic systems (Sect. 12.4.2.2). A compensating trisomic with two translocation chromosomes is prone to irregularities of segregation, and is difficult to apply in any program.

The segregational behaviour of compensating trisomics is not like that of a primary, secondary, telocentric or tertiary trisomic, nor entirely that of a disomic. If the segregation is such that the complex is maintained intact from one generation to the next, it behaves in principle like a disomic but with more complex linkages. However, the extra material usually prevents male transmission of the compensating complex, but female transmission is not necessarily reduced. The compensating complex carries duplications for several genes, and usually at least one allele is dominant. When the recessive allele is in the normal chromosomal type, in an F2 there is an excess of recessives because of reduced male transmission of the trisomic type. A complication is the ample opportunity to produce new genetically unbalanced types by irregular segregation of the complex. Among these, tertiary trisomics are the most common. Even the less complex telocentric compensating trisomics suffer from instability (de Vries 1985).

6.2.2.4.2 Alien Addition

6.2.2.4.2.1 Types, Origin

In an alien addition a chromosome of another species is present in addition to a full genome of the host species. In a sense it may be considered a trisomic because the extra chromosome is homoeologous at least to some extent with one of the chromosomes of the recipient: otherwise no transfer and/or acceptance would be possible. When evolution and species differentiation have been accompanied by the formation of translocations in one or both species involved, the alien extra chromosome may be a translocation chromosome and the addition in a sense is a tertiary trisomic. It might also be an inversion chromosome relative to the homoeologous host chromosome.

Because during the construction and maintenance of an alien addition, it is often present as a univalent, there is a probability that it suffers centromere breakage, and as a result telocentrics and isochromosomes of alien additions are not uncommon. Basically, therefore, there are five types of alien addition: the normal, the telocentric, the isochromosome, the tertiary and the inversion addition. The first three are the most common.

When an alien addition is present once, it is called a monosomic addition, when it is present twice, it is a disomic addition.

Alien additions will not readily arise spontaneously, but have to be constructed. The complete series of *Secale cereale* (rye) and *Secale montanum* additions to hexaploid wheat have been constructed as well as the additions of several of the related *Aegilops* species, and even of the much less closely related barley (*Hordeum vulgare*). They are somewhat less common for other allopolyploids like oats, cotton and tobacco, and especially for diploids, but several additions of wheat chromosomes to rye have been reported by Schlegel et al. (1986, 1988).

The usual way to make additions is to double the chromosome number of a hybrid between the recipient (host) species and the donor species of the addition. This amphidiploid is subsequently backcrossed to the host. After the first backcross the entire genome of the host is present twice, but that of the donor only once. The latter will be univalent at meiosis. Subsequent backcrossing to the host, or selfing, will produce complete genomes of the host combined with none, one, two or at most a few of the host. If there are two or more, these will be reduced to one or none by another backcross or selfing. The monosomic additions are most readily selected when they carry a dominant marker. They are difficult to maintain because the univalent will tend to be lost. Making a disomic addition requires male transmission in addition to female transmission. The latter is not very frequent, but male transmission is even worse. Yet, in most cases, in allopolyploids disomic alien additions have been obtained. Even these, however, are not completely stable because the pairing and chiasma formation are not always normal in the foreign background. Their maintenance by selfing is not guaranteed. Different additions and different genetic background have different meiotic stabilities.

Provided it is technically possible, a *somatic hybrid* between a somatic (diploid) protoplast and a (haploid) tetrad protoplast (Pirrie and Power 1986) directly replaces the making of a sexual hybrid, doubling it and backcrossing it to the recipient. The subsequent steps for isolating an addition are the same as in the completely sexual approach.

One step further is the *asymmetric fusion* between one normal (recipient) protoplast and a donor protoplast irradiated with several hundreds of Grays of ionizing radiation. The high doses necessary to limit transmission to one single chromosome, normal or rearranged, are so detrimental for the donor nucleus that no viable (asymmetric) hybrid is obtained (*Lycopersicum*: Wijbrandi et al. 1989; *Brassica*: Yamashita et al. 1989). Elimination of a smaller part of the donor genome is hardly better than no elimination. In programs of fusion with irradiated donor protoplasts one of the aims has been to completely destroy the donor nucleus, possibly releasing the DNA that, by transformation, could be inserted into the recipient genome, hopefully including the desired gene (Gupta et al. 1984; Dudits et al. 1987). No definite results have been obtained so far.

Another approach again to the isolation of monosomic additions or even substitutions from a somatic fusion product is by treating it with a spindle-disturbing substance, which may result in somatic chromosome segregation and aneuploid cell progeny. Examples are: simple in vitro culture with an extended callus phase; further, the addition of colchicine, benomyl, para-fluorophenylalanine, chloramphenicol, amiprophos-methyl, CIPC to the medium. However, additions or substitutions from somatic cells have not been reported as yet.

A somatic approach to the direct transfer of single chromosomes between species is the separation of condensed metaphase chromosomes from cells of the donor species, and sorting them in a flow cytometer (de Laat and Blaas 1984) or other sorting devices to accumulate the chromosome involved. The chromosomes can be isolated as such, or a dividing cell can be treated with amiprophos-methyl. This separates the chromosomes which subsequently form micronuclei, which can be sorted in a flow cytometer (de Laat et al. 1987; Ramulu et al. 1988). Separation of chromosomes by pulsed-field gel electrophoresis has been quite successful in yeast, but has not yet appeared possible with the much larger plant chromosomes. Fusion of isolated chromosomes with recipient protoplasts in polyethylene glycol (Szabados et al. 1981; Griesbach et al. 1982; Ramulu et al. 1988) is possible, but electroporation of protoplasts may turn out to be more promising. The host cell must be in mitosis or nearly so because differences in condensation caused by differences in the cell developmental state are destructive for the chromosome. Selective systems for isolating cells with the extra chromosome are a prerequisite as long as the success of transfer is too limited for selection on a cytological basis. Occasionally, visually selected fusion cells can be isolated and grown separately or in a nurse culture (Hein et al. 1983). Microinjection of chromosomes into cells has appeared possible in animals, but plant cells (protoplasts) with chromosomes injected into them have not yet been found to proliferate.

6.2.2.4.2.2 Relevance

Alien additions play an important role in chromosome manipulation, both in the transfer of alien genes, where they are an intermediate step (Sect. 10.4.4.2) and in systems for maintaining male sterile lines for hybrid varieties (Sect. 12.4.2).

Disomic additions would, in principle, be interesting in allopolyploids to directly introduce an alien dominant gene, for instance disease resistance into the host. Both the generally negative effect of an extra pair of chromosomes and the usually insufficient stability have made such attempts unsuccessful (Wienhues 1966; cf. Sect. 10.4).

6.2.2.4.2.3 Characteristics, Identification and Consequences

In the *somatic karyotype* the monosomic or disomic additions are only distinguished from other types of polysomics when the chromosome can be identified, either by specific gross structural characteristics, by its C-banding pattern, or by molecular probes. Rye chromosome additions to wheat are usually readily recognized because they have pronounced terminal C-bands which are absent from wheat chromosomes. Occasional variants of rye chromosomes, however, may not have sufficient heterochromatin to differentiate them readily from some of the wheat chromosomes. With careful banding or with specific rye DNA probes, identification remains possible.

In *meiosis* the alien addition forms a univalent when present singly, and a bivalent when disomic. As mentioned above, alien chromosomes may fail to behave regularly in meiosis and may form univalents even when present twice, more frequently than the host chromosomes.

Monosomic alien additions are variably and often poorly transmitted to the progeny; disomic additions, on the other hand, are transmitted considerably better, but still not perfectly. The inheritance is not trisomic (for monosomic additions) because the extra chromosome does not pair at random with the homoeologues of the host species and usually does not simply replace it in segregation. Any dominant or co-dominant genes in the extra chromosome will be expressed as long as the chromosome is present. The correlation between the presence of the extra chromosome and the expression of the dominant allele locates the gene in the chromosome. This has been a very important method to locate specific genes, especially co-dominant genes coding for enzymes that differ slighty between the host and the donor of the addition.

Monosomic addition chromosomes may, through centromere breakage of the univalent, form telocentrics and isochromosomes. These are found regularly, although not always at high frequencies, in the progeny of especially monosomic additions.

Alien additions, when crossed with homoeologous nullisomics or monosomics, can subsequently be made to replace the homoeologous chromosome of the host. This then represents a homoeologous alien substitution. The way this is done and used in chromosome manipulation is discussed in Section 10.4.4.1.

Chapter 7

Diagnosis: Identifying Cytogenetic Causes of Variants of the Karyotype and the Generative Cycle

7.1 Diagnosis: The Collection of Specific Information: Context, Objectives and Means

When the role of cytogenetics in plant breeding was discussed in Section 1.2 the importance of *information* (*general* as well as *specific information*) was mentioned. The present chapter on the diagnosis of variants of the karyotype and of the reproductive cycle deals with methods for collecting specific cytogenetic information about specific material, by making use of their typical karyological and meiotic characteristics. The analysis should include any additional, relevant information available: the context in which the original deviation was found, the history of the material, morphology and biochemistry of the plant, segregation of morphological, biochemical or molecular markers, recombination, fertility, etc. It may even be necessary to carry out additional experiments, and although it may seem superfluous, it is important to check for experimental errors, in classification, administration, pollination, etc. Together these are the diagnostic *tools*.

In many cases the plant on which (or on the progeny of which) the original observation was made, and which prompted the analysis, is not available. It may have died, or has been discarded. Yet, especially when deviations of the generative cycle are the objects of a diagnostic analysis, this would be the plant of which karyotype and meiosis would be the most informative. It is then necessary to resort to sister plants, or selected progeny that show the same deviant behaviour or characteristics, or from which the deviation in the original deviant can be reconstructed.

Different *types of abnormalities* for which a cytogenetic diagnosis can be considered are listed in Table 7.1.

Most of these concern the generative reproduction. Three phases can be distinguished: (1) *meiosis*, resulting in haploid spores; (2) the *haplophase*, where *gametes* are formed; (3) *fertilization* and the development of *embryo* and *endosperm*, resulting in mature seeds. The haplophase might be considered a separate (haploid) somatic phase, and embryo development the initial stage of the diploid somatic phase, but both can also be seen as part of the generative cycle.

Table 7.1. Abnormalities of karyotype and generative reproduction that would prompt a cytogenetic diagnosis

a. Karyotype abnormality observed incidentally
b. Meiotic abnormality observed incidentally
c. Repeatedly occurring somatic malformation or other unexpected segregant in a progeny
d. Repeatedly occurring karyotype abnormality within a progeny
e. Repeatedly occurring meiotic abnormality within a progeny
f. Reduced fertility
g. Unexpected or disturbed segregation ratio
h. Altered recombination frequency

It should be noted that diagnostic difficulties may lead to diagnostic errors. These occur more frequently than often assumed, and are made not only when the criteria used are insufficiently discriminating. Misclassification in karyotype analysis, for instance, is not uncommon. As a result, plants with a rearrangement may be classified as normal, and, exceptionally, the reverse. Detailed observations, if necessary, involving the measurement of the chromosomes of several cells and applying a chromosome banding system, would often be a solution when immediate recognition of potentially deviant chromosomes is doubtful, but is usually considered too laborious and, consequently, too costly for most purposes. More discriminating molecular techniques are usually not even considered. Errors are the less probable, the more striking the morphological change caused by the rearrangement, and especially when the aberration is numerical. In species with large chromosome numbers, however, even the simple counting of chromosomes may result in errors.

A second source of error has its origin in disturbance of meiosis, less frequently of mitosis, as a consequence of the initial aberration. Sometimes new variants result which are sometimes readily recognized, but sometimes not. One example is the origin of a duplication from a simple terminal translocation heterozygote (Sects. 5.2.2 and 5.4.1.4). The number of chromosomes is not increased. When the classification is based on the morphological characteristics of the chromosome that carries the translocated segment, the duplication is not immediately distinguished from the translocation. In other cases, the duplication may not be distinguished from one of the normal chromosomes. When the probability of secondary variants is real, a more careful classification is required than normally considered necessary.

Observations on male meiosis are not sufficient when deviations in maternal meiosis are expected. Female meiosis, however, is much less accessible to analysis than male meiosis. A number of techniques have been developed, some rather recently, specifically for the study of female meiosis in plants (Jongedijk 1987).

Haplophase and Endosperm Development. The importance of the *haplophase* in plants is considerable, but the only aspect (briefly) considered here is its role in determining the competitiveness of the gamete in fertilization: *certation*.

Table 7.2. Context in which the original observation is made

a. Nothing particular, the deviation was a surprise
b. The deviant plant was found in a population scored for another segregating, karyological deviation
c. The plant is a hybrid
1. Within species
2. Between species
d. The plant has been treated:
1. For chromosome doubling
2. For haploidization
3. For induction of mutations
4. For affecting its physiology
e. The plant was found in a program of chromosome manipulation

There are, normally, very specific relations between the numbers of genomes of the parental plant, embryo and endosperm. Deviations from these relations are often detrimental for the proper interaction between these tissues. They may result in disturbance of the *development of the endosperm* and failure of normal viable seed to develop. This phenomenon has received considerable attention. It is important in artificial polyploidy, more specifically in newly developed allopolyploids, and in some species hybrids. Developmental abnormalities of the endosperm and possible causes will be briefly considered when relevant.

Context in Which the Abnormality Is Found. The *context* in which the deviation is found can be relevant for its identification (Table 7.2). It is often the first source of information available, but it may be misleading. In the present discussion the context will be considered after other observations have given an impression of what the potential cause of the observed irregularity is, and mainly as a check of its probability.

The following situations may be distinguished:

1. No abnormality is expected, and it comes as a surprise. Occasionally, an environmental effect may be involved, for instance affecting fertility; even mutagenetic effects are not excluded. Several pesticides are mutagenic and have been shown to inadvertently induce chromosomal aberrations, but naturally occurring poisons, such as colchicine from *Colchicum autumnale* in pastures, hardly have an opportunity to cause gene or genome mutations (Sybenga 1956). Spontaneous mutations are rare but do occur, including chromosomal aberrations. In a normal wheat field about 1 out of 100 plants carries a chromosomal deviation, most frequently monosomy ($2n - 1 = 41$). A major reason why other crop plants, especially diploids, tend to show lower frequencies of spontaneous aberrations is their lower tolerance rather than a lower initial frequency. Aberrations involving the karyotype or meiotic behaviour are usually not primarily detected because of their deviating plant morphology, although they are often at least slightly off-type. They are primarily found in special cytological checks, or when they happen to be used as

a parent of a hybrid and either in the F1 or in later generations abnormalities are observed.

2. The presence or absence of a segregating chromosomal construction is scored on the basis of the most differentiating karyological or meiotic characteristic (Sect. 8.3.2). A complication arises when the meiotic behaviour of the variant is not normal and novel abnormalities appear, sometimes even in considerable frequencies. These then must be distinguished from the original deviation and identified. The secondary deviants are sometimes predictable and can then be identified simply on the basis of the information given in Chapters 5 and 6. Occasionally, the secondary variants are not predictable or not readily distinguished from the original deviation or a possible alternative, then a diagnostic approach is necessary. Entirely unexpected variants may also occur, then the information available from the "context" is misleading.

3. The deviant plant is a hybrid, or is found among the progeny of a hybrid, between genetically distant forms of one species or between different species. Abnormalities are not unexpected, but their character is not predictable. The closer the parents are related, the fewer the complications, but even between cultivars of the same species, karyotype differences are found. Differences between species, in addition to differences in the karyotype, may also involve genes that affect the reproductive processes, including meiosis.

4. A treatment has been applied, for instance for chromosome doubling, for haploidization, to induce mutations. In vitro culture is a treatment.

5. A special karyological construction has been or is being made from the combination of different specific variants, with the purpose of isolating a specific new chromosomal construction. (Chaps. 10, 11 and 12). The progenies raised during the process must be scored for this particular type, and in principle all that is required is to look for the simplest and most discriminating somatic or generative characteristic available. In this case, however, the desired type may not be expected to be particularly frequent and several other, undesired deviations may have been induced simultaneously. Some of the unwanted changes may be mistaken for the desired variant.

In principle, the information provided in Chapters 4, 5 and 6 is sufficient to identify unintended, extra variants and for distinguishing them from a desired form, but in the present chapter this identification is approached systematically.

7.2 The Diagnostic Strategy

It may be repeated that it is good or even essential to check first the *experimental* and *administrative procedure*: can the deviation be due to errors in pollination, in the use of insufficiently checked parents, in the choice of parents, etc.? Are the aberrant plants the result of contamination?

The information available at the start of the analysis consists essentially of the observation of the *specific* morphological, karyotypic, generative, including meiotic and segregational *abnormality* (Table 7.1), the *context* in which it was found (Table 7.2) and the genetic *history* of the plant or the population, insofar as it can be reconstructed. This history (Does the plant have a species hybrid among its ancestors? Have any of its ancestors been irradiated? Is it from a highly inbred line of a cross-breeder? etc.) is important initial information.

Although the most logical *first check* would seem to be one of the *karyotype*, for practical reasons it is often a check of *meiosis*. Meiotic observations are generally the most informative observations and include important elementary karyotype information: chromosome number and gross structural morphology. There are several cases in which the structure of the chromosomes is not normal, but where no irregularities are observed in the gross karyotype morphology. Then the identification of the cause of the abnormality observed starts with a meiotic analysis. It is also possible that a mutant of meiosis or another phase of the generative reproduction is involved. Many of such different cases can be distinguished by their specific meiotic behaviour. It is only natural, of course, to use somatic karyotype information when it is readily available or when meiotic observations are difficult to obtain. However, when meiosis can be studied, it is the most direct and most informative source of information. On the basis of meiotic observations, a decision can be made as to which aspects of the karyotype should be analyzed in detail, if any. When for practical reasons a karyotype analysis precedes the meiotic analysis, meiotic analysis can profit from the information obtained. This would be the case, for instance, when the original deviant plant is not available, and new progeny is grown. Then a rapid karyotype check of the seedling would be useful as a preliminary to meiotic analysis, if the deviation suggests that a karyotype abnormality may not be excluded.

Diagnosis thus starts with a check of the karyotype (in meiosis or directly in somatic mitoses) for *karyotype variants*. For each karyotype variant, different possibilities of *meiotic behaviour* are discussed separately (Table 7.3), which can be compared with the observations made. If necessary, a more *detailed karyotype analysis* may be made. Next, the compatibilty of the results of the meiotic analysis with the *context* and the genetic *history* of the material is considered, followed by the *expected consequences* for fertility, recombination, possible abnormal segregation ratios and segregants. These can be compared with observations on the material.

When a satisfactory conclusion cannot be obtained on the basis of meiotic observations, an analysis of the *haplophase* and the early *embryonic development* and, possibly even additional experimentation, may be necessary (test crosses, etc.). The latter will be desired especially when not only the character of the aberration but also the specific chromosomes involved must be known.

Table 7.3. Aspects of karyotype, meiotic behaviour, haplophase, embryonic development and their consequences, important for determining the cause of a karyotypic or generative abnormality

I. Karyotype
  a. Chromosome number as in normal parent(s)
    1. Chromosome structure not obviously different from normal parent(s)
    2. Chromosome structure suggests change
  b. Chromosome number different from normal parent(s)
    1. Aneuploidy
      a. Hyperploidy, polysomy: one or more extra chromosomes, normal or structurally deviant
      b. Hypoploidy: one or more chromosomes less in a normal or structurally deviant karyotype
    2. Euploidy
      a. Gametic chromosome number
      b. Additional genomes
        1. Triploidy
        2. Double the original genome number
        3. Other ploidy level

II. Meiosis
  a. Types of configurations
    1. Diakinesis/metaphase
      a. No abnormalities seen
      b. Univalents
      c. Heteromorphic or otherwise abnormal bivalents
      d. Multivalents
        1. Trivalents
        2. Quadrivalents
        3. Higher multivalents
    2. Anaphase I and later meiotic stages
      a. Anaphase I
        1. Chromatid bridges with fragments
        2. Chromatid bridges without fragments
        3. Chromatid loops
        4. Unequal chromatids in the same chromosome
        5. Laggards
      b. Anaphase II chromatid bridges and laggards
      c. Abnormal cellular processes
        1. First division restitution
        2. Second division restitution
        3. Other abnormalities
  b. Distribution of characteristic configuration(s) over cells
    1. Maximum of one per cell
      a. Always one
      b. Less than one per cell, distribution skew, suggesting one per cell as a maximum
    2. Zero to many per cell
      a. Distribution does not deviate from random
      b. Distribution skew

III. Haplophase and early embryonic development
  a. Certation
  b. Endosperm failure

IV. Consequences of abnormal generative behaviour
  a. Altered fertility
  b. Altered recombination
  c. Unusual segregation
    1. Segregation ratio
    2. Abnormal or unexpected segregants

7.3 The Diagnostic Use of Meiotic Behaviour and Configurations: Types and Distribution; Causes and Consequences

7.3.1 Stages

The main "*tools*" for the diagnosis of an unidentified abnormality in the karyotype or in the generative cycle, are the specific characteristics of the aggregates of chromosomes (*configurations*: univalents, heteromorphic bivalents, different types of multivalents) observed at diakinesis-metaphase I of meiosis. They result from chromosome pairing and chiasma formation during the first meiotic stages. In addition, the characteristics of individual chromosomes or chromatids at later stages (from anaphase I onward: chromatid bridges with and without fragments, chromatid loops, unequal chromatids of one chromosome) provide information on the properties of the chromosomes involved. If possible, the information from different stages is combined. A special category includes genetically conditioned abnormalities of meiotic behaviour and of meiotic cell development (asynapsis, desynapsis, different types of nuclear restitution, failure of cytokinesis).

Chronologically, meiotic pairing comes first. However, the pairing stages are generally not readily accessible. Moreover, the difficult technique, and the absence of clear morphological markers, make it advisable to study other stages first and analyze pachytene and especially SCs only when strictly necessary. Similarly, because metaphase I configurations are better accessible and more conspicuous and in essence provide not much less information than diplotene and diakinesis configurations, the latter will be considered only when necessary.

7.3.2 Basic Configuration

The configuration as observed in a particular cell at metaphase I is not necessarily the most characteristic configuration for the chromosomal constitution. For two reasons the observed configuration may be different, usually less complex or smaller than the most critical configuration.

1. It has resulted from the breakdown of larger pairing configurations as a consequence of limited chiasma formation. An interchange heterozygote, for instance, may have a quadrivalent in some cells, but a trivalent with a univalent, or two bivalents, in other cells, when certain chromosome segments fail to have chiasmata (Sect. 5.4.1).

2. The configuration observed in a specific cell is one of different alternative pairing configurations, other alternatives being combinations of a larger number of chromosomes. Tertiary trisomics, for instance, may form quinquivalents in some and trivalents with bivalents in other cells, with the same number of chiasmata (Sect. 6.2.2.3).

Thus, when metaphase I cells are analyzed, after a first check of the karyotype, the next check is on the presence of *other*, larger *configurations* in the same cell or in sister cells. However, it is possible that different configurations in different cells have an independent origin in an independent, simultaneous event. The best way to distinguish the different events is by first identifying a "primary" configuration and subsequently determining if any smaller configurations are present that cannot be accounted for by the "primary" deviation. However, when such secondary phenomena occur, and especially when different "primary" deviations occur simultaneously, the identification may become complex.

7.3.3 Distribution of Configurations

The meiotic configurations associated with the different karyotype and cell-biological variants have been discussed in Chapters 5 and 6. With respect to these configurations and other relevant chromosome characteristics, two aspects are important:

1. Their *type* (number of chromosomes involved, the way these are associated, chromatid size, etc.).
2. Their *distribution* over cells.

A test of the distribution of the configuration over cells is a relatively simple check to distinguish between two different situations: Do all chromosomes of the complement have a similar probability of forming part of the configuration or are these probabilities significantly different per chromosome? This differentiates between the situation in which an overall cellular or genomic effect is present, and that in which an abnormality involves only one or a few chromosomes.

The check on distribution consists of first estimating the average frequency of the configuration; next to construct an expected (usually binomial) distribution on the basis of the average just estimated, and finally to compare the expected with the observed distribution in a χ^2 test. Tests on distribution should be applied with caution. There may be simple disturbing factors that invalidate the conclusions.

7.4 The Diagnostic Protocol

The diagnostic protocol follows Table 7.3. The first subdivision is according to the gross karyotype. Next, the possible different meiotic diakinesis-metaphase I configurations are considered. Here, the variants are listed which could cause these types and distributions of configurations, on the basis of the karyotype as given. For each possible aberration suggested by the diakinesis/metaphase observations, the expected consequences for anaphase I and later are given and, if relevant, also for pachytene. Further, the details of the corresponding karyotype, and the consequences for fertility, segregation and recombination are shown (Table 7.1). These can be compared with the observations made, which should lead to a reasonably reliable diagnosis.

7.4.1 No Chromosome Number Deviation

The difference between the two options (see Table 7.3): I.a.1 (*Chromosome structure* obviously *normal*) and I.a.2 (*Chromosome structure deviating*) depends strongly on the resolving power of the gross karyotype analysis made at meiosis or during a simple somatic karyotype check. Therefore, when the first option is discussed, the possibility must be left open that a difference in chromosome structure has been overlooked. The two categories will be treated together, but both must be considered.

7.4.1.1 Meiosis: Diakinesis/Metaphase I: No Obvious Deviations

Abnormalities in segregation, reduced fertility, appearance of abnormal plants, etc. may well occur without gross karyotype and meiotic abnormalities in the deviant plants or their parents. This does not necessarily mean that the origin is not to be found in cytogenetic irregularities, but it will be difficult to trace it with cytogenetic approaches. There may be slight deviations, not observed due to statistical reasons or because the analyses as described below do not have sufficient resolution. If feasable, a thorough karyotype analysis may occasionally give some clue. Making a hybrid with a normal stock and analyzing meiosis and segregation in the F1, following the protocol given below, may work. The different possible situations will not be discussed. There may be other reasons for unexpected abnormalities to appear. Suddenly reduced fertility, for instance, may result from the sensitivity of special genotypes to unusual environmental conditions. There may be desynapsis or irregular post-meiotic processes. The origin may, for instance, be in an alien cytoplasm into which a nuclear genotype was brought to produce cytoplasmic male sterility and which has subsequently been used for the development of non-hybrid varieties, after a general restorer has been introduced. If the restorer is sensitive to unusual temperatures, sterility may suddenly appear. Analysis of mitochondrial DNA may give a clue, but there is little reason to look further for typical cytogenetic approaches.

7.4.1.2 Univalents (Table 7.3: II.a.1.a)

Only univalents are considered that are not the incidental result of breakdown of larger configurations or that are alternative pairing configurations for more complex basic associations.

Univalents are formed by a variety of mechanisms and may be found at high frequencies, even including all chromosomes in the meiocyte, or at lower frequencies, down to only one, and then accompanied by normal bivalents. They may occur in rather constant or in variable numbers. The reasons for univalents to be formed may be chromosomal (due to properties of individual chromosomes), genomic (for instance, lack of homology between entire genomes) or cellular (genetic or environmental conditions affecting cellular processes: for instance, asynapsis and desynapsis). An even number of univalents is observed, i.e. the "configuration" is a pair of univalents.

Always one pair per cell (Table 7.3: II.b.1.a).
Monosomic alien substitution: One normal chromosome is absent, the homologue has been replaced by a chromosome from an alien species with insufficient homology to form a bivalent with the (original) single chromosome. When homoeologous, they may pair occasionally and form a bivalent in a small percentage of the cells. In addition, the substitution may not be genetically perfect and affect the physiology of the cell, resulting in a slight increase in the spontaneous univalent frequency. Then the distribution of the univalents over cells deviates significantly from random, with the class of one pair predominating.

Karyotype details: C- or other chromosome banding may demonstrate a morphological difference between the two non-homologous chromosomes, even at meiosis. This is the case, for instance, with monosomic rye substitutions in wheat ($2n = 6x + 1 - 1 = 42$, cf. Sect. 10.4.4.2 and Fig. 4.3C, where the difference in banding pattern between the different species is clearly visible). *Anaphase I*: The univalents lag at the equator and get lost, or segregate at random to the poles or break at the centromere and produce telocentric chromosomes, or isochromosomes, or centromere translocations between the two chromosomes involved. *Consequences*: Reduced fertility, reduced recombination and abnormal segregants among the progeny, resulting from aneuploidy, occasionally including telocentrics, isochromosomes or, rarely, centromere translocations. *Context*: The monosomic substitution would be expected in a chromosome manipulation program (Table 7.1e), where it would be considered a desired intermediate product rather than an aberration. The unexpected appearance (Table 7.1a) of a monosomic (or heterozygous) substitution has been observed in breeding programs involving certain wheat varieties in which chromosome pair 1B has been substituted by rye 1R. When these are crossed with normal wheat, two univalents are formed, one of which is 1B, the other 1R. The origin of most single-chromosome substituted wheat varieties is a hybrid between wheat and the wheat-rye amphidiploid triticale made to introduce disease resistance from rye into wheat.

The derived monosomic substitution may have been initially encountered in almost any of the ways listed in Table 7.1: as a karyotype abnormality, a meiotic abnormality, unexpected segregants (aneuploids) in the progeny, reduced fertility, reduced recombination, etc. There are few reports of alien substitutions outside wheat (Sect. 10.4.4.1). They have not been described with certainty for diploids, because they are expected to be imbalanced. The *identification* of the substitution, if not possible by a karyotype analysis, or when not known from the history of the material, requires complicated test crosses (Sects. 6.2.2.4.2 and 10.4.4.1).

Often one pair of univalents per cell, distribution skew, suggesting one per cell as a maximum (Table 7.3: II.b.1.b).

1. Clear cases have been reported by Parker (1975) for *Hypochoeris radicata* (2n = 16), and by Tease and Jones (1976) for *Crepis capilaris* (2n = 12). The *karyotype* is completely normal, one pair of chromosomes that could be individually identified even without C-banding was almost consistently univalent. *Pachytene* did not show reduced pairing. *Anaphase I* predominantly, but not always normal, univalents usually segregating to different poles. *Consequences* few; there was some effect on fertility. The mechanism has not been explained. *Context* spontaneous, discovered by incident. For plant breeding, the case is of little practical interest because the phenomenon has not been reported for cultivated plants and the effect is limited.

2. *Disomic* (or homozygous) *substitution* of an alien chromosome. Pairing and chiasma formation of the alien pair is not as effective in the genetic background of the host as in its own background. *Karyotype* details: with detailed banding or molecular probes the difference between substitution and normal (absent) chromosomes can sometimes be made visible (see above: always one pair of univalents). *Pachytene*: Usually no special features. *Anaphase I*: The univalents may lag and get lost or segregate at random to the poles, occasionally the centromeres split. *Consequences*: A slight reduction in fertility, and the appearance of occasional aneuploids for the chromosome involved, including rare derived types (telocentrics, isochromosomes). The *identification* of the substitution, if not possible by karyotype analysis or from the history and context, requires complicated test crosses (Sect. 10.4.4.1). A monosomic substitution by a chromosome of a donor related to the host may give a similar univalent distribution. *Context*: Derivatives of chromosome manipulation programs.

3. *Translocation homozygote*. When one translocation chromosome has become exceptionally small, it may (in the homozygote) form significantly fewer chiasmata than normal. *Karyotype*: The translocation is occasionally recognizable because of conspicuous changes in the size of the trans-

located chromosomes and/or an altered C- or other banding pattern. The chromosomes may even be recognized at a first inspection of meiosis, but not frequently. If the overall chiasma frequency is somewhat below normal, as for instance in inbreds of outbreeding species, where occasional other univalents occur in addition to the pair considered, the condition may still be recognized as caused by a single chromosome pair by a deviation in the univalent distribution from binomial: the class with one univalent pair per cell is in excess. However, this is not a very sensitive test. An example of a rye interchange homozygote with relatively frequent formation of a single univalent pair is given by Sybenga (1958). *Pachytene*: Not necessarily different from normal, except possibly an occasional pair of univalents when the translocated chromosomes are exceptionally small. *Anaphase I*: Occasional laggards, random segregating or amphitelically splitting chromosomes, or infrequent centromere breaks. *Consequences*: Slightly reduced fertility; occasional aneuploid segregants in the progeny, occasionally with morphological consequences, sometimes up to several percent, mainly monosomics in allopolyploids, trisomics in diploids; reduced crossing-over in the small translocation chromosome. *Context*: Spontaneous at very low frequency; more frequent in generations following mutagenic treatment; in chromosome manipulation programs.

These asymmetric translocation homozygotes may have practical consequences when the small chromosomes carry (semi)dominant dose-sensitive genes affecting morphological characters.

A test cross with a normal individual will reveal the translocation as a heterozygote in the F1 hybrid, and a series test crosses with a tester set may *identify* the *chromosomes* involved.

Univalents: *Varying number* per cell, *distribution random* over cells (Table 7.3: II.b.2.a).

The random distribution suggests that all chromosomes have approximately the same probability of forming univalents. There are basically two different possibilities: (1) a genetic or externally induced *cellular* effect (asynapsis, desynapsis) or (2) a *genomic* cause (lack of homology). The difference between the two cases is not simply established without knowledge of the history of the material or detailed experimentation. The history of the material and, if necessary, a detailed *karyotype* analysis will usually distinguish between effects of hybridity and those of desynapsis. In a hybrid, either the genomes may be insufficiently homologous or genetic imbalance may cause chromosome pairing or chiasma formation to be disturbed. This results in desynapsis even when there is sufficient homology for pairing. There may be sufficient genetic variation to cause variable behaviour between plants or hybrid populations. There may also be increased sensitivity to environmental factors causing variation in the frequency of univalents.

The possibility that a spontaneous interspecific hybrid is involved cannot really be excluded when the species is an outbreeder, with potential alien pollen parents present in the neighbourhood, or when the origin of the plants is uncertain.

In the case of asynapsis or desynapsis the expression may be complete (all chromosomes are univalents) or partial (not all chromosomes are univalent, or even only a few). In both cases the distribution of univalents is, in principle, random. Asynapsis and desynapsis may be monogenically (usually recessive) or polygenically conditioned. *Inbreeding* of outbreeders usually leads to a lower level of chiasma formation than normal in the outbreeder and is expressed as a higher univalent frequency. It is genetically determined (Rees and Thompson 1956; Sybenga 1958) and might be considered a mild form of polygenic desynapsis. If the history of the material or the segregation does not give a clue as to which mechanism is involved, a simple genetic experiment may be necessary. It could consist of crossing the deviant plant (if fertile, otherwise a number of sister plants) and studying the F2. In case the F2 segregates similar deviants, monogenic synapsis or desynapsis is involved. In case no clear segregation is observed, a polygenic system may be assumed. Environmental effects can be checked by changing the environment.

Consequences: Reduced fertility, reduced recombination, disturbed segregations and abnormal segregants among the progeny, primarily aneuploids, in part derived aneuploids (telocentrics, isochromosomes and centromere translocations caused by centromere breakage) are common consequences of univalent behaviour at *anaphase I*. In special situations a high degree of asynapsis or desynapsis will result in nuclear restitution. When followed by parthenogenetic embryo development, seed with the original unreduced chromosome number is produced with almost the same genetic composition as that of the parent.

Univalents: *Distribution skew* (Table 7.3: II.b.2.b), with the possibility of more than one univalent per cell, still considering the normal chromosome number, is only realized when different causes of univalent formation are combined. These combinations will not be considered separately here.

Initial observation (Table 7.1), reason for a meiotic analysis: unexpected segregants (c), including karyotype aberrations (d), reduced fertility (f) or altered recombination (h).

7.4.1.3 Heteromorphic or Otherwise Abnormal Bivalents (Table 7.3: II.a.1.b)

Heteromorphic bivalents are composed of two chromosomes of different shape and/or size. As with univalents, the heteromorphic bivalent can be the breakdown product of a larger complex. In that case there should be larger configurations in other cells, and other breakdown products in the same cell. This is not considered here.

Bivalents composed of two different chromosomes are not always visibly heteromorphic. They may be symmetric and either apparently normal or abnormally shaped. Regarding heteromorphic or otherwise abnormal bivalents, there may be many in the same cell, but often there is only one. True heteromorphic bivalents are always heteromorphic, or replaced by two univalents. The degree to which the heteromorphism is recognized depends on the morphological difference between the two chromosomes involved.

Heteromorphic bivalents: Always one or a *maximum* of *one* (Table 7.3: II.b.1.a).

1. *Deficiency* and *duplication heterozygotes*. In heterozygotes of *deficiencies* and also of *duplications* (tandem, inverted or within-chromosome displaced), when large enough to result in chromosomes of visibly different size, heteromorphic bivalents or their breakdown products are consistently formed from pachytene to metaphase I (Sects. 5.1 and 5.2). Duplications displaced to other chromosomes, large enough to be morphologically visible and especially when distally located, will often form occasional multivalents, and will, therefore, primarily be discussed further in later sections. Both for deficiencies and duplications there is usually only one heteromorphic bivalent per cell. Deficiencies and duplications large enough to produce heteromorphic bivalents can usually be recognized in the *karyotype*, either in mitosis with the usual somatic markers, or at pachytene when distinct patterns of chromomeres are available, as in maize. They are rare. Polymorphisms for small heterochromatic C- or other types of bands, however, are common and often recognizable in meiosis. They are not considered deficiencies or duplications in the regular sense. When apparently only heterochromatic bands are involved and no phenotypic effects are observed, it may be considered a polymorphism. With more pronounced differences between the component chromosomes, at *anaphase I* the chromatids of the separating chromosomes may have different lengths as a result of chromatid exchange proximal to the rearrangement (Sects. 5.1.4 and 5.2.4). *Consequences*: There may be a selection *against transmission* of the deviant chromosome *through* the *male* line. *Segregation*, therefore, tends to be disturbed in an F2. Deficiencies tend to be less readily transmitted and to have a more pronounced phenotypic effect than duplications. A genetic analysis may occasionally identify the rearrangement. The absence of genes in a deficiency can give the impression that they are recessive mutations that segregate abnormally.

A special form of deficiency, not infrequent in allopolyploids, is the monotelocentric, where one telocentric chromosome replaces a metacentric chromosome, one entire chromosome arm is absent (Sect. 6.2.1). It is readily recognized both at mitosis and meiosis. At anaphase I the sister chromatids are not different, because there is no chromatid exchange in the deficient arm. An *isochromosome* (Sect. 6.2.1) replacing a normal metacentric chromosome combines a large deficiency with a duplication and forms a very characteristic configuration as an alternative for the heteromorphic bivalent: ring univalent. The heteromorphic bivalent may have the isochromosome in the form of a small ring associated with a normal chromosome. This may make it appear like an inversion heterozygote at diakinesis/metaphase I (Sect. 5.3.3). The isochromosome will form ring univalents as an alternative configuration, which an inversion univalent will not. Telocentrics are occasionally observed as deficiencies for one arm in diploids (rye: Melz and Winkel 1986; maize: Weber 1983), but in allopolyploids they have been reported repeatedly, and isochromosomes as well.

Context: Deficiencies (and duplications) large enough to form a recognizable heteromorphic bivalent have been found (Table 7.2) infrequently spontaneously and after a mutagenic treatment; telocentric chromosomes as deficiencies (monotelosomics) and isochromosomes are not uncommon in the progeny of allopolyploids with regular univalent formation, where they result from centromere breakage. Producing deficiencies and duplications may occasionally be objectives of breeding programs.

The *initial observation* (Table 7.1) may be a karyotype deviation (a), a meiotic abnormality (b: the heteromorphic bivalent); reduced fertility (f); disturbed segregation (g); or altered recombination (h).

2. *Pericentric* and *paracentric inversion* heterozygotes (Sect. 5.3). A pericentric inversion, if very asymmetric with respect to the centromere, will consistently form a heteromorphic bivalent at meiotic metaphase. A paracentric inversion heterozygote or a pericentric inversion, symmetric with respect to the centromere, will not have an obviously changed morphology of the chromosomes, except when clear chromosomal markers (C-bands, for instance) are available. However, if the inverted segment is large enough to contain a chiasma, and a chiasma is also formed in one of the end segments, a typical "frying pan", "spoon" (or "6-" or "9-shaped") bivalent appears at diakinesis metaphase I. The chiasma closing the ring is proximal to the chiasma which keeps the two chromosomes together (Sect. 5.4.1.4). This is an abnormal bivalent type which, unlike the heteromorphy, does not necessarily occur in all cells. When only one such inversion occurs, the distribution over cells is very skew, clearly suggesting that only one is possible. When there are more, the distribution is less skew. With a chiasma in both end segments, a homomorphic but still abnormal "figure-8" bivalent is formed. Both configurations are relatively rare and but can be occasionally observed in species hybrids (Levan 1941; de Vries 1989). If a pericentric inversion is asymmetric with respect to the centromere, detection is also possible at *anaphase I* because of different sister chromatids, provided a chiasma had formed in the pairing loop. This, however, is easily overlooked. For paracentric inversions anaphase bridges are the most distinguishing feature (Sects. 5.3.3, 7.4.1.5.2 and 7.4.1.5.3).

Pachytene pairing analysis is usually the best method to identify inversions, but not always readily performed. It distinguishes pericentric from paracentric inversions only when the centromeres can be made visible. Pericentric inversions can be recognized in the *karyotype* when they are asymmetric or when the C- or other banding pattern has changed. For paracentric inversions only the latter is available.

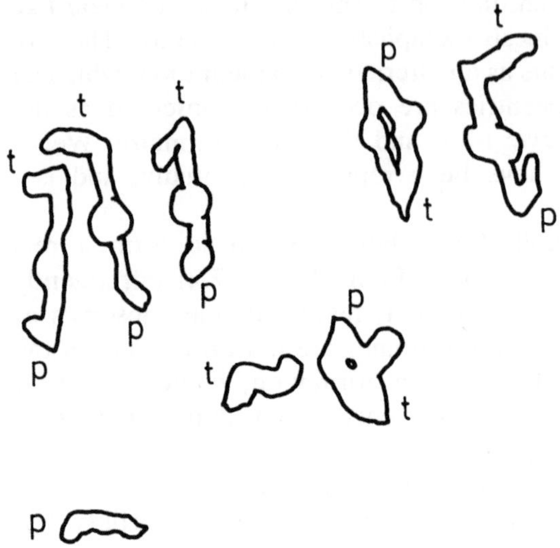

Fig. 7.1 Heteromorphic bivalents in the hybrid between *Lolium temulentum* (*t*) and *L. perenne* (*p*); 2n = 14. (After Sybenga 1975; drawn after Rees and Jones 1967)

Consequences (Sect. 5.3.4): Reduced fertility when the pairing loop is large enough for frequent chiasma formation, resulting in unbalanced duplication-deficiency gametes. There will always be locally reduced recombination, which may be one of the reasons to suspect a meiotic abnormality. When the deficiencies are small, viable progeny may be formed that contain them, in addition to the accompanying duplications, both always of the same type. These result in abnormal segregants. Without chiasmata in the loop, inversions may well go undetected.

Context: For plant breeding, inversions are occasionally important, primarily when they reduce recombination in specific chromosome segments in intra- or interspecific hybrids, which may prevent gene transfer. They have been found spontaneously and after mutagenic treatments in some material but are generally rare.

3. *Interspecific hybrids* may form any of the heteromorphic bivalents mentioned above when the rearrangement involved is present. In addition, bivalents in *hybrids between species* with *chromosomes* of *different* size are heteromorphic. The number per cell varies depending on the number of chromosomes sufficiently different to be distinguished at meiosis. Often all chromosomes are involved. Examples are the hybrids between certain *Allium* species (Jones and Rees 1968); between *Lolium temulentum* and *L. perenne* (Fig. 7.1; Rees and Jones 1967; Evans and Macefield 1972), between *Gossypium arboreum* and *G. raimondii* (Endrizzi et al. 1985), and between species of the *Aloeaceae* (Brandham 1990). In the latter case the heteromorphism is due to the presence of pericentric duplications which lead to occasional E-type or L-S bridges at anaphase I (Sects. 5.2 and 7.4.1.5.2). Karyotype analysis will readily

reveal the cause of heteromorphic bivalents formed by chromosomes of different size. The heteromorphism itself does not affect fertility when pairing is near normal, which does not appear to be uncommon as long as no complete homologues are present also. At pachytene occasionally loops are visible of stretches of possibly non-coding DNA, and in other cases the extra material of the larger chromosomes is simply not observed in the synaptonemal complex in the electron microscope or the pachytene structure visible in the light microscope (Rees and Jones 1967). Large and small chromosomes segregate at random at anaphase I. When viable progeny is obtained, random segregation may be found, but it is not unexpected that combinations of chromosomes from the same parent are more viable and consequently overrepresented in the progeny.

The *consequences* are limited except when E-type bridges are formed, which interfere with the proper course of meiosis and lead to reduced fertility and possibly aneuploidy. In other cases decreased fertility and reduced recombination will be the consequences of gene imbalance. This, rather than the meiotic behaviour, could be the reason for the recovery of low numbers of viable recombinants, including those resulting from chromosome recombination (segregational dysgenesis).

Context: These heteromorphic bivalents are only found in special *interspecific hybrids* and, at lower frequencies in their descendants. The initial observation may be on the *karyotype*, either in the hybrid or in its descendant or in *meiosis*, or it may be a reduction in fertility or apparent recombination.

7.4.1.4 Multivalents

7.4.1.4.1 Trivalents (Table 7.3: II.a.1.c.1)

Always one *trivalent* (Table 7.3: II.b.1.a) is not expected when the chromosome number is normal: it would be accompanied by a univalent. The consistent formation of two trivalents or their breakdown products can theoretically be realized, but is not sufficiently relevant for the present purpose. The combination of consistently one trivalent with a univalent and no larger configurations in other cells is a combination of different, independent abnormalities. An example is a *translocation monosomic* with an *alien addition*, as could be constructed for instance in wheat. Such trivalents would have a distribution of one per cell, possibly with breakdown products in some cells. A second example is a *primary trisomic* combined with a *monosomic* for another chromosome. This combination results from crossing a nullisomic-tetrasomic with a normal plant, as discussed below under "quadrivalents". Both types could be recognized in a detailed C- or other banded karyotype. *Anaphase I* may show erratic univalent behaviour and possibly irregular trivalent orientation. The *context* would be primarily an experiment such as a program of chromosome manipulation. They would not occur spontaneously or after mutagenic treatments. The *consequences* are reduced *fertility*, appearance of abnormal segregants (*aneuploids*) in the progeny because of the loss of the univalent or incorrect segregation of the trivalent.

Distribution skew with a maximum of *two trivalents* would be possible when the potential hexavalent of a double interchange heterozygote involving two chromosome pairs systematically falls apart into two trivalents or their breakdown products. It is improbable that in a large population of meiocytes a larger configuration would never be observed.

7.4.1.4.2 Quadrivalents (Table 7.3: II.a.1.c.3): Chromosome Number Normal; Karyotype Visibly Changed

Always one quadrivalent (Table 7.3: II.b.1.a) or its breakdown products or alternative pairing configuration (mainly trivalent with univalent or two bivalents). The quadrivalent is a ring, a chain, or a branched configuration. There are three main types: (1) translocation heterozygote; (2) displaced duplication heterozygote; (3) nulli-tetra substitution.

1. *Translocation heterozygote*. Most will be interchanges (Sect. 5.4.1), which may show a structural alteration of the *karyotype*, but not necessarily. Depending on the size of the interchanged segments and the non-translocated arms, the frequency of quadrivalents (chains, rings, branched types) will vary. Translocation heterozygotes can be found in diploids and allopolyploids. Heterozygotes are not common in normal inbreeding populations, and in outbreeders they are expected to segregate. *Anaphase I* may show occasional 3:1 instead of 2:2 segregation and differently sized chromatids after the formation of intersitial chiasmata when the chromosome segments involved were of different size.

Consequences: Usually (not always) reduced fertility; altered recombination; altered segregation; unusual segregants, usually aneuploids. All these factors might suggest a cytogenetic diagnosis, in addition to (in certain cases) karyotype alteration and meiotic abnormality.

Context: Translocation heterozygotes may be found to occur spontaneously at a low frequency, in mutation and in vitro culture programs more frequently, as well as in hybrids, including intra- and interspecific hybrids. In the latter a possible disturbance in fertility can often be ascribed to genetic imbalance in addition to the meiotic behaviour of the translocation. Interchanges may "float" in populations as a balanced polymorphism (Sects. 5.4.1 and 12.3).

2. *Duplication heterozygote* (displaced duplication: Sect. 5.2). Some types of duplication, especially the simple terminal duplications derived from an apparently terminal translocation (Sects. 5.2.2 and 5.4.1.4) can form configurations similar to those normally formed by translocations. They may also have karyologically recognizable different chromosomes, and do not show an

increase in chromosome number. If a duplication cannot be distinguished from a translocation somatically, the special configuration types formed and especially their relative frequencies can help to distinguish them from a translocation. This is not too difficult when the parental translocation is known and can be compared with the presumed duplication. In other cases the distinction may be more difficult. Again, the relative frequencies of the different configurations are often the best aid in distinguishing the two types. These are discussed in Sections 5.2.4 and 5.4.1.4. *Anaphase I*: Unequal chromatids are not uncommon. *Consequences*: Reduced fertility, reduced recombination, aberrant segregation and aneuploid segregants. Homozygotes tend to be rare. *Context*: Infrequently spontaneous; not frequent in mutation and in vitro programs or in hybrids; more common in the progeny of certain translocation heterozygotes and combinations of translocations, and in chromosome manipulation specifically set up to produce duplications (Sect. 11.2).

3. *Nulli-tetra substitution* (only in allopolyploids; cf. Sears 1954). One pair of chromosomes has been eliminated and a homoeologous set is present fourfold. Occasionally, the chromosomes can be recognized in the karyotype, but in many cases this is not so. Again, there is always one quadrivalent per cell, or its alternative pairing configuration (two bivalents) or breakdown products. Because the tetrasomic quadrivalent behaves somewhat differently from the translocation or duplication quadrivalent, whereby two bivalents are more frequent than the quadrivalent, and because four structurally identical chromosomes are present, whereas one normal pair is absent, a provisional distinction between the nulli-tetra combination and a translocation or duplication heterozygote can usually be made. The history of the material will aid in the identification, and there may be a distinct morphological effect usually absent in the translocation heterozygote. *Anaphase I* tends to be regular, although the quadrivalent may at times segregate differently from 2:2. The *consequences* are sightly reduced fertility, occasional aneuploidy in the progeny. *Context*: A nulli-tetra combination will hardly be found outside a program specifically set up to make one, or among its progeny. If there is any doubt, a test cross with a normal plant will give the clue: A nulli-tetra x normal combination, not considering abnormal progeny, will consistently give a monosomic-trisomic which forms a trivalent with a univalent. The nulli-tetra combination reproduces true to type by selfing. A translocation heterozygote, not considering abnormal progeny, will give 50% normal progeny and 50% heterozygotes and will segregate upon selfing unless it is a thoroughly balanced permanent heterozygote. The duplication will behave comparably, but (male) transmission tends to be reduced, and homozygotes usually have low viability.

Quadrivalents: *Distribution random* (Table 7.3: II.b.2). The frequency may vary from few to many, the chromosome number is normal, but at the tetraploid level. The typical case is the (natural) autotetraploid. In addition to quadrivalents of various types and their breakdown products (trivalents with univalents, open bivalents), there will be alternative pairing configurations:

bivalents, often rings. For several reasons the frequency of quadrivalents in autotetraploids can be quite low (Sect. 6.1.2.2.2.3). Yet the distribution over cells will be binomial unless the system of quadrivalent suppression is different for different chromosomes. Other exceptions are autoallopolyploids and certain tetraploid hybrids. In such cases, when no other distinctive characteristics are available, conceivably the distinction from allopolyploids with one or two translocations forming relatively few multivalents could be difficult. The pattern of inheritance, although not free of pitfalls, can give a clue. For further details, the reader is referred to Sections 6.1.2.2.9.3.2.2, and 11.3.1.

At *Anaphase I* the segregation of the chromosomes over the two poles may be irregular, especially when linearly orientated multivalents and univalents are present. The *consequences* are tetrasomic segregations (Sect. 6.1.2.2.2.4) and (occasional) aneuploidy. *Context*: Natural and induced autotetraploids and their hybrids are encountered in several cultivated species where the autopolyploid nature is well established. Problems are expected only with unknown tetraploids and tetraploid hybrids between superficially studied species.

7.4.1.4.3 Higher-Order Multivalents: Chromosome Number Normal, Chromosome Structure Normal or Deviant (Table 7.3: II.a.1.d.3)

They must be even-numbered multivalents, or unevenly numbered multivalents accompanied by a univalent, or, exceptionally, combinations of unevenly numbered multivalents. There are two main causes: (1) higher-order autopolyploidy, or autotetraploids with translocations; (2) multiple translocation heterozygotes.

1. Higher-order natural *autopolyploidy* is rare, especially in cultivated plant species. The chromosome number as such may help in the identification. Polyploid hybrids will be discussed briefly with genome analysis in Section 9.3.2.3. The distribution of the multivalents over cells in higher autopolyploids is in principle binomial, which may distinguish this case from multiple translocations. *Anaphase I* is not necessarily abnormal when multivalent formation is reduced, but may well be so with frequent multivalents, especially because large configurations tend to segregate less regularly than smaller configurations. The *consequences* are polysomic inheritance, and the formation of relatively frequent aneuploids.

Heterozygotes for *translocations in autotetraploids* produce configurations of up to eight chromosomes (octovalents), with a skew distribution and a maximum of one per cell. Breakdown products are frequent. There will be quadrivalents formed by the non-translocation chromosomes, in addition to the larger configurations. Fertility is reduced, in the progeny there is segregation for the translocation. They may be expected occasionally spontaneuously in natural and induced autotetraploids. In programs of allopolyploidization of autopolyploids (Sect. 12.2.2.1) they are produced on purpose.

2. Heterozygotes of *translocations* involving *more than two chromosomes*. In some species large complexes of translocation chromosomes occur regularly as a natural condition (Sects. 5.4.1.3 and 12.3). In this case, although usually fertility is not perfect, it is adequate for reproduction, and abnormal segregants are not really frequent. If in a wild species such complexes are systematically found, there is a real probability that balanced complex translocation heterozygosity is present, but the frequency of occurrence of such forms is low. No cultivated species are known to carry balanced translocations, but (as yet unsuccessful) attempts have been made to create them (Sect. 12.3). Smaller breakdown products are expected, but unlike in polyploids, these are not based on alternative pairing modes and will be of various types and compared to polysomic multivalents, less frequent. The distribution of the multivalents is skew. *Anaphase I* in non-balanced complex translocation heterozygotes tends to be irregular to a variable degree. Further *consequences* are comparable with those of single translocations, but usually with stronger effects on fertility, recombination, segregation of deviants, etc. *Context*: Translocations, involving more than two chromosomes in diploids, may appear spontaneously but are relatively rare. In a few species of ornamentals (*Paeonia*, *Oenothera*, *Rhoeo*, see Sect. 12.3) balanced complex translocation heterozygosity involving a few to all chromosomes of the complement may be encountered. In *Rumex* (Smith 1969) and *Viscum* (Barlow and Wiens 1976) sex-linked translocation complexes have been described (Sect. 12.3). Polyploid hybrids may show large translocation complexes more frequently (Sect. 9.3.2.2). Combinations of translocations occur spontaneously and, more frequently, after mutagenic treatment or in vitro culture.

For the systematic, unevenly numbered, higher-order multivalents with a univalent, not occurring as a regular breakdown product, complex explanations are necessary. The consequences and the context are comparable with those of the translocations involving two chromosomes, with extra complications caused by the univalent.

7.4.1.5 Chromosome Number Normal; Diakinesis-Metaphase I Normal

7.4.1.5.1 Unequal Sister Chromatids (Table 7.3: II.a.2.a)

Chromosomes with unequal sister chromatids at anaphase I have been discussed in the previous sections as the consequence of a number of diakinesis/metaphase I configurations. They are the result of chromatid exchange between two chromosome arms that differ in size. It is possible that the corresponding metaphase configuration has escaped attention: heteromorphic or abnormal bivalents resulting from deficiencies, duplications, pericentric inversions, chromosome size difference between species (Sect. 7.4.1.3) and, less probably, translocation or duplication multivalents. Whenever chromatid differences at anaphase are observed, it is necessary to go back to diakinesis/

metaphase I and it may be expected, when sufficiently large differences show up at anaphase I, that these will also be visible at earlier stages.

7.4.1.5.2 Chromatid Bridges (Table 7.3: II.a.2.b)

a) *Maximum one per cell*, distribution *skew* (Table 7.3: II.b.1), suggesting that one chromosome is involved. There is one main cause of a single chromatid bridge per cell at anaphase I: a chromatid exchange in the pairing loop of a heterozygous *paracentric inversion*. It is accompanied by an acentric fragment which in all cells has the same length for a particular inversion (Fig. 5.6A). The frequency of the bridge depends on the formation of chiasmata in and proximal to the loop. The inversion gives unusual bivalent shapes in diakinesis to metaphase I (Sect. 5.3.4), which are often not clear enough to help much in the diagnosis. Gross karyotype abnormalities are not visible. *Consequences*: Reduced fertility, the degree depending on the frequency of the bridge. Few viable duplications and deficiencies are recovered, and consequently few abnormal segregants. *Recombination* is often strongly *reduced* in and around the segment involved. These two consequences suggest a diagnostic analysis. In certain material (e.g. maize) the bridge may be the start of a breakage-fusion-bridge cycle (Sect. 5.3.4) and may then occasionally result in variegation of the endosperm. Special orientation of the bridge in embryo-sac mother cells may exclude the recovery of deficiency and duplication chromatids in the egg and prevent a reduction in fertility (Sect. 5.3.4).

With two paracentric inversions, two bridges may be formed. When bridge formation is rare, the distribution may deviate so little from random that it is difficult to distinguish from the following case.

b) *Zero to a few or even several bridges per cell*; *distribution random* over cells, suggesting that each chromosome may be potentially involved with equal probability (Table 7.3: II.b.2.a). There are three main types: (1) U-type exchange; (2) premeiotic errors; (3) E-type or L-S and other special bridges.

1. The *U-type exchange* (Jones 1969) is randomly distributed over cells, and is a typical cellular, rather than a chromosomal aberration. All chromosomes may be potentially involved. The bridges are assumed to be due to errors in chromatid exchange, which would normally lead to chiasmata. It is found in unbalanced genotypes such as inbred lines of outbreeders, interspecific or other wide hybrids and possibly after special treatment (Sect. 3.2.2). There is also an *acentric fragment*, but its *size varies*. It tends to reflect the size of double the segment distal to the normal position of the chiasmata. The frequency of the event per cell varies and is usually small compared to the total number of chiasmata per cell. There are no special *karyotype* conditions connected with the phenomenon. The *consequences* are primarily reduced fertility and possibly abnormal segregants, but these have not been reported. The *context* is: wide hybrids, genetic or environmentally conditioned imbalance.

2. *Premeiotic errors*, probably mainly errors of DNA replication; distribution over cells random. These are a collection of unspecified abnormalities, including spontaneous inversions originating before pairing. The *size* of the *fragments varies*. The cause may further be asymmetric exchange unrelated to chiasma formation but representing random chromosome breaks. Chromatid bridges and short side arms occasionally give the impression of being double, for which no adequate explanation has been found (John and Lewis 1965). *Stickiness* of chromosome ends also results in bridges, but *not* accompanied by an *acentric fragment*. The cause here is usually not a premeiotic error, but an abnormality of the chromosome condensation process, for instance caused by irradiation, extreme temperatures, chemicals, genetic imbalance. The frequency of aberrations caused by premeiotic errors may be low, but if stickiness occurs, many chromosomes per cell may be involved. There is no relation to specific *karyotype* characteristics. The distinction from U-type exchanges is that the latter are accompanied by acentric fragments, and that they follow the distribution over chromosomes of recombinational events. The *consequences* are primarily reduced fertility. The *context* is usually a drastic stress condition: irradiation, high temperature, chemical treatment or a genetic imbalance (hybrid, special genotype).

3. *E-type*, *L-S* or other special *bridges*. The E-type or L-S bridge has a typical chromosomal-genomic basis. The most extensive report (Brandham 1990) is of hybrids between species of Aloeaceae with pronounced differences in chromosome size, resulting in heteromorphic bivalents in diakinesis-metaphase I. The phenomenon is rare and has been considered earlier with heteromorphic bivalents, which represent one way of identifying them (Sects. 5.2.4 and 7.4.1.3.2).

7.4.1.5.3 Chromatid Loops (Table 7.3: II.a.2.c)

One per cell as a *maximum*. The distribution suggests that one chromosome pair is involved (Table 7.3: II.b.1). The primary cause of chromatid loops, connecting the sister centromeres of one specific anaphase I chromosome, is chromatid exchange in the pairing loop of a paracentric inversion, in addition to a chiasma in the interstitial segment, between the centromere and the inversion (Fig. 5.6A). They are often difficult to detect; fragments have the same length; if an inversion is involved, there should be more cells with a bridge and a fragment than cells with a chromatid loop and a fragment.

7.4.1.5.4 Anaphase II Bridges (Table 7.3: II.a.2.e)

Maximum one per cell; non-random distribution suggesting one as a maximum per cell (Table 7.3: II.b.1).

A chiasma in the paracentric inversion loop together with a chiasma in the interstitial segment results in a bridge at Anaphase II. Usually, anaphase I can be studied also, and should give the typical paracentric inversion phenomena.

More than one per cell, distribution random (Table 7.3: II.b.2.a), suggesting cellular irregularities causing novel asymmetric rearrangements and stickiness. Although anaphase II can be disturbed independently of anaphase I, there is usually considerable correlation.

7.4.2 Chromosome Number Not Normal; Aneuploidy: Hyperploidy

There are several types of hyperploids, and different numbers of extra chromosomes. Some are structurally abnormal.

7.4.2.1 *Trisomics*: There is One Extra Chromosome

7.4.2.1.1 Univalents (Table 7.3: II.a.1.a)

a) *Aways one* (Table 7.3: II.b.1.a). Four situations must be considered: (1) a single B-chromosome; (2) a monosomic alien addition; (3) a trisomic in species with low multivalent association; (4) a restructured extra chromosome.

1. Single *B-chromosome*. In plants, single B-chromosomes without translocations involving A-chromosomes will always be univalent. In species where they have been studied, their morphology is known and tends to be different from that of A-chromosomes. Plant morphology is usually not affected. For a detailed karyotype analysis, C-banding, in situ hybridization with specific probes, and RFLP analysis, if available, are very effective in identifying B-chromosomes. If other tests fail, the characteristic accumulation system of B-chromosomes (Sect. 3.1.4.1.2.3) may help in their identification. *Consequences*: Accumulation of the extra chromosome in the progeny unless there is strong selection for productivity. The univalent may misdivide, and telosomics and isochromosomes of B-chromosomes and other derivatives are not uncommon. There are few other consequences. *Context*: Programs involving B-chromosomes; programs involving little known species that happen to carry B-chromosomes.

2. Monosomic *alien addition*. It is not probable, with the history of the plant known, that this possibility would not have been considered at the beginning: alien additions do not arise spontaneously. In some genotypes especially of the Triticinae, when there is sufficient homology with a chromosome pair of the recipient, a trivalent will be observed with low and variable frequency. The alien addition is common in chromosome manipulation programs or varieties resulting from them in allopolyploids and in diploids (Sect. 10.4.4.2). The plant morphology may be typical of special alien additions. In a

number of cases a detailed *karyotype* analysis, including C-banding and in situ hybridization combined with knowledge of the origin of the alien addition, can identify the specific chromosome involved. In the Triticinae, where they have been produced repeatedly in polyploids and even in diploids, *C-banding* will often identify them. *Consequences*: At *Anaphase I* the univalent tends to get lost, and male gametes with the extra chromosome are less competitive in fertilization than normal gametes. Yet occasionally the extra chromosome is transmitted through the pollen and disomic additions may result. The univalent may misdivide and produce telocentrics and isochromosomes in the progeny. Dominant or hyperstatic genes in the alien chromosome may be expressed, which are not normally present in the recipient species. *Context*: The monosomic addition is practically always the result of some manipulation carried out with the purpose of introducing a gene from a different species, where the addition is an intermediate step, made on purpose or unintentionally. Occasionally, however, it results from hybridization involving alien substitutions. In the latter case, when the substitution is unknown, the alien addition may be unexpected.

3. *Primary* or derived *trisomics* in species with a restricted polysomic multivalent association will (practically) consistently produce a single univalent at diakinesis-metaphase. Such species are discussed in Sections 6.1.2.2 and 11.3.1.2.3. They are not very common. SC analysis may show trivalent pairing. A *karyotype* analysis, when sufficiently discriminating techniques are available, is, as usual, a good approach. The *consequences* are as discussed for primary trisomics (Sect. 6.2.2.1.5): trisomic segregation. This may be the best way of identification. There is variable transmission, reduced through the male; the chromosome may misdivide and produce secondary products. The *origin* is rarely spontaneous, somewhat more frequent in in vitro culture, but frequent in the progeny of triploids. The *context*, therefore, will often be a program of autopolyploidy breeding or gene transfer between diploid and tetraploid forms of those rare species, where multivalents are (practically) absent.

4. *Restructured chromosomes* derived from normal chromosomes (tertiary or similar trisomics with more than one translocation or inversion) may be so different from primary trisomics that they have lost the possibility of pairing with the normal complement. Hardly any instances are known and the chance of encountering one without expecting it is small. The trisomic (and tetrasomic) barley reported by Tsuchiya (1969) with fully independent pairing of a pair of restructured chromosomes is an example (see also Ramage 1991). The plant phenotype may show special characteristics. Pachytene pairing in SCs may give some indication. The *consequences* are limited to what is normal for trisomics: abnormal segregation, but not trisomic; variable transmission, usually low through the male; origin of telocentrics and isochromosomes. The context is restricted to programs involving rearranged chromosomes. In in vitro cultures with long callus phases such trisomics may arise incidentally, but have not been convincingly reported.

b) *Zero* to many univalents *per cell*, *random distribution* (Table 7.3: II.b.2.a): not expected when no further complex configurations are observed in other cells. *One* to *many per cell*, no other configurations and a skew distribution suggest one of the possibilities discussed above combined with partial desynapsis, which may or may not be the consequence of the presence of the extra chromosome. *Karyotype*, *Anaphase I*, *consequences* and *context* are as for the alien addition, and for univalents in normal karyotypes.

7.4.2.1.2 Heteromorphic Bivalents (Table 7.3: II.a.1.b)

Without other configurations in sister cells, these are not expected in the category "one extra chromosome". Accompanied by a univalent it could be any combination of cases discussed in Sections 7.4.1.3 and 7.4.1.1.2, and not to be considered here in detail. When their identity cannot be derived from the history of the material, or from a detailed karyotype analysis, special test crosses may be required.

7.4.2.1.3 Multivalents (Table 7.3: II.a.1.c)

Multivalents are characteristic of trisomics, and in addition to trivalents, higher forms may occur. There is almost without exception a maximum of one per cell, and in other cells breakdown or alternative pairing products are found.

Trivalents
A single and apparently *homomorphic trivalent*, formed at a variable frequency and with only breakdown configurations in sister cells, suggests *primary trisomy*. The trivalent may have any of the shapes discussed in Section 6.2.2.1. The plant will usually have a characteristic morphology (Sect. 6.2.2). A detailed *karyotype* analysis, if possible including banding, will show that the morphology of the extra chromosome is like that of one normal pair, except for occasional polymorphisms. At *Anaphase I* the trivalent often, but not always segregates 2:1; the univalents behave as discussed earlier. *Consequences*: Variable transmission, usually reduced, especially through the male; trisomic segregation; trisomic gene segregation; telocentrics and isochromosomes in the progeny; somewhat reduced fertility as a result of genetic imbalance rather than meiotic chromosome behaviour. The *context* is: spontaneous, infrequent in diploids, more frequent in allopolyploids, especially in desynaptics; somewhat more frequent after mutagenic treatment; common in the progeny of triploids; occasionally in the progeny of other trisomics. If plant morphology does not give a clue and a karyotype and molecular analysis cannot be carried out with sufficient resolution, a genetic analysis is the best way of identification.

Abnormal Trivalents
Single *ring trivalents* are almost exclusively formed by secondary trisomics (isochromosome trisomics: Sect. 6.2.2.2). Alternative pairing configurations are a (*ring*) *univalent* with a homomorphic bivalent at diakinesis-metaphase I or a *heteromorphic trivalent* with occasionally a small ring at one side, formed by the isochromosome. The isochromosome is occasionally recognized by its gross morphology; in other cases, a detailed *karyotype* and/or molecular analysis is required. *Anaphase I* orientation may result in a large, probably lethal deficiency in one daughter cell and an extra normal chromosome in the other cell. The univalent, although potentially a ring univalent, behaves like other univalents. The *consequences* are loss of the extra chromosome, primary trisomy, telocentrics, reduced fertility, abnormal segregation, but not trisomic (Sect. 6.2.2.2). *Context*: Spontaneous, very infrequently; somewhat more frequently after mutagenic treatment; more frequently in the progeny of other trisomics, especially primary trisomics, and in progeny of triploids. There is considerable variation per chromosome and per species in the frequency of isochromosome formation.

Heteromorphic trivalents composed of two normal chromosomes and one telocentric chromosome result from telocentric trisomy (Sect. 6.2.2.2). The telocentric is usually univalent in at least 1/3 of the cells. It is readily recognized in the karyotype, even in meiosis. Anaphase, consequences and context are very similar to the previously discussed non-primary trisomics. It will not or very infrequently form isochromosomes or centromere translocations.

Heteromorphic trivalents are also formed when two telocentrics of a heterozygous *Robertsonian* or *centromere split* (Sect. 5.5.3) associate with the corresponding metacentric chromosome. The metacentric is in the center of the trivalent. The chromosome number is one larger than normal. The total amount of chromatin has not significantly changed and no gene imbalance effects are expected. When one of the telocentrics is small, a heteromorphic bivalent with a telocentric univalent will be observed regularly. The trivalent in most cells and the heteromorphic bivalent with telocentric univalent in others, in addition to a karyotype analysis, will readily distinguish it from other aberrations. In *Anaphase I* the orientation usually leads to balanced daughter nuclei, but occasionally trisomy for one of the telos is recovered in the progeny. The *consequences* are usually limited to (rather infrequent) formation of aneuploids, primarily telocentric trisomics. The *context* is (rarely) an interspecific hybrid or its derivatives, quite infrequently but not negligibly the origin is spontaneous and hardly more frequently it arises after a mutagenic treatment. Although centromere split is relatively common in univalents at meiosis, the simultaneous recovery of the two telocentrics of the same metacentric chromosome in the same plant is rare even in the progeny of plants with high univalent frequencies. By combining the two separately derived telocentrics, a Robertsonian split can be simulated (Sect. 5.5.1).

Other *heteromorphic* or otherwise abnormal *trivalents* are formed in combinations of extra chromosomes with the rearrangements causing heteromorphic bivalents (Sect. 7.4.1.3), inversion trisomics, etc. Tertiary trisomics may form heteromorphic trivalents, but are expected to show at least some larger configurations in other cells.

Quadrivalents and *Higher-Order Multivalents* (Table 7.3: II.a.1.c.2, II.a.1.c.3) Trisomics do not normally form *quadrivalents* as the largest configuration, unless as a combination of different aberrations. These have to be identified separately on the basis of the information given for other trisomics and other aberrations.

Larger Multivalents
One *chain quinquivalent* (chain-of-five) as the maximum configuration, no other multivalents in the same cell, and a quadrivalent with a univalent or a trivalent with a bivalent in other cells are characteristic of the combination of a translocation and an extra chromosome. There are three main categories: (1) *tertiary trisomic* (Sect. 6.2.2.3); (2) *translocation trisomic* (Sect. 6.2.2.3); and (3) *compensating trisomic* (Sect. 6.2.2.4.1). They can be distinguished on the basis of the types and relative frequencies of the alternative pairing configurations.

1. The *tertiary trisomic* forms a chain quinquivalent, or a trivalent with a bivalent, or two bivalents with a univalent as alternative pairing configurations in a theoretical random pairing ratio of 3:2:1. At diakinesis-metaphase I lack of chiasmata may reduce the size, and consequently the observed frequency of the larger configurations, but there will never (or very exceptionally) be a ring quadrivalent.

2. The *translocation trisomic* again has three pairing configurations: a quinquivalent, a trivalent + bivalent combination and a quadrivalent with a univalent. At diakinesis-metaphase the quinquivalent and the trivalent are chains, the quadrivalent can be a ring (the translocation complex). Chiasma failure reduces the frequency of larger configurations. This is the only type that can form a ring quadrivalent.

3. The *compensating trisomics* are a heterogeneous category, which may, for instance, have a metacentric translocation chromosome together with another metacentric translocation chromosome as the compensating combination, or a telocentric chromosome combined with a metacentric translocation chromosome. In the latter case characteristic heteromorphic configurations are observed and in the karyotype the telocentric will be readily recognized. There are no ring quadrivalents and the only two alternative pairing configurations are a quinquivalent and a trivalent/bivalent combination in a 2:1 ratio. Compensating trisomics containing only metacentric translocation chromosomes have a maximum of seven associated chromosomes. Lack of chiasmata as well as pairing preferences may result in univalents.

In the karyotype, the numerical deviation is readily established. To identify the extra chromosome and the rearrangements, detailed *karyotype* analyses or test crosses are necessary. The *Anaphase I* behaviour tends to be somewhat disturbed in most of these trisomics because of the irregular orientation of the large multivalents and the occurrence of univalents. The *consequences*, therefore, are usually the loss of the extra chromosome in the first two cases, because this is tolerated. The compensating trisomic formally segregates 1:1, but in practice the compensating gamete is formed somewhat less frequently and functions less efficiently. Male transmission of the gamete with the extra chromosome is low, often approaching O, which is the reason for its use in special programs (Sect. 12.4). Fertility is reduced, partly due to segregational irregularities and partly perhaps to gene imbalance. There is reduced recombination in certain chromosome segments, and new types of abnormal segregants are found in the progeny. For details, see Section 6.2.2.4.1. The *context* can be spontaneous in the progeny of translocation heterozygotes (Points 1 and 2). The compensating trisomic can arise from certain combinations of translocations. Neither of the three is expected in normal material. Gene segregation is different for the three types (Sects. 6.2.2.3.4 and 6.2.2.4.2.3).

One *quinquivalent* per cell, accompanied by quadrivalents in the same cell, is possible when in natural autotetraploids an extra chromosome is present: *pentasomy*. It is not at all infrequent in autotetraploids, and readily recognized once it has been established that the species or form is autotetraploid. The *frequency* of quinquivalents depends on the frequency of quadrivalent formation, which, as shown before, can vary considerably. Translocations in autotetraploids produce larger configurations. *Anaphase I* combines the irregularity of the tetraploid with that of the extra chromosome. The *consequences* are limited: slightly decreased fertility, frequent loss of the extra chromosome, slightly altered segregation for specific genes. The *context* usually is spontaneous with reasonable frequency in autotetraploids, slightly increased after mutagenic treatment and in vitro culture.

Zero to several quinquivalents per cell, without larger configurations in the same or in sister cells are not expected when a single chromosome is extra. One *multivalent larger than quinquivalent*, when an extra chromosome is present, is possible in diploids, allopolyploids and autopolyploids, for instance when more than two chromosomes are involved in a heterozygous translocation and the extra chromosome is homologous to one of them. Different possibilities can be conceived. They occur in the progeny of combinations of translocations, are usually rather unstable at meiosis, and have reduced fertility. They are readily recognized on the basis of chromosome number and meiotic behaviour, but identification of the chromosomes involved, as usual, requires additional information.

7.4.2.1.4 Anaphase I and Anaphase II Aberrations

Anaphase bridges, fragments, loops and unequal chromatids (Table 7.3: II.a.2) with otherwise normal metaphase I in trisomics are rare, unless as combinations of aberrations discussed above. These must be identified separately.

7.4.2.2 Two or More Extra Chromosomes (Table 7.3: I.b.1.a)

The case of two extra chromosomes is of sufficient specific interest for a discussion. It also gives a basis for the identification of more than two extra chromosomes.

7.4.2.2.1 Only Normal Bivalents (Table 7.3: II.a.1.a)

Includes their (exceptional) breakdown products: two univalents. Three cases are distinguished:

1. A pair of *B-chromosomes*. If known for the species, these are readily recognized, and their presence should not be unexpected from the context. When wild ancestors or distant relatives are involved, B-chromosomes may appear as a surprise. Further identification, including a karyotype analysis, is as for B-chromosomes discussed above in the section on trisomics. *Anaphase I*: Normal. In pollen or embryo-sac division often non-disjunction. *Consequences*: Upon selfing increase in the number of chromosomes in the progeny; occasionally, slight effects on meiotic behaviour. *Context*: In wild relatives of some cultivated species, and their hybrids.

2. Disomic *alien addition*. The context would most probably give a clue. In addition, a detailed karyotype analysis might show the presence of two deviating chromosomes which are apparently equivalent with chromosomes from another species. There may be a slightly increased frequency of cells with one pair of univalents when the alien chromosome is not quite adjusted to the genetic environment. *Anaphase I*: Mostly normal, but the occasional univalents may lag, or split precociously, or break in the centromere. *Consequences*: Morphological effects, expression of alien genes, slightly reduced fertility, some segregation of trisomics and euploids. *Context*: In chromosome manipulation programs of allopolyploids, rarely diploids.

3. *Tetrasomy* with failing multivalent formation. Tetrasomy is rare in most diploids, but can occur in allopolyploids. It does not usually appear spontaneously, but is expected in the progeny of a trisomic. As in autotetraploid species that fail to form multivalents (Sects. 6.1.2.2.2 and 11.3.2.3), the same can occur when a chromosome is tetrasomic in an allopolyploid. *Hexasomy* in autotetraploids with exclusively bivalent pairing will also result in one extra bivalent. Both possibilities are rather remote. *Anaphase I*: Normal. *Consequences*: Some specific morphological effects; slightly reduced fertility; tetrasomic inheritance for the linkage group involved in a tetrasomic allopolyploid, hexasomic for the linkage group in a hexasomic autotetraploid. *Context*: Chromosome manipulation programs; autotetraploids.

Distinction between the three types is based on the karyotype, including molecular markers if available; plant morphology; to some extent on meiotic behaviour; pollen mitosis (B-chromosomes); segregations.

7.4.2.2.2 Univalents (Table 7.3: II.a.1.b)

Always *one pair* (Table 7.3: II.b.1.a), very infrequently two. The apparent case of two non-homologous *alien addition* chromosomes, practically restricted to allopolyploids, most probably the result of a cross between two different alien addition lines. A *karyotype* analysis should be able to confirm this conclusion when sufficient details of the chromosomes can be made visible. *Anaphase I*: The two univalents will tend to lag, divide precociously, or break (infrequently) at the centromere, or segregate at random to the poles. *Consequences*: Morphological effects; reduced fertility; segregation of different, new combinations of aneuploids for the chromosomes involved and euploids; infrequent recovery of the same combination in the progeny. *Context*: Chromosome manipulation programs.

Other possibilities are not excluded, but negligible in frequency.

Often, not always *one pair* of univalents; distribution *skew*, suggesting one pair to be involved primarily (Table 7.3: II.b.1.b).

Most probably a disomic addition with reduced chiasma formation as a result of insufficient adaptation of the chromosomes to the cellular environment. The consequences and context are evident: the normal univalent behaviour at Anaphase I, and its consequences for the progeny, are as discussed above (Sect. 7.4.2.2.1). Various other univalent distributions can occasionally be encountered, suggesting that in addition to the extra chromosomes other irregularities occur, including asynapsis, desynapsis, or hybridity.

7.4.2.2.3 Heteromorphic Bivalents (Table 7.3: II.a.1.c)

Consistently one heteromorphic bivalent, or its breakdown product: two morphologically different chromosomes (Table 7.3: II.b.1.a), no other abnormalities in sister meiocytes. The most obvious explanation is one pair of alien addition chromosomes, one of which is a *telocentric*. This telocentric should also be visible in the somatic *karyotype*. *Anaphase I* behaviour and *consequences* can readily be derived from the previous sections on alien additions. It is not an unexpected aberration in the progeny of a disomic or monosomic alien addition. It is not impossible to conceive of other situations where a single heteromorphic bivalent could be seen, for instance involving an isochromosome clearly different from an asymmetric normal metacentric, but most of these are not of practical interest.

7.4.2.2.4 Multivalents (Table 7.3: II.a.1.c)

A *trivalent* is not an obvious configuration. A trivalent in combination with a univalent can be a primary trisomic together with an alien addition, as is the result of crossing a tetrasomic with a disomic alien addition. It is not expected outside the context of a chromosome manipulation program and would be a planned result.

Two trivalents without univalents, no larger configurations, but alternative pairing configurations and breakdown products (bivalents and univalents) in other cells is a more probable condition, even in diploids. It is the typical configuration of a *double primary trisomic*. Detailed *karyotype* analysis would confirm trisomy for two normal chromosomes. At *Anaphase I* only some unusual segregation and univalent behaviour are expected and the *consequences* include a reduction in the number of extra chromosomes in the progeny, often to zero, with infrequent production of tetrasomics and telocentrics and isochromosomes, and trisomic segregations. The *context* is spontaneous in the progeny of a (possibly not previously detected) triploid or desynaptic. In chromosome manipulation the triploid could have been constructed with the aim of yielding primary trisomics.

Quadrivalents: Consistently *one* per cell, or its alternative pairing (two bivalents) or breakdown products is the typical configuration of the *tetrasomic*. It is more common in allopolyploids than in diploids, where the excessive amount of extra genetic material is often not well tolerated. The *karyotype* should show four identical chromosomes of one type; *Anaphase I* will probably show 2:2 segregation for the chromosomes involved in most cells, but 3:1 in some. The consequences are limited: mainly some reduced fertility, altered segregation pattern for the genes in the quadruple chromosomes, some loss of the extra chromosomes in the progeny. There are clear morphological effects. The *context* is the progeny of a trisomic, or a chromosome manipulation program. It is not expected to be found spontaneously as it is a rather indirect product of a triploid or a desynaptic (Sect. 6.2.2.1.2).

A *quadrivalent* may also result when a heterozygous translocation is combined with a disomic alien addition, and it is possible to conceive more such combinations. It will not be difficult to recognize them, especially as their origin will usually be known. Independent segregation of the two separate events is one characteristic. A *hexavalent* may be formed when the translocation involves three chromosomes or when a translocation is combined with tetrasomy for one of the chromosomes in the translocation. A hexavalent in combination with randomly distributed quadrivalents points to an autotetraploid with two extra chromosomes: a hexasomic.

7.4.3 Hypoploidy (Table 7.3: I.b.1.b)

One or more chromosomes less than in the normal karyotype.

7.4.3.1 Monosomics

One chromosome less than normal; the most common type of hypoploidy.

7.4.3.1.1 Univalents (Table 7.3: II.a.1.a)

Always one per cell (Table 7.3: II.b.1.a), occasionally two more: *primary monosomy* (Sect. 6.2.1). One normal chromosome is absent, the most common aneuploid in allopolyploids, less common in diploids. The *karyotype* shows only the absence of one chromosome. *Anaphase I*: Lagging, random segregation, precocious split or centromere break of the univalent. *Consequences*: Slight but characteristic morphological changes. Occasionally, when dose-dependent genes occur in the particular chromosome (Rht1 and Rht2 dwarfing genes in semi-dwarf wheat varieties, see Sect. 6.2.1.3), striking rogues may appear in the field. In meiosis in some cases increased desynapsis, but often no abnormality observed, or a few occasional univalents, hardly more than normal. In the progeny characteristic monosomic segregation of the chromosomes with primarily monosomics, several disomics and few nullisomics in the progeny. No recombination in the chromosome involved. *Context*: Spontaneous, not infrequent: in a normal field of wheat about 1 out of 100 plants is karyologically abnormal, usually monosomic (Riley and Kimber 1961). The frequency can be considerably increased after certain treatments (Sect. 10.4.4.1). Frequent in regenerants from callus and common in chromosome manipulation programs of allopolyploids.

XO/XX sex determining systems as are present in several insect taxa have not been reported for plants. They would also systematically show one univalent (X-chromosome) in male meiosis. There is one chromosome less than in the homogametic sex, which has two X-chromosomes.

One univalent to many per cell; *distribution random* (Table 7.3: II.b.2.a) is not well conceivable, unless for statistical reasons not distinguishable from the next case: *distribution skew* (Table 7.3: II.b.2.b). This occurs when monosomy for specific chromosomes reduces chromosome pairing and chiasma formation. In wheat monosomy for chromosome 3D (Sears 1954) results in mild desynapsis with several univalents. Nullisomy has a stronger effect. These extra univalents are approximately randomly distributed, but the combined distribution in the monosomic is in principle skew.

7.4.3.1.2 Heteromorphic or Otherwise Abnormal Bivalents

These must be accompanied by a univalent. It is most probably a combination of a monosomic with an independent aberration causing the heteromorphic bivalent.

7.4.3.1.3 Multivalents

Trivalents (Table 7.3: II.a.1.c.1)
Consistently one trivalent per cell or its obvious breakdown products (Table 7.3: II.b.1.a) is the consequence of *translocation monosomy*, or *tertiary monosomy*, the second most important type of monosomy. One chromosome of the set of four associated with a heterozygous translocation is absent. It occurs in diploids more frequently than primary monosomy because the absence of short translocation chromosomes is tolerated better than that of normal chromosomes (Sect. 6.2.1). *Karyotype* details: Possibly one abnormal chromosome recognizable, expected to be the larger translocation chromosome. *Identification* of the chromosome lost is possible in test crosses with a translocation tester set, a trisomic set or a monosomic set when either the monosomic or the trisomic are male transmissible. *Anaphase I*: Occasional, irregular segregation. *Consequences*: Fertility usually reduced and morphology abnormal to variable degrees. Segregation abnormal, which can be used for locating genes. Recombination reduced in the chromosomes involved. *Context*: Spontaneous in the progeny of translocation heterozygotes; used in chromosome manipulation programs.

Higher-order multivalents: Infrequent. Possible as a combination of translocation heterozygosity with primary or tertiary monosomy; tertiary monosomy derived from a translocation involving more than two chromosomes. Anaphase I, consequences and context: combination of the separate cases.

7.4.3.2 Hypoploids: Two Chromosomes Less than Normal (Table 7.3: I.b.1.b)

7.4.3.2.1 Only Normal Bivalents

Nullisomy is only expected in allopolyploids. *Anaphase I* regular. *Consequences*: Plant phenotype abnormal, expression of specific recessive alleles in homoeologous chromosomes, fertility reduced, no segregation and recombination for genes in chromosome involved, no unusual segregants. Most nullisomics have a slightly abnormal meiosis. See further under: univalents, random, zero to many per cell: nullisomics with desynaptic effects.

7.4.3.2.2 Univalents (Table 7.3: II.a.1.a)

Always two, occasionally more univalents (Table 7.3: II.b.1.b): there is most probably a double primary monosomic: two different normal chromosomes are absent (Sect. 6.2.1). *Karyotype*: In (C-)banded preparations the chromosomes can possibly be seen to be present in pairs, except two, which are single. *Pachytene*: Often two univalents, occasionally only bivalents with one bivalent irregularly and non-homologously paired, and possibly a disturbance in pairing of other bivalents. *Anaphase*: Two univalents that behave as expected: lagging, random inclusion in daughter groups, precocious split or centromere break. *Consequences*: As single monosomic, but with more different combinations of phenomena. *Context*: In allopolyploids rarely spontaneous, potentially more often after special treatment; in chromosome manipulation programs; not in diploids.

Zero or more univalents per cell. *Distribution random* (Table 7.3: II.b.2.a): probably a nullisomic with desynaptic effects (see above and Sect. 6.2.1). *Distribution skew*: (Table 7.3: II.b.2.b): Double monosomic with desynaptic effects (see above and Sect. 6.2.1). For statistical reasons it is often not possible to distinguish between the two distributions.

7.4.3.2.3 Heteromorphic Bivalents and Multivalents

Heteromorphic bivalents and multivalents occur primarily in combinations with the previous cases and other rearrangements. These and other special combinations will not be discussed here, but their identification can be derived from combinations of the aberrations discussed above.

7.4.4 Euploidy (Table 7.3: I.b.2.a)

7.4.4.1 Gametic Chromosome Number ("Haploidy")

7.4.4.1.1 Only Normal Bivalents (Table 7.3: II.a.1.a)

Exclusively bivalents, very infrequently a univalent pair as in normal diploids, is not expected when a monohaploid is involved. Plants with the gametic chromosome number, when derived from allopolyploids, are in reality polyhaploids and may occasionally have a very regular bivalent formation, although usually not quite without univalents. Cotton haploids (*Gossypium hirsutum*) are an example, and here the bivalents are hetermorphic (Sect. 6.1.1). Exclusively bivalents are expected and found in (di)haploids of autotetraploids, also when they form exclusively bivalents at the tetraploid level.

7.4.4.1.2 Univalents (Table 7.3: II.a.1.a)

Univalents without any other configuration, although theoretically expected, is in practice exceptional. Most haploids or polyhapoids form at least occasionally one or more bivalents or even multivalents (Sect. 6.1.1.3). *Consistently one* (pair) is also not expected to be realized except if in the previous case (Sect. 7.4.4.1.1) one of the chromosomes happens to have been substituted by a chromosome from a more distantly related species, i.e. with an alien disomic or monosomic substitution in the original polyploid.

One to *several* pairs of *univalents. Distribution random.* Although not considered to be the typical meiotic condition for haploids, it is not uncommon for several *di-* or *polyhaploids* derived from stable allopolyploids with somewhat better pairing differentiation than discussed above. *Karyotype*: Not much extra information. *Pachytene*: Relatively regular pairing. *Anaphase I*: variable. *Consequences*: Reduced stature; low fertility, or completely sterile, genetic imbalance of gametes and zygotes; segregation of many abnormal or unusual types in the progeny, including aneuploids. *Context*: Infrequently spontaneous; more frequent in special programs (induced parthenogenesis by chemicals, foreign pollination, etc., anther, microspore or ovule culture).

7.4.4.1.3 Heteromorphic or Otherwise Abnormal Bivalents (Table 7.3: II.a.1.c)

These represent a normal condition in haploids, and hardly need further discussion. As mentioned for bivalent-forming dihaploids of upland cotton, the chromosomes of the two parental species may have different sizes, resulting in heteromorphic bivalents. In monohaploids the bivalents are composed of largely non-homologous chromosomes which may well be of quite different size and there may be other differences that result in abnormal bivalent forms.

7.4.4.1.4 Multivalents, Anaphase and Cellular Abnormalities

In mono-, di- and polyhaploids these may be common as a result of pairing in duplications and between chromosomes in different genomes that carry chromosomal rearrangements of different types. These have been briefly mentioned earlier and further discussion is not necessary.

Aneuploids are rare in monohaploids but not really uncommon in di- and polyhaploids. Their characteristics and identification can be derived from the information provided and will not be discussed here further.

7.4.4.2 Triploidy (Table 7.3: I.b.2.b)

The designation "triploid" may be confusing. The most straightforward case is that of an autotriploid of a diploid species (Sect. 6.1.2.2.1). This can be the triploid hybrid between a diploid and its autotetraploid form or a spontaneous triploid appearing among the progeny of a diploid. The hybrid between a diploid and a related autotetraploid or allotetraploid species can also be designated as a triploid. In the latter case the same combination of three genomes is found in the trihaploid derived by reduction from an allohexaploid.

In several cultivated species different levels of ploidy occur within the genus. In the present context it is simplest to consider the basic genome as the unit, rather than the "standard" chromosome number, as in the previous sections. This means that for the autotetraploid ($4x = 48$) potato the triploid number is considered to be $3x = 36$ and not $6x = 72$. For trihaploids derived from allohexaploids, the reader is referred to the (tri)haploid (Sect. 7.4.4.1).

7.4.4.2.1 Univalents

Univalents (Table 7.3: II.a.1.a) without other configurations except bivalents, including the extreme case where (almost) no bivalents occur, is a real possibility. The case where, in some cells, trivalents are observed is not considered in this section. Consistently one univalent per cell or one set of three (Table 7.3: II.b.1.a) is not expected.

Many univalents, random distribution: There are two situations: (a) combined with a bivalent as an alternative pairing configuration for a trivalent; (b) as a result of breakdown of trivalents and bivalents due to lack of chiasmata. Three general cases are distinguished: (1) *autotriploid* with exclusive bivalent/univalent pairing. The univalents represent the equivalent of one complete genome (or slightly more) and are present in the basic chromosome number. (2) *Desynaptic autotriploid*. Trivalent frequency so low that (practically) no trivalents are observed, and the majority of configurations are open bivalents and univalents. (3) Spontaneous or artificial *triploid interspecific hybrid* between diploid and allotetraploid species or *trihaploid* derived from allohexaploid. These can have any univalent distribution between (1) and (2). The AAB combination of genomes resembles (1). In the ABC combination, there are mainly univalents due to insufficient homology, possibly reinforced by desynapsis. *Distinction*: Simple on the basis of previous knowledge of the history. *Anaphase* I segregation is usually irregular, depending on univalent frequency. *Pachytene* pairing, especially in SC preparations may give an impression of the extent of the homology, but caution is required because of the possibility of extensive non-homologous pairing. *Consequences*: Low fertility. *Context*: Rarely (but not impossible) spontaneous as natural or incidental autotriploids or triploid hybrids; more frequently trihaploids from auto- or allohexaploids; as artificial hybrids and in chromosome manipulation programs.

7.4.4.2.2 Heteromorphic or Otherwise Abnormal Bivalents (Table 7.3: II.a.1.b)

At the triploid level, but without trivalents, these are expected only in the same cases as the univalents in the previous section. Heteromorphic bivalents may then be the consequence of differences in chromosome size between genomes, and may have a frequency corresponding to one genome, or less when the bivalents have broken down because of low chiasma frequencies.

A single heteromorphic or abnormal bivalent per cell will be an incidental deficiency, duplication, inversion, etc., as in forms with the normal chromosome number, and can be identified in the same way as discussed there.

7.4.4.2.3 Multivalents

7.4.4.2.3.1 Trivalents (Table 7.3: II.a.1.c.1)

Trivalents are the typical configuration for autotriploids and triploid hybrids between related species (Sect. 6.1.2.2.1.3). When no other configurations than trivalents, bivalents and univalents are observed, the latter two in equal numbers or with univalents in excess, the conclusion is justified that one of these two cases is involved. With low trivalent frequencies, the distribution of the trivalents over cells should still be random, even when rarely more than one is found. In triploid hybrids with low chiasma frequencies, the univalent frequency may be high and the bivalent and especially the trivalent frequency low.

Usually the distinction between an autotriploid and a triploid hybrid can be made on the basis of the history of the material. Occasionally, the triploid has an unknown origin. It is not entirely excluded that trivalents in triploids are due to other reasons than triploidy alone, but this must be considered an exception. Additional aberrations will usually be recognizable by the appearance of configurations larger than trivalents.

When one of the *trivalents* is *heteromorphic*, a *karyotype* analysis may show the presence of a deviating chromosome, for instance a telocentric, a deficiency, a duplication or an inversion chromosome. If this does not give a clue, the identification of the aberration is similar to that of heteromorphic bivalents at the diploid level. A complete genomic set of heteromorphic trivalents is most probably due to chromosome size differences between species involved in a triploid interspecific hybrid or trihaploid. It is improbable that this would not have been known to start with.

7.4.4.2.3.2 Higher-Order Multivalents (Table 7.3: II.a.1.c.2,3) (in Triploids)

Multivalents larger than trivalents are expected to occur as *one per cell*, at most, and with a distribution that suggests that this is the maximum. Quadrivalents, when not found in aneutriploids and not accompanied by larger configurations in other cells, are not simply due to translocations as in diploids, because these should at least occasionally form larger multivalents. When the quadrivalent is indeed the largest configuration, and this has been checked in a large number of cells, the origin must be complex. It could be the combination of a tetrasomic and a disomic, which might possibly be seen in the *karyotype* if sufficient details can be made visible. If there is *one extra chromosome*, the quadrivalent indicates tetrasomy, and it should be found in a maximum of one per cell. With two extra chromosomes a maximum of two quadrivalents should be seen in cases of double trisomy.

Without extra chromosomes *quinquivalents* are not expected when no hexavalents occur in other cells. *Hexavalents* result from translocations and some forms of duplication. Many variants of meiotic behaviour with and without extra chromosomes at the triploid level can be conceived, but will not be discussed.

In all these cases triploidy as such dominates the *pachytene* and *anaphase* I behaviour as well as the *consequences* and the *context*.

Similarly, *anaphase* I (and later) abnormalities (Table 7.3: II.a.2) may occur in triploids, but they will have the same origin as in diploids, with possibly some complications due to triploidy as such. In some triploids apomixis or very specialized meiotic and fertilization systems result in apparent normal fertility; this is an interesting phenomenon, but rare and it will not be discussed further here.

7.4.4.3 The Doubled Chromosome Number (Table 7.3: I.b.2.b.2)

"Double" the number of chromosomes refers to twice the standard number of chromosomes for the species considered. For diploids this is the tetraploid chromosome level, for instance in the form of autotetraploidy. Further, it could be a spontaneous or constructed tetraploid hybrid resulting from the combination of unreduced gametes. For allopolyploids the situation is comparable, but at a higher ploidy level in relation to the basic chromosome number. For natural or induced, established autotetraploids the doubled level of ploidy is octoploid. The autotetraploids themselves are considered elsewhere (Sects. 6.1.2.2 and 11.3.1.2) and above in Section 7.4.1.

7.4.4.3.1 Only Normal Bivalents

There are basically two possibilities:

1. The plant is an autopolyploid with an exclusive bivalent formation. It is an exceptional but possible and important situation. In the *karyotype* the four homologues of each chromosome are identical and at *pachytene*, if analyzable, quadrivalents may or may not be present. *Anaphase* I is regular; fertility normal or slightly reduced because of gene dose effects. *Consequences*: Segregation tetrasomic with or without double reduction: a good diagnostic characteristic, if markers are available. These markers are duplex if the plant observed is the first somatically doubled generation, and heterozygous. No other abnormalities are expected. Different special mechanisms assumed to cause the absence of quadrivalents are discussed in Sections 6.1.2.2.2.3 and 11.3.1.2.3. *Context*: Programs of induction of polyploidy; infrequently genetic manipulation programs involving an in vitro phase; rarely spontaneous.

2. The plant is an *allopolyploid*, the result of spontaneous hybridization combined with the functioning of unreduced gametes. It is not expected to be at all frequent, but cannot be excluded when different related species are grown together and when the species involved tend to form functional unreduced gametes regularly. The possibility of an admixture should not be overlooked. A taxonomic description may be expected to recover alien in addition to "standard" parental characteristics. A detailed *karyotype* analysis may possibly show chromosome characteristics from different species and not four identical copies of each chromosome type. The presence of potential parents in the vicinity of the parent of the deviant should be checked. *Pachytene* is not expected to show many, but possibly a few quadrivalents. *Anaphase I* is normal. *Consequences*: Fertility normal, except when disturbed by genetic imbalance, which would certainly be the most common situation; segregation disomic. *Context*: Infrequently spontaneous; further: in hybridization programs, generative or somatic.

7.4.4.3.2 Univalents (Table 7.3: II.a.1.a)

The consistent or regular presence of *one pair of univalents* and no other configurations except bivalents is exceptional and would have an origin comparable to the univalent pair discussed for normal diploids. A smaller or larger number of *randomly distributed* univalent pairs, again with only bivalents, points to asynapsis or desynapsis in a spontaneous or induced autotetraploid with infrequent quadrivalent pairing. When, as a result of pronounced asynapsis or desynapsis, the frequency of univalents is large, the formation of quadrivalents may be suppressed merely because of this. Alternatively, a new allotetraploid (see above) may be involved. The same approach as for the previous cases can be used for the identification of the type of polyploid. *Anaphase I* segregation is irregular. *Consequences*: Fertility low; further: similar to desynapsis with normal chromosome number, with additional complications due to the presence of additional genomes. *Context* is as above.

7.4.4.3.3 Heteromorphic or Otherwise Abnormal Bivalents

These bivalents combined with a doubled chromosome number, and no other meiotic abnormalities, must be considered a combination of different, independent events and can be treated as such.

7.4.4.3.4 Multivalents

Trivalents as the maximum configuration are not expected, but could result from special rearrangements combined with suppression of quadrivalents.

Quadrivalents are the typical configurations for autotetraploids and for tetraploid hybrids between related species. A single quadrivalent, however, found consistently or occasionally, but then in other cells replaced by its alternative pairing configurations or breakdown products, is not probable. The origin could be heterozygosity for a translocation or a duplication in an autotetraploid with suppressed multivalent formation. In a tetraploid species hybrid, the translocation or duplication would have to occur as a heterozygote within one of the two parents. In somatic fusion programs it could be present in either one of the parents from the start, or may have originated as a result of in vitro culture.

Randomly distributed quadrivalents with a variable frequency typically point to *autotetraploidy* or to a tetraploid hybrid between related species. *Morphology*: "Gigas" and other polyploidy characteristics. *Karyotype*: Four identical chromosomes of each type, or slightly variable when a hybrid is involved. *Pachytene*: If analyzable, several to many quadrivalents. *Anaphase I*: Slightly irregular. *Consequences*: Reduced fertility, degree depending on anaphase behaviour and physiology; segregation of off-type aneuploids; tetrasomic inheritance. Further characteristics are dealt with in Sections 6.1.2.2.2 and 11.3.1.2.2. *Context*: *Spontaneous*, for instance after regeneration in in vitro programs.

7.4.4.3.5 Anaphase and Cellular Irregularities

Anaphase I irregularities combined with a doubled chromosome number without other meiotic abnormalities are comparable to the cases in which single quadrivalents and heteromorphic bivalents occur as mentioned above. Hybrids between closely related tetraploids which form quadrivalents in combination with anaphase I bridges due to paracentric inversion heterozygosity are found, for instance, in cultivated tetraploid *Tradescantia* spp. and their hybrids (Sect. 12.2.2.2).

7.4.4.4 Aneuploidy in Combination with Doubled Chromosome Number

Trisomics (one chromosome less) and pentasomics (one chromosome more) are common in established, induced or even in natural autotetraploids. At meiosis single *trivalents* without an accompanying univalent and *quinquivalents* respectively are expected. For details, see Section 7.4.2 and 11.3.1. When a spontaneous polyploid species hybrid is involved, aneuploidy has apparently resulted from a meiotic irregularity at the origin of the unreduced gamete forming the polyploid. Especially after asymmetric, but also after symmetric protoplast fusion, normal chromosome aneuploids are expected, but here hyperploidy can also involve any kind of spontaneously rearranged chromosome (Wijbrandi et al. 1989). For the exact identification the same methods as used in allopolyploids or autopolyploids should be used: karyotype analysis, if necessary test crosses. Occasionally, B-chromosomes may be present and mistaken for extra normal chromosomes. They are not really common, but as in any other case of aneuploidy (Sect. 7.4.2) it is useful to check for their presence in the original material. They will not arise de novo.

7.4.4.5 Higher Ploidy Levels

A spontaneous origin of ploidy levels higher than the doubled number is not excluded in generative reproduction. Their identification follows the lines of doubled chromosome numbers (Sect. 7.4.4.3).

An interesting origin is protoplast fusion. Wijbrandi et al. (1989) reported that symmetric and asymmetric fusion between *Lycopersucum esculentum* (tomato) and irradiated *L. peruvianum* protoplasts resulted in euploid and aneuploid regenerants at the tetraploid and hexaploid level. The genomes and even the individual chromosomes involved could be recognized by RFLP and other marker analyses, both directly and in segregating progenies. The many accompanying rearrangements in irradiated protoplasts in asymmetric fusion and those formed in the callus phase disturb the analysis considerably. Instead of RFLP and other marker analyses, karyotype analysis may be possible in favourable material, and test crosses followed by analysis of segregation are effective when sufficient markers are available.

Abnormal cellular processes (Table 7.3: II.a.2.f) in standard karyotypes, especially first division restitution, resulting in unreduced gametes which may develop parthenogenetically, are somewhat more common in autotetraploids than in diploids (Sect. 12.5). They may occur in, but also outside, programs for the induction of apomixis or variants thereof and can often be readily recognized as variants of the meiotic process. The origin of unreduced gametes is usually meiotic restitution (Sect. 11.3.1.2.1.2). It can be a genetically conditioned phenomenon. In addition to restitution, leading to unreduced gametes, several other types of abnormalities of meiosis can be encountered, of which desynapsis is the most common. The *consequences* are usually low fertility and abnormal segregants. The *context* is variable: special genotypes, including interspecific hybrids, abnormal environmental conditions. They tend to be readily recognized, although identification of their origin if not obvious from the context, may require special experimentation.

Haplophase (Table 7.3: III.a) and *endosperm development* (Table 7.3: III.b) abnormalities are as discussed for natural or induced, established autotetraploids. The reader is referred to Section 6.1 and to the induction of allopolyploidy in Chapter 11 (Sect. 11.3.2.2).

Chapter 8

Estimating, Recording and Manipulating Recombination

8.1 Chromosome Recombination and Exchange Recombination

Recombination is the formation, in the gametes of a plant or animal, of *new combinations* of different alleles of two or more genes. *New* refers to the original combination of alleles present in the genomes of the parental gametes from which the individual concerned has been formed. The subject is discussed in most textbooks on general genetics. The *observation* of recombination is based on the expression of the relevant alleles in the phenotype (in the broadest sense) of the progeny. Due to dominance and epistasy not all recombinational events are immediately detectable in the progeny phenotype. Analysis of recombination, therefore, is usually indirect and requires special experimental design. Recombination is the basis of most plant breeding programs, and it is appropriate to analyze its components, the ways it can be manipulated to serve its purpose optimally, and its actual role in plant breeding.

Recombination has two components: *chromosome recombination* and *exchange recombination* or *crossing-over* (Sect. 3.2.4.1).

8.1.1 The Role of Chromosome Number in Recombination

8.1.1.1 Chromosome Recombination in Diploids as the Basis of Mendelian Inheritance

The Mendelian "law" of independent assortment of genes is based on the phenomenon of independent orientation of bivalents at meiosis in disomics (diploids and allopolyploids). The consequence of independent orientation of bivalents is independent segregation of the homologous chromosomes derived from the parents, and similar independent segregation of alleles of genes in these chromosomes. The segregation of these genes, in the progeny obtained after selfing of a heterozygote (Mendelian inheritance) as well as in a test cross have been briefly discussed in Section 3.2.4. The free recombination, resulting from the independent assortment of genes in different chromosomes, implies 50% recombination, but at the same time maintenance of the original combination in the other 50% for each pair of alleles. The higher the number of chromosomes in a diploid, the more segments can recombine with this frequency of 50%: the overall level of recombination is directly proportional to the number of chromosomes.

8.1.1.2 Variation in Chromosome Recombination

8.1.1.2.1 Ploidy Level

8.1.1.2.1.1 Haploids

In haploids, chromosome recombination is excluded. The few vegetatively reproducing haploids are not expected to segregate except by mutation. In several insects and spider mites haploidy of the males (arrhenotoky) is part of the reproductive system, there is no meiosis and all their sperm have the same genotype. When a single male fertilizes a female, segregation among her progeny results exclusively from recombination during her own meiosis.

8.1.1.2.1.2 Polyploids

In *allopolyploids* the number of chromosomes increases proportionally with the level of polyploidy. The direct consequence is a higher level of chromosome recombination and, because each homoeologous group (Sect. 6.1.2.3) consists of more than two chromosomes, homozygosity of all copies of a particular gene as a result of inbreeding is attained later compared to the diploid. However, heterozygosity for segments in one pair will only lead to observable segregation when in the homoeologous pair both chromosomes have the recessive allele. This obviously reduces the effect of the larger number of chromosomes on increasing recombination. On the other hand, the interaction between different alleles of the corresponding genes in homoeologous chromosomes in allopolyploids is quite complicated. Therefore, the effects of recombination tend to be more complex than predicted by simple dominance models. In addition, an increase in recombination in allopolyploids compared to diploids is of a different kind than an increase in recombination due to increased chromosome number with an equal genome number in diploids.

For *autotetraploids* the situation is different again. Heterozygosity is maintained much longer than in diploids and even allopolyploids (Sect. 6.1.2.2.2.4) and segregation continues for many more generations. The allelic interactions are more complex than in diploids. However, since the variation between alleles derived from a single species (as in an autopolyploid) may be expected to be less pronounced than the variation between alleles derived from different species as in allopolyploids, the interactions between the alleles are somewhat less complex than in an allopolyploid. Both chromosome recombination and recombination, resulting from crossing-over, continue much longer in autopolyploids than in diploids and allopolyploids. There is, however, a limit to the increased variation in expression of new allelic combinations, as in allopolyploids. In general, continued segregation is considered a drawback rather than an advantage because it postpones stabilization, and in addition, due to the complex mode of segregation, recombination in autopolyploids is much more difficult to exploit.

8.1.1.2.2 Chromosome-Associated Restrictions of Chromosome Recombination

8.1.1.2.2.1 Affinity

Exceptionally, alleles of genes in different chromosomes do not segregate independently. The phenomenon has been described for yeast and mice and later for *Drosophila* (Michie and Wallace 1953) as "affinity" or linkage between genes in different chromosomes. Causal mechanisms have not been found. Affinity has been searched for in plants, especially maize, but has not been reported to have been observed unequivocally.

8.1.1.2.2.2 Permanent Translocation Heterozygosity

In several species of plants and animals translocations have been reported to "float" in populations and even have been established permanently as heterozygotes (Sects. 5.4.1 and 12.3). Genes in different chromosomes of a heterozygous translocation complex are linked, because chromosome recombination is either excluded or does not come to expression as a result of lethality of the recombined products. Linkage between genes within homologous chromosomes is usually even closer than without translocations. The usefulness of such translocation heterozygotes in nature is assumed to be at least partly in keeping specific gene blocks together. A second favourable consequence is the maintenance of heterozygosity. It has been attempted to introduce permanent translocation heterozygosity in cultivated plant species, but as yet without practical results (Sect. 12.3). These are typical examples of restriction of chromosome recombination.

8.1.1.3 The Relative Importance of Chromosome Recombination in Plant Breeding

In breeding programs of predominantly self-fertilizing species, usually several generations of selfing follow after hybridization. In diploids and allopolyploids, in each generation 50% of the chromosome segments that were heterozygous in the previous generation become homozygous and then are not available for recombination. With so few chromosome segments recombining, with necessarily maximally 50% recombination, combined with a rapid decrease in the number of heterozygous segments, for practical purposes exchange recombination tends to play a limited role in plant breeding. Most of the recombination is chromosome recombination, usually at most doubled by exchange recombination. In order to make better use of recombination by crossing-over, heterozygosity must be maintained over longer periods, or restored regularly by hybridization. This makes the program so much longer that most breeders do not tend to make use of this possibility (Stam 1977; Sybenga 1989).

8.2 Exchange Recombination: Crossing-Over

The basis of exchange recombination has been discussed in Section 3.2.4.1.2. Chromosome recombination results in 50% recombination (Sect. 8.1.1), a level equalled by exchange recombination only when the markers are far enough apart to have consistently at least one exchange event between them. Since this is normally not the case for markers positioned closer together than at or near the opposite chromosome ends, the majority of genes do not have the opportunity to recombine frequently.

8.2.1 Estimates of Crossing-Over

8.2.1.1 Segregation of Genetic Markers

Genetic markers used to study segregation in a recombination analysis, from which crossing-over percentages are to be derived, can be of different types. They may be alleles of genes of which the expression can be recognized in the phenotype: macro- or micromorphology, including cellular morphology or chromosome behaviour, but also isozymes, etc. They may further be specific non-coding DNA segments, recognized with the use of molecular probes, for instance repetitive DNA segments, and coding or non-coding segments recognized in an RFLP analysis (Helentjaris 1987). They can also be chromosome structural rearrangements visible in the somatic karyotype, at meiosis or because of their genetic consequences.

The segregation of two markers may be independent or correlated. In most normal situations by far, correlated or interdependent segregation (linkage) indicates that the markers are situated in the same chromosome. Exceptions may be due to heterozygosity for chromosomal rearrangements (Sects. 5.4.1.4 and 12.3), or to selection of specific recombinants or alleles either at the gametic or at the plant level. Rarely, the reason may be special chromosome behaviour of non-rearranged chromosomes at meiosis (affinity: Sect. 8.1.1.2.2.1). The frequency of *recombination* between markers in the same chromosome, as observed in segregating populations, is an indication of the frequency of *crossing-over* between them, but it is not a direct estimate, and it is lower. Deriving crossing-over frequencies (often expressed as percentages) from recombination estimates is discussed in detail in books on general genetics or cytogenetics.

The reason why recombination and crossing-over are not interchangeable has been briefly discussed in Section 3.2.4.1, where it was shown that two cross-overs between the markers will cancel each other's effect. It is also shown that as a consequence the maximum possible recombination is 50% even though the number of cross-overs does not have a theoretical limit. For low crossing-over frequencies the correspondence between recombination and crossing-over is close, because the probability is small that two cross-overs occur between the two markers, thus reducing the recombination frequency. This is enhanced by interference that reduces the probability of cross-overs being positioned close together. With larger crossing-over frequencies recombination increases, but so does the probability that two cross-overs occur in the same interval.

Estimating crossing-over from recombination between segregating markers requires the presence of proper markers, and, as mentioned above, is indirect. Rarely, these markers will be the genes that have to be recombined for a practical purpose. When they are, an estimate of the frequency of crossing-over is not of primary interest because recombination has already taken place, and only the fact that they do recombine (or do not) is relevant. There are mainly three reasons of interest in crossing-over frequencies:

1. They give an impression of the overall level of recombination in the material, useful for roughly estimating the size of populations required for a reasonable chance to find recombination between a desired gene or gene complex and an undesired linked gene or complex.
2. They give an impression of the distribution of recombination: are there important segments practically without recombination and others with a relatively high level?
3. They are the basis of genetic (chromosome) maps showing the relative positions of markers and useful genes in linkage groups in chromosomes, and the level of recombination between them. These maps also show where regions with limited crossing-over occur.

8.2.1.2 Chiasma Frequency, Recombination Nodules and Chiasmate Chromosome Arm Association Frequency

Chiasmata are the result of the same primary exchange event as crossing-over. They can, therefore, be used to estimate the frequency and distribution of crossing-over in specific chromosome segments. In fact, chiasmata are more directly related to crossing-over than recombination between genes, because they represent exchange events directly, whereas recombination is the combined result of an a priori unknown number of cross-overs in the segment considered. In addition, it is occasionally possible (rarely in plants) to trace the course of the chromatids in subsequent chiasmata and analyze chromatid interference.

In order to estimate crossing-over frequencies from chiasma frequencies, markers are again necessary. In this case these are chromosome structural markers: centromeres, telomeres, distinct C-bands, deficiencies, breaks of chromosomal rearrangements if recognizable, and occasionally a distinct gene, usually a multicopy gene made visible by in situ hybridization. In order to be used as markers in chiasma frequency studies, it is not necessary that they are heterozygous as is required in segregation analyses. Some of these markers can be used as segregational markers in genetic experiments, provided they are polymorphic, and then an analysis of crossing-over in specific chromosome segments can be combined with a cytological analysis. The use of *chromosomal markers* in genetic segregation experiments is the basis of locating genes on and within chromosomes, which is necessary for constructing genetic chromosome maps. The frequencies of chiasmata between these markers provide additional information that cannot, or only less directly, be obtained from genetic segregation analysis alone.

In principle very promising, but in practice quantitatively applicable only in a few plant species, is the determination of the frequency and location of late *recombination nodules* in synaptonemal complexes, studied in the electron microscope. There is little doubt that they represent the sites of exchange of non-sister chromatids at the original location. Especially in combination with centromeres, NORs, occasionally heterochromatic segments and other systematic or incidental chromosome structural markers (translocations, inversions, etc.), they permit the exact physical localization of recombination events. Disadvantages are the difficulty of applying the technique and the great amount of work required for a satisfactory quantitative analysis. This section further deals with chiasmata only. The combination of an analysis of chiasmata with a segregation analysis is considered in Section 8.3.

The stage in which chiasmata can be studied in most detail is diplotene. There are very few plant species where diplotene is at all accessible to a chiasma frequency and localization analysis and a quantitative analysis is almost never possible. Stack et al. (1989) report on a method of interpreting diplotene bivalent shapes in terms of the number of chiasmata they contain. In insects, and especially several grasshopper species, diplotene is much more favourable, and several detailed studies have been performed (Sect. 3.2.2). In some of these species it has been possible to prepare metaphase configurations in such a way that the core of the chromosomes becomes clearly visible after incubation in silver nitrate (Rufas et al. 1987; Santos et al. 1987; Suja et al. 1991). The exact course of the chromatids in the chiasmata and the exact location of the point of exchange can be observed in the structures that would otherwise hardly be distinguished as separate chiasmata. These cores are difficult to demonstrate in plants, but do exist (*Vicia*; de Jong, pers. comm.).

In most plant species, therefore, chiasma frequency and distribution must be studied at diakinesis-metaphase I on chromosomes that are too condensed to show any details of the chiasmata themselves. It is often not even possible to distinguish one chiasma from two or more. In most material knowledge of the specific changes in bivalent morphology (Sect. 3.2.2.2) will make it possible to make a reasonably reliable estimate of the number and location of the chiasmata, but this is by far not as accurate as diplotene and metaphase core observation. Yet, like Stack et al. (1989) for diplotene in *Lilium*, Kumar et al. (1990) succeeded in developing systematic methods for establishing the number of chiasmata in diakinesis and metaphase I configurations in *Pisum*, on the basis of their morphology. In this case the configurations were quadrivalents in autotetraploids, in which on first inspection chiasmata are difficult to score. In a number of species there is a genetically and environmentally determined difference between plants or varieties in the degree chiasmata can be recognized. An example is rye (Fig. 3.10), where in some genotypes or under some conditions chiasmata can be distinguished with reasonable certainty, whereas in others this is impossible or almost so (Sybenga 1975). Several species with small chromosomes are even more difficult.

When it is not possible to distinguish between one or more chiasmata in a chromosome arm, all that can be seen is whether or not there is *chiasmate association* between the homologous arms, or, in multivalents, the homologous segments. The chiasmate association frequency of a chromosome segment is transformed into the approximate number of chiasmata involved by applying the same mapping functions used for transforming recombination frequencies into crossing-over frequencies (Sybenga 1975; see Sect. 8.2.1.4). Chiasma frequencies are transformed into crossing-over frequencies by dividing by two: in each chiasma only two of the four chromatids participate in exchange.

Occasionally, there is some uncertainty about the character of chromosome association at metaphase I. Chromosome stickiness keeps chromosomes together at anaphase I without chiasmata necessarily being involved. It results in anaphase bridges and is usually caused by the action of factors disturbing normal chromosome behaviour (heat, irradiation, and other environmental stress factors, as well as gene imbalance). It is probable that in a few cases non-chiasmate associations, possibly related to stickiness, but not persisting until anaphase I, keep metaphase chromosomes together. These too are formed primarily under abnormal conditions (Orellana 1985). For the present purpose they are neglected unless it can be shown that they play a significant role.

8.2.1.3 Mapping Functions

The first to formulate the mathematical relation between crossing-over and recombination, considering only double cross-overs, was Haldane (1919). The function was based on the Poisson distribution of cross-overs, which is correct only in the absence of interference. Interference requires special adaptations to the distribution of chiasmata, as it reduces the frequency of double cross-overs in a non-random way. In addition, interference differs in intensity and range between species and possibly even between genotypes. Kosambi's mapping function (Kosambi 1944) uses an 'average" measure of interference for correcting the frequency of double cross-overs in the segment considered. It is in common use mainly because it is simple to apply and reasonably realistic for average intervals. It has the form of: $y = ½(\text{th } 2x)$, where y is the recombination frequency (or chiasmate association frequency) and x the crossing-over frequency (or chiasma frequency, respectively). More complex mapping functions have been used in human genetics, which fit the special characteristic of human genetics better than Kosambi's function. Mapping functions are not necessary for short intervals because, in part as a result of interference, double cross-overs do not occur. For large intervals (recombination approaching 50%) no accurate transformation of recombination frequency into crossing-over frequency is possible because of the large inherent error of the recombination estimates. The use of large segregating populations would seem to reduce the error of recombination estimates close to 50%, but then variation due to different causes (environmental, physiological within plants, etc.) becomes important and increases the error again.

In addition to recombination frequencies, meiotic observations can be used to study other genetic parameters like interference and distribution of genetic exchanges. These can best be studied when separate chiasmata can be distinguished. Interference affects the distribution of the numbers of chiasmata over chromosomes and chromosome arms, and this can be used to make estimates. These subjects have been considered by Sybenga (1975) and will not be discussed here.

8.2.1.4 Models for Deriving Chiasmate Association Frequencies from Configuration Frequencies

One can determine the genetic length of chromosome segments directly from chiasma counts in specific chromosome segments. It is also possible to derive map distance purely from the frequency of chiasmate association observed at meiosis. Frequencies of chiasmata and of chiasmate association have been reported for numerous chromosome segments when markers were available, such as large deficiencies or duplications and in translocation heterozygotes. It should be remembered that the chiasmate association frequency is twice the recombination frequency.

In several cases the borders of the different segments are not clear, nor can the different segments always be identified. This is especially so in large configurations with segments of comparable size. Mathematical analytical models for estimating how the chiasmate association frequencies are distributed over the segments of the configurations can provide some help.

Even for a bivalent it is often not possible to distinguish the two arms at meiotic metaphase I, nor, when the chromosomes are strongly contracted, to count the number of chiasmata in the arms. A relatively simple model can be used to estimate the chiasmate association frequencies of the separate arms in this case. It is based on a few assumptions that may be correct in most cases, but that may not be so in others. One general assumption is that chiasma formation in different segments is independent, i.e. that there is no interference.

Several such models have been discussed by Sybenga (1975) and one will be briefly considered here: that for the normal bivalent.

When two homologous chromosomes have paired to form a bivalent, chiasmata may be formed in both arms, or in one and not in the second, or in the second and not in the first, or in neither of the two arms. In the first case, a ring bivalent is formed, in the second and third an open bivalent and in the fourth case a pair of univalents. When the arms cannot be unequivocally recognized, the two types of open bivalents cannot be distinguished. The assumption that both arms will then have the same chiasma frequency is not necessarily correct. There are three types of recognizable configurations of which the frequency can be determined: ring bivalents, open bivalents and pairs of univalents. The sum of the frequencies equals 1, so there are two degrees of freedom from which two parameters can be estimated. When it is assumed that there is no interference across the centromere, these two parameters may be the chiasmate association frequencies of the two arms. On the other hand, if the assumption of no interference is not accepted, three parameters would be required, including one for the estimate of interference. Callan and Montalenti (1947) made the assumption that the two arms of the metacentric chromosomes of *Culex pipiens* had the same association frequency and they used one degree of freedom to estimate the average arm association frequency and the other to estimate interference. Presently, the main interest is in the chiasma frequencies of the two arms, and the assumption of no or very restricted interference across the centromere is accepted. This is realistic for chiasmata not close to the centromere, and this is common in many instances.

With chiasmate association frequencies a and b for the arms A and B, the frequencies (Fig. 8.1) of the configurations are:

ring bivalents: $r = a \,.\, b$
open bivalents: $o = (1 - a) \,.\, b + (1 - b) \,.\, a = a + b - 2a \,.\, b$
univalent pairs: $u = (1 - a) \,.\, (1 - b)$

Two of the three equations are independent. From the relations:

$a \,.\, b = r$ and $a + b = o + 2r$

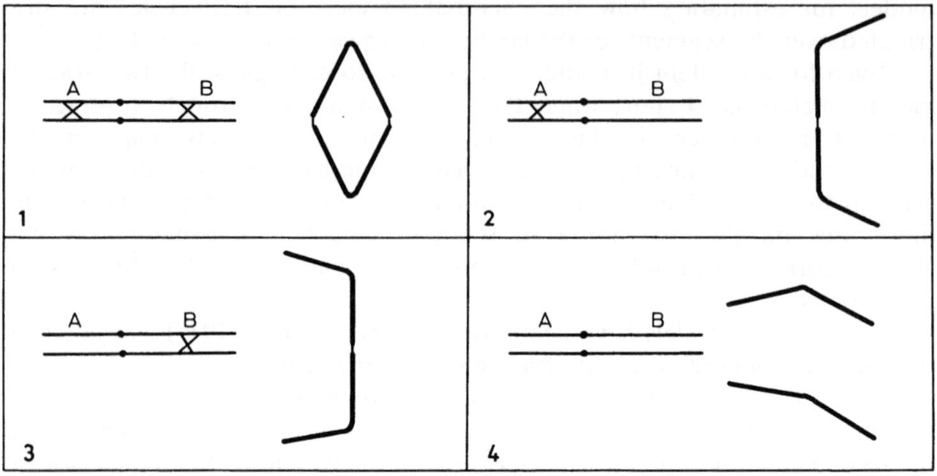

Fig. 8.1 Diagram of the metaphase I shapes **(1–4)** of a bivalent of a metacentric chromosome pair, arms *A* and *B* with chiasmatic association frequencies *a* and *b* respectively. Rings are formed when both arms have at least one chiasma (frequency $a \cdot b$); open bivalents when one has a chiasma, the other does not: $(1 - a) \cdot b + (1 - b) \cdot a$; univalents: $(1 - a) \cdot (1 - b)$. (After Sybenga 1975)

a quadratic equation can be derived:

$x^2 - (o + 2r)x + r = 0$, where the two roots for x represent a and b, respectively:

$$x_{(a,b)} = \frac{o + 2r \pm \sqrt{\{(o + 2r)^2 - 4r)}}{2}.$$

It is clear that real roots for a and b are obtained only when the discriminant is positive. There are cases where this is not so. By adjusting the expression for the discriminant by introducing the univalent frequency, which equals $u = 1 - r - o$, it can be shown which these cases are. The discriminant now becomes $(o + 2r)^2 - 4r = 4r\,(r + o - 1) + o^2$ or $o^2 - 4r\,.\,u.$, from which can be concluded that a relative excess of the combination of configurations where both arms are associated (r = rings) with configurations where both arms are not associated (u = univalents) can prevent real roots from being obtained. This is the opposite of (positive) interference, where the combination of one arm associated and the other not (open bivalents) is in excess. The excess of rings and univalents represents *negative interference*. In the quadratic equation the discriminant determines the difference in chiasmate association frequency between the arms. Therefore, when there is a difference, negative interference may be present, but it is not observed. It simply reduces the difference estimated, which is then not a good estimate. The difference, on the other hand, is increased by positive interference and it is correctly estimated only in the absence of interference.

The model of the bivalent has been discussed in some detail because the models for more complicated configurations are based on the same principles. The *interchange heterozygote* is one of the most frequently observed rearrangements with complex configurations in which the different end segments cannot readily be distinguished, and where a model helps in estimating chiasmate association frequencies. It will not be discussed, but the reader is referred to the original literature (Sybenga 1975).

8.2.1.5 Variation in Crossing-Over

Crossing-over and its corollary chiasma formation vary in two ways: *frequency* and pattern of *distribution*. In the latter again two components can be distinguished: *interference* and *localization*. Both frequency and distribution vary between taxa in a sometimes characteristic way which must have a background in genetic differences. In addition, there is genetic variation within species and even within populations. Genetic variation in crossing-over and chiasma formation is quite common. It can be monogenic in the form of varying degrees of asynapsis (failure of pairing) or desynapsis (failure of chiasma formation or chiasma maintenance after sufficient pairing). Genetic variation in crossing-over and chiasma formation can also be polygenically conditioned (Sect. 7.4.1.1).

Similarly, localization is under genetic control. In a hybrid between the related subspecies *Secale cereale dighoricum* and *S. cereale turkestanicum*, which both have a rather distal chiasma localization, Jones (1967) reported that the chiasmata had a nearly random distribution. In the F2 different types segregated suggesting that several, although not many genes were involved. A pronounced difference in chiasma localization is observed between *Allium cepa* (distal to subdistal) and *A. fistulosum* (extreme proximal). In the hybrid the distribution of the chiasmata is nearly random, apart from effects of interference, and in the F2 segregation of different distribution types can be observed (Fig. 8.2, after Darlington 1965), accompanied by great variation in chiasma frequency. Here, too, a limited number of genes is probably responsible. How such genes operate is not known, but possibly through chromosome pairing. Within species variation in chiasma formation is in part concentrated in specific chromosome segments. An example is the analysis of variation in crossing-over in the loop of an inversion heterozygote in different genotypes of barley (Säll et al. 1990), where part of the variation was apparently due to chiasma formation specifically in the loop area.

In addition to genetic causes, several environmental agents affect crossing-over and chiasma formation, some directly, some by affecting the degree of heterochromatinization. There is little recent information. A relatively old, brief review is given by Sybenga (1972). In Section 8.4 variation in crossing-over and the ways it can be manipulated will be briefly considered again, and more extensively in Chapter 10.

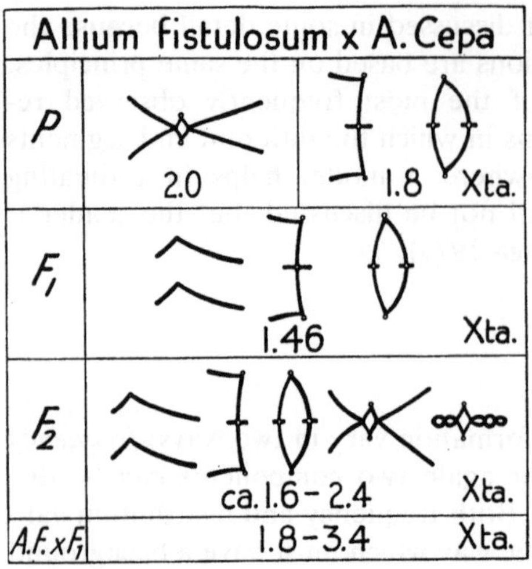

Fig. 8.2 Genetically determined chiasma localization in *Allium* (Darlington 1965; data from several authors combined). In *A. cepa* the chiasmata have a somewhat distal distribution, in *A. fistulosum* a pronounced proximal distribution with, on average, slightly more chiasmata. In the hybrid (F_1) the distribution tends to be more random, but somewhat distal as in *A. cepa* and the chiasma frequency is reduced. In the F_2 different types segregate, some with distal, some with proximal and some with completely random distribution. In the last case the chiasma frequency may be greatly increased; in other cases it is lower than in the lowest parent

Chromosomal markers used in gene localization studies may have an effect on crossing-over. This may disturb the analysis of genetic distance between such markers and gene loci. A few cases are mentioned in the sections on recombinational chromosome mapping, and should serve as a warning regarding the possibility that the map distances estimated may not always be fully representative of the situation in the absence of such markers. Similarly, when rearrangements occur in breeding material, they may well affect the frequency of recombination, usually by reducing it. There are exceptional cases where recombination can be increased in specific segments of translocation heterozygotes. These will be mentioned briefly in Section 8.4 when the manipulation of recombination is discussed.

8.3 Recording Recombination: Genetic Chromosome Maps

8.3.1 Recombinational Maps of Genes and Other Markers; Three-Point Test

The frequency of recombination between two loci can be considered to represent a distance between these loci. When the recombinational (or genetic) distances are known between more than two loci, these loci can be placed on a map, on which the recombination frequencies are represented by map distances. Knowing that genes are located in a linear order in the chromosomes, this map can represent the genes on the chromosomes, but it is necessarily one-dimensional: there is a straight line with genes at specific distances.

The construction of a genetic chromosome map requires not only that the distances are known, but also the order. Establishing genetic distance and order is most accurately and most conveniently performed when three genes (or other markers) are involved: *three-point test*. There are modifications of the three-point test in which more than three markers are involved, but these present a number of technical difficulties. Chromosome morphological markers may well be included in a three-point test, as long as two forms ("alleles") of each of the three markers can be seen to segregate. In a standard three-point test a triple heterozygote is test-crossed with a recessive or co-dominant (the most common type for chromosomal and biochemical genetic markers) triple homozygote. The F2 has a much more limited resolution. In a test cross, $2 \times 2 \times 2 = 8$ classes can be recognized, with seven degrees of freedom from which seven independent parameters can be estimated. Without linkage the eight classes are expected to occur at equal frequencies. The three monofactorial segregations can be analyzed (expected to be 1:1 in a test cross) and the three recombination frequencies between the markers can be estimated. One degree of freedom is then still available. The largest recombination frequency is assumed to represent the largest distance and is between the genes farthest apart. The third gene must be located between them.

When there is simultaneous recombination between the gene in the middle and both genes at the sides, there is double crossing-over, which restores the original allelic combination between the outer genes (Sect. 3.2.4.1.2.1; Fig. 3.13). As a consequence, these outer genes do not show recombination. To obtain a realistic estimate of the genetic distance between the two outer genes, the double cross-overs have to be added to the number estimated directly. After this correction the product of the recombination frequencies within in the two segments must equal the double cross-over frequency between the outer markers, if crossing-over in the two segments is independent. If this is not so, there is *interference*. Interference is the rule rather than the exception. In Table 8.1 an example of a three-point test is given, including as estimate of interference. This uses the last degree of freedom in the analysis.

The result of a three-point test is the most elementary *genetic map*: three markers, their order and the distances between them. A combination of different three-point tests including one or, preferably, two of the markers from previous tests extends the map to all markers involved. Special statistical methods are available to cope with the inevitable variation between recombination estimates from different experiments. However, as long as only genes are involved and no chromosomal markers, even when the chromosome

Table 8.1. Three-point test in maize

All three factors involve properties of the seed: the segregation can be read on the cob of a selfed F_1 plant.

Parents: P_1 *c* (colourless aleurone) – *sh* (shrunken seed) – *wx* (waxy endosperm)
P_2 *C* (coloured aleurone) – *Sh* (smooth seed) – *Wx* (normal endosperm)

$$F_1: \frac{c - sh - wx}{C - Sh - Wx}, \text{ test-crossed with } \frac{c - sh - wx}{c - sh - wx}$$

Types in test cross with numbers of seeds found (representing gametic ratios):

C Sh Wx	*C Sh wx*	*C sh Wx*	*c Sh Wx*	*C sh wx*	*c Sh wx*	*c sh Wx*	*c sh wx*	Total
2238	672	19	98	107	39	662	2198	6033

Monofactorial segregations: $C : c = 3036:2997$, $Sh : sh = 3047:2986$ — slight shortage of recessives; $Wx : wx = 3017:3016$

Crossing-over $C - Wx$:
$$C\,wx = 672 + 107$$
$$c\,Wx = 98 + 662$$
$$1539 \quad \text{crossing-over } \frac{1539}{6033} \times 100 = 25.51\%$$

Crossing-over $C - Sh$:
$$C\,sh = 19 + 107$$
$$c\,Sh = 39 + 98$$
$$263 \quad \text{crossing-over } \frac{263}{6033} \times 100 = 4.36\%$$

Crossing-over $Wx - Sh$:
$$Wx\,sh = 19 + 662$$
$$wx\,Sh = 672 + 39$$
$$1392 \quad \text{crossing-over } \frac{1392}{6033} \times 100 = 23.07\%$$

The greatest crossing-over percentage corresponds to the greatest distance, and this must be between the outer two loci. The order, therefore, is *C-Sh-Wx*. The sum of *C-Sh* and *Sh-Wx* is 27.43 which is more than *C-Wx* estimated directly (25.51). The difference is a result of double crossing-over.

Double crossing-over:
$$C\,sh\,Wx = 19$$
$$c\,Sh\,wx = 39$$
$$58 \quad \text{percentage } \frac{58}{6033} \times 100 = 0.96\%$$

The product of 23.07% and 4.36% = 1.01% is the expected double crossing-over frequency. The difference is due to interference. The coincidence value *c* can be calculated as $\frac{0.96}{1.01} = 0.95$ and the interference equals $1 - 0.95 = 0.05$.

The distance *C-Wx* can be estimated directly when double crossing-over is taken into account, i.e. the double crossing-over frequencies count twice:

$$C\,wx = 672 + 107 + 2 \times 19$$
$$c\,Wx = 98 + 662 + 2 \times 39$$
$$1655 \quad \text{crossing-over } \frac{1655}{6033} \times 100 = 27.46\%$$

(Results from a student course, Department of Genetics, Wageningen.)

involved in known, the map does not have any other relation with this chromosome other than that it is situated somewhere on it. All genes between which linkage has been established (together forming a *linkage group*) can be brought into one recombinational map.

8.3.2 The Location of Genes on Chromosomes

For genetic chromosome mapping, i.e. to correlate the genetic map with the chromosome on which it is located, preferably in relation to specific chromosomal landmarks, but also for several other reasons, it is of interest or even necessary to know the chromosome on which a gene is located. This, of course, makes sense only when the chromosome involved can be identified not only as the carrier of linkage groups, but also morphologically in a cytological preparation. This is possible on the basis of specific characteristics such as size and centromere location, a banding pattern, the presence of a NOR, etc., or other special markers. Chromosomes can also be marked by specific rearrangements (translocations, inversions). Absence of a chromosome, as in monosomics, can mark it in genetic experiments, and the monosomic can be identified and maintained. Different types of trisomics, or even special constructions like substitutions, compensating trisomics, etc., can also make chromosomes recognizable.

In order to be used in a genetic linkage analysis, the chromosomal marker must be polymorphic, i.e. it must occur in different forms. Like a gene, it can then segregate in homozygotes of two types (like recessive and dominant for a gene) and heterozygotes. The simplest alternative forms of a marker are absence and presence. Specific rearrangements (deficiencies, duplications, translocations, centric split, inversions: Sect. 8.3.2.2) and aneuploids (monosomics and different forms of trisomics: Sect. 8.3.2.1) can be used in linkage analyses, and scored as absent, heterozygous or homozygous for the rearrangements. Nullisomics, monosomic, disomic, trisomic, tetrasomic are the different categories for the classification of segregating aneuploids. The segregation can be scored in the same segregating population as any gene, and on the basis of their co-segregation their linkage relationships can be determined. When there is linkage, the genes involved are on the marked chromosome.

In the following sections methods to correlate genes with chromosomes are discussed (Sects. 8.3.2.1 and 8.3.2.2) and in subsequent sections the location of genes within chromosomes (Sect. 8.3.3).

8.3.2.1 Aneuploids

8.3.2.1.1 Monosomics

In the few cases where monosomics survive in diploids, they are very effective for gene localization. There are two ways to use them: (1) at their origin, then their generative reproduction is not necessary; (2) in a segregating progeny derived from a monosomic parent heterozygous for the genetic marker. Here, generative reproduction is required. In the first case the monosomic must be formed with sufficient frequency, and it is necessary to have a method to identify the chromosome involved, either on the basis of its morphology or because of a unique origin from, for instance, a known rearrangement causing frequent univalent formation and subsequent recovery of the monosomic. The best-known examples are those of maize (Weber 1983, 1991) and tomato (Khush and Rick 1966; see Sect. 6.2.1).

A few genes can be localized in monosomics when their expression in the hemizygous state deviates from normal. This is occasionally the case. Usually, gene localization is simplest when the carrier of an aberration has a dominant allele and is crossed with a normal pollen donor homozygous for the recessive allele. When in the progeny the monosomics consistently have the recessive phenotype, and the normal plants the dominant phenotype, the gene must be located on the monosome. In principle the same method can be applied with a monosome that can be propagated generatively with maintenance of its monosomy (Fig. 8.3). The use of rearranged monosomes, for instance in translocation (or tertiary) monosomics, makes it possible to identify the location of the gene in specific chromosome segments, especially when several rearrangements involving the same chromosome are available (Khush 1973). For general application monosomics are not sufficiently common in diploids.

In *allopolyploids*, monosomics are the favoured material for locating genes on chromosomes. They can be obtained relatively readily and in some important crop species the complete series is available. The number of morphological markers is usually small in polyploids, because in order to be expressed, the recessive allele must be present in all genomes simultaneously. For enzyme markers and other polymorphisms like RFLPs this is not necessary, and this is one of the reasons why in polyploids the genes analyzed are most frequently biochemical and molecular markers. Especially in wheat (Sears 1954) monosomics have been used extensively, not only for gene localization, but also for making special constructions, primarily chromosome substitutions.

There are several ways to determine if a gene is located in the monosome in an allopolyploid (Fig. 8.3).

1. When nullisomics are viable, the simple expression of a recessive (null) allele in the nullisomic locates the gene in the chromosome involved. When a dominant allele is present in more than one genome, nullisomy for any single chromosome will not lead to suppression of the dominant phenotype. An example is the red seed factor on the group 3 chromosomes. In the variety Chinese Spring it is present only in one genome and nulli 3D shows the recessive phenotype. In other varieties the dominant allele is also present in the homoeologous chromosomes, and nulli 3D does not show the recessive phenotype.

2. Genes with a hemizygous deviating expression will be recognized directly in monosomics.

3. In the case the monosomic series has the dominant allele, which must be considered the more probable situation because new mutant marker alleles tend to be recessive, a simple approach is sufficient. The monosomic is crossed

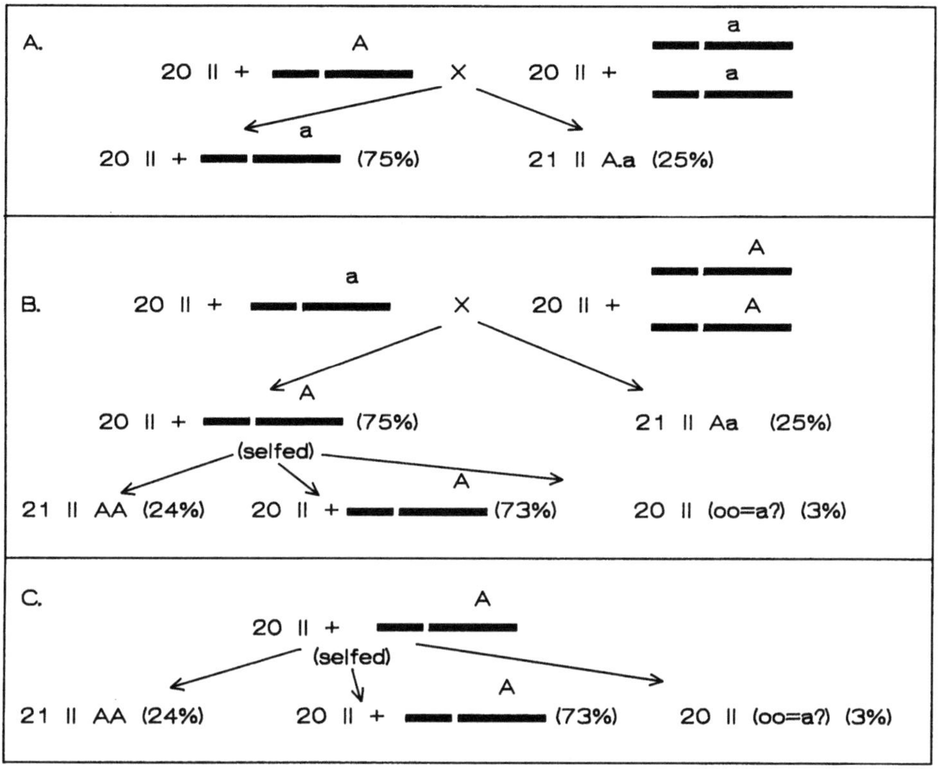

Fig. 8.3 Locating genes on chromosomes by using monosomics. **A** The monosomic has the dominant allele (*A*.) and is crossed with a homozygous recessive disomic (*aa*). In the F_1 all monosomics have the recessive phenotype (*a*.), all disomics the dominant phenotype (*Aa*) if the gene is located on the chromosome involved. If not, the F_1 has the dominant phenotype. **B** The monosomic has the recessive allele (*a*.) and is crossed with a homozygous dominant disomic (*AA*). In the F_1 the monosomics are selected; they have the dominant phenotype (*A*.). In the F_2 no recessives segregate if the gene is on the chromosome involved. The nullisomics may have the recessive phenotype, but these are rare. **C** The monosomic has the dominant phenotype (*A*.) and is selfed; as in **B**, recessive types may segregate spontaneously when the nullisomics have the recessive phenotype, without previous introduction of the recessive allele, but they tend to be infrequent

with the normal stock with the recessive allele and in the F1 the monosomics will show up directly as recessives, when the gene is located on the chromosome involved. Biochemical markers, for instance isozymes, are usually semi-dominant, i.e. both alleles can be recognized simultaneously and then it is not necessary that either allele is present in a specific parent.

4. A slightly more complex method is necessary when the monosomic contains the recessive allele. It is crossed with a normal pollen parent with the dominant allele. Because of reduced male transmission of gametes with an incomplete set of chromosomes, the monosomic is generally used as the

female in hybridization. In the F1 the monosomics are selected and selfed. There is a normal 3:1 segregation if the gene is not in the chromosome involved. If it is located in the monosome, however, the single chromosome in the F1 has had only the dominant allele of the normal parent and in the F2 no recessives will appear. There is also no recombination between genes in this chromosome. Only when the nullisomic has the phenotype of the recessive allele, will some segregation be seen, but there is a large deficit of these "recessives" because of the reduced appearance of nullisomics. Unrau (1950) reported an example for wheat, where the allele for red glumes is dominant over that for white glumes. The series of monosomics tested all had white glumes and were crossed with a normal stock with red glumes. The F2 involving all monosomics except 1B segregated 3:1. In the F2 of mono-1B 6.6% had white glumes and all appeared to be nullisomics. This method has been used to locate several genes in wheat, including genes for disease resistance.

It has been suggested to use monosomics formed spontaneously in callus culture for the localization of gene loci, especially of isozymes. The loss of a gene function simultaneously with the loss of a particular chromosome or chromosome segment can be readily detected even without regeneration. This would be especially convenient with asymmetric protoplast fusion, where one of the two fusion partners has been damaged seriously by ionizing irradiation. Only part of the genome is transmitted, and when with the loss of specific chromosomes specific functions have disappeared, it may be assumed that the genes for those functions are located in the chromosomes lost. The method resembles the use of man-mouse cell hybrids for gene localization in man, where a natural loss of human chromosomes is accompanied by the loss of specific markers. In the case of asymmetric hybrids, however, it has appeared that simultaneously with the loss of chromosomes so many rearrangements tend to take place that the method is not reliable (Wijbrandi 1989). The use of chromosome elimination in certain hybrids (*Hordeum vulgare* × *H. bulbosum*, for instance) to locate genes on chromosomes would seem to be more promising, but in practice it does not appear to function satisfactorily.

Telocentrics, replacing normal metacentric chromosomes (monotelocentrics), are effective in locating genes in chromosomes if the loci are in the lost arm. They are more specific than complete monosomics because they determine the arm involved, and are usually better tolerated. They will be discussed later with the localization of genes within chromosomes. When the gene is on the arm still present, its segregation can be sufficiently disturbed by reduced male transmission for detection of its location.

8.3.2.1.2 Primary Trisomics

The typical characteristics of the genetic segregation of genes located on chromosomes of which more than two copies are present (Sect. 6.2.2.1.4) can be used to locate genes on chromosomes. Primary trisomics give the most straightforward information. They are found spontaneously very infrequently in diploids. Among the progeny of desynaptics they are more frequent, but desynaptics themselves are rather rare. Treatment with certain spindle poisons (Sect. 6.2.2.1.2) results in (infrequent) formation of trisomics and other aneuploids. Usually, the best source of trisomics is a triploid (Sect. 6.1.2.2.1.4). The main disadvantages of triploids as the origin of primary trisomics are that they are sometime difficult to make even by crossing diploids with tetraploids. In addition, they tend to have very low fertility.

Trisomic segregation, in the duplex often close to 12:1 (Sect. 6.2.2.1.5), and always sufficiently different from the normal disomic 3:1 ratio, and the possible role of double reduction have been discussed i.. Section 6.2.2.1.4. Crossing a wild-type primary trisomic with a stock in which the recessive allele is present leads automatically to the duplex condition of the trisomics in the progeny. Examples of trisomic segregations, some including double reduction, are given in Table 6.10. It is always necessary to check the normal disomic ratio of the gene involved, in the same genetic background as that of the trisomic. This avoids the risk that a disturbed segregation ratio in the diploid is erroneously interpreted as a trisomic ratio. It is also necessary to adjust the size of the segregating population to the low frequency of recessives. In small populations the trisomic ratio is low enough to cause the incidental absence of recessives. This could lead to the conclusion that in the population analyzed the recessive allele in not present, with the possible consequence that this population is discarded.

8.3.2.1.3 Telocentric Trisomics

Genes located on the chromosome arm which is present in three copies in telocentric trisomics do not segregate in the typical trisomic way (Sect. 6.2.2.2.4). Yet their segregation is sufficiently abnormal to make it possible to decide whether or not genes are positioned in the extra arm. Rhoades (1936) analyzed the progeny of a test cross involving the telocentric of the short arm of chromosome 5 of maize and the genes *bm* (brown midrib) and *pr* (purple aleurone). The presence of the extra arm was, in this particular case, visible in the plant morphology (short, broad leaves) which made cytological checks unnecessary. The genetic make-up of the segregating female parent is shown in Fig. 8.4A, and the segregation in the test cross by a *bm bm pr pr* pollinator is given in Table 8.2. Using the segregating trisomic as the male parent would have been less favourable because the extra chromosome has a reduced rate of transmission through the pollen parent. With complete recovery of the telocentric 50% is expected to be found in the progeny. The telocentric is often univalent at meiosis, which purely because of chance pairing is expected to occur in one-third of the meiocytes. In addition, a single short arm has a greater than average probability of not forming chiasmata even after pairing.

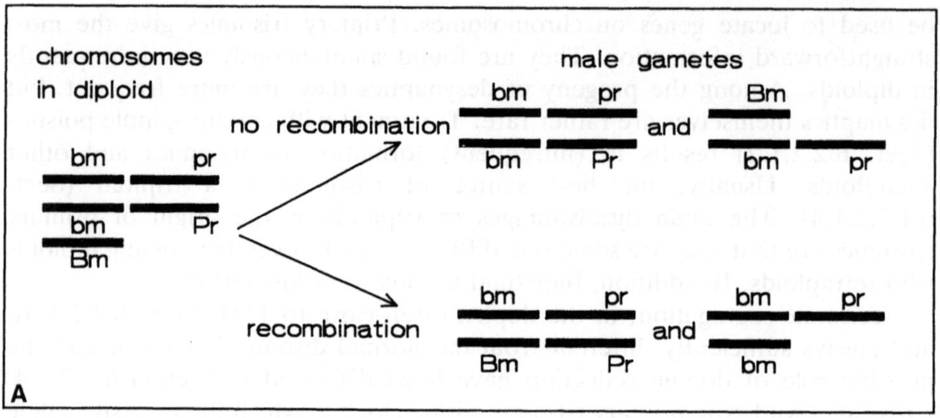
chromosomes
in diploid
male gametes
no recombination
bm
pr
bm
Pr
Bm
and
recombination
A

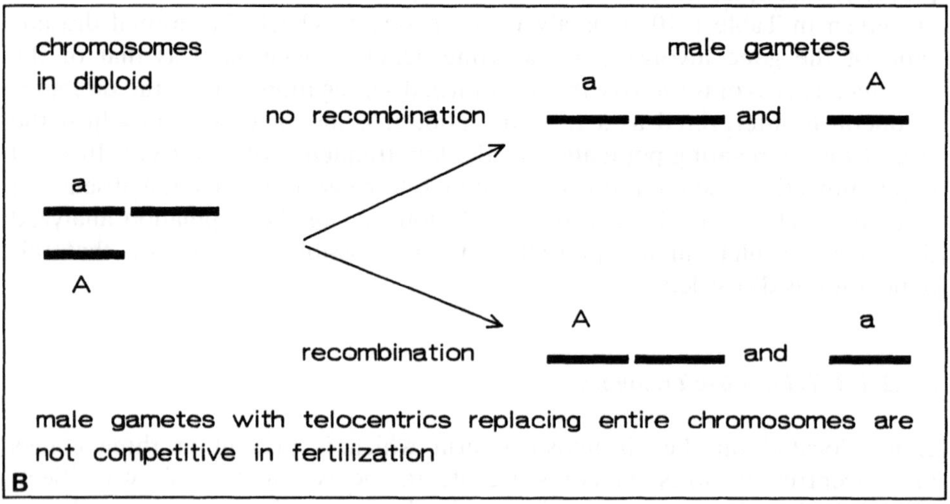
chromosomes
in diploid
male gametes
no recombination
a
A
and
recombination
male gametes with telocentrics replacing entire chromosomes are not competitive in fertilization
B

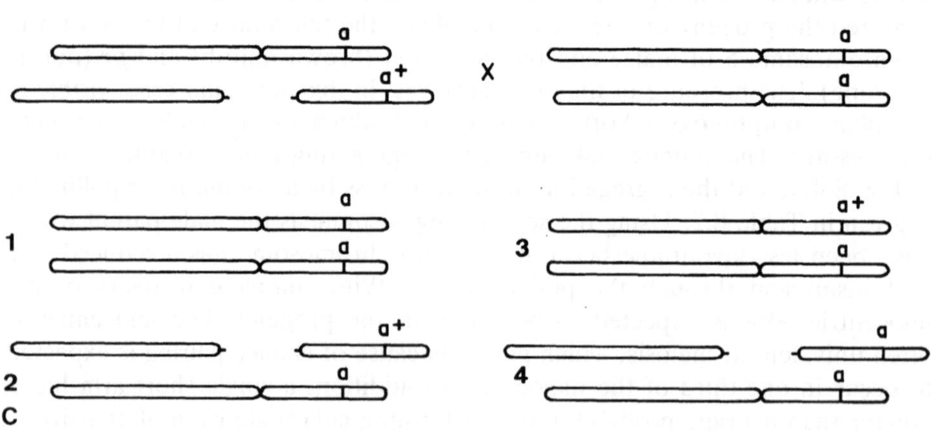
a
a+
X
1
2
3
4
C

Table 8.2. Segregation in a test-cross progeny of a telotrisomic for the short arm of chromosome 5 of maize. (Rhoades 1936)

Genes	2n (normal)	2n + telo	total	ratio
Pr:*pr*	63:64	31:35	94:99	1:1
Bm:*bm*	1:171	85:0	86:171	1:2

bm: brown midrib; *pr*: purple seed. The 1:2 ratio for *Bm*:*bm* is a consequence of the loss of the extra chromosome.

Consequently, the telocentric is often lost during anaphase or is split into chromatids that fail to function properly at the second meiotic division.

In the example given, transmission of the telocentric was 30%. Segregation of *pr* was apparently quite normal, and it can be concluded that it is not located in the trisomic arm. The recessive allele *bm* was not present in the trisomic population, whereas the dominant allele *Bm* was present only once in the disomic population. This makes it clear that the gene is located on the short arm of chromosome 5. The single *Bm* plant among the disomics is the result of recombination between the centromere and the locus, which is apparently quite restricted. A small number of recessives among the trisomics resulting from recombination could have been present, but were not observed. Recombination in trisomics is more complex than in disomics, but in the present simplex condition the situation is relatively simple.

The situation would have been more complicated if a recessive allele had been present in only one of the two normal chromosomes of the (duplex)

Fig. 8.4 Three ways of using telocentric chromosomes derived from metacentrics to map the centromere in relation to gene loci. **A** The *telotrisomic* in maize of Table 8.3 used by Rhoades (1936) to determine the location of gene loci on chromosome arms and to estimate locus-centromere distance. The normal chromosomes have the recessive alleles of *bm*, the telo has the dominant allele (*Bm*). The normal chromosomes are heterozygous for *pr*, which is not present in the telo. **B** *Monotelosomic*: one telocentric replaces a normal chromosome, one complete arm is absent, usually manageable only in allopolyploids. The telo carries the dominant allele *A*, the normal chromosome the recessive allele *a*. When used as the male parent in a test cross, pollen with the telo and the dominant allele cannot compete with pollen with the normal chromosome and the recessive allele. Only by crossing-over will the dominant allele be transferred. The frequency of recovery is a measure of crossing-over and thus of centromere-locus distance. A cytological check is advisable when pollen competition is not complete (cf. Sears 1969). **C** *Centromere split* (Robertsonian split) as centromere marker. *Top* The heterozygote has a locus marked by the recessive allele (*a*) in the normal chromosome and the dominant allele (*a*+) in the split chromosome. The centromere is marked by the split. The heterozygote is test-crossed with a double recessive (*aa*), structurally normal type. In the progeny (*1–4*) both segregate and can be classified: the parental types (*1* and *2*) and the recombinant types (*3* and *4*). Recombination is a measure of locus-centromere distance (example given in Sect. 8.3.3.1.1)

heterozygous parent, especially for estimating recombination. The probability that a recessive allele will be recovered among the trisomics is then reduced, because it would require that the recombinant telocentric occurs in combination with two normal chromosomes with the recessive allele. It would also be difficult to conclude how often a dominant allele among the disomic progeny is the result of recombination. However, for simply demonstrating that the gene is located in the arm involved, it would be sufficient to conclude that among the trisomic progeny the frequency of the dominant phenotype is significantly greater than the expected 50%.

8.3.2.1.4 Tertiary and Other Complex Trisomics

Tertiary trisomics and compensating trisomics can be used to locate genes on chromosomes like primary trisomics, but their main interest is in gene localization in specific segments of chromosomes, rather than locating genes on a chromosome. Their segregation is not like that of the primary trisomic, but usually sufficiently different from the disomic ratio (Sects. 6.2.2.3.4 and 6.2.2.4.1) to locate genes on the rearranged chromosome. It is not possible to decide directly whether the gene is located on the segment with the centromere, derived from one chromosome or on the translocated segment derived from the other chromosome. When such trisomics are formed anew with sufficient frequency, they can give the same information as when available as an existing stock. Certain translocation heterozygotes produce tertiary trisomics quite frequently and certain other complex rearrangements frequently produce compensating trisomics (Sybenga et al. 1990).

A special category of aneuploids with extra chromosomes is represented by the alien additions. The correlation between the presence and absence of a marker (including desired genes to be transferred from a donor species to a cultivar; cf. Sect. 10.4.4) and an alien addition locates the marker in the specific addition chromosome. This can be combined with a deficiency analysis (Sect. 8.3.3.2.1) to determine the location of the marker within the chromosome: Hu and Quiros (1991) report on the monosomic addition of the *Brassica oleracea* chromosome 4C to *B. campestris* (2n = 20 + 1). Six RFLP and isozyme markers in this chromosome were followed in the addition line. In 55% of the plants one to five markers were spontaneously missing, which was accompanied by deletions of varying size. In the addition line with chromosome 5C, similar results were obtained for three genetic markers. Because the chromosomes involved are rather small and do not contain many morpholgical markers, the deficiency analysis in this case is rather coarse.

Trisomics and additions can be used by the same methods in diploids and allopolyploids.

Autopolyploids present considerable difficulties for gene localization with the use of aneuploids. Chromosomes less (trisomics in autotetraploids) and more (pentasomics in autotetraploids) do have different segregation ratios than regular tetrasomics, but the deviation is not always easy to interpret. Whenever possible, it is advisable to try to derive a diploid from the autopolyploid and use any one of the methods described for diploids. Trisomics at the diploid level have been successfully employed for gene localization in potato, which is normally an autotetraploid (Ramanna and Wagenvoort (1976).

8.3.2.2 Chromosomal Rearrangements

In diploids and allopolyploids the use of chromosomal rearrangements is quite effective in locating genes on chromosomes. Like the rearranged trisomics mentioned above (Sect. 8.3.2.1.4), some rearrangements, in addition to simply serving to locate genes in chromosomes, can be used to define the position of genes within chromosomes. Gene localization with rearrangements may be based on the visible presence or absence of specific chromosome segments in the rearranged chromosomes by which the expression of the gene is affected (deficiencies or duplications). In other cases there may be changes in linkage relationships (inversions, translocations; Sects. 5.3.4 and 5.4.1.4).

8.3.2.2.1 Deficiencies and Inversions

Instances are known where a deficiency is accompanied by the loss of specific genes. An old example is that of McClintock (1931) who induced deficiencies in stocks with dominant alleles of maize by applying ionizing irradiation. Deficiencies for the terminal segments of the long arm of chromosome 6, which could be recognized in pachytene bivalents by the absence of prominent knobs and chromomeres, were accompanied by the loss of specific genes. In 1968 Khush and Rick used irradiated pollen to pollinate multiple recessive stocks of the commercial tomato, and when recessive progeny appeared, checked them for deficiencies. Several genes could be accurately located on and even within chromosomes (Sect. 8.3.3).

Inversions are not very common in cultivated plant species, either because they are simply rare or because they are not readily detected in cases where chiasma formation is localized. Linkage of a segregating inversion, either pericentric or paracentric, with a gene shows that the gene is located in the inversion chromosome, and the recombination frequency may even give a rough estimate of the distance between gene and inversion. The inversion can be scored on the basis of its meiotic behaviour or the partial sterility it causes, but this distinguishes only the heterozygote from the two types of homozygote. Recognition of the inversion by its effect on chromosome morphology (most readily but not necessarily always possible with pericentric inversions) makes it possible to establish all three possible classes, and improves the resolution of the linkage analysis. This, however, is rarely possible.

Heterozygosity for pericentric inversions may occasionally result in viable deficiencies (Sect. 5.3.4) which, when accompanied by the loss of specific genes, indicate the location of that gene in the deficient segment (Burnham 1962). Since such deficiencies are rarely viable in the homozygous state, it requires a test cross with a recessive allele in a normal chromosome to demonstrate the loss of the gene. This appears similar to the method briefly described above which uses deficiencies to locate genes in chromosomes. It is only incidentally of any use. Inversions have an effect on recombination between genes in the inversion chromosome.

8.3.2.2.2 Translocations, Duplications and Robertsonian Splits

Translocations have been used extensively to locate genes on chromosomes, and in some instance they are also useful for locating genes within chromosomes. The problem with translocations, including the most frequent type, the interchange, is that two chromosomes are involved and that it is not readily decided on which of the two chromosomes the gene is positioned when a linkage analysis shows it to be linked with the translocation. Even without scoring the translocation as such, a change in the linkage relationships between two or more genes in the homozygote or, more readily, the fusion of two linkage groups into one in the segregating progeny of a heterozygote, indicates that the genes involved are located on the chromosomes of the translocation. The reason for this is given in Section 5.4.1.4. It is necessary to repeat the test with other translocations involving one of the same chromosomes which was present in the first translocation tested. This is one of the applications of a translocation tester set (Sect. 5.4.1.3).

There are different ways to classify the translocation. When the heterozygote is significantly less fertile than both homozygotes, it can be rather simply recognized in a segregating progeny either by recording pollen stainability or by seed set. Not always, however, is the heterozygote sufficiently less fertile than the two types of homozygote to make it readily identifiable. In barley (Kramer 1954) and maize (Burnham 1962), the reduced fertility of the heterozygote is sufficient for its classification. In rye, fertility is hardly or not reduced because the orientation of the translocation quadrivalent is predominantly alternate, which results in genetically balanced spores. If partial sterility is an insufficient diagnostic characteristic, the specific meiotic configuration of the heterozygote (Sect. 5.4.1.4) distinguishes it readily from the two homozygotes and thus can also be used for classification, but this is a very laborious approach.

When, like in the two cases just mentioned, only the heterozygote (NT) can be distinguished from the two other types (NN + TT), but when for the gene studied the dominant (*AA* + *Aa*) versus the recessive (*aa*) phenotype is recorded, the analysis requires a special variant of a linkage analysis (Kramer 1954; Burnham 1962). As long as it is sufficient to establish that there is

Table 8.3. Examples of joined segregations of interchanges, scored in root tip mitoses, and genetic markers. (de Vries and Sybenga 1984)

				Dominant all elc			Recessive all elc			
Gene	Transl.	Chrom.	Plants	TT	NT	NN	TT	NT	NN	r
an	T282W	5R/7R	320	7	160	–	144	9	–	5.0%
an	T303W	5R/7R	429	–	45	187	–	149	48	21.7%
ti	T282W	5R/7R	349	–	171	1	–	2	175	0.1%
ti	T273W	1R/5R	184	–	76	5	–	9	94	7.6%
mo	T305W	2R/5R	60	–	33	1	–	1	25	3.3%
mo	T273W	1R/5R	92	–	17	29	–	22	24	55.4%

Conclusion: *an* on 5R or 7R (is 7R)
ti on 5R (on 5R or 7R, and on 1R or 5R)
mo on 2R (on 2R or 5R, and not on 1R and 5R)

Test-crosses with either the normal or the interchange homozygote as the recurrent parent. TT: Interchange homozygote; NT heterozygote; NN normal karyotype. For the genes, the recessive and the dominant alleles are indicated; r: recomlination.

linkage, this complication can be avoided. The subject is treated again later in this chapter.

When the karyotype has changed sufficiently to classify the progeny plants of a test cross or an F2 on the basis of chromosome morphology, all three genotypes, homozygote normal (NN), heterozygote (NT) and homozygote translocated (TT), can be distinguished, and this gives the best resolution possible. Due to background variation in the somatic chromosome morphology, slight changes in the karyotype are hard to distinguish unequivocally without careful measurement when distinct banding patterns are not available for the segments involved. Karyotype measurements are very time-consuming (Sect. 4.2.3.2).

Examples of gene localization and linkage analysis involving interchanges that can be classified in mitosis have been given by de Vries and Sybenga (1984; Table 8.3).

As with pericentric inversions, there is a special variant in the use of translocations to localize genes on chromosomes that makes use of the fact that some translocation heterozygotes have viable deficiency-duplication gametes and progeny. Very rarely, the deficiency produces the phenotype of the null-allele of a gene in the segment involved. This is distinguished from a recessive mutation by the repeated appearance of the "mutant" among the progeny of the heterozygote. When homozygous deficiencies are not viable, a cross with a recessive mutant will give a comparable result. It is not a very useful approach for a systematic program of gene localization, but in maize and barley a few examples are known in which, more or less by accident, genes have been located in the deficient segment resulting from adjacent segregation of the interchange complex (Burnham 1962).

Another special variant in the use of translocations to localize genes on chromosomes and to some extent even their location within chromosomes are translocations between B-chromosomes (Sect. 3.1.4.1.2.3) and (normal) A-chromosomes. When the translocated B-chromosome does not have a stable accumulation mechanism, it may be lost, and a deficiency results, possibly accompanied by the loss of a marker gene. If the accumulation mechanism is still intact, accumulation may occur with the consequence of duplication of the A-chromosome segment, possibly including specific genes. This has been used to duplicate genes when this is desired (maize: Shadley and Weber 1986; cf. Sect. 11.2.2), but it also locates the gene in the A-chromosome segment. In addition to maize (see also Roman 1947; Birchler 1983), A-B translocations have been reported for a few more plant species (*Lolium*: Evans and Macefield 1977; rye: Pohler and Schlegel 1990), but there not yet used for gene localization.

In *Robertsonian splits* or centromere splits a metacentric chromosome has been replaced by the corresponding two telocentrics. It is recognizable in mitosis and meiosis and can thus be used in a genetic analysis involving genetic markers. Semisterility is not normally observed. When there is linkage between the marker and the rearrangement, the marker is assumed to be located on the chromosome concerned. Since a linkage analysis is involved and not merely a disturbed segregation, the Robertsonian split, which is a centromere marker, is also useful for determining the genetic map distance between the genetic marker and the centromere. This will be discussed again in Section 8.3.3.1.1.

8.3.3 The Physical and Recombinational Location of Genes Within Chromosomes

8.3.3.1 Natural Chromosome Morphological Markers

8.3.3.1.1 The Centromere

The first step in the development of a genetic chromosome map from a genetic map is to find suitable chromosomal markers. The most obvious is the centromere, which is present in every chromosome and a good marker in monokinetic systems.

Insufficient genetic variation is available in the morphology or other classifiable behaviour of centromeres of higher plants for direct use in linkage analysis. Unlike in the *Ascomycetes*, tetrad analysis is not possible because the products of meiosis do not stay together. The few exceptions (*Salpiglossis variabilis* and a few genera of *Ericales* and *Juncales*) do not include cultivated species, and the species concerned are not particularly favourable for genetic analysis. It is possible, however, to use exceptional mutants of meiotic behaviour that create conditions under which an analysis resembling tetrad analysis can be performed. Such mutants are not common, but when they occur in a species, they can be used for any gene desired. Rhoades and

Dempsey (1966) used the "elongate" factor in maize, which inhibits the normal course of the second meiotic division: second division restitution (SDR). The resulting diploid spores (half-tetrads) carry both chromatids of the same chromosome. After crossing-over between the locus and the centromere, the diploid spore has become heterozygous, because the two chromatids are different. Test-crossing the F1 plants with a double recessive diploid produces simplex triploids. Because the elongate factor does not cause SDR consistently, some diploid progeny are formed also, but these can be recognized. Although the triploids are rather sterile, they are sufficiently fertile to be test-crossed with a double recessive diploid. The simplex heterozygous parents in the test-cross populations can be distinguished from the duplex heterozygotes and the nulliplex homozygotes. Their frequency is a measure of the recombination between the centromere and the locus.

Several useful variants of this method in which meiotic restitution is used to analyze gene-centromere genetic distance have been reported, such as the *ps* (parallel spindle) gene in potato (Watanabe and Peloquin 1989) but they are very specialized and will not be discussed. Desynaptics with high frequencies of first division restitution can in principle also be used but do not have normal recombination patterns (Jongedijk et al. 1991a).

There are different ways to use telocentrics derived from metacentric chromosomes to map centromeres in relation to genes. Three are considered (Sybenga 1972):

1. The use of telocentric trisomics (Fig. 8.4A) has been discussed above. If transmission of the extra chromosome through the pollen is rare, the trisomic simplex heterozygote with the dominant allele in the extra chromosome can be used as the pollinator in a test cross. All progeny with the dominant allele are the result of recombination between the marker and the centromere. If male transmission of the telocentric is not negligible, and the plant morphology cannot be used as a marker of trisomy, cytological classification of the progeny is necessary.

In the case of a trisomic, the three chromosomes must make a choice between two potential pairing partners, and in only 2/3 will the telocentric pair with one of the normal chromosomes, including the critical region between centromere and locus. In the simplest case, where both normal chromosomes carry the recessive allele and the telocentric the dominant allele, recombination is readily recognized, and the observed recombination frequency should be multiplied by 3/2 to find the frequency representative of the distance between centromere and locus. With other combinations of alleles and chromosomes, the analysis is much more complicated. The most favourable combination mentioned cannot be constructed in a simple way and whether or not it is present in the material at hand must be decided from the segregation. There is still another complication, which is generally present when polysomics are used: the choice of pairing partner may not simply result in 2/3 pairing of the telocentric. It may be useful to check meiosis first, before applying a correction of the observed recombination frequency. The best correction factor is obtained when meiotic association of the arm concerned can be determined. The simple frequency of association of the telocentric with a normal chromosome is not sufficient. It must be determined in relation to the frequency of chiasmate association between the corresponding arms of the two normal chromosomes. There is often a considerable reduction in chiasma frequency between locus and centromere, here at least in part because of pairing problems in the trisomic arm.

2. A telocentric replaces a normal chromosome: one arm is deficient (Fig. 8.4B). It is well tolerated in allopolyploids (Sears 1954) and used in wheat and cotton, for instance to determine recombination between the centromere and a gene (Endrizzi and Kohel 1966; Sears 1969). In diploids where hypoploids are tolerated (Khush 1973; Weber 1983; Melz and Winkel 1986) monotelocentrics tend to be viable. There is crossing-over in the disomic arm only. In a test cross with the heterozygote as the female, the progeny is expected to segregate 1:1 for the normal and the telocentric chromosome, and also for the two alleles of a gene in the disomic arm. From the combined segregation the frequency of recombination between the gene and the centromere can be estimated. Telocentrics replacing a normal chromosome are poorly transmitted through the pollen in competition with normal pollen. With the heterozygous parent as the pollinator, and abundant pollination, only pollen with the normal chromosome functions in fertilization, and the allele carried by this chromosome is transferred. Genes close to the centromere will rarely or never show crossing-over between centromere and locus, and the original allele in the normal chromosome will be recovered. Recombination will be observed when the gene is more distally located: the simple frequency of recovery of the allele from the telocentric is a direct estimate of recombination between centromere and locus, and classification of the telocentric is not necessary.

As shown by Endrizzi and Kohel (1966) for cotton and Fu and Sears (1973) for wheat, there may be a considerable reduction in crossing-over around the centromere when telocentrics are used in recombination analyses. This may be compensated in more distal regions such that the reduction is not found for more distal genes (Dvorak and Appels 1986).

3. The third method makes use of the substitution of a normal chromosome by the two corresponding telocentrics (Robertsonian split, Fig. 8.4C). There is usually regular segregation of the trivalent at meiosis (Sect. 5.5.5.5.3) resulting in balanced spores, and as a result 1:1 segregation of the normal chromosome and the combination of the two telocentrics. Recombination between a locus (in either one of the arms) and the centromere is analyzed by crossing a recessive or co-dominant stock with a telocentric substitution with the other allele, and test-crossing the F1 with the recessive. In the segregating test-cross generation the chromosomal condition must be scored by karyotype analysis. Recombination is readily determined. In the example from Sybenga et al. (1990) for two seed storage protein genes in rye (Sec-1 and Sec-3) this method has been applied. The segregation for the centromere split (NT = heterozy-

gote; NN = normal) and the *Sec-3* gene coding for high molecular weight (HMW) secalins was as follows:

NT/$Sec\text{-}3^+$	NT/$Sec\text{-}3^-$	NN/$Sec\text{-}3^+$	NN/$Sec\text{-}3^-$	Total
19	6	3	23	51

Recombination 9/51 = 17.6%.

For the centromere split and the locus of *Sec-1* (omega secalins) the segregation was:

NT/$Sec\text{-}1^+$	NT/$Sec\text{-}1^-$	NN/$Sec\text{-}1^+$	NN/$Sec\text{-}1^-$	Total
50	21	23	51	145

Recombination 44/145 = 30.3%.

In a more complex situation of this analysis the shorter of the two telocentrics was involved in a reciprocal translocation in the satellite. Recombination between the *Sec-3* locus in the *other* arm and the centromere was reduced to practically 0. In a meiotic analysis it could be shown that the chiasma frequency in the entire arm with *Sec-3* was not reduced and that some proximal chiasmata still occurred. There must have been only a slight distal shift of chiasma localization, comparable with that observed by Fu and Sears (1973) mentioned above, but apparently much stronger than without the translocation. It is not clear whether in this case it was due to the telocentric constitution, the translocation in the other arm or to genetic effects of another character.

In the first two cases mentioned above using telocentric chromosomes to estimate the genetic distance between locus and centromere, the arm in which the gene was located was automatically determined. In the last case, however, this is not so. When two markers segregate, in addition to the centromere marker, a three-point test can be performed, and the order of the genes can be determined. This establishes whether the genes are located in different arms or in the same arm; in which arm, is still not possible to determine.

In the rare case where segregation is not regular and (telocentric) trisomics appear, it is in principle possible to use the frequency of rare recessive trisomics as an estimate of locus to centromere distance, based on the phenomenon of double reduction (see Sybenga 1972). It has been used in human genetics with spontaneous trisomics and could be applied in all cases where spontaneous trisomics are found. In the case of telocentrics, it also gives information on the arm in which the gene is located. For the present purpose, this is too unusual and too complex to be discussed.

8.3.3.1.2 Heterochromatin, Chromomeres and Banding Polymorphisms

In McClintock's 1931 and later experiments with the deficiencies mentioned earlier (Sect. 8.3.2.2.1), the exact locations of the border(s) of the deficiencies were established by using knobs and chromomeres. With natural polymorphisms for such chromosome morphological markers, including prominent C- and N-bands, linkage studies involving known genes and these markers give an indication of the position of the genes with respect to the chromosomal markers. This does not, however, show the precise position of the genes in the chromosomes, and the chromosomal markers are just landmarks in chromosomal gene mapping. Polymorphisms, in regard to size or absence or presence, have often been reported for large as well as small chromomeres, for prominent and for minor C-bands, N-bands, etc. They occur between different genotypes and especially between related species and are useful markers for recombination analyses (Loidl 1987). One disadvantage is the large amount of cytological work required. In wheat, Curtis and Lukaszewski (1991) mapped the centromere, 11 C-bands and two seed storage protein genes in chromosome 1B of tetraploid wheat. For these markers polymorphism was found in hybrids between different accessions of Langdon durum wheat and *Triticum dicoccoides*. In the short arm recombination was almost completely restricted to the satellite. In the long arm, too, recombination was mostly distal.

8.3.3.1.3 The NOR and Other Multigene Loci: In Situ Hybridization

NOR-associated heterochromatin, if present as in many Triticeae, is variable. In addition, there is considerable molecular variation in the spacer regions between the r-DNA genes, and this has been used as a genetic marker in combination with other markers. There are examples of such analyses in the Triticinae, for instance by Dvorak and Appels (1986). The NOR is visible as a morphological structure without scorable variation in the chromosome, and the way it is scored is molecular.

The coincidence of a genetic molecular marker with a structure recognizable in a chromosome like the NOR is not common, but has a parallel in the situation where different alleles of multicopy genes (besides the NOR: histone genes, storage protein genes, t-RNA genes) can be scored on the basis of biochemical variation and where, in the same material, the locus can be made visible in the chromosome by in situ hybridization (Gustafson et al. 1990; Fig. 4.6B). For plant breeding this is of considerable potential interest, but neither the genetic variation nor in situ hybridization techniques have been sufficiently developed for general application in genetic chromosome mapping. The fine resolution possible with in situ hybridization in animal material, especially at somatic prophase and pachytene (Moens pers. comm.; Moens and Pearlman 1990), is very promising. A specific low copy number (120 copies) X-chromosome satellite DNA of only about 1000 bp each could be seen to be located in the chromatin loops extending from the bivalent cores. The different copies were separated by spacer DNA which had not hybridized with the probe, and each copy could be seen separately to be labelled after hybridization with the labelled probe. Several loops were involved. The number of copies could be counted and agreed with the number established by biochemical methods. It could also be seen that the X-chromosome had two chromatids, each with its own set of chromatin loops. This approach would be possible only in cultivated plant species with relatively short chromosomes, but there it could be of considerable interest in gene localization. In situ hybridization itself has been used on several occasions, but not yet in a genetic mapping analysis in combination with other loci because this requires hybridization polymorphism.

8.3.3.2 The Break Points of Rearrangements as Genetic and Chromosomal Markers

The use of chromosome structural rearrangements in the localization of genes on chromosomes may, at the same time, give an indication of their location within the chromosomes. When linkage studies are involved, the exact localization is not determined, but the recombination frequency between locus and rearrangement gives at least some indication of its position. This particular aspect will be discussed in Sect. 8.3.4 (recombinational chromosome maps). Here, the more direct physical localization, insofar as it has not already been discussed earlier, will be considered.

8.3.3.2.1 Deficiencies

Deletion of single bands and small groups of bands in the salivary chromosomes have been used in *Drosophila melanogaster* already in the 1930s to make very detailed genetic chromosome maps. The disappearance of a genetic function with an X-ray induced deficiency for a single band or, for a series of larger overlapping induced deficiencies always including a specific band, has made it possible to map genes on specific polytene chromosome bands (Slizynska 1938; see Sybenga 1972). Such a detailed mapping is not possible in plants, but since the early 1930s (McClintock 1931) chromosomal rearrangements, including deficiencies induced by X-irradiation of pollen, and involving smaller and larger chromosomeres and knobs have been used, especially in maize. Maize has been analyzed in detail genetically, and its pachytene chromosomes have a specific chromomere pattern. Not only were genes located on chromosome segments, but also the recombination percentages and from these the map distances between these genes and chromosomal markers could be determined.

The following example is from McClintock (1931). A large deficiency of about 60% of the long arm of chromosome 6 included the locus of the gene *pl*. At pachytene the break of the deficiency could be located in the chromomere pattern. Closer to the centromere was the recessive allele *y* of the gene for white endosperm. A plant of the composition *y def*/*Y pl* was the pollen parent in a test cross with *y Pl*/*y Pl*. Pollen transmission of *y* was impossible as long as it was in the same chromosome as the deficiency. Non-recombinants give exclusively *yyY* or yellow triploid endosperm, and all white seed (*yyy*) must result from recombination between *y* and the deficiency. There were 100 white seeds among a total of 635, or 15.9% recombination. The normal recombination percentage between *pl* and *y* is 28%, so *pl* must be located half-way between *y* and the deficiency, provided the deficiency does not seriously disturb crossing-over in the interstitial segment.

This method and that in which developing embryos are irradiated shortly after fertilization, and where recognizable deficiences are coincidental with the disappearance of genetic functions, are also appropriate for other species. The latter approach, for instance, has been used on a considerable scale by Khush and Rick (1968) in tomato. Established deletions recognizable in C-banded metaphase chromosomes have been mapped and used to map genes on chromosomes in wheat by Curtis et al. (1991). Wheat, as an allohexaploid, has considerable tolerance for deletions. An interesting variant of this type of deficiency mapping makes use of chromosomal instability induced by the presence of certain alien chromosomes. In wheat, chromosome 6Bs of *Triticum speltoides* (syn. *Aegilops speltoides*) induces chromosome instability, resulting in chromosomal aberrations, including terminal deficiencies. Kota and Dvorak (1988) found that the loss of recognizable chromosome segments was accompanied by the loss of specific genetic markers. This locates the markers (genes or otherwise) on the segment lost.

8.3.3.2.2 Translocations, Duplications and Inversions

Like centromere (or Robertsonian) splits (Sect. 8.3.3.1.1) and deficiencies (Sect. 8.3.3.2.1), the segregation of translocations with known break points can be combined with the segregation of genetic markers. There are some complications, however, that must be taken into account. These are mainly due to the fact that the segregating parent contains the translocation in a heterozygous condition, which has consequences for its meiotic behaviour (Sect. 5.4.1.4).

There are three major effects:

1. Reduction of crossing-over in the vicinity of the break point, which coincides with the point of pairing partner exchange in the pairing configuration. It apparently results from the disturbance of pairing in this region as can be seen by variable and reduced pairing in the SC (de Jong et al. 1989) and in light microscope preparations of pachytene, for instance in maize (Burnham 1962). The decrease is possibly accompanied by an increase further away in the chromosomes as a result of decreased interference (Sybenga 1972, 1975).

2. Changes in interference relations, due to the special spatial relations between the pairing chromosomes around the point of pairing partner ex-

change. In special cases this may have an opposite effect compared to the decrease mentioned above. It then results in an increase in the probability of chiasma formation in chromosome segments that without the translocation do not have a chiasma (Parker 1987: the plant species *Hypochoeris*; Arana et al. 1987: the grasshopper *Eyprepocnemis*). This occurs possibly mainly in chromosomes that normally have only one chiasma, but in the quadrivalent may have one on either side of the point of partner exchange.

3. A different interference effect, sometimes seriously disturbing genetic analyses, has been reported by Sybenga and Mastenbroek (1980) for rye. The same phenomenon had appeared earlier in meiotic analyses of translocation heterozygotes (Sybenga 1975). In the example of rye, the translocation was T282W between chromosome arms 7RS and 5RL. The locus of the gene *Br/br* for brittle stem is situated in the short (non-translocated) arm of 5R and the locus of *An/an* (the main anthocyanin factor) on the long (non-translocated) arm or 7R. The heterozygote for the three factors (*an* – T282W – *br*) was test-crossed with a homozygous interchange stock (TT), double recessive for *an* and *br*. The following segregation was obtained:

An/NN/*Br*: 158; *An*/NN/*br*: 2; *an*/NN/*Br*: 0; *an*/NN/*br*: 9
An/NT/*Br*: 6; *An*/NT/*br*: 1; *an*/NT/*Br*: 1; *an*/NT/*br*: 143

The recombination frequencies between the three markers were: *an* – T282W: 5%; *an* – *br*: 1.25%; T282W – *br*: 5.63%.

The largest recombination frequency is between *br* and T282W, although it is known that T282W is in between the two other markers. Taking the known order as *an* – T282W – *br*, the frequency of double cross-overs can be determined as 6 + 9 = 15 or 4.68%. This is not only significantly more, but very much more than the expected frequency of 5% × 5.63% = 0.281% and clearly indicates strong negative interference at a considerable distance across the break point. The crossing-over frequency between *an* and *br* can be re-estimated as 10.63%. The negative interference is even too strong to be accounted for by exclusive formation of double cross-overs. It is necessary to assume a predisposition of the chromatids in the multivalent to one specific orientation at anaphase II. In addition to, or rather as a consequence of, negative interference across the break point, there is positive interference between adjacent segments with respect to the break point.

It is clear that these effects around the point of partner exchange of an interchange have considerable consequences for estimates of mapping distance between genetic or other markers and an interchange, particularly because the effects are quite variable.

In spite of these complications, translocations have been used in genetic chromosome mapping with considerable success. An example of an early genetic chromosome map segment in maize is given in Fig. 8.5 based on data from McClintock (1931) and others (Sybenga 1972; see also Burnham 1962). A more recent example involving two translocations and a centromere split, and

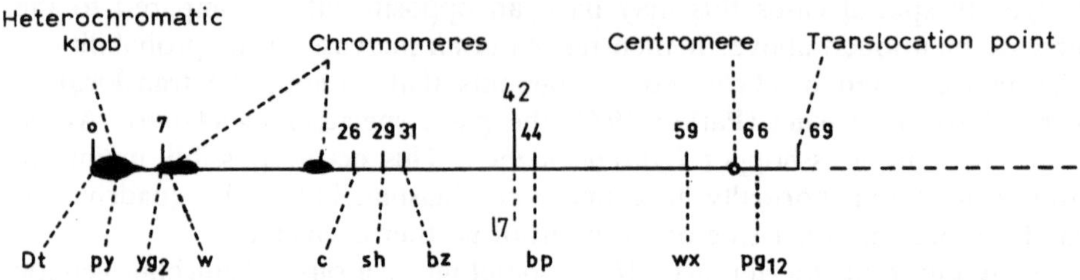

Fig. 8.5 Map of the short arm of chromosome 9 of maize, based on data of Creighton and McClintock (1931) and others (Burnham 1962; Sybenga 1972); chromosomal markers have been included: a heterochromatic knob at the end, two conspicuous chromomeres and the centromere. There is a break point of an interchange with chromosome 8 in the other arm, just on the other side of the centromere.

According to more recent information (Coe et al. 1990), several more markers have been located around the centromere and the distances are slightly different: at map distance 50 from the end of the short arm: *lo*2 (lethal ovules); at 54: *w*11 (white); at 56: *wx*1 (waxy endsperm); at 59: *d*3 (dwarf); (centromere); at 61: *pg*12 (pale green); at 62: *ar*1 (argentia); at 63; *v*1 (virescent); at 64: *ms*2 (male-sterile); at 66: *gl*15 (glossy).

In addition, for several less specifically located loci the general region and the order have been determined: there is another dotted locus near the tip (*Dt*5); there are four loci around 30 (*l*6: luteus; *dek*12: defective kernel; *hcfl*13: high chlorophyll fluorescence; *v*28: virescent). Between 40 and 55 there are *G*6: golden sheath; *Mr*: mutator of R-m; *bafl*: barren stalk fastigiate; *ga*8 gametophyte factor; *Les*8: lesion; *Rf*2: fertility restorer; *Zb*8: zebra cross-bands; *da*1 dilute aleurone; *mal*1: multiple aleurone layering).

With large deficiencies produced by A–B translocations it could be shown that about 40% of the physical length is found between about 52 and the centromere, less than 15% of the genetic map, which indicates greatly reduced recombination in the proximal region.

In the distal, approx. 25% of the map, 8 RFLP markers are reported and 6 between about 50% and the centromere. The region inbetween had not yet been marked by RFLP. (Coe et al. 1990)

also incorporating data from earlier sources, is given by Sybenga et al. (1990) for the short arm of chromosome 1R of rye, carrying the satellite. Dvorak and Appels (1986) and Lawrence and Appels (1986) report 26% recombination between the NOR, using spacer DNA polymorphism between the rDNA genes as marker, and the gene for seed storage protein *Sec-1*. Singh et al. (1990) recorded 26.1 ± 4.3 cM (centiMorgan) between *Sec-1* and the centromere and 5.5 ± 1.7 cM between *Sec-1* and a rust resistance locus, which in turn was 16.0 ± 4.8 cM from the telomeric heterochromatin, in wheat-rye translocations in a wheat background. Gustafson et al. (1990) located *Sec-1* in the satellite of 1R by in situ hybridization using a molecular clone of a segment of *Sec-1*. Its exact location was difficult to determine because of the small size of the satellite and the comparatively large area covered by the hybridization signal, in addition to variation in chromosome contraction. To assign a definite location to such a variable signal, the average position from several different observations must be taken. De Jong et al. (1989) showed that in the SC the break point of translocation T248W between chromosome arms 1RS and 6RS is just distal to the NOR with a short pairing region in between which often failed to be paired. Sybenga et al. (1990) report about 20% recombination between T248W and *Sec-1*. Light microscope MI analyses indicated 40% chiasmate association or also about 20% recombination between T248W and the chromosome end (Sybenga 1975). Similar MI observations did not show chiasmata between the break point of interchange T850W (1RS/4RL) and the end of chromosome arm 1RS and, consequently, there should be little or no recombination in this segment. The break point of T850M is just proximal to the terminal C-band of 1RS. There was, however, 2% recombination between T850W and *Sec-1*, which must have taken place in the interstitial segment, where it may well have been lower than in the same segment in a normal chromosome. The map of the satellite of 1RS can thus be constructed as:

NOR – (≈2 cM) – T248W – (23 cM) – *Sec-1* – (2 cM) – T850W
– (0 cM) – C-band.

Between the NOR and *Sec-1* the locus of Pg1 could be mapped. The results of Singh et al. suggest a much greater distance between *Sec-1* and the telomere (16.0 + 5.4 cM), which may be due to the different situation in the rye translocation in the wheat background.

This example demonstrates the complexities of detailed chromosome mapping using different material and different approaches. Yet it seems that the combination of translocation breaks as studied in SCs with segregating genes may have a higher resolution at the chromosomal level than in situ hybridization in condensed mitotic metaphase chromosomes.

Translocations between A- and B-chromosomes (Sect. 3.1.4.1.2.3) have been used very successfully for gene localization in maize (Beckett 1991). The approach differs from that discussed above and includes deficiencies and aneuploidy rather than linkage relationships. It will not be discussed here.

8.3.3.3 Rearranged Aneuploids: Tertiary and Other Complex Monosomics and Trisomics

When direct localization of genes with specific markers is not possible, indirect approaches may still lead to satisfactory results, for instance using tertiary and more complex trisomics such as *compensating trisomics*. These can also be used to locate genes in specific chromosome segments. In addition to existing tertiary and other trisomics, newly formed trisomics can be very effective for gene localization within chromosome segments, provided they are formed sufficiently frequently. In one of the analyses leading to the map of the satellite of 1RS of rye (Sect. 8.3.3.2.2), the new formation of compensating trisomics was used (Sybenga et al. 1990), but this and similar examples will not be discussed here.

8.3.4 Recombinational Chromosome Maps: An Example

An old example of a recombinational chromosome map is shown in Fig. 8.5 (see above; Burnham 1962; Sybenga 1972). It represents the short arm and a short segment of the long arm of chromosome 9 of maize, and is based on data published by Creighton and McClintock (1931), McClintock (1931) and others. Deficiency mapping (Sect. 8.3.2.2.1) showed that *Dt* should be at the very end of the large terminal heterochromatic knob because it could be removed without removing the bulk of the knob. In the short segment between the knob and a slightly proximal conspicuous chromomere, the genes *py*, *yg*2 and *w* were located. Between a second conspicuous chromomere and the centromere, five genes are located and shown with their genetic distance. On the other side of the centromere, one more gene and the break point of a translocation were mapped. More recent additions to the map, including RFLP markers (cf. also Helentjaris 1987), are listed in the legend to Fig. 8.5.

It is clear that for detailed gene mapping of the chromosomes, several advanced and more classical techniques (segregation of genes and chromosomal markers scored in the karyotype, meiotic chiasma counts, in situ hybridization, SC analysis) are available, but that some of these require specialization and much labour. As yet in situ hybridization in condensed somatic metaphase chromosomes has less resolution than some of the other techniques. As shown above, techniques which make use of less contracted chromosomes, for instance pachytene bivalents with the chromatin intact and extending in loops from the chromosome core, permit a much finer localization of a gene locus by in situ hybridization. A prerequisite remains that the chromosome involved can be identified.

8.4 Manipulating Recombination

One objective of the manipulation of recombination, especially of recombination by crossing-over is to increase recombination in order to break a close linkage between a desired allele of one gene and an undesired allele of another gene within a cultivated species. A special application is the transfer of a desired gene into a cultivated species from a related wild or other cultivated species, where it is linked to chromosome segments that are not to be transferred, usually because the genes they contain do not satisfactorily substitute for the genes they replace in the recipient. In such situations crossing-over is normally insufficient and special techniques of transfer are required. These will be discussed in some detail in Section 10.4.

In addition to stimulating recombination, it is occasionally useful to prevent recombination in specific chromosome segments in order to maintain specific allelic combinations. This includes restricting exchange recombination, keeping allelic combinations intact within chromosomes, or restricting chromosome recombination, by combining different chromosomes into single linkage groups, which also restricts recombination within linkage groups. This will be discussed in Section 12.3.

In principle, chromosome recombination can be increased by replacing metacentric chromosomes by the corresponding telocentrics, which increases the number of chromosomes by a factor of two. This is not effective for the transfer of genes from one species or form to another, and it is potentially of practical importance only for modifying overall levels of recombination. In the genera *Gibasis* and *Cymbispatha* of the Commelinaceae, several species have acrocentric, even practically telocentric chromosomes, whereas in related genera they are metacentric (Jones 1974, 1976). There must be an adaptation in the system of chiasma distribution to assure that each (smaller) chromosome always has at least one chiasma. This is not necessarily an immediate natural reaction to centromere split, because in rye, for instance, our own unpublished results show that centromere-split homozygotes tend to fail to have a chiasma in short telocentrics in a large proportion of the cells, which results in insufficient fertility. Even in cultivated species with sufficient chiasmata, it is not certain that centric split would have any advantage.

More common in nature are systems that restrict recombination between chromosomes. One system could be centromere fusion, the opposite of centromere split, mentioned above. It is not known to have occurred with positive effects in plants. However, translocation heterozygosity, which has a similar effect, is quite common. In many cases interchanges "float" in the population, but exactly what their main advantage is, is uncertain. Crossing-over is restricted to the distal segments, apparently because this is favourable for the mechanical behaviour during metaphase I/anaphase I orientation at meiosis and because it restricts the breaking up of proximal gene blocks. This subject, including attempts to develop artificial systems in crop plants, is discussed in Section 12.3. In many animals, especially arthropods, and in some dioecious plant species, translocation heterozygosity involves the sex chromosomes (Smith 1969; Barlow and Wiens 1976), such that sexual dimorphism ensures permanent heterozygosity, but it is restricted to one sex.

Chapter 9

Genome Analysis: Identification of and Relations Between Genomes

9.1 Genome Analysis: Different Concepts

Originally, the use of the term *genome analysis*, as introduced by Kihara (1930; see also Lilienfeld 1951), was restricted to the identification of the diploid species which were combined in allopolyploids. It does not refer to the analysis of (single) genomes in terms of DNA and gene composition or of gene arrangement as considered in Section 8.3, nor the morphological structure of the chromosomes (karyotype analysis: Chap. 4). Gradually, the meaning of the term was broadened to include the identification of individual chromosomes constituting the different genomes of allopolyploids, in terms of their homoeology relations with corresponding chromosomes in the different constituting species. Subsequently, quantitative meiotic pairing (affinity) between different, related genomes, not combined in one allopolyploid, was included in genome analysis. In Section 9.2 genome analysis in the original sense, including a discussion of methods to identify the chromosomes belonging to specific genomes and to homoeologous groups will be considered. In Section 9.3 the analysis of quantitative pairing relations between genomes, and its interest for evolutionary and taxonomic studies follow.

9.2 Genomic Composition of Allopolyploids

9.2.1 Identification of the Progenitors

There are a few complications. Potential progenitors may have become extinct, or have not yet been discovered or changed in the course of time. Also, the genomes as present in the allopolyploid may not have remained quite the same. One mechanism of change is simply mutation, which in an allopolyploid is more readily tolerated than in a diploid, because of the presence of a related genome which still has an intact allele of a gene mutated in the other genome. In addition to gene mutation, chromosome structural rearrangements may have become established. Further, allopolyploids with a somewhat different origin may have hybridized. This may have led to combinations of chromosomes other than those present in any original diploid. This has possibly taken place in wheat, where the origin of the B-genome is still uncertain.

There are basically three steps in the identification of the diploid progenitors of an allopolyploid.

1. *Selection*, on the basis of karyotype, morphology, histology, anatomy and biochemistry, of the species that are the most probable progenitors. Analytical procedures ("extrapolated correlates", Anderson 1949) have been developed that make a good selection possible merely on the basis of morphology and anatomy. Biochemical methods can refine the approach considerably. Johnson (1967) found that the proteins of *Aegilops squarrosa* and *Ae. caudata*, which had been suggested as potential progenitors of allotetraploid *Ae. cilindrica*, when mixed gave the same electrophoretic pattern as that of *Ae. cilindrica*. Separately, the electropherograms did not add up to that of the allotetraploid.
2. Analysis of *meiotic pairing* in the hybrids between the allopolyploid and its proposed progenitors. If the right diploid species has been selected, it must pair with one genome of the allopolyploid, and the other genomes of the allopolyploid form univalents. There may be an occasional quadrivalent if a translocation has been established either in the polyploid or in the diploid progenitor. This approach has been in use since Kihara carried out his first analysis of the Triticinae in the 1920s. When more than one species is eligible, the one with the most regular meiotic behaviour in the hybrid is considered the best candidate.
3. *Resynthesis* of the allopolyploid from its putative progenitors. It should closely resemble the polyploid species, and the hybrid between the synthetic and the original allopolyploid should be fully fertile, with limited segregation in the F2, only involving simple allelic differences.

In addition to the possibility that the allopolyploid has diverged from the original by mutation or introgression from related forms, it is possible that the true diploid progenitor was not available. An old example is the analysis of tobacco, *Nicotiana tabacum* L. (2n = 4x = 48) by Goodspeed and Clausen (1928). On the basis of morphological studies, it was decided that *N. sylvestris* and *N. tomentosa*, both 2n = 24, were the best candidates. The F1 hybrids between *N. tomentosa* and *N. sylvestris* with *N. tabacum* had 12 bivalents and 12 univalents in meiosis. Both diploid species had apparently a genome in common with the allopolyploid, and this was not the same genome, because the F1 between the diploids had only univalents, and only infrequently an occasional bivalent. Doubling the chromosome number of this diploid hybrid resulted in a plant closely resembling *N. tabacum*. Apparently normal pollen was produced, but embryo sac development was abnormal and only few seeds developed. Selection for increased fertility had some results, but not enough. Later, it appeared that *N. tomentosiformis*, related to *N. tomentosa*, produced amphidiploids with *N. sylvestris* that were immediately fertile, and gave fertile hybrids with *N. tabacum*. *N. otophora*, however, did the same. One of these two species, in combination with *N. sylvestris*, must have been the progenitor of tobacco, but it is not certain which one. No chromosomal rearrangements were detectable in the F1 between the resynthesized allopolyploid and tobacco, but genetic segregation was abundant in the F2, suggesting that the new and old allotetraploids differed in numerous genes.

Hexaploid bread wheat and the tetraploid macaroni wheats are another example of early genome analysis. Kihara (1930–1937) was the first to determine the genomic constitution of the different species. The A genome is found in the diploid *T. monococcum*, *T. boeoticum* and other species of the "Einkorn" group. The A and B genomes are combined in the tetraploid "Emmer" wheats, including the hard macaroni wheat, *T. durum*. The B genome has been derived from the genus *Aegilops* (also considered a section of the genus *Triticum*), but the exact origin is not certain. It was first believed to be *Ae. speltoides*, but for several reasons this could not be maintained. It may well be a composite genome, to which different diploid species have contributed. In the hexaploid bread wheats (*Triticum aestivum*) a third genome has been introduced, the D genome (from "Dinkel", spelt). McFadden and Sears (1944) found that the D genome had its origin in *Aegilops squarrosa*. Crossing cultivated tetraploid wheats with *Ae. squarrosa* and doubling the chromosome number of the hybrid resulted in a fertile spelt-type wheat, which gave a fertile hybrid with *T. aestivum*.

The tetraploid cotton species *Gossypium hirsutum* (the most widely grown cotton: upland cotton) and *G. barbadense* (Sea Island, Egyptian, Pima cottons) both have 2n = 52 and are constituted by the A genome from an "Old World" *G. herbaceum*-type diploid cotton and the D genome of the diploid American wild species *G. raimondii* (Endrizzi et al. 1985). Hexaploid oat (*Avena sativa*, 2n = 42) has the A, B, and C genomes (Rajhathy 1983), and the wild species *A. fatua*, *A. sterilis* and *A. byzantina* have the same basic genome composition. The tetraploid *Avena* species have AC and AsB (Nishiyama et al. 1989; Rajhathy 1991). Already in the 1930s the diagram in Fig. 9.1. was constructed for three *Brassica* allopolyploids (Morinaga 1934; U 1935). The genomic constitution of many polyploid perennial grasses of the Triticeae have been and are still being analyzed in Logan (USA), see for instance Wang (1989).

9.2.2 Assigning Chromosomes to Genomes and Homoeologous Groups

The chromosomes of an allopolyploid can be classified:

1. According to the *genome* to which they belong;
2. According to their *homoeologous group*.

In wheat, for instance, there are three different genomes, A, B and D, with seven chromosomes each, numbered 1–7. There are three chromosomes

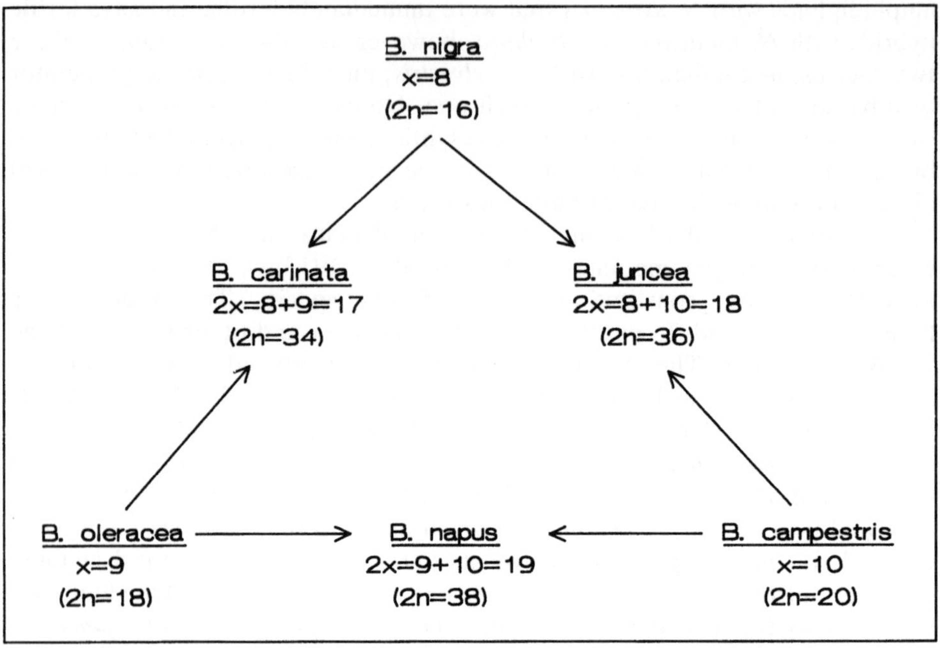

Fig. 9.1 The relations between some of the diploid and allotetraploid *Brassica* species. (After U 1935)

in each homoeologous group; for group 1 these are: 1A, 1B and 1D, and for homoeologous group 2: 2A, 2B and 2D. The 21 wheat chromosomes together are classified as: 1A, 1B, 1D; 2A, 2B, 2D; . . . ; 7A, 7B, 7D. The chromosomes within a homoeologous group are related, but not closely, and they do not normally pair in meiosis. The mechanisms responsible for this pairing differentiation, which may vary considerably between allopolyploids, are discussed in Sections 6.1.2.3. and 11.3.2.2.

To assign chromosomes to their genomes or homoeologous groups, they must be identified. Usually, it is not sufficient to distinguish them individually in the karyotype by C- or other types of banding, or simply to identify them according to a gene or the linkage group they carry. Sometimes the chromosomes of the different genomes are sufficiently different to distinguish the genomes. In allotetraploid cotton (*G. hirsutum*) there is sufficient difference in chromosome size between the genomes to identify the genome to which a particular chromosome belongs. Within the genomes, the identification is somewhat more difficult. In wheat, the chromosomes of the B genome can readily be recognized on the basis of their banding pattern, and even within the B genome individual chromosomes can be identified. The A and D genomes of wheat and even their individual chromosomes can also be distinguished, but less simply (Fig. 4.3.C). For most other allopolyploids a detailed C-band analysis has not yet been carried out or is simply not possible.

However, even when genomes can be recognized by C-banding, it must still be determined which of the different genomes is which.

To assign chromosomes to their genomes and homoeologous groups, special chromosomal markers are usually required: monosomy, trisomy for complete chromosomes or specific arms (telosomy) offer the best opportunities. The way these chromosomal modifications are obtained is discussed in Sections 6.2.1.2 and 6.2.2.1.2.

In order to establish which set of chromosomes together form one specific genome, a hybrid is made between the allopolyploid of which the chromosome to be identified is marked and one of the known ancestral species of the allopolyploid. The chromosome to be classified can be a nullisomic, a monosomic or a monotelosomic. In the F1 the monosomics (or telosomics, respectively) are selected. When the marked chromosome is one of the chromosomes belonging to the ancestral species with which the hybrid was made, there will be one normal bivalent less. It is replaced by a univalent in case of a nulli- or monosomic, and a heteromorphic bivalent when it was a telocentric. In case the monosomic is a chromosome of another genome, the full number of normal bivalents will be observed and there is a univalent less, or the telo is univalent. A translocation in the parental species or the allopolyploid complicates the issue, but only slightly. Then, if the marked chromosome is one of the genomes corresponding to the diploid hybridization partner, and if it is also one of the chromosomes involved in the translocation, instead of a quadrivalent, a trivalent may be formed or a heteromorphic quadrivalent containing a telocentric.

By crossing the parental diploids with each line of a complete monosomic or telosomic series, the genome to which each of them belongs can be determined. It is assumed that the marked chromosome has been identified and can be carried over to subsequent generations. To identify the marked chromosomes several methods are available: from the karyotype (Chap. 4), the linkage group (Chap. 8), or a tester set (Chap. 6). Or it is simply registered as "this particular chromosome" and given an arbitrary name or number for registration.

To identify the *homoeologous group* to which a series of marked chromosome belongs the parental species are not required. Again, use is made of a monosomic (or, preferably, but not always possible, a nullisomic) series, and a tetrasomic or at least a trisomic series. The nullisomics (or monosomics) are crossed with the tetrasomics (or trisomics). In the hybrid progeny plants with the apparently normal chromosome number are selected. The combination of a nullisomic with a tetrasomic immediately yields a monosomic/trisomic combination, but more combinations are formed, when monosomics and trisomics are used. This is the rule rather than the exception.

The critical type combines monosomy for one and trisomy for the other chromosome. It can be recognized by having one univalent and a trivalent in meiosis, in addition to normal bivalents. Of course, when the original

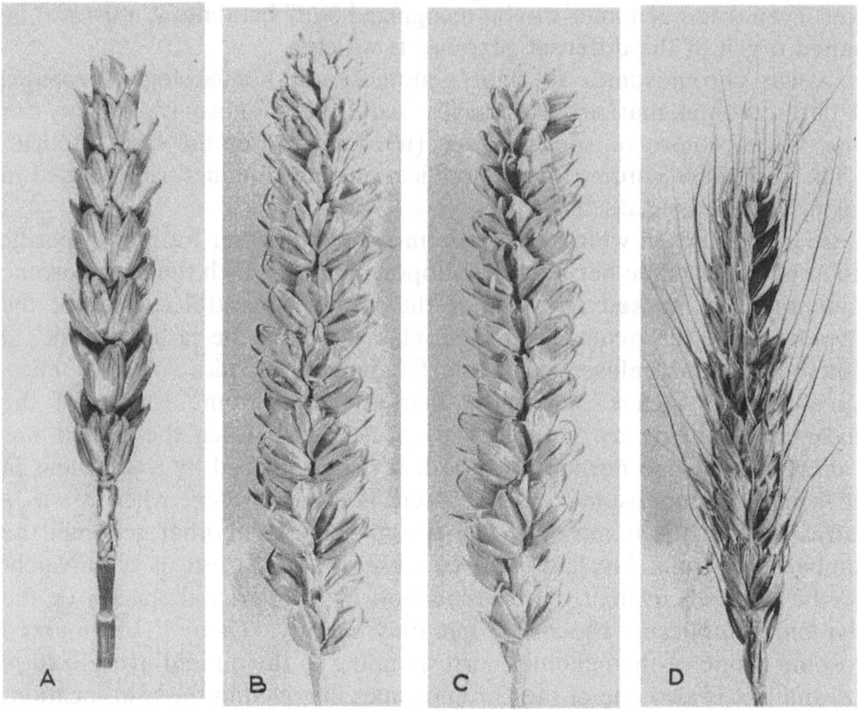

Fig. 9.2 Nulli-tetra compensation in wheat: **A** nulli 2D; **D** tetra 2A; **B** normal wheat; **C** nulli 2D – tetra 2A. The close similarity between **B** and **C** shows that there is good complementation, suggesting that 2A and 2D are largely homologous genetically. (Courtesy of E.R. Sears, see also Sybenga 1972)

tetrasomic and nullisomic involve the same chromosome, a normal type appears. The mono/trisomics are selfed, and in the progeny, plants are again selected that have the normal chromosome number, but in meiosis have one quadrivalent: these are nullisomic for one chromosome and tetrasomic for the other. The essence of the analysis now is to compare the morphology and fertility of the nulli-tetrasomic plants with those of normal plants. In most cases the nulli-tetrasomics will be far inferior to normal plants and not better than the original nullisomics. However, in a few combinations tetrasomy compensates, at least partly, for nullisomy, the plant morphology will look more normal, and the fertility is partly restored. In wheat, Sears (1954) observed that sets of chromosomes could be found that would compensate in nulli-tetra combinations: *nulli-tetra* compensation (Fig. 9.2). It was assumed that these chromosomes belonged to the same homoeologous group. By making a critical set of combinations of nullisomy with tetrasomy, Sears (1954) could classify all seven groups of three homoeologous chromosomes.

An alternative method to identify the chromosomes of homoeologous groups is genetic, and in principle quite simple when a sufficient number of genes have been mapped on their respective chromosomes: chromosomes that carry homoeologous linkage groups must belong to the same homoeologous group.

In both cases, there is a possibility that in the course of evolution chromosomal rearrangements and mutations have occurred, which may considerably complicate the analysis.

The same methods, or modifications thereof, can be used to identify homoeologous chromosomes of genomes that are not combined in an allopolyploid, but belong to different species. By crossing disomic additions of rye to wheat with nullisomics (or, if necessary, monosomics) of wheat, Koller and Zeller (1976) obtained homozygous substitutions of specific rye chromosomes for wheat chromosomes. In some combinations telocentric chromosomes were used instead of complete chromosomes. It appeared that specific rye chromosomes could compensate well for specific wheat chromosomes, which apparently were their homoeologues except for chromosomes 4 and 7. The telo of the short arm of chromosome 4R of rye substituted much better for 4A and 4D of wheat than the complete chromosome 4R. On the other hand, most of the long arm of 4R of rye compensated for arms 7AS (short arm of 7A), 7BS and 7DS. Yet substitution of the complete 7R for 7A and 7B gave better fertility than the nullisomic. The conclusion was that there is an ancient translocation in rye between chromosomes 4R and 7R. The reasonable compensation of 7R for 7A and 7B must be due to the very limited effect of the segment translocated from 4R to 7R. This cytological conclusion was confirmed by the localization of the gene for 6-PGD (6-phosphogluconate-dehydrogenase), using wheat aneuploids, including specific additions, and triticale and rye (Hsam et al. 1982). In wheat the loci were found on chromosomes 6A, 6BL and 7BL and in rye on 4RL and 6RL. It is not uncommon that for similar enzymes gene loci in different chromosomes are present, in this case chromosomes of groups 6 and 7. There is, apparently, homoeology between 6RL and the group 6 chromosomes of wheat, but 4RL is homoeologous with one arm of the group 7 chromosomes. Similar analyses have been performed for different species of the genus *Aegilops* and hexaploid bread wheat, combining compensating chromosome combinations and gene loci. RFLP analysis can give similar information on the genetic structure of the individual chromosomes of different species.

9.3 Relations Between Genomes: Homology, Homoeology, Affinity

9.3.1 Diploid Hybrids

The extent of pairing between the chromosomes of different species may be considered an indication of their genetic relationship. In addition to a theoretical interest, there are two main reasons for a practical interest: to transfer genes between species it is useful to know whether there is a reasonable probability of exchange. Further, if there is exchange, how frequent is it, how is it localized, and how large must a segregating population be to isolate a sufficient number of recombinants? No examples are yet available of estimates of the required population size. There is a second reason for interest, but here the opposite of recombination is required: when an allopolyploid between two species is planned, it is useful to know if the pairing affinity between the chromosomes of the two species is low enough for regular disomic segregation.

The estimation of the genetic distance between species on the basis of chromosome association at meiotic metaphase I of the hybrid is not without pitfalls, nor is it simple to determine the level of chromosome association most favourable for practical applications. For the construction of allopolyploids, in general, species with little chromosome pairing affinity, provided the genomes are not too far apart genetically, will tend to form more stable allopolyploids than species with a strong tendency to pair. Yet relatively good pairing in the diploid hybrid is not always accompanied by poor meiotic behaviour in the polyploid. The homoeologous chromosomes in the dihaploid of allotetraploid cotton *Gossypium hirsutum* or the equivalent, the diploid hybrid between *G. herbaceum* (A genome) and *G. raimondii* (D genome), pair almost as well as in the diploid species in spite of a considerable difference in chromosome size (Sect. 6.1.2.3). In the allotetraploid *G. hirsutum*, however, all homoeologous pairing between the A and D genomes is suppressed by the availability of complete homologues (Endrizzi et al. 1985). Complete differentiation, apparently, is not always necessary.

With respect to the transfer of genes from a wild into a cultivated species, a high level of recombination between the homoeologous chromosomes would seem to be desired. However, this will not only lead to the introduction, by recombination, of the desired alien gene into a chromosome of the cultivated species, it will also introduce several undesired genes. Their removal requires an extensive series of backcrosses. A moderate level of recombination applied in cycles of alternating backcrossing and strong selection may sometimes be more effective. Further details on gene transfer between different species and on the construction of allopolyploids will be discussed in Sections 10.4 and 11.3.2.2.

With respect to estimating the level of metaphase I chromosome association itself, caution is necessary (Sect. 8.2.1). Under some conditions achiasmate associations may be mistaken for chiasmata. In normal material distal chiasmata are usually not found within the distal heterochromatin, where this is present, as can be seen from the shape and location of the C-bands at the chiasma (Jones 1978, see also Sect. 3.2.2.2 and Fig. 3.11). In desynaptic material and in interspecific hybrids, however, several of the metaphase I associations are in the heterochromatin and do not show the typical chiasma appearance. Orellana (1985) showed that in rye-wheat hybrids, many of such apparently terminal chiasmata were lost between diakinesis and full metaphase

I, whereas the frequency of associations (one-third of the total associations) that had the typical chiasma structure with the C-bands at the sides did not decrease. The frequency of unequal chromatids at anaphase, which is the result of crossing-over between a chromosome with terminal heterochromatin of rye, and a chromosome without terminal heterochromatin of wheat, corresponded with the frequency of metaphase I association between such chromosomes with the typical chiasma characteristics. Orellana (1985) concluded that the frequency of genetic exchange between wheat and rye is much lower (about one-third) than expected on the basis of metaphase I association frequencies. This agrees with the low frequency of genetic exchange observed in such material.

In addition to the apparent difficulties in recognizing true chiasmate associations in diploid and polyploid interspecific hybrids, there is the possibility that as a result of genetic imbalance in the hybrid the frequency of chiasmata is reduced relative to the actual pairing affinity. This is a form of desynapsis, which prevents an evaluation of the true pairing relations. To estimate recombination, this is not important, but for theoretical purposes and for the construction of an allopolyploid, it is of some consequence. Similarly, when genes are present that reduce homoeologous pairing specifically, without significantly reducing homologous pairing (the *Ph* gene in wheat and comparable systems, see Sect. 6.1.2.3.4), pairing in diploid hybrids is not a reliable criterion of homoeology. In spite of these drawbacks, it does give valuable information if proper care is taken (Wang 1989).

9.3.2 Quantitative Models for Polyploid Hybrids

Because estimates of the frequency of chiasmate association in diploid hybrids give only a limited impression of the pairing relations between the species, which may be further distorted by desynapsis and non-chiasmate associations, polyploid hybrids have certain advantages. With more than two genomes combined in the same genetic background, competing for pairing partners, it may be assumed that their relative pairing (or affinity) relations are more accurately expressed than in diploid hybrids.

Kimber and coworkers in Missouri (USA) have made great contributions to the development of mathematical models to estimate pairing affinity relationships between genomes in polyploid hybrids, based on diakinesis and metaphase I configuration frequencies. In addition, a few other groups have designed comparable models, sometimes using different approaches which, in some instances have consequences for the estimates obtained.

9.3.2.1 Triploid Hybrids

The triploid interspecific hybrid is the simplest polyploid hybrid, but in many respects also the most useful. The models contain a number of simplifications.

One is the assumption, common to all current models, that pairing starts at or near the ends and that there are no interstitial pairing initiation points. This is not an entirely realistic assumption, as is clear from synaptonemal complex studies where parallel alignment (Loidl and Jones 1986; Loidl et al. 1990) and several points of pairing partner exchange have been observed repeatedly in several species and hybrids. Usually, however, it is an acceptable simplification, especially when chiasma formation is mainly terminal or subterminal. The consequence of this assumption is that there are nine pairing modes (three in each arm, in all combinations). In the triploid, six result in a trivalent and three in a bivalent with a univalent (Table 6.1).

In autotriploids, the nine pairing modes have an equal probability of occurring. In hybrids, however, there may be differences in affinity between the three genomes, resulting in preferences for specific pairing combinations, and consequently differences in the frequencies of the nine configurations.

With three genomes, say A, B and C, there are three ways in which the chromosomes can start pairing. Between the ends the chromosomes are free to exchange partner, which results in trivalents. For each chromosome arm, the representative of genome A can pair with that of B, and that of C is free; or, the arm of genome A can pair with that of C and that of B is free. Finally, the arm of B can pair with that of C and that of A is free. Each combination of two chromosomes (AB, AC and BC) can have its own affinity, expressed as the probability of being paired. With decreasing affinity between two genomes, the affinity between the others will increase relatively and the frequency of bivalents with a univalent will increase at the expense of the frequency of trivalents. When the chromosomes cannot be distinguished, the mere increase in bivalents does not contain sufficient information for deciding *which* pairing combination is increased, or what the differences in affinity between all of the three genomes are. Since pairing preference of one combination is at the expense of other combinations, it would be sufficient to estimate two affinity parameters, as then the third can be derived immediately, but even this is not realized because in the current models only one estimate can be made.

Different solutions have been proposed for this dilemma. In the model of Alonso and Kimber (1981) the relative affinity between the two most closely related genomes, say A and B, is estimated (x) and it is assumed that the third (C) is equally distant from the first two, i.e. the affinity between A and C is equal to that between B and C and represented by y, where $y = (1 - x)$.

In the models proposed by Sybenga (1988) and Crane and Sleper (1989a), even though the fact that only one affinity parameter can be estimated is recognized, all three affinities are separately represented in the models. In the first this leads to a range of possibilities for each affinity factor, here estimated as *preferential pairing* factors (p1, p2 and p3, in which the sum equals 0). In the Crane and Sleper (1989a) model optimization is applied, which ideally leads to one best estimate for all three, but in practice also to a range, simply because there are not enough degrees of freedom for more.

These affinity or preferential pairing coefficients represent relative affinities. Potentially, there is also a possibility to estimate the *level* of the affinities. In the model of Alonso and Kimber (1981) the chiasma frequency or rather the average frequency (c) of the chromosome arms to have at least one chiasma, is a measure. Here, a complication arises: in addition to reduction of chromosome pairing when genomes are not closely related, there may be genetic reasons for a general decrease in chiasma formation not related to low affinities. Then the average level of affinity is confounded with chiasma frequency.

One important difference between the different models is in the way chiasma formation is considered. The situation is simple when the chiasma frequency is high and all chromosome arms have at least one chiasma. Then affinity is apparently strong, but may still be different for different combinations. There are only trivalents and ring bivalents, the latter accompanied by a univalent. The ratio of the frequencies of these two comfiguations is a measure of possible differences in affinity between the genomes, and can be derived directly from the observations. The trivalents are represented mainly by chain-shaped trivalents with occasional other types, arising when interstitial chiasmata are formed (Sect. 6.1.2.2.1.3).

With lower chiasma frequencies a trivalent may fall apart into an open bivalent and a univalent. Ring bivalents similarly are reduced to open bivalents, and the difference between trivalent and bivalent pairing disappears when arms fail to have chiasmata. In both the model of Alonso and Kimber (1981) and Sybenga (1988), the ratio (r) of trivalents/ring bivalents is the single basis for the relative affinity estimates. In the first model, the average arm association frequency c is only the second parameter estimated as a measure of the degree of affinity. The reasoning is based on that of Driscoll et al. (1979) and the pairing diagram of Table 6.1, and can be summarized as follows: After trivalent pairing, a trivalent is seen at metaphase I when both paired arms have one or more chiasmata; the frequency is c^2, multiplied by the frequency of trivalent pairing. After bivalent pairing a ring bivalent is formed with the same frequency c^2, but multiplied by the frequency of bivalent pairing. Open bivalents and univalent pairs are derived from rings as well as from trivalents and have a frequency of $2.c(1 - c)$ and $(1 - c)^2$ respectively. This implies that the frequencies of open bivalents and univalents do not contain information with respect to preference for pairing, which is present only in the relation between trivalents and ring bivalents.

Pairing between the two most closely related chromosomes has the frequency x, and the frequency with which either one pairs with the third has a frequency y. When there is pairing, it is either of type x or type y and thus $(x + y) = 1$. The frequencies of the three pairing modes $(x + y + y)$ together add up to $(x + 2y)$. The most closely related chromosomes pair with a frequency of $x/(x + 2y)$. *Ring bivalents* with a univalent are formed when *both* arms of two chromosomes pair with one another. This occurs between the most closely related chromosomes with a frequency $[x/(x + 2y)]^2$. In addition, ring bivalents are formed when pairing occurs in both arms of the two less homologous combinations: $2[y/(x + 2y)]^2$. This adds up to $(x^2 + 2y^2)/(x + 2y)^2$.

Trivalents are formed when the most closely related chromosomes pair in one arm, and the other arm of either one pairs with the third chromosome. There are four possible combinations of this type, together with a frequency $4xy/(x + 2y)^2$. However, there are also two possibilities for y/y associations forming trivalents, frequency: $2[y/(x + 2y)]^2$; together: $(2y^2 + 4xy)/(x + 2y)^2$.

Open bivalents with a univalent have fewer possibilities. There is one x-type combination: $x/(x + 2y)$ and two y-type combinations: $2y/(x + 2y)$. These combinations add up to 1. The fact that both arms can have the association is accounted for by the factor 2 in the chiasma formula involving c (see above and below). The formula for the sets of three univalents is simply 1. The final formulae are as follows:

Sets of three *univalents*:	$(1 - c)^2$ (x and y not involved)
Open bivalents with univalent:	$2c(1 - c)$ (x and y not involved)
Ring bivalents with univalent:	$c^2(x^2 + 2y^2)/(x + 2y)^2$
Trivalents:	$c^2(2y^2 + 4xy)/(x + 2y)^2$

The affinity relations x and y occur only in the ring bivalent and trivalent formulae. As proposed by Sybenga (1988), the ratio r of the trivalent frequency and the ring bivalent (with univalent) frequency in combination with the equation $y = (1 - x)$ should give a direct solution of x. Thus, $r = (4xy + 2y^2)/(x^2 + 2y^2)$ or:

$$x = (2r \pm \sqrt{(2r - 2r^2 + 4)})/(3r + 2).$$

Alonso and Kimber (1981) give a more complex formula from which x can be derived also. In addition to x, they estimate c as an estimate of the overall level of affinity, but the third degree of freedom is not used. In order to accomodate this extra degree of freedom, but also in order to obtain values for x when the observations are in conflict with the assumptions of the model (for instance when r is larger than 2, which would give a negative discriminant in the formula for x, or when rearrangements occur resulting in a number of univalents smaller than the number of bivalents, or larger configurations than trivalents), Alonso and Kimber (1981) do not estimate x directly but prefer to solve x by optimization.

In the models of Sybenga (1988) and Crane and Sleper (1989a) the remaining two degrees of freedom after estimating an affinity parameter are used to estimate the chiasma frequencies of the two arms. This is a valid procedure especially when one of the two arms of a chromosome fails to have a chiasma relatively more frequently than the other simply because of its length or other intrinsic factors independent of pairing affinity. This confounds *level* or *degree* of affinity (in contrast to *relative* affinity x) with chiasma frequency, but in view of the type of observations (metaphase I configurations)

Table 9.1. Meiotic configuration frequencies in three triploids; taxonomic nomenclature from original publication. (Yen and Kimber 1989)

Triploid[a]	Cells	Univ.	Bivalents Open	Bivalents Ring	Triv	Quadr. Open	c[b]	x[c]	SSD[d]
A obs.	?	7.03	0.66	6.33	0.00	0.00	0.951		
calc.		7.03	0.65	6.33	0.00	0.00		1.00	0.00
B obs.	60	2.52	0.42	2.03	4.48	0.03	0.968		
calc.		2.64	0.43	2.27	4.98	0.00		0.50	0.52
C obs.	40	5.98	1.02	4.92	0.98	0.05	0.926		
calc.		6.08	1.08	4.85	1.02	0.00		0.95	0.02

[a] A: hybrid between an autotetraploid and a well-differentiated other species (*Triticum tauschii* x *Secale cereale*, genomes DDR); B: autotriploid (*T. speltoides*, genomes SSS); C: hybrid between an autotetraploid and an apparently not closely related species with similar genome symbol (*T. speltoides* x *T. longissimum*, genomes SSlSl). In B and C two cells had a quadrivalent, possibly due to a translocation.
[b] c Average metaphase I association frequency per chromosome arm.
[c] x Relative affinity between *closest* genomes.
[d] SSD = sum of squares of the difference between observed (obs.) and calculated (calc.) frequencies the basis of the model and the value of x as estimated by optimization.

this is inevitable. The models of Driscoll et al. (1979) on which the later and more detailed model of Alonso and Kimber (1981) is based, neglect this difference on the assumption that the difference in chiasma frequency between arms is small compared with the differences caused by differences in affinity. Or: once two chromosomes pair at any locus, they do so sufficiently to form a chiasma, independently of whether this is in a short or a long arm. The correctness of this assumption has not been satisfactorily proven, even though some indications have been obtained that it may be correct under certain conditions. In diploids there are real differences between chromosome arms in frequency of chiasmate association, and the same may be expected to be true for polyploids.

The uncertainty introduced by combining three different genomes with two independent relative affinities, where only one independent parameter can be estimated, is circumvented by combining two identical genomes with a different, third genome (Yen and Kimber 1989, 1990a,b). The triploid hybrid is produced by crossing an autotetraphoid of one species with the diploid of a second. In Table 9.1 (Yen and Kimber 1989) the meiotic analysis of three different triploids is shown. In one triploid the single unrelated genome does not show any association with the other two, which are fully homologous as demonstrated by the high c-value. The relative affinity (x) between these two genomes is equal to one. This implies that $y = 0$. The second example is an autotriploid: again, c is high, and, because the affinity between the three genomes is equal, x is 0.5. In the third example there is some trivalent formation, indicating that there is some affinity between the two species. The two homologous genomes still cause the c-value to be high, but x is somewhat

Table 9.2. Meiotic pairing of a set of three homoeologous chromosomes in a triploid hybrid. The genomes are 1, 2 and 3

Arm B \ Arm A	1 − 2 (1/3 + *p*1)	1 − 3 (1/3 + *p*2)	2 − 3 (1/3 − (*p*1 + *p*2)
1 − 2 (1/3 + *p*1)	$(1/3 + p1)^2$	$(1/3 + p1)(1/3 + p2)$	$(1/3 + p1)(1/3 - p1 - p2)$
1 − 3 (1/3 + *p*2)	$(1/3 + p1)(1/3 + p2)$	$(1/3 + p2)^2$	$(1/3 + p2)(1/3 - p1 - p2)$
2 − 3 (1/3 − (*p*1 + *p*2)	$(1/3 + p1)(1/3 - p1 - p2)$	$(1/3 + p2)(1/3 - p1 - p2)$	$(1/3 - p1 - p2)^2$

For each arm of each set of three chromosomes pairing is possible between 1 and 2, between 1 and 3 and between 2 and 3. The other arm is unpaired. With random pairing each combination is expected with a frequency of 1/3. Preferential pairing causes an increment *p* of the combination involved. The other combinations must together decrease with the same amount; their *p* value is *negative*. There are three different combinations, each with its own *p*: *p*1, *p*2 and *p*3. The sum of the three *p* values is 0, when two are given, the third is fixed: only p_1 and p_2 are required in the formulae.
The total pairing configuration frequencies are:
trivalents: $2/3 - 2p1^2 - 2p2 - 2p1p2$
bivalents with univalent: $1/3 + 2p1 \quad 2p2 + 2p1p2$
To obtain the metaphase configuration frequencies, these must be multiplied by the chiasmatic association factors *a* for one arm, *b* for the other: $a \cdot b$ for trivalents and ring bivalents with univalent. The frequency of open bivalents with a univalent is $\{(1 - a) \cdot b + a \cdot (1 - b)\}$ and $(1 - a) \cdot (1 - b)$ for sets of three univalents. In the latter two expressions the pairing formula is not present. To estimate the pairing parameters, the ratio trivalents/ring bivalents gives all the information available in the observations.

lower than 1. This is probably the most useful application of triploid models, as it gives an estimate of relative affinity between two species, which is often difficult to obtain in a diploid hybrid. A disadvantage is that $y = 0$ does not necessarily mean that in the diploid the two species do not pair. In fact, the reciprocal triploids and the diploid should all three be analyzed for a good affinity estimate.

In the model of Sybenga (1988) not only chiasma frequencies (in fact, bound arm frequencies) are estimated for both arms, but also different affinity parameters for the three combinations of the three chromosomes. These are given as *preferential pairing* values added to the one-third for each combination expected when pairing is random (Table 6.1). There are three *p* values, one for each of the three possible combinatons: p_1, p_2 and p_3 with $p_1 + p_2 + p_3 = 0$. When one *p* value is positive (pairing frequency > one-third) this must be compensated by negative values for one or both of the other combinations, i.e. they pair with a frequency of less than one-third. The overall level of pairing is partly expressed in the average chiasmate association frequency (*c*).

The pairing matrix for the three pairing combinations for both arms is given in Table 9.2. The diagonal represents bivalent pairing with a total of $1/3 + 2p_1^2 + 2p_2^2 + 2p_3^2$. The remaining are trivalents: $2/3 - 2p_1^2 - 2p_2^2 - 2p_3^2$. For the metaphase I configuration frequencies the pairing frequencies must be multiplied by the chiasmate association frequencies *a* (for arm A) and *b* (for arm B) or *c* when the arms are considered to be equal. Trivalents and ring bivalents (with univalent) are both formed with a frequency *a.b* x the pairing

formula. Like in the Alonso and Kimber (1981) model, the pairing formulae do not appear in the open bivalent (with univalent) and sets-of-three-univalents formulae. From the ratio trivalents/ring bivalents the single pairing affinity estimate must be derived. Since there are three different preferential pairing factors with a sum of 0, two estimates are required, or certain assumptions must be made. With the restrictions of Alonso and Kimber (1981), $p_2 = p_3 = 0.5(1 - p_1)$ and $p_1 = 1/3\sqrt{[2(2-r)/(r+1)]}$.
In this situation it is possible to transform p into x.

The importance of the preferential pairing model is in the opportunities it offers to see what the different possibilities for the three preferential pairing estimates are for each specific observed (or fictitious) value of r, and in the possibility to differentiate between the two arms of the (average) configuration with respect to chiasmate associaton frequency. It can be derived that, for any value of p_1 (within the acceptable range of $-1/3$ and $+ 2/3$):

$$p_2 = (-p_1 + \sqrt{\{[2(2-r)/3(r+1)] - 3p_1^2\}}/2 \text{ and}$$
$$p_3 = (-p_1 - \sqrt{\{[2(2-r)/3(r+1)] - 3p_1^2\}}/2 = 0 - p_1 - p_2.$$

The discriminant D in the formula for the roots of the quadratic equation that can be derived for the chiasmate arm association frequencies a and b equals $o^2 - 4u(t + r)$ and:

$$a = (2t + 2r + o + \sqrt{D})/2;$$
$$b = (2t + 2r + o - \sqrt{D})/2;$$

where t, r, o and u are the frequencies of trivalents, ring bivalents, open bivalents and sets of three univalents, respectively.

When $r = 2$, all p values are 0 and there is no difference in affinity between the three genomes. It is essentially an autotriploid when either a or b or both are large, but when the chiasma frequency is low and desynapsis can be assumed to be excluded, it is a triploid hybrid with equal and great distance between the three genomes. With r between 1 and 2 all p values are within the limits of $-1/3$ and $+2/3$. In this range each of the three genomes must necessarily be involved in pairing with both other genomes, but certain combinations will be preferred. With values of r lower than 1 it is also possible that the chromosome of one genome is always in the center and the other two always at the ends of the trivalent. This has its consequences for recombination between the three genomes. Alonso and Kimber (1981) do not accept this possibility on the assumption that when there is sufficient homology for two genomes to pair with the third, they must also be sufficiently homologous to pair with one another. When pairing preference is indeed a matter of DNA homology, this would be a reasonable assumption. However, pairing initiation is more probably dependent on active pairing initiation sites ("zygomeres", Sybenga 1966) of which several occur within each chromosome. Their specificity may be DNA sequence-dependent, but they may have a DNA-independent regulation system. In Fig. 9.3 three sets of zygomeres are shown diagram-

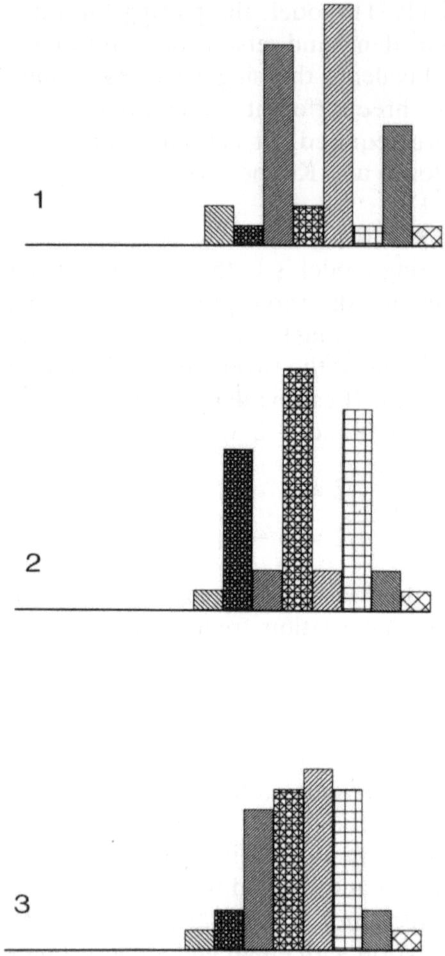

Fig. 9.3 Diagram of hypothetical zygomere patterns in a pro-terminal segment in three homoeologous chromosomes. The size of the *column* represents the activity, the type of marking the specificity. Only the activity varies in this example. Chromosomes **1** and **2** could be homoeologues between which pairing is not possible as long as full homologues are present, but where very limited pairing is possible in the dihaploid. In the presence of chromosome **3**, both could pair with **3** but still not effectively with each other. Chromosome **3** would always be in the center of a trivalent of three metacentric chromosomes with such zygomere patterns in both arms. (Sybenga 1988)

matically. Sets 1 and 2 have only relatively inactive sites in common, but in set 3 active zygomeres are present that occur also in sets 1 and 2. Trivalents are possible only with set 3 in the center if pairing between sets 1 and 2 is excluded.

In an example of a hybrid between allotetraploid *Trifolium repens* (2n = 32) and diploid *T. niger* (2n = 16), given by Sybenga (1988), a value of r = 0.397 was found. Considering that none of the p values can exceed 0.667 nor be smaller than 0.333 and that their sum equals 0, the possible ranges are rather limited. At one extreme $p_2 = p_3$, and the conditions of the Alonso and Kimber (1981) model are fulfilled, with $p_1 = 0.505$, and p_2 and p_3 both -0.253. Slight changes in p_1 have large consequences for the other two preferential pairing parameters, implying a large error in their estimates. The limit of $-1/3$ is reached quite rapidly for p_3, so the range for the positive preferential pairing parameter is short. Apparently, considerable variation in the preferential pairing parameters is compatible with the observations.

In the example of the triploid hybrid between tetraploid *T. repens* and diploid *T. niger*, the frequency of trivalents is low enough to permit the conclusion that the *niger* chromosome is always in the center and that the two *repens* genomes do not exchange material. It is not excluded that the two *repens* genomes also pair and recombine with each other. This would require the assumption that the presence of the *niger* genome disturbs the system of pairing differentiation between the *repens* genomes. The consequence would be that not only are genes from *niger* introduced into *repens*, but also that the *repens* genomes exchange segments between them. This could ultimately result in a partial autotetraploid character of the *repens* genomes. In the progeny this can lead to quadrivalents and substitutions between entire *repens* chromosomes. The consequence is a segmental allotetraploid. For taxonomic studies this is not of much importance, but in a breeding program, it may cause unexpected difficulties. It is important to recognize this possibility and to backcross the hybrid derivatives in early stages in order to reconstitute the original genomic constitution (if desired).

The model permits the estimate of one more pairing parameter, of a different character (Sybenga 1988). This will not be discussed here.

Parallel alignment of all three chromosomes with numerous exchanges of pairing partners (Loidl and Jones 1986; Loidl 1990) results in a trivalent frequency of over twice the ring bivalent frequency, the expected maximum with pairing initiation starting at the chromosome ends. Such high frequencies of trivalents have been observed repeatedly, although not frequently (for instance: Wang and Berdahl 1990). They are not necessarily accompanied by high quadrivalent frequencies in the corresponding autotetraploid (Sect. 11.3.1.2.3).

In contrast to the Alonso and Kimber (1981) and Crane and Sleper (1989a) models, no optimization is applied, and no abnormalities in the observations can be accomodated (less univalents than bivalents, higher multivalents, all pointing to translocations). If they occur, adjustments are necessary. Neglecting such irregularities implies that for two sets of chromosomes (in the case of a translocation) the observations are misleading. The question then arises whether no result is really worse than an invalid result.

Special adaptations of the model have shown their use in special situations, but do not need discussion here. Nor will the model of Crane and Sleper (1989a) be discussed, which can give insight into the ranges of possible relations between genomes, but requires a complex optimization program.

There is one more model (Jackson and Casey 1982; Jackson and Hauber 1982) that has found rather wide application in polyploids. It was originally set up primarily to compare meiosis in hybrids with meiosis in autopolyploids, without estimating pairing parameters. The assumption about pairing is similar to that in the models discussed. It makes use of a theoretical distribution of chiasmata over chromosome arms that is somewhat artificial. To estimate pairing parameters the models discussed here are more adequate.

9.3.2.2 Tetraploid Hybrids

Tetraploid hybrids between allotetraploids and between diploids and allohexaploids are made rather frequently both for the transfer of genes and for the study of taxonomic genome relations. Quantitative estimates of pairing and chiasma parameters are of interest in both cases. In programs of gene transfer, for instance, the specific pairing and genetic exchange relations between all four genomes are important for estimating the overall exchange frequency between recipient and donor. They are also important for checking if any exchange recombination takes place between the different genomes of the recipient, which is usually not desired. It is quite well possible that the genotype of the hybrid promotes homoeologous pairing between the parental genomes, and it is important to be able to recognize the level at which this occurs.

The analysis of affinity has been carried out in material with normal chromosomes, where overall levels are considered, and in material with marked chromosomes, especially telocentrics, which enables the recognition of the behaviour of specific chromosome arms.

9.3.2.2.1 Tetraploid Hybrids with Unmarked Chromosomes

Affinity or preferential pairing estimates are primarily based on the frequency of multivalent formation in relation to chiasma frequency. This implies that all the uncertainties encountered with multivalent formation in autotetraploids will also operate in tetraploid hybrids: reduced quadrivalent pairing by shifting the point of pairing partner exchange or variation in its initial position; localized pairing initiation; localized chiasma formation, etc. Increased multivalent formation may result from multiple pairing initiation.

Further, between four different genomes six different pairing combinations are possible, each with a specific probability, depending on the affinity relations between the genomes. These probabilities should, in principle, be estimated independently. The metaphase I observations, when individual chromosomes cannot be distinguished, are ring quadrivalents, chain quadrivalents, trivalents with a univalent, ring bivalents, open bivalents and sets of two or four univalents. These six classes represent five degrees of freedom from which, in theory, five different parameters can be estimated. With a set-up comparable with that of the triploid hybrids, however, only one degree of freedom can be assigned to the pairing system. This is clear from the model for the autotetraploid (Sect. 6.1.2.2.2) and Table 9.4 where the quadrivalent pairing frequency f is the only parameter specifically available for the pairing pattern. In hybrids this is the same. It is impossible to estimate all five independent affinity parameters, and special approaches and restrictions in the models are necessary to make a distinction possible between different pairing modes. In spite of these limitations, different attempts have been made to construct mathematical models for the quantitative analysis of pairing affinity parameters for tetraploid hybrids. It should be noted that the faith with which the conclusions have been accepted is not always justified.

In a rather unspecific way, affinity between homoeologous chromosomes can be measured by the method developed by Gaul (1958), which entails no more than estimating the number of pairs of chromosomes that are capable of pairing and forming at least one chiasma. In hybrids between allopolyploids where several more or less related genomes are combined and partial desynapsis reduces full expression of homologies, and in the case of limited affinity, this may be an interesting approach. The number p of chromosome pairs capable of pairing and genetic exchange is given by.

$$p = \frac{(X^2 + X - B)}{[C.(2X - B)]},$$

where X equals the total number of chiasmata, B the total number of observed pairs of chromosomes and C the number of cells analyzed. Applied to the hybrid between *Triticum aestivum* (2n = 42) and *Agropyron intermedium* (2n = 42) it appeared that each of the three wheat genomes could pair and form chiasmata with one of the *Agropyron* genomes. A somewhat more recent application of the same formula, slightly modified, by Thomas and Kaltsikes (1977) on colchicine-treated meiosis in a pentaploid *Triticum* hybrid led to the conclusion that primarily the association of homologues was reduced by colchicine. The method has not had wide application because of the limited specificity of the results.

With the newer models it is possible to analyze whether pairing between the four ho(oe)ologues of the tetraploid is random as in an autotetraploid or not. If it is not, this is attributed to differences in affinity, or preferential pairing (Driscoll et al. 1979; Jackson and Casey 1982; Jackson and Hauber 1982). In this respect they do not in their intention differ from the autotetraploid model of Sybenga (1975), but the information they provide is less detailed than in the latter. All models assume the restriction of one point of partner exchange, coinciding with the centromere. As discussed with the autotriploids, this is a simplification that in some materials may not be acceptable. In some instances, however, it is.

Most models consider pairing first, and are an extension of the trisomic and triploid model of Sybenga (1965) and the tetraploid model of Sved (1966), Sybenga (1975), and used by several later authors.

Jackson and Casey (1982) assume that pairing follows the affinity relations, and that the formation of chiasmata subsequently determines which configurations are formed and in which relative frequencies. The pairing model is based on straightforward autotetraploid pairing, i.e. there are twice as many quadrivalents as pairs of bivalents. The way in which the final configuration frequencies are determined is based on the average chromosome arm association frequency P (identical to c of Alonso and Kimber 1981; Kimber and Alonso 1981), and a binomial distribution of chiasmata, with a maximum of four chiasmata per chromosome. Because chiasmata are not distributed binomially, this is not realistic, and in a subsequent paper Jackson and Hauber (1982) introduce a correction factor for chiasma distribution. On the basis of the formulae provided in these papers, it is possible to construct expected configuration frequencies for a tetraploid on the basis of associated arm frequencies, assuming random chromosome pairing. These can be compared with the observed frequencies and a conclusion about the presence or absence of preferential pairing can be made. The same, however, can be done much simpler by the use of Sybenga's (1975) f factor, which gives an estimate of the difference between the autotetraploid random pairing expectation and the observations, and assumes a potential difference between arms in chiasma frequency (Table 9.3). In addition, even with a correction factor, chiasma distribution remains an uncertain factor in the models, and not really necessary.

More sophisticated models, for instance those of Kimber and Alonso (1981), estimate distinct affinity factors, for instance x and y as discussed for the triploid hybrid (Sect. 9.3.2.1).

Sved (1966; see also Sybenga 1975) was the first to consider quantitative differences in affinity and the effect of different levels of affinity on multivalent formation and marker segregation in amphidiploids. In Fig. 9.4 one set of four chromosomes of a tetraploid hybrid is shown. Chromosomes 1 are homologous, and chromosomes 2 are homologous, but 1 and 2 are homoeologous. Arms A and B pair independently. Homologous pairing (A1-A1, A2-A2, B1-B1, B2-B2) has a frequency $(1 - a)$ for the A arms and $(1 - b)$ for the B arms. Homoeologous pairing (A1-A2 and B1-B2) has a frequency a and b respectively. With random pairing $a = b = 2/3$.

Without desynapsis and with full pairing between homologues and between homoeologues, the frequencies of homologous bivalents, homoeologous bivalents and quadrivalents can be expressed in terms of a and b. *Homologous bivalent pairs* are formed with a frequency $(1 - a)(1 - b)$. With an allelic constitution of a gene m as in Fig. 9.4, the anaphase I daughter cells will *both* receive *Mm* and so will the gametes, and thus no segregation results. *Homoeologous bivalent pairs* are formed with a frequency $\frac{1}{2}ab$. The gametes resulting from such bivalents segregate $\frac{1}{4}MM : \frac{1}{2}Mm : \frac{1}{4}mm$, i.e. there are $\frac{1}{4}$ double recessive gametes.

The remaining configurations are *quadrivalents* with a frequency $a + b - 3/2ab$. These segregate $1/6\,mm$ gametes. The total frequency of *gametes with two recessive alleles* equals $1/6a + 1/6b - 1/8ab$.

Table 9.3. Formulae to estimate expected quadrivalent and bivalent pairing frequencies, and the chiasmatic association frequencies in the two arms in autotetraploids (or tetrasomics) when the chiasma frequency is too low to make all arms associate at meiotic diakinesis or metaphase I (Sybenga 1975; cf. Table 6.3). The estimated quadrivalent pairing frequency is *f*, which should be lower than the observed multivalent frequency. The theoretical maximum of both is 0.667. The chiasmate arm association frequencies are a and b after bivalent pairing, a′ and b′ after quadrivalent pairing

---

rq ring quadrivalents	*cq* chain quadrivlents
t trivalents with univalent	*r* ring bivalents
o open (rod) bivalents	*u* univalents (not associated with trivalents)

Frequencies expressed in terms of sets of four chromosomes.

$$rq = fa'^2b'^2$$
$$cq = 2f(a'b'^2 + a'^2b' - 2a'^2b'^2)$$
$$t = 4f(a'b' - a'b'^2 - a'^2b' + a'^2b'^2)$$
$$r = (1 - f)ab$$
$$o = f(a' + b' - 4a'b' + a'b'^2 + a'^2b') + (1 - f)(a + b - 2ab)$$
$$u = f(1 - a' - b' + a'b'^2 + a'^2b' - a'^2b'^2) + (1 - f)(1 - a - b + ab)$$
$$f = \frac{(t + 2cq + 4rq)^2}{16rq}$$

a′ and b′ can be estimated from:

$$a'b' = \frac{4rq}{t + 2cq + 4rq} \quad \text{and} \quad (a' + b') = \frac{2cq + 8rq}{t + 2cq + 4rq}$$

The averages for the arm association frequencies in bivalents and quadrivalents together can be derived from a quadratic equation derived from

$$\bar{a}\bar{b} = \frac{4rq + 2cq + t + 2r}{\text{total}} \quad \text{and} \quad (\bar{a} + \bar{b}) = \frac{2cq + 2t + 2o}{\text{total}}$$

From $fa' + (1 - f)a = \bar{a}$ and $fb' + (1 - f)b = \bar{b}$ the estimates for a and b can be derived. These are very indirect and may have a great error.

Deviation of *f* from 2/3 (0.667) indicates preferential pairing, or any other reason for quadrivalent pairing other than 2/3 as discussed in the text. The significance of the deviation can be tested by reconstructing the numbers of configurations from a′, b′, a, b, a, and b, and $f = 0.667$, and testing these in a χ^2 test against the numbers observed

---

These formulae can be combined to relate the frequency of quadrivalents to gene segregation and also to permit an estimate of preferential pairing in both arms independently. Sved (1966) used this model to analyze a set of observations (Table 9.4) on doubled *Gossypium* hybrids published by different authors and collected by Phillips (1964). The original hybrids were triploid hybrids between allotetraploid *G. hirsutum* (AADD) and diploids closely related to the D genome. The amphidiploids were, in fact, hexaploids (A A D1 D1 D2 D2), but the two A genomes always formed a bivalent, and were not considered; only segregation of genes in the D genomes were considered. With the model, the maximum frequency of multivalents is two-thirds

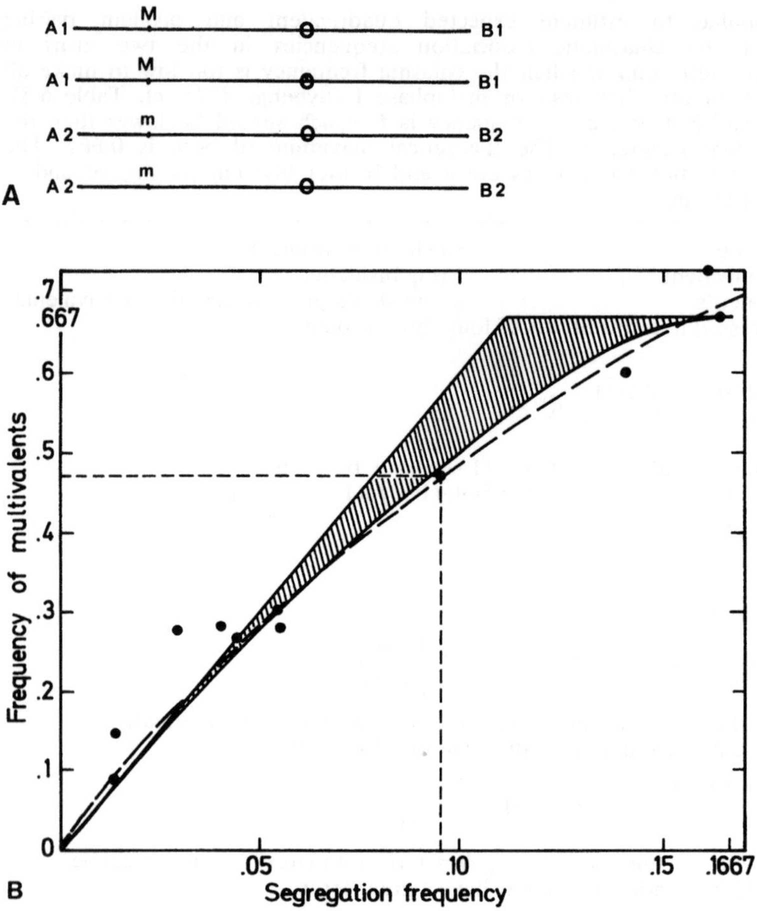

Fig. 9.4 The relation between multivalent frequency and gene segregation in tetraploid hybrids. **A** Two pairs of homologous chromosomes (arms *A* and *B*), with varying degrees of preferential pairing between the pairs. Pairing starts from the ends and all paired arms have chiasmata. One marker locus: *M*/*m*. **B** The relation between multivalent frequency and segregation of alleles in the gametes. The *right-hand border* of the *shaded area* gives the relation when both arms vary identically. The *left-hand border* when one arm consistently pairs preferentially compared to the other. The *points* are examples of polyploid *Gossypium* hybrids (cf. Table 9.3). (Sybenga 1975; after Sved 1966)

and the maximum segregation ratio is one-sixth. This represents full homology as in an autotetraploid. The left border of the shaded area in Fig. 9.4B gives the line for $a = 2/3$ with variable b (from 0 to 2/3) or $b = 2/3$ and variable a. The right border gives the expected values for $a = b$, both between 0 and 2/3.

Most observations fit this line reasonably well and thus permit an estimate of the average pairing differentiation between the chromosomes of the different D genomes. A few points fall slightly to the right of the line, but this

Table 9.4. The relation between gene segregation and multivalent frequency in synthetic *Gossypium* allohexaploids (2n = 6x = 78). Two genomes are fully homologous and form bivalents consistently. Multivalents are formed only by the other four. (Philips 1964; Sved 1966)

Hexaploid	Genomes	Segregation/Multivalents				
		No. loci	Av. segr.	Average per cell	Average per set	No. cells
G. hirsutum × *arboreum*	$A_2A_hD_h$	4	5.1:1	8.68	0.668	115
G. barbadense × *arboreum*	$A_2A_bD_b$	5	6.1:1	7.80	0.600	107
G. hirsutum × *raimondii*	$A_hD_hD_r$	10	9.5:1	6.16	0.474	108
G. hirsutum × *harknessii*	$A_hD_hD_{2-2}$	5	17.1:1	3.65	0.281	220
G. hirsutum × *armourianum*	$A_hD_hD_{2-1}$	5	17.4:1	3.96	0.305	258
G. hirsutum × *aridum*	$A_hD_hD_4$	4	21.3:1	3.48	0.268	279
G. hirsutum × *lobatum*	$A_hD_hD_7$	4	23.7:1	3.66	0.282	283
G. hirsutum × *thurberi*	A_hD_hD	3	32.9:1	3.61	0.278	44
G. barbadense × *gossypoides*	$A_bD_bD_6$	4	71.6:1	1.13	0.087	109

may be due to sampling error. In the hybrid *G. hirsutum* with *G. raimondii*, for example, an average segregation ratio of 9.5:1 or 0.095 is observed combined with a quadrivalent frequency of 0.474. The combination of the segregation and quadrivalent frequency formulae gives a quadratic equation in a and b from which a and b can be solved. The discriminant appears to be slightly negative, corresponding to a position on the graph just to the right of the curve where $a = b$ and the conclusion is that $a = b$. A separate estimate of this "differentiation" factor from the quadrivalent frequency yields 0.566 and from the segregation ratio 0.463. The correspondence is reasonably good, with the differentiation estimate derived from the quadrivalent frequency being slightly larger. This suggests that there is some multivalent breakdown due to other reasons than mere affinity differences. The difference with the two-thirds expected with random chromosome pairing is large for both estimates, suggesting that the *G. raimondii* genome used in the hybrid is not exactly the same as the present D genome of upland cotton. Since ten loci were involved, most of the chromosomes are represented. The results for the different doubled hybrids are rather different (Fig. 9.4 and Table 9.4), showing that there is considerable variation in the affinity relations. It may be noted that these doubled hybrids are closely related to the 2:2 hybrids of the Kimber and Alonso (1981) model discussed below.

In the models discussed above (Sved 1966; Sybenga 1975; Jackson and Casey 1982; Jackson and Hauber 1982; and the model of Crane and Sleper 1989b briefly referred to later), pairing is completed before chiasma formation. When pairing is incomplete, this is confounded with incomplete chiasma formation. They first consider the different types of arm combinations in pairing between four chromosomes, and assign chiasmate associations to the paired arms. The possibility that the failure of pairing resulting from insufficient affinity is the cause of chiasma failure, is recognized, but not explicitly taken into account in the models.

Other models, however (Driscoll et al. 1979; Kimber and Alonso 1981), assume that the primary cause of reduced chiasma formation in hybrids is insufficient pairing, which is due to insufficient affinity between genomes or to differences in affinity between genomes competing for pairing partners. As soon as there is pairing, chiasma formation follows with the same effectiveness as in normal homologues, which is near 100%, and the same for all arms, long or short. Here again, pairing and chiasma formation are confounded, but the basis is slightly different. This approach is compatible with the hypothesis that pairing initiation and chiasma formation are closely related. The principal difference between the two approaches is not really in the suggested difference in concept. The pairing-first models are not in conflict with the concept that pairing is incomplete because of limited affinity. The operational difference between the models is that Driscoll et al. (1979) and Kimber and Alonso (1981) directly assign chiasmata (or rather chiasmate associations) as representatives of pairing events to different combinations of arms. The difference in approach between the models leads to slight differences in the outcome of calculations, which cannot be neglected.

The model of Driscoll et al. (1979) merely compares presumed autotetraploid behaviour with observed meiotic behaviour, and as such is not of interest for estimating specific pairing parameters. However, it forms the basis of the models of Kimber and Alonso (1981) which make it possible to estimate specific affinity parameters, and as such a discussion of its main characteristics is relevant. The autotetraploid pairing-first model (Table 6.3), in which twice the number of quadrivalents as pairs of bivalents appears, is not the starting point. Instead, chiasmate arm associations (not chiasmata as in Jackson and Casey 1982) are assumed to be directly distributed randomly over the arm pairs available.

There are four chromosomes, with four pairs of arms (Fig. 6.3). Differences between arms in chiasma frequency are not taken into account for the reason mentioned above. Without chiasmate association, all four chromosomes are present as *univalents*, as in any pairing model. When *one chiasma* is formed, any pair of homologous arms may be involved. There are 12 possible combinations to choose from, six for arm A and six for B. There are always one open (rod) bivalent and one pair of univalents.

When all four available arm pairs are associated, there are three possible combinations. Starting with A1/A2, the other A arm pair association must necessarily be A3/A4. The other two of the four associations can be in three combinations: B1/B2 with B3/B4 gives *two ring bivalents*, and B1/B3 with B2/B4 as well as B1/B4 with B2/B3 each give a *ring quadrivalent*. This is the same result as that of the other pairing models for autotetraploids with four arms associated.

With three arms associated, there are nine different possibilities:

A1/A2 B1/B2 B3/B4: rbiv + obiv
A1/A2 B1/B3 B2/B4: cq
A1/A2 B1/B4 B2/B3: cq
A1/A2 A3/A4 B1/B2: rbiv + obiv
A1/A2 A3/A4 B1/B3: cq
A1/A2 A3/A4 B1/B4: cq
A1/A2 A3/A4 B2/B3: cq
A1/A2 A3/A4 B2/B4: cq
A1/A2 A3/A4 B3/B4: rbiv + obiv

where rbiv is a *ring bivalent*, obiv an *open bivalent* (rod) and cq a *chain quadrivalent*. Together this adds up to 6 cq + 3 (rbiv + obiv) = 2:1, the same as with the first autotetraploid pairing model.

With two arms associated, the expected configuration frequencies start deviating from the autotetraploid pairing-first model of Sybenga (1965, 1975) and others. The first chiasmate association can choose between 12 different combinations: 6 for arm A and 6 for arm B. Again assuming it is A1/A2, the second chiasmate association has only one out of seven pairs available:

A1/A2 B1/B2: rbiv + 2u
A1/A2 B1/B3: triv + u
A1/A2 B1/B4: triv + u
A1/A2 B2/B3: triv + u
A1/A2 B2/B4: triv + u
A1/A2 B3/B4: 2 obiv
A1/A2 BC/BD: 2 obiv

Here, triv is a *trivalent* and u a *univalent*. Together there are 1/7 (rbiv + 2u), 2/7 (2 obiv) and 4/7 (triv + u), or 0.143 (rbiv + 2u), 0.286 (2 obiv) and 0.571 (triv + u). The total number of possibilities is 12 × 7 = 84. This can also be derived from the combination of six possibilities for arm A plus six for arm B, i.e. total 12 for the first chiasma, combined with 12 for the second chiasma. Of the resulting 144 combinations 5 × 12 are not available for the second because these are occupied by the first, and 7 × 12 = 84 remain.

The pairing-first model gives a different result: there are again six combinations for each of the two arms, and for each of the nine pairing modes (six quadrivalents and three pairs of bivalents), or a total of 12 × 9 = 108. When checking the combinations, it appears that 18 of these occur three times among the 108, and these are considered an essential part of the model. If from the 18 triple combinations only 18 (single) combinations are considered real possibilities, 2 × 18 = 36 should be removed, and again 84 combinations would remain and there would be no difference between the two approaches. All together, the pairing-first model gives: 1/9 (rbiv + 2u), 4/9 (2 obiv) and 4/9 (triv + u) or 0.111 (rbiv + 2u), 0.444 (2 obiv) and 0.444 (triv + u), i.e. sufficiently different from the first model to cause concern for some ranges of chiasmate association frequencies. It is not easy to decide which of the two models is best.

Kimber and Alonso (1981) have adopted the model of Driscoll et al. (1979) as the basis for their affinity model. They assume that affinity determines qualitatively (choice of partner) and quantitatively (frequency of association) which arm combinations will be formed and, further, that between each set of four chromosomes one, two, three or four associations can form. Following Driscoll et al. (1979), the frequencies of the different types of chiasmate associations are given in terms of the average chiasmate association frequency c (Table 9.5), where the coefficients are derived as shown earlier. They sum up to unity.

Superimposed over the "random" arm association frequencies are expressions for the effects of differences in affinity. As in the triploid model, there is basically only one independent affinity factor, x with a complementary affinity factor $y = (1 - x)$. This is not sufficient to estimate the five independent affinities possible for six combinations. In their tetraploid hybrid model Kimber and Alonso (1981), therefore, restrict their attention to a limited number of relations, comparable to their approach to the triploid hybrid, but with more, different alternatives. In the model of Crane and Sleper (1989b), which is based on a tetraploid pairing-first model, the same alternatives are considered, but in addition two parameters for the two arm association frequencies a and b are introduced. This model is greatly over-parameterized and the equations involved can only be solved by complex iterative optimization. In addition to giving specific optima, the Crane and Sleper model (1989b) gives possible alternatives and indications of the most probable ranges of the affinity estimates. Although the basic model is not particularly complex, the optimization procedure, although robust, is far from simple. It will not be discussed.

In the model of Kimber and Alonso (1981) four different genomic relations are considered:

1. All genomes have equal affinities, and can be represented as 4:0. This can be an autotetraploid, genomes AAAA or, when the genomes are

Table 9.5. The expected frequencies of the meiotic configurations formed by four homologous chromosomes on the basis of the frequency of arm association c. The univalents include those which in combination with a trivalent form one set of four chromosomes. (Driscoll et al. 1979; Kimber and Alonso 1981)

Univalent	$1.1\,c^4 - 2.29\,c^3 + 5.14\,c^2 - 8\,c + 4$
Open bivalent	$-1.90\,c^4 + 6.48\,c^3 - 8.57\,c^2 + 4\,c$
Ring bivalent	$0.19\,c^4 - 0.38\,c^3 + 0.86\,c^2$
Trivalent	$3.43\,c^4 - 6.86\,c^3 + 3.43\,c^2$
Chain quadrivalent	$-2.67\,c^4 + 2.67\,c^3$
Ring quadrivalent	$0.67\,c^4$

not closely related but equidistant with respect to affinity: AA′A″A‴. The model is, in fact, the autotetraploid of Driscoll et al. (1979) discussed above. Since all affinities are equal, no specific affinity component is introduced into the formula. In the extreme situation where (practically) no affinity exists, the formula 1:1:1:1 (genomes ABCD) gives a better description than 4:0.

2. Two genomes are closely related (affinity x) and more distant (affinity y) from two more that are also closely related (also affinity x): the 2:2 model with genomes AAA′A′, AA′BB′ or AABB, etc.
3. Two genomes are closely related (affinity x) and the other two more distant from one another and from the other two (affinity y): 2:1:1 (genomes AABC, AA′BC, etc.).
4. Three genomes are relatively closely related (affinity x) and one is more distant from the other three (affinity y): 3:1 (genomes AAAB, AAAA′ or AA′A″B, etc.).

A restriction of the possibilities to discrete cases is necessary for the construction of the models, which contain only one basic affinity factor, x. These four are not unrealistic. In intermediate situations the results will show between which of these models the observations fit best.

There are four final affinity models consisting of sets of formulae expressing the frequencies of the different classes of configurations in terms of x, y and c. The value of c can be derived directly as the frequency of associated arms. It is not possible to design formulae to solve x and y directly, as the number of observed classes of configurations (and thus the number of corresponding formulae) is larger than the number of parameters to be estimated. The equations are not suitable for solving by a maximum likelihood approach. Arbitrary values for x and y are introduced into the formulae, together with the estimate of c, and by optimization the best-fitting values for x and y are derived. These are subsequently introduced into the expressions to give the best-fitting theoretical configuration frequencies for each model. These frequencies are compared with the original frequencies by calculating the sums of squares of the differences between observed and expected frequencies. The model giving the best fit is considered to approach the real situation best. When different formulae give equal fit, the real situation is considered to be halfway between the two.

The models are too complex and the formulae too long to be given here in full detail. For application, the reader is referred to the original literature (Kimber and Alonso 1981).

The 2:1:1 model has different x and y arm associations than the previous model. Only one is an x-type combination, the other five are y-type associations.

It is clear that these models have their limitations: no distinction is made between the two arms of the chromosomes with respect to chiasma frequency; only two of the potentially six pairing parameters are estimated; the pairing model is not necessarily the best; optimization with an under-parameterized model carries the risk that factors neglected in the model seriously affect the esimates. When translocations occur, these are usually simply neglected except that larger configurations are not included. However, there are no satisfactory alternatives yet, except, perhaps, the Crane and Sleper (1989b) model mentioned. Probably, the Crane/Sleper conclusions are slightly more meaningful than those based on the Kimber/Alonso model.

The practical importance of these models for plant breeders is in the possibility they offer for estimating the level of recombination between different genomes in a polyploid hybrid. This is of interest not only for estimating the required population size for the transfer of specific (or less specific) genes from one species to another. It is also important for getting an impression of the rate of introduction (or loss) by recombination of genes from genomes which are present in the polyploid hybrid, but not as donors of the genes for which the hybrid was made. No quantitative estimates, based on such analyses are yet available, neither of the required population size for transferring a gene, nor of the probability of introducing undesired genes, or losing genes.

9.3.2.2.2 Tetraploid Hybrids with Marked Chromosomes

One of the first quantitative analyses of genome relations with the use of marked (telocentric) chromosomes was that of Riley and Chapman (1966), who analyzed a wheat *Aegilops speltoides* hybrid ($2n = 21 + 7 = 28$) with the genomic composition ABDS. The action of the *Ph* gene on chromosome 5BL was partly neutralized by the presence of the S (*speltoides*) genome, such that considerable homoeologous pairing occurred. In the wheat variety used two different chromosomes had been replaced simultaneously by their telocentric and among the hybrids, plants with two telocentrics instead of the two normal chromosomes were selected. Emphasis was on the group 5 chromosomes, and homoeologous and non-homoeologous combinations (involving group 3 and group 6 together with group 5 telocentrics) were made. Because of the presence of the S genome, all four homoeologous chromosomes could, in principle, pair and form chiasmata resulting in configurations of up to four chromosomes. Arm 5B with the *Ph* gene was always present.

No pairing was observed between non-homoeologous chromosomes. The telocentrics of the group 5 chromosomes paired in all combinations: 5A with 5B, 5A with 5D and 5B with 5D. Usually, the telocentrics paired with each other forming a small bivalent. Occasionally, a triradial trivalent was observed, including one normal chromosome, or a chain quadrivalent with the telos at the ends. In the combinations 5AL with 5BL and 5AL with 5DL there was relatively little association between the telocentrics, but in the 5BL and 5DL combinations they were associated in about half of the cells analyzed. Apparently, 5B and 5D in this analysis had a much closer affinity than either one with 5A. Since there were quadrivalents, including both telos, 5S of *Ae. speltoides* must have been involved in pairing also.

The different affinities can be expressed quantitatively. There are six two-by-two combinations, with frequencies:

$$5A - 5B = a;\ 5A - 5D = b;\ 5B - 5D = c;$$
$$5A - 5S = x;\ 5B - 5S = y;\ 5D - 5S = z.$$

The few triradial trivalents were included with the telocentric bivalents. The estimates $a = 0.08$; $b = 0.08$ and $c = 0.52$ could be obtained directly. For x, y and z an indirect approach is necessary: association between a telocentric and an unmarked chromosome can be of four types: in the presence of a 5AL and a 5BL telocentric, with normal chromosomes 5D and 5S, for instance, there can be heteromorphic association between 5A and 5D, between 5A and 5S, between 5B and 5D or between 5B and 5S. The frequencies T of association events can be given as:

$$T(5A/5B) = b + x + c + y;$$
$$T(5A/5D) = a + x + c + z;$$
$$T(5B/5D) = a + y + b + z.$$

These include marked and unmarked associations. The association of two telos counts double because two are involved. Thus, $T(5A/5B) = 0.01 + 0.54 + 2 \times 0.17 = 0.89$. Similarly, $T(5A/5D) = 0.81$ and $T(5B/5D) = 0.46$. With the directly obtained values for a, b and c substituted, the following equations can be given:

$$0.89 = 0.08 + x + 0.52 + y, \text{ or } 5A - 5B: 0.08;\ 5A - 5S: 0.10;$$
$$0.81 = 0.08 + x + 0.52 + z, \text{ or } 5A - 5D: 0.08;\ 5B - 5S: 0.19;$$
$$0.46 = 0.08 + y + 0.08 + z, \text{ or } 5B - 5D: 0.52;\ 5D - 5S: 0.11.$$

There is considerable variation in affinity and it is somewhat surprising that under the conditions of relaxed pairing restriction the supposedly closely related B and S genomes pair less frequently than the B and D genomes. The affinity relations do not fully correspond to nulli-tetra compensation, nor with later results by other authors on other chromosomes. Apparently, species differentiation is expressed differently under different conditions, for different parameters and possibly even for different chromosomes.

Specific relative affinities may be estimated quantitatively by using telocentrics, but the results cannot be entirely compared with homologous pairing between complete chromosomes. In the same marked hybrid the association between the homologous long arm group 5 telos was only 86% against about 100% for the long arm in a normal chromosome pair. If this is representative of the average long arm, the 52% association between 5BL and 5DL telos corresponds to about 60% of homologous affinity.

This analysis has been described in some detail to show the approach. The conclusion that the B genome appears to be closely related to the D genome is in agreement with the presumed taxonomic relations, but the relation between the B and S genomes is not readily explained. Belfield and Riley (1969) later extended the analysis with several more combinations and concluded that for other chromosomes no striking differences in affinity between the genomic combinations could be observed.

Alonso and Kimber (1983) developed models of chromosome pairing for telocentrics, comparable to those for triploids and tetraploids as discussed above. Again, hybrids of wheat with *Ae. speltoides* were studied, where the suppression of homoeologous pairing is released. The conclusion was that the 2:2 model was the best, supposedly with B and S forming a relatively closely related pair of genomes, and A and D another related pair, although probably somewhat less closely, but both more distant from B and S. This is not in correspondence to the results given above of Riley and Chapman (1966) and Belfield and Riley (1969) but could be confirmed in N-banded preparations (Jauhar pers comm). The reason is not immediately clear.

9.3.2.3 Higher Polyploids and Aneuploids

Models for pentaploid hybrids have been developed by Espinasse and Kimber (1981), again based on the autopolyploid models of Driscoll et al. (1979) and applies to several hybrids between hexaploid allopolyploids, primarily wheat, and allotetraploid relatives of wheat. The results are less readily interpreted than for the lower ploidy levels, but several conclusions could be drawn. The models are even more complex than those for tetraploid hybrids and will not be discussed here.

Thomas and Kaltsikes (1977) applied a modification of the formula of Gaul (1958) to analyze the effect of colchicine on chromosome pairing and chiasma formation in the pentaploid hybrid ($2n = 35$) between tetraploid wheat ($2n = 28$) and hexaploid triticale ($2n = 42$).

Similar approaches are possible for *hexaploid hybrids*. The models of Jackson and Casey (1982) and Jackson and Hauber (1982) include pentaploids and hexaploids but have similar disadvantages as those for lower ploidy levels: highly artificial model for chiasma distribution, in principle autopolyploidy as the basis and no distinction between arms. The latter, of course, also applies to the Kimber *cum suis* models.

For *aneuploids* special adaptations of the models as used for euploid polyploid hybrids can be designed, as briefly mentioned, with the telocentrics (Kimber et al. 1981). These will give information on specific chromosomes; however, they will not be discussed here.

Amended models for triploids, tetraploids and pentaploids have been developed by Chapman and Kimber (1992, where further references).

Chapter 10

Manipulation of Genome Composition: A. Gene Transfer

10.1 Objectives

In most plant breeding programs the manipulation of the genetic composition of the variety under construction consists of the selection of a desirable combination of alleles derived from related cultivars (the primary gene pool). Less frequently, a slightly more distant form of the same species or occasionally even a different but closely related species (the secondary gene pool) is used. The combination of selected alleles results from simple recombination in hybrids between the selected parental forms. The emphasis may be on the introduction of specific alleles of specific genes or on more complex combinations of less clearly defined gene complexes. This is the simplest form of genetic manipulation.

It is not uncommon that more difficult objectives must be realized for which simple selection after hybridization within the primary or with the closely related secondary gene pool is insufficient. For instance, the transfer of genes or complexes of genes from one species to another across effective barriers against genetic exchange (the tertiary gene pool: Harlan and DeWet 1971) requires special measures whereby cytogenetic approaches can sometimes give a solution. Some of these involve "chromosome manipulation" (Thomas 1981; Riley and Law 1984). These approaches are the subject of the present chapter. Other forms of cytogenetic manipulation involve gene dose effects and specific combinations of genes and genomes (Chap. 11) and the manipulation of genetic systems (Chap. 12). In addition to practical chromosome manipulation, there is a wide field of application for cytogenetic techniques or chromosome manipulation in research (Riley and Law 1984), which is referred to somewhat superificially in several sections of different chapters of this book.

An alternative for gene transfer across regular recombination barriers is *gene mutation*. This is not always successful when very specific, especially dominant alleles are desired. In programs to introduce genes from other forms, mutations have an application in removing epistatic genes, or the original allele when this is dominant and not removed by homologous exchange (Sect. 10.2). There is a role for cytogenetics in programs of mutation breeding when side effects include chromosomal aberrations. These have been considered in Chapter 7. Mutations will not be further discussed here.

10.2 Molecular Versus Generative Gene Transfer

Gene transfer from outside the primary or readily accessible part of the secondary gene pool is the field in which in vitro cell biological and molecular approaches offer very promising possibilities. Their application has been limited till now, and in several instances cytogenetic approaches are still the only potentially successful possibility available. For a comparison of cell biological and molecular techniques, on the one hand, and generative techniques, on the other hand, for gene transfer and other aspects of genetic manipulation, see Sybenga (1989).

The great interest in gene transfer by molecular techniques (Gasser and Fraley 1989) has considerably reduced the interest in cytogenetic methods for transferring genes between species. Another reason for the decreased interest in cytogenetic techniques for gene transfer is that disappointing experience (especially the rapid breakdown of introduced disease resistance) has made plant breeders hesitant. There have not been sufficient practical results from molecular and cellular techniques for a similar disappointment, but a first indication that here too disappointment may reduce the initial optimism is the limited application of the first practically useful molecular transfer, that of herbicide resistance. This does not mean that positive results are not obtained in other fields. One important difference with respect to the final chance of success between molecular/cell-biological and cytogenetic techniques is the very large input available for the former and the very modest facilities for the latter approach. With the development of several new (including molecular) methods to monitor the transfer of recessive and hypostatic genes (Sect. 10.3), the interest in the potentially very successful methods of generative gene transfer may be expected to increase.

10.3 Identification of Transferred Chromatin

During the process of gene transfer by whatever method, a dominant gene can simply be followed by its expression. The transfer of recessive alleles and hypostatic genes presents serious problems, not only because the primary transfer is difficult to monitor, but also because usually a series of backcrosses is necessary to recover as much of the original genome as possible. Replacement of a dominant allele in the recipient by a recessive alien allele by homoeologous recombination enables its expression in the homozygote. When a series of backcrosses is required, these have to be alternated with selfing to recover the types homozygous for the new gene. For hypostatic alien genes the situation is more complex. Not only for tracking the alien gene during transfer but, more urgently, for its expression in the final cultivar, the epistatic gene has to be removed. This is possible by deletion or mutation. Introduction of a recessive allele by translocation instead of homologous recombination similarly requires the removal of the original dominant allele of the recipient before expression of the alien allele is possible. It is not unexpected, therefore, that in practical programs of alien gene transfer almost exclusively dominant and epistatic genes have been involved.

One advantage of the use of recombinant DNA techniques in gene transfer is that the detection of the presence or absence of the gene is not dependent on the phenotype, but can be scored in the bulk DNA by molecular means. The same technique enables the detection of genes transferred by other means, provided the corresponding DNA is available in the form of a molecular probe. In species where molecular transformation rather than the isolation of the gene DNA presents the problem, the availability of a DNA probe of the gene for direct *molecular identification* of genes transferred by other means between cultivars and especially between species can be a great help.

In addition, there has been considerable progress in the use of molecular markers that are closely linked to the gene to be transferred. These are much simpler to find and to clone than a gene that has not yet been identified on a molecular basis. Especially Restriction Fragment Length Polymorphism (RFLP) markers are useful to monitor the presence of genes in all forms of gene transfer. This will extend the range of application of gene transfer by cytogenetic techniques to recessive alleles and genes that for other reasons are not readily scored in the plant phenotype. It seems reasonable to expect that in the future molecular, cell biological and generative cytogenetic techniques can be combined profitably to provide very effective facilities for gene transfer and, possibly also, other ways of genotype manipulation. In several, but still a limited number of crop species, neutral molecular markers are available on a sufficient scale: maize (Evola et al. 1986); tomato, tobacco (Tanksley et al. 1988); wheat (Chao et al. 1989). Within the same species, some chromosomes are more densely marked than others.

When large numbers of RFLP markers are available in critical chromosome segments, it is possible to establish the approximate size of the segment transferred. This may be important when, in addition to the target gene, less desirable genes are simultaneously transferred and not readily separated from the target gene by backcrossing (linkage drag, hitch-hiking). Recipient plants with small segments transferred from the donor can then be selected directly after transfer. The transfer (by generative recombination) of the gene for tobacco mosaic virus resistance (Tm2) from *Lycopersicon peruvianum* to the cultivated tomato (*L. esculentum*) could be followed in detail by Young and

Tanksley (1989) in the high density RFLP map of the short arm of chromosome 9 by straight forward linkage analysis. The longest segment transferred was 51 cM (centiMorgan) long, approximately the entire arm, the shortest segment only 4 cM. Reducing the size of an introduced alien segment by repeated backcrossing without selection is quite slow. Young and Tanksley (1989) estimate that with the aid of RFLP markers two generations are sufficient to isolate as short a segment as backcrossing with random recovery would do in 100 generations.

Instead of using molecular probes as markers of the genes themselves or as linked markers in genetic segregations, it is in principle possible to use them directly in in situ hybridization, provided a probe of sufficient length is available and the transferred segment is at least several kb long. It may mark a segment or a gene transferred from one species or genotype to another. The unplanned introgression of large chromosome segments (often entire arms or whole chromosomes) of rye into wheat (see also below) has been made visible by the use of a molecular marker specific for rye, which hybridized only to the rye segment (Fig. 4.6A; cf. Lapitan et al. 1986). Somewhat different is the technique described by Le et al. (1989), where no specific probes are required. It involves the preparation of labelled, total DNA from one species, mixing it 1:1 with unlabelled DNA of the other species and hybridizing it to preparations of the material to be tested. The unlabelled DNA partly saturates the DNA of one species and the labelled DNA, although having some homology with the chromosomes of this species, it is concentrated on the chromosomes of its own type. The difference in label on the chromosomes in the preparation is sufficient to distinguish between chromosome segments derived from the different species. A comparable approach was used by Heslop-Harrison et al. (1990) to check wheat for rye chromosome segments.

Although single copy genes cannot yet be made visible by in situ hybridization in plants, low copy number repetitive genes have been located in somatic chromosomes. Multiple copies of originally single copy genes are often introduced by molecular transformation and can then be made visible (Mouras et al. 1987). They occur naturally in several instances: t-DNA, seed storage protein genes (e.g. Ambros et al. 1986; Gustafson et al. 1990; see also Sects. 4.2.4.2 and 8.3.3.1.3). The relative (and unexpected) inaccuracy of gene localization by in situ hybridization in condensed chromosomes is a drawback, which may possibly be overcome by using prophase or pachytene chromosomes (Sect. 8.3.3.1.3).

C- and N-bands can be acceptable markers of specific chromosome segments and, although in many cases not sufficiently specific, have been useful in chromosome segment transfer in the Triticinae. The segments must be rather large to be recognized. Examples will be given in the discussion of the techniques of gene and chromosome segment transfer.

Finally, monitoring of transferred segments, provided they are of sufficient size and sufficiently differentiated from the host, is possible by testing meiotic pairing with marked, known chromosomes. Using telocentrics of chromosome

7 of wheat, Eizenga (1987) could conclude that a segment of chromosome 7 of *Agropyron elongatum*, carrying gene Lr19 for resistance to *Puccinia recondita* rust, was on chromosome 7A and not on 7D as assumed earlier. The segment had been transferred by homoeologous recombination and was apparently sufficiently differentiated to show preferential pairing.

10.4 Different Approaches to Generative Gene Transfer

The way introgression of alien genes into the genome of a recipient is realized by generative means depends on the possibilities of generative recombination, and here the degree of chromosome pairing and exchange is crucial. Kimber (1984) distinguishes three levels of chromosome pairing affinity between the donor and recipient and for each level a different approach for transferring genes is appropriate.

1. *High pairing affinity*: Normal recombination is sufficient. Backcrosses may be necessary to separate desired genes from undesired linked genes.
2. *Medium affinity*: The affinity is not sufficient for direct transfer by genetic exchange between chromosomes of the donor and the recipient. Artificial increase in the effective affinity is necessary and sometimes possible by manipulating the genetic system which regulates chromosome pairing and the level of exchange.
3. *Low affinity*: If sufficient pairing cannot be induced, translocation of a chromosome segment with the desired gene to a chromosome of the recipient is the only solution. Here, molecular transformation would be an alternative where possible.

The three levels of affinity are not well separated and the course taken depends on the genetic and economic possibilities the material offers and on the attitude and the facilities of the breeder. In many practical programs careful planning does not precede selection of recombinants and trial and error is the common approach.

The actual situation is usually still more complicated. For instance, chromosomal rearrangements may block the separation of undesired from desired genes, even in hybrids where recombination is otherwise quite satisfactory.

10.4.1 Gene Transfer by Direct Recombination

With sufficiently high levels of chromosome pairing affinity, artificial introgression of an alien gene from a (not necessarily closely) related species into a chromosome of a cultivated species can often be realized simply by hybridization followed by straight recombination and recovery of most of the original genome by cycles of backcrossing. Even then, there are bottlenecks. Some are rather obvious, but nevertheless repeated failure often precedes their recognition unless the material is analyzed carefully at the start of the program.

First, the hybrid may be difficult to make, but with the proper choice of compatible genotypes and with modern techniques of embryo rescue, success can be substantially increased. Once the hybrid has been made, several barriers to gene transfer to the cultivated species may be encountered apart from limited recombination. One is low fertility in backcrosses. The reason may be physiological, due to genetic incompatibility between the combined genomes, resulting in disturbance of any of the many steps in gamete [prefertilization: premeiotic, meiotic or post-meiotic, including embryo sac and pollen (tube) development] and embryo and endosperm development (post-fertilization barriers). Female fertility may be better than male fertility, but not necessarily so.

When this form of incompatibility is not so strong as to result in sterility, it may still deregulate some of the subtle meiotic processes in the hybrid and lead to unusual patterns of recombination: *disruptive mating*. This may prevent genetic exchange in chromosome segments that otherwise would be accessible to recombination. It may also break up linkage blocks that could be better maintained. At the same time, normally close linkages may be broken between desired and undesired genes that can be of considerable practical interest (Khush 1984, see below). Although not in all respects favourable, disruptive mating can have its merits.

Fertility may be disturbed because of meiotic errors resulting from chromosomal differences: not only the lack of homology with the consequence of univalent formation, but also chromosomal rearrangements. Chromosomal rearrangements, when heterozygous in a hybrid, can have serious effects in addition to disturbing fertility: they may prevent recombination between genes that must be separated.

Although found with relatively low frequencies, *inversions* may be an obstacle to gene transfer. They affect only one chromosome at a time. In interspecific hybrids between different *Allium* species made to transfer genes from wild sources into the onion, *Allium cepa*, *pericentric* inversion heterozygosity has been observed that practically prevents recombination between the chromosome segments involved (Levan 1941; de Vries 1989; Albini and Jones 1990). *Paracentric* inversions have been observed in several hybrids between wheat and related species. However, when their identification is merely on the basis of the presence of anaphase bridges, an alternative explanation can be hybrid dysgenesis leading to U-type chiasmata or premeiotic errors. Paracentric inversions can be distinguished from other meiotic disturbances causing anaphase bridges with fragments on the basis of fragment length (Sects. 5.3.4 and 7.4.1.5.2). This is identical in all cells in paracentric inversion heterozygotes, but (more) variable when the cause is different.

When a single inversion occurs in a single chromosome of a heterozygote made to transfer a gene from one species or cultivar to another, the probability that this gene will be located in the inverted segment is not large. If it does occur, there are two possibilities:

1. The entire inversion is transferred to the recipient cultivar and maintained there in a homozygous state. If no deleterious alleles of other genes are co-transferred, and no indispensable alleles lost, this may be an acceptable procedure. A possible negative effect of the co-transferred genes may be compensated by adjusting the genotype. Such an adjustment, however, reduces the genetic variation available to the original breeding program. In addition, in later programs involving this new cultivar, the inversion will again be encountered.

2. It is attempted to remove the gene from the inversion by reciprocal double crossing-over. Then it is useful to have an impression of the expected frequency of this event in order to know how large a segregating population must be to obtain at least one recombinant. An approximate estimate of reciprocal double exchange in the inversion can be obtained from a quantitaive meiotic analysis. It should be noted, however, that the probability of transferring a specific gene depends not only on the frequency of reciprocal double exchange in the inversion, but also on the position of this gene in the inverted segment. In the central region of this segment the probability of transfer is much greater than near the ends. Double reciprocal exchange tends to be a very rare event.

Estimates of double reciprocal exchange are obtained as follows. In paracentric as well as pericentric inversions the frequency at diplotene and often also at metaphase I of configurations of the typical "pretzel" type corresponds to the frequency of one or more chiasmata in the inverted segment. With paracentric inversions the frequency of single anaphase I and II bridges corresponds to the frequency of single and disparate double exchanges. Double bridges are formed by complementary double exchanges. The difference between the frequency of "pretzels" at diplotene or diakinesis-metaphase and the frequency of bridges at anaphase I and II corresponds to the frequency of reciprocal double exchange, which does not produce bridges, and which is the category desired. It should be noted that the estimate of the difference is inaccurate. The probability that triple exchanges complicate the analysis is negligible, as even double exchanges tend to be quite rare.

Burnham (1962) gives examples of reciprocal double crossing over in *Drosophila* and maize where genes were transferred. In maize a genetic analysis is possible, and at anaphase I and II bridges can be scored, but analysis at metaphase I is unreliable. In spite of the small size of the chromosomes of maize, exchange recombination is not infrequent. The following example is from Rhoades and Dempsey (1953). A colourless aleurone, liguleless (*a1 lg2*) type was crossed with a wild-type homozygote for a paracentric inversion in the long arm of chromosome 3. The partially sterile F1 with the composition:

```
               (inversion)
                 Lg  A1
---------X===========X---------
             1   2   3                    (segments)
---------X===========X---------
                 lg  a1
```

was backcrossed as pollen parent to the double recessive parent. Recombination between the two genes must be the result of reciprocal double exchange, in segments 1 and 2, or 2 and 3. In all other cases bridges are formed that do not produce male-viable spores. This automatically selects for reciprocal double exchanges, in addition to no-exchange types. Parental types were found in 1410 (*Lg2 A1*) and 1215 (*lg2 a1*) plants and recombinants in 6 (*Lg2 a1*) and 8 (*lg2 A1*) plants. The recombinant frequency was 0.5%, compared to 28% in normal karyotypes. A comparable frequency of the parental types will also have been recombinants transferring both genes together. In the reciprocal backcross, with the female as the heterozygote, the frequency of recombinants was seemingly much higher, but the excess was due to recovery of deficiency/duplication gametes through the egg.

Comparable results have been reported for pericentric inversions, but the cytological situation is somewhat different. Single exchanges and disparate double exchanges cause the sister chromatids to be different (Sect. 5.3.3). Both reciprocal and complementary double exchanges cause the sister chromatids to be equal. The frequencies of these two events may be assumed to be the same. If unequal anaphase chromatids can be scored reliably, the difference between the frequency of "pretzels" at diplotene (or the corresponding metaphase I figures) and the frequency of unequal anaphase chromatids represents the reciprocal and complementary double cross-overs in the inverted segment. Half of this difference is due to the desired reciprocal exchanges. Examples of gene transfer from a pericentric inversion chromosome to a normal chromosome and vice versa are given by Burnham (1962).

When, as briefly discussed above, the entire chromosome segment including the target gene, the rearrangement, and several linked loci is simply transferred without being detected, and it is subsequently made homozygous in the host, it does not result in meiotic abnormalities. Then, the rearrangement may not be recognised as the reason for the failure of recombination between the target gene and other genes. Occasionally, a rearrangement can be recognized in the karyotype, especially when detailed banding patterns can be made visible, but unless there is a reason, such a check will not often be carried out. At later stages of the breeding program slight negative effects may appear to be associated with the introduced gene. These may be due to pleiotropic effects of the gene itself or to the action of linked genes or gene complexes. Cycles of backcrosses with the original cultivar in an attempt to remove the effects of linked genes are often not effective (Young and Tanksley 1989, see above), especially when they are not carefully monitored.

A check for the presence of chromosomal rearrangements and an analysis of their meiotic characteristics, for instance in a test cross with normal material, and especially with respect to recombination, is useful in such cases.

In many of the large number of interspecific hybrids between grasses of the Triticinae studied by the groups of Kimber in Missouri (see Chap. 9) and of Dewey in Logan (Utah), *translocations* have been found to be relatively frequent (Dewey 1984). In Section 5.4.1.4 their effects on recombination and segregation, and in Section 9.3.2 their effects on chromosome pairing affinity estimates, especially in polyploid hybrids, are mentioned. For gene transfer, translocations may be quite bothersome. Not only in interspecific hybrids, but also within species, translocations are not uncommon. In wheat, many varieties have been found to differ from each other in one or even more translocations. These are disturbing when they cause desired and undesired genes important in breeding programs to be closely linked, and normal backcross programs fail to break such linkages. In many cases translocations without an apparent undesired linkage have been carried over unknowingly into new cultivars without causing serious problems. In later breeding programs these may cause unforeseen complications.

As explained in Section 5.4.1.4 all genes in all chromosomes involved in a translocation or set of translocations are linked in the heterozygote. Especially genes in interstitial segments are difficult to separate from the translocation and from other genes near the break points. There are examples of considerable recombination in interstitial segments (de Vries 1983), but these are an exception. In interstitial translocations double recombinants are required to separate the gene from the translocation, as with inversions. In addition, even with reasonable frequencies of crossing-over, the accompanying reduction in fertility of translocation heterozygotes is a disturbing factor. When no striking deleterious effects accompany genes linked to translocations, chances are that, as with inversions (see above), the entire translocation is co-transferred with the gene when the gene is located in the complex. Unlinked translocations may be carried over by chance and made homozygous in the final cultivar when no extensive series of backcrosses is involved.

10.4.2 Manipulation of Recombination

In many interspecific hybrids recombination is reduced compared to the species themselves, apart from the effect of the rearrangements. One reason, of course, is insufficient pairing affinity. However, even with reasonable or even high average levels of pairing and genetic exchange, as in normal nonhybrid material, the *pattern* of *crossing-over*, expressed as *chiasma localization*, can practically exclude specific chromosome segments from recombination. In several cereals chiasmata are localized distally, but in variable degrees and with genetic variation. In the genus *Allium*, to which onion, leek and several more cultivated and wild species belong, some species have random chiasma distribution, but several have proximally localized chiasmata and others terminal localization (Jones 1983). In both of the latter two cases large chromosome segments are practically free of recombination. This pattern can be occasionally altered by using specific genotypes (Sect. 8.2.1.5; Jones 1967, 1984) or artificial conditions promoting exchange recombination. Khush (1984) reports success for rice with hybridization with exotic genetic sources that lead to upsetting the meiotic pairing and exchange system in the hybrid (disruptive mating, see above). This results in breaking up repulsion phase linkages in addition to the transfer of a gene to a hom(oe)ologous chromosome. These linkages would normally prevent the introduction of desired alien genes without linked undesired genes (linkage drag, hitch-hiking). The fact that abnormal genetic exchange patterns will not only affect the chromosome segment with the target genes, but the entire genome, implies that recovery of the original genome with only one or a few genes from the donor requires many generations of backcrossing.

Change of the genetic exchange pattern in species hybrids with less drastic consequences has been reported by Jones (1967) for two subspecies of rye, *S. cereale dighoricum* and *S. cereale turkestanicum*. Both had distal chiasma localization, but in the hybrid chiasma formation was close to random, except for normal, positive interference. It was possible to isolate progeny lines with the original parental chiasma distribution in addition to lines with random distribution. Apparently, there was segregation of genes for chiasma localization. In hybrids between *Allium* species with different localization patterns, the regulating systems are also disturbed, and near-random chiasma distribution is observed. In their progeny different localization and chiasma frequency patterns appear (Fig. 8.2). Within species, the possibilities for adjusting the chiasma pattern are limited, but some genetic variation in chiasma distribution patterns is common within species. Whether the genes involved act through chromosome pairing or directly on chiasma formation is usually not clear.

Seemingly conflicting is the use of desynapsis in increasing crossing-over in specific segments. Prakken (1943) reported on monogenic recessive conditional and partial desynapsis in rye, expressed only under draught stress, where the original system of localization had been altered. Somewhat unexpected is the high cross-over frequency in asynaptic or possibly desynaptic maize where Rhoades (1947) found a considerable increase instead of a decrease, and especially double cross-overs in certain chromosomal regions were much more frequent than in controls. This was not only true for the haploid gametes which could have been the result of selection of cells with high levels of crossing-over. Diploid gametes formed after restitution by failure of meiosis I, which would be expected to be from low chiasma frequency meiocytes, where restitution is expected to be most frequent, showed a similar increase. Jongedijk et al. (1991a) report on variable recombination in desynaptics of diploid potato where some segments had reduced and other segments increased recombination. The subject of level and distribution of recombination in desynaptic plants is briefly discussed again in Section 12.5.

There are no reports of attempts to exploit desynapsis for purposes of breaking specific linkages, but there are certainly opportunities.

The possibility to enhance recombination in specific chromosome segments by altering the interference pattern has been implied in a few cases discussed above. One component of interference is the pairing pattern. It has appeared possible, exceptionally, to alter the pairing pattern around chromosomal rearrangements such that very close linkages are broken. Whereas in some translocation heterozygotes a reduction of pairing around the break point results in a reduction in crossing-over in the region concerned, at most accompanied by a slight increase some distance away, in other translocation heterozygotes there is a clear increase already at short distances from the break point (Arana et al. 1987; Parker 1987). This is striking especially in short chromosomes with normally only one chiasma, but which in interchange heterozygotes tend to have two. This is a complex approach to gene transfer and referred to here only to demonstrate the wide potential for manipulation of exchange recombination. In addition, there is no certainty that in this case chiasma distribution is involved, i.e. that chiasmata are formed in segments where they are practically absent normally, or that the frequency is merely increased. In the first case the shift in chiasma location can be used to break close linkages, in the second case a larger segregating population would have the same effect of recovering rare recombinants with less complications.

A second example of increased crossing-over in specific chromosome segments caused by chromosomal rearrangements is the combination of a telocentric with an interchange heterozygote. In interchange T240W (3R/5RL) of rye (*Secale cereale*), the gene for brittle stem (*br*), located in the short arm of 5R, cannot be separated from the interchange by normal crossing-over in the heterozygote. A telocentric 5RS combined with the interchange, however, appears to pair with the interchange chromosome 5RL/3R much more effectively than a complete chromosome 5R. As a consequence, recombination between the locus of *br* in the telocentric and the translocation break point is sufficiently frequent for the recovery of recombinants (own unpublished results). A comparable effect has been observed for the tertiary trisomic for 5RS/3R, derived from T240W, where the extra chromosome recombines readily with one of the normal chromosomes (de Vries 1984), whereas in the interchange heterozygote no recombination between the same markers and the translocation has ever been observed even in very large populations (de Vries and Sybenga 1984). Such constructions may be useful for breaking linkages between rearrangements and genes and between normally closely linked genes.

Knowing that specific environmental conditions alter the frequency as well as the localization of chiasmata, simple treatment of heterozygotes with extreme temperatures, irradiation or chemicals (colchicine, for instance: Curtis and Feldman 1988) may increase the probability of obtaining otherwise rare recombinants, even when the overall chiasma frequency is decreased. There are no indications for specific effects on specific chromosome segments, but if a rare recombinant has to be obtained, it may pay to check if an externally applied agent changes the pattern of chiasma distribution sufficiently to potentially result in crossing-over in unusual segments. Also, with a reduction in the average level of recombination, the pattern of distribution can be significantly altered. The necessity to maintain a reasonable level of chiasma formation, not only for recovering any recombinants et all, but also for maintaining a satisfactory level of fertility, is obvious.

In addition to genetic and environmental conditions, B-chromosomes may quantitatively alter chiasma distribution in plants (review by Jones and Rees 1982; more recently on *Crepis*: Parker et al. 1990). It is not certain whether the effect of B-chromosomes on cross-over distribution is sufficient to use them to replace simple selection in larger populations.

In most breeding programs it is not recognized that the level of recombination between genes within chromosomes is relatively low, and usually insufficient opportunity is given for exchange recombination. The level of crossing-over, as long as it is not extremely low, is usually less important than localization, because somewhat lower levels are readily compensated by selecting in a larger population. Only when population size is a bottleneck (tree breeding; small progenies due to F1 sterility) does the level become important.

10.4.3 Modification of Chromosome Pairing Affinity

There is no strict boundary between levels of pairing that are sufficient for genetic exchange and those that are not. The choice of approach depends on the situation, and especially on the possibilities to alter the homoeologous pairing level.

The same problems as discussed above in Section 10.4.1 for intraspecific hybrids with sufficient pairing affinity will be encountered here, and sometimes even more pronounced, but they will not be discussed again in the same detail: difficulty in making the hybrid, sterility of the hybrid and the first backcross generations, difficulty in recognizing the gene to be transferred, barriers to recombination other than homoeologous pairing affinity such as restricted pattern of crossing-over and chromosomal rearrangements.

Artificially induced homoeologous recombination was first used by Riley et al. (1968) to transfer dominant stripe (yellow) rust resistance from *Aegilops comosa* to bread wheat. Homoeologous recombination could not very well be induced by nullisomy for chromosome 5B of wheat, because the effect of the absence of 5B was compensated by the presence of the homoeologous chromosome in *Ae. comosa*. By crossing the amphidiploid of wheat and *Ae. comosa* with *Ae. speltoides*, genes that lifted the natural block against homoeologous recombination were introduced from the last species. In a series of backcrosses of the highly infertile triple hybrid with wheat, and selecting for stripe rust resistance, wheat with a recombined wheat/*Ae. comosa* chromosome was recovered which was completely balanced. In the intermediate stages a serious breakup of the wheat genome had occurred, both because of chromosome segregation in the complex hybrid and because of intergenome crossing-over. The end result after repeated backcrossing was stable again. Exactly when recombination had taken place during the various intermediate stages was not discussed, nor the size of the segment finally transferred.

This method and several variants using special combinations of homoeologous pairing promoting mutants, have been applied successfully on many occasions, primarily in wheat, where cytogenetics, genetics and taxonomy of related species are well developed (Islam and Shepherd 1991; Koebner and Shepherd 1986, 1987; Rogowsky et al. 1991). The use of neutral molecular or other markers closely linked to the gene to be transferred is a necessity in the case of hypostatic genes, but this approach has not yet been applied widely. In allopolyploids the same is true for recessive genes when in more than one genome a dominant allele occurs of which only one is replaced by the alien allele. As in the case of straightforward recombination, the removal of extra dominant or hyperstatic genes by mutation or repeated introduction of the desired gene in different genomes is required. Such complications, of course, are just as troublesome for molecular transformation as they are in straightforward and induced homoeologous recombination, and other ways of gene transfer as discussed below.

In wheat several ways are available to modify homoeologous pairing (Sects.6.1.2.3, 11.3.2 and 12.2): nullisomy for chromosome 5B, deficiency for 5BL or a segment of 5BL, mutations (less effective than a deficiency) of the *Ph1* gene on chromosome 5BL and *Ph2* (weaker than the *Ph1* gene) on 3DS, inducing different levels of homoeologous pairing (Sears 1984). In addition to such mutants and deficiencies with relatively strong effects, each chromosome appears to have its individual reaction, and the genotypes of both wheat and the donor species have considerable effects. The effects of complete genomes and individual chromosomes have been studied in detail for instance in the combination of rye and wheat, and in both species major and minor genetic variation is present that affects homoeologous pairing (Dvorak 1977; Sears 1984; Miller and Reader 1985; Gupta and Fedak 1987; Cuadrado and Romero 1988; Naranjo et al. 1989).

Combinations of deficiencies for pairing promoting chromosome segments, mutations of pairing restricting genes and specific genetic background can result in almost any level of homoeologous pairing between not too distant genomes. The level can thus be adjusted to the requirement of the experiment, such that there is a good chance to transfer a gene without too much redundant alien material, while at the same time the disturbance of the rest of the recipient genome remains limited. Outside wheat, such regulatory systems have not been well developed.

A single homoeologous exchange tends to transfer a considerable piece of chromosome, which may not be desired. By repeating induced homoeologous

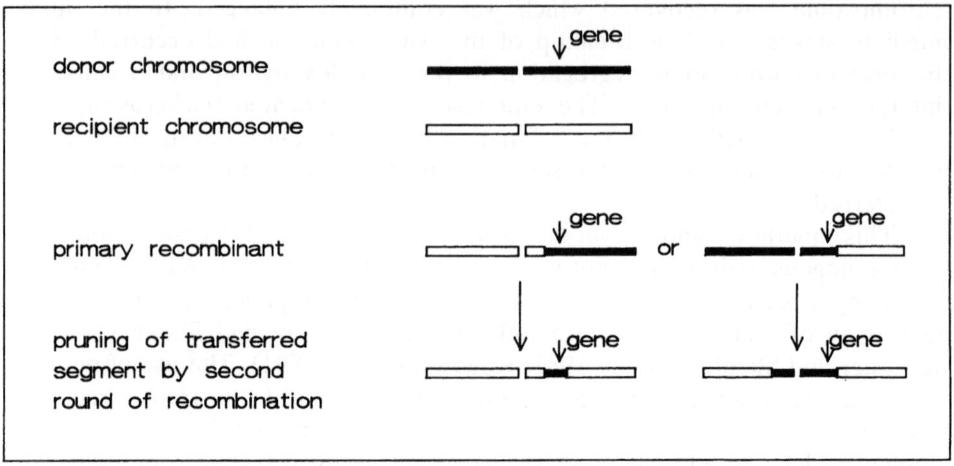

Fig. 10.1 Reducing the segment transferred by homoeologous exchange in subsequent generations. The position of the gene is relatively proximal in the donor chromosome, which is unfavourable for transfer. Two situations are considered:

In one situation the recombined chromosome recovered in a disomic has a large distal segment derived from the donor, but the centromere and the other arm are from the recipient.

In the second situation the primary recombinant has the centromere and one arm of the donor and a large segment, without the centromere, from the recipient.

In a second round of recombination in the first situation the distal segment of the donor chromosome is replaced by a homoeologous segment of the recipient and only a small interstitial segment remains of the alien chromosome. In the second situation a large segment of the other arm is replaced, still leaving a relatively large, but not necessarily harmful segment with the centromere of the donor

genetic exchange, the size of the alien segment can be "pruned" to a suitable size (Koebner and Shepherd 1986, 1988). When the gene involved has a proximal location (Fig. 10.1), the distal alien segment transferred together with the gene can be replaced by homoeologous recombination. When the location of the gene is distal, and the proximal chromosome segment is the alien segment, the gene can be transferred directly into a recipient chromosome with only a small distal segment. In subsequent cycles the segment transferred can be reduced again. This should be done with a system that does not produce the maximum homoeologous recombination. Drastically upsetting the original genome by the introduction of a strong system (nullisomy for chromosome 5B, some strong *ph* mutants, hybridization with *Ae. speltoides* in wheat, for instance) should be avoided. There are mutants of the *ph* gene on chromosome 5B that have an intermediate effect, and these are preferred. Also, in cases where straightforward recombination between different species is possible, it is necessary to restrict the size of the segment transferred. It may be assumed, however, that the closer the donor and the recipient species are taxonomically, the less deleterious an alien fragment will be.

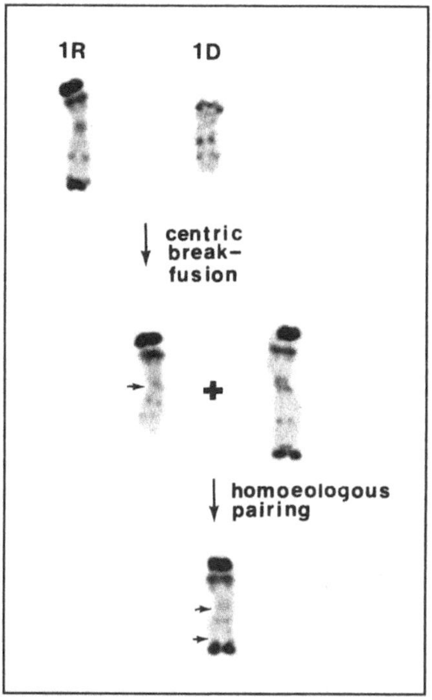

Fig. 10.2 Transfer of HMW glutenin subunits 5 and 10 (*GluD1*, important for bread-making quality) from chromosome *1D* of wheat to *1R* of rye in hexaploid triticale Rhino. First, a centric break-fusion translocation was induced in wheat with 20″ + 1R′ and 1D′, followed by a transfer of this 1Rs.1DL chromosome to Rhino by hybridization and selection. The recipient Rhino line had a 5D(5B) substitution, inducing homoeologous pairing. This led to pairing between 1RL of the normal chromosome 1R of Rhino, with 1DL of the 1RS.1DL construct. Exchange recombination resulted in a predominantly 1R chromosome with an interstitial 1DL segment (carrying the wheat *GluD1* gene). This segment could be morphologically identified. The breaks are indicated by small arrows. (Courtesy of A. Lukaszewski)

An example is the transfer of an interstitial segment of chromosome 1D, containing the gene *GluD1*, to rye chromosome 1R in triticale, replacing the corresponding rye gene *Sec-3*. *GluD1* is an important gene for bread-making quality, normally not present in triticale (Fig. 10.2). Lukaszewski et al. (pers. comm. 1991) first made a monosomic substitution in wheat, containing 20 normal wheat chromosome pairs, one chromosome 1D and one chromosome 1R. The univalents produced a 1RS.1DL centromere translocation (see Sect. 10.4.4.2.2.2). This chromosome was transferred to the triticale variety Rhino, which had a 5D(5B) substitution, and consequently a high level of homoeologous pairing and recombination. Pairing between the long arm of the normal chromosome 1R and the 1DL arm of the 1RS.1DL translocation chromosome led to the exchange of a terminal segment, resulting in a "rye" chromosome 1R with an interstitial segment from 1D (Fig. 10.2). Rogowsky et al. (1991) report detailed monitoring by molecular means of the reduction by homoeologous recombination of the rye 1RS arm translocated to 1D of wheat (see also Islam and Shepherd 1991).

This "pruning" of alien chromosome segments, although rather laborious, is especially important when more genes than one are introduced in sequence. Maintaining several large alien segments would seriously contaminate the recipient genome, and would not only make it unbalanced, but also hardly accessible for further breeding. An advantage of molecular single gene transfer is that little excess material is carried with the gene. Homoeologous recombination used properly can have comparable effects. In other species than wheat these possibilities have not been extensively explored, and even in wheat most projects do not involve such refinements.

In practice it is often quite difficult to separate the desired from the undesired gene. An example is the apparently close linkage between rust resistance and the undesired yellow starch colour in wheat given by Eizenga (1987). A chromosome segment with both genes could be readily transferred by homoeologous recombination from *Agropyron elongatum* to chromosome 7D and even to 7A of wheat, but the two genes could not be separated. The cause of the close linkage has not been reported. Special cases of co-transfer of genes with potentially undesired effects together with alien genes of commercial interest are those where "gametocidal" or "cuckoo" chromosomes are involved. These result in preferential transmission of the transferred segment in heterozygotes (Marais 1990). When homozygous, the effect disappears, but will return in later breeding programs. The effect of a homozygous "cuckoo" segment on field performance of a variety carrying it, is not well known. There may be ways to "prune" such segments by increasing the homoeologous chromosome pairing affinity, or by altering the pattern of recombinational exchange if possible. In exceptional cases induced translocation may be an alternative, but the probability of transferring the right segment is small. Alternatively, mutation or deletion of the undesired gene, or even adjustment of the background genotype may be tried. It is clear that in addition to technical feasibility the economics of the project plays a role, in addition to the importance of the effect of the associated gene. Not in all cases is the effect of gametocidal or "cuckoo" genes detrimental (Sects. 6.2.1.3 and 10.4.4.2.2).

When genes are introduced into chromosomes of an allopolyploid by homoeologous recombination, it is expected that most transfers will take place between the alien genome and the most closely related genome of the recipient. This, however, is not necessarily consistently so. Eizenga (1987) reports that the transfer of the *Lr19* gene for *Puccinia recondita* resistance, transferred from *Agropyron elongatum* into wheat (earlier considered to have been transferred consistently to chromosome 7D in numerous repeated transfers) had in fact been to chromosome 7A in at least one case. In most cases, however, the transfer of genes from *A. elongatum* is to the D-genome of wheat, in accordance with the close relationship between the two genomes (Knott 1988).

In material other than wheat subtle regulations of homoeologous pairing are usually not available, although there is considerable genetic variation in homoeologous pairing that can be exploited much more extensively than has been done. Especially in allopolyploids without effective genetic systems for reinforcing pairing differentiation, genetic variation of a more quantitative nature can be important for gene transfer.

There are more bottlenecks in interspecific gene transfer than limited fertility, difficult recognition of the gene to be transferred and restricted recombination. *Expression* of the alien gene in the final cultivar is essential, but it may appear to be suppressed or abnormal. As indicated earlier, when the alien allele is recessive, homoeologous recombination will make it replace the dominant allele and then it can be expressed in the homozygote; epistatic genes must be removed (by mutation or deletion) before a hypostatic gene can be expressed. Persistent deviations in expression but also close linkage with an undesired gene, resembling pleiotropy, can occasionally be solved by selection of the proper background genotype where expression of the undesired effects are reduced or compensated. Traditional practical introgression programs have paid insufficient attention to possibilities to find solutions for such complications.

The use of translocations directly induced in the interspecific hybrid in the somatic phase, for transferring chromosome segments when recombination is not possible, is not especially attractive, even though not in principle impossible. Lapitan et al. (1984) found several translocations between rye and wheat chromosomes in plants regenerated after prolonged in vitro culture of scutellar tissue of wheat-rye hybrids. None were homologous and their value is dubious. Translocations have been more successful when induced in the plant in additions and substitutions (Sect. 10.4.4.2.2).

10.4.4 Indirect Gene Transfer Via Addition or Substitution

10.4.4.1 Isolation of Addition and Substitution Lines

In order to circumvent the complications encountered with direct introgression from a distant species, indirect methods have been designed, primarily using the generative phase. The principal difference to the techniques discussed above is the introduction of an intermediate step consisting of first transferring a single entire chromosome from the donor to the recipient and subsequently transferring the target gene or a chromosome segment with this gene to the recipient. This eliminates the complication of disturbing the host genome by random introgression of numerous genes from several chromosomes simultaneously with the desired gene.

The first step is usually the creation of an *addition line* which has one pair or, if this cannot be realized, a single chromosome from the donor species in addition to the complete genome of the recipient. In many cases it is advisable to use a (monosomic) substitution instead of an addition. The extra chromosome (pair) or monosomic substitution must carry the gene to be transferred. If the allele in this chromosome is dominant or epistatic, its expression shows the presence of the correct chromosome. As in the cases described in the previous section, if it is a recessive or hypostatic allele, special methods are necessary to detect it. For simplicity, by far the most extensive use of this technique has been made for the transfer of dominant alleles, primarily for disease resistance. Usually, addition chromosomes can be maintained complete through several generations. Occasionally, however, deletions may occur, not only including heterochromatic segments, but also segments with genes which are to be transferred to the recipient species (Hu and Quiros 1991: additions of *Brassica oleracea* to *B. campestris*: cf. Sect. 8.3.3.2). Loss of dominant resistance genes can be traced by checking resistance (which should be done regularly). Deletions involving genes with incomplete penetrance, or hypostatic and recessive genes or alleles, can sometimes be a problem. Molecular and other co-dominant markers are not always known for the particular addition chromosome.

The construction of an addition line has been considered in Section 6.2.2.4.2.1. First, the recipient (say genome A) and the donor (genome B) are crossed and the hybrid is doubled (AABB). This amphidiploid is backcrossed to the recipient, resulting in a population (in this case an allotriploid) with a double set of chromosomes from the recipient and a single genome derived from the donor (AAB). Selfing or intercrossing these plants results normally in the complete transmission of the recipient genome, but of only a few chromosomes of the donor. Plants with one extra chromosome, showing the presence of the desired gene are selected. They are recognized by the expression of the gene or, still exceptionally, by the presence of specific linked marker alleles. A cytological check on the presence of an extra chromosome, if possible, identified by its shape, size or C-banding, is usually carried out. It may also be a single arm (telocentric) or an isochromosome formed by centromere misdivision of the univalent in the triploid or monosomic stages. An isochromosome is especially favourable because the double arm gives a double probability of transfer.

This is the monosomic addition, which can be made disomic by selfing or intercrossing and selection on the basis of the chromosome number, if possible, completed with a C-band or other test. When it is not a priori known on which chromosome the gene involved is positioned, its expression in combination with counting the chromosome number is the usual selective criterion.

In addition to generative methods for producing additions and substitutions, there are, potentially, a few ways to produce additions and substitutions on the basis of somatic cell genetic techniques. These have also been discussed in Section 6.2.2.4.2.1, but their importance is still limited.

The *disomic addition* itself has been suggested to be directly useful when the gene concerned is dominant, and transfer of single genes or small chromosome segments from this chromosome would not be necessary. It has appeared, however, that the disomic addition is usually not entirely stable and that gradually the extra chromosomes are lost because, for the plant, this pair is redundant when the selection pressure for the new gene is low. In addition, the phenotypic effects of an extra pair of chromosomes tend to be so pronounced that even with extensive selection for a tolerant genetic background, no satisfactory result is obtained (Wienhues 1966, 1973). Intermediate forms such as disomic compensating trisomics (Sects. 6.2.2.4.1 and 12.4.2.2) still carry too much excess genetic material and are not really very stable. The idea of directly exploiting additions has practically been abandoned, and has made place for methods that transfer the desired gene, if inevitably with a small segment of alien chromosome, into a chromosome of the recipient.

By far the largest number of gene transfers from an alien into a cultivated species by using additions or substitutions has been realized in wheat. Yet, in a number of other allopolyploids and even diploids, alien additions are available. There are several reasons why the use of additions has been limited in these cases: direct transfer by homoeologous recombination is relatively simple; or there is insufficient genetic and cytogenetic knowledge of the species; or specific monosomic addition and substitution lines and other material are not available and difficult to construct; or there is insufficient financing; etc. In cotton, monosomics, trisomics, telosomics and alien additions are available (Endrizzi et al. 1985), but usually not necessary because direct gene transfer is successful, either by recombination or by (spontaneous or induced) translocation. In oats (*Avena sativa*) the situation is similar, but some instances of applicable gene transfer have been reported (Thomas 1968; Aung and Thomas 1978); in *Brassica napus* alien additions from *B. nigra* have been constructed (Jahier et al. 1989), in tobacco additions from *N. glutinosa* were made by Gerstel already in 1945; Jena and Khush (1989) report alien additions from *Oryza officinalis* to *O. sativa*, and more examples could be found. In diploids additions are also possible. Examples are rye (Schlegel 1982; Schlegel et al. 1986; Schlegel and Kynast 1988); sugar beet (de Jong et al. 1985; Heijbroek et al. 1988); Welsh onion (*Allium fistulosum*: de Vries et al. 1991). In all these cases successful transfer of a gene from the alien chromosome to the host has been reported.

It should be noted that in many cases it is not necessary to make special constructions or apply special techniques or intermediate forms like additions to transfer single chromosomes from one species to another. Even monosomic substitutions are frequently obtained simply by first backcrossing an interspecific hybrid to the recipient species and then selecting progeny with the gene to be transferred. In many cases a spontaneous addition is first obtained and, when sufficiently viable, subsequently a substitution.

10.4.4.2 Gene Transfer from Alien Addition or Substitution Chromosomes

There are three main routes leading to the transfer of a gene from the addition chromosome to the recipient genome. The route chosen depends on the possibilities the material offers. Because steps involving monosomics, additions and substitutions are often involved, the transfer of genes from alien addition or substitution chromosomes to recipient chromosomes has been applied primarily to allopolyploids and especially to wheat, where a large array of cytogenetic types and methods is available in diverse genetic backgrounds. A detailed recent review is given by Islam and Shepherd (1991). The three alternatives discussed here are: homoeologous recombination, radiation-induced translocation and meiotic centromere translocation.

10.4.4.2.1 Homoeologous Recombination

Homoeologous recombination between an alien chromosome present as an addition or a substitution, directly, reinforced or induced, is not essentially different from the situation where an entire alien genome is present, except that there are less complications. A homoeologous substitution is much more favourable than an addition because of both the alien and the host chromosome only one copy is available, which avoids (preferential) pairing between two homologous host chromosomes. It appears that where homoeologous recombination is possible, direct transfer involving entire genomes and followed by a series of backcrosses has usually been preferred over the more complex way via the construction of additions and substitutions. As long as dominant alleles are transferred and the selection for recombinants is relatively simple, this procedure may be acceptable, in spite of the obvious possibility that several undesired genes are introduced simultaneously in intermediate stages. During subsequent cycles of backcrossing, the hybrid is gradually transferred into the equivalent of a substitution and finally the proper segment is all that remains of the host species. With less simple methods of selection it will pay first to create the substitution. When this is too difficult, of course, there is no choice but to skip the substitution.

There are several examples, old and new. Knott (1988) reported on the homoeologous recombination between a chromosome of *Agropyron elongatum* (substituted for one chromosome 7D of wheat), and the remaining chromosome 7D, transferring stem rust resistance to wheat. The substitution was spontaneous, originating in the backcross progeny of the hybrid between the two species selected for rust resistance (see also Islam and Shepherd 1991). The complication that induction of homoeologous pairing is not restricted to the specific chromosome pair involved, but involves all chromosomes of the host at the same time, is common in wheat. It is much less so in species where the initial pairing differentiation is more pronounced and not genetically enhanced, such as cotton, to some extent oats and allopolyploid Brassicas. The unintended induction of (homoeologous) pairing between the recipient genomes in the addition or substitution is hardly less disturbing than inducing homoeologous recombination in the hybrid.

10.4.4.2.2 Translocations

10.4.4.2.2.1 Reciprocal Translocations

As early as 1956, Sears published a report on the transfer of dominant rust resistance from diploid *Aegilops umbellulata* to allohexaploid bread wheat via the addition of the *Aegilops* chromosome carrying the gene. The start was a hybrid of *T. aestivum* with the amphiploid of *T. dicoccoides* and *Ae. umbellulata*, followed by two backcrosses to wheat. Every generation was checked for the presence of the resistance. This resulted in a wheat plant with an extra *Aegilops* chromosome carrying the resistance. The addition had negative effects on the plant phenotype and productivity. In the progeny of this plant an isochromosome arose by centromere breakage of the univalent chromosome in meiosis. This isochromosome was used as the donor of the segment with the resistance gene. The advantage over a normal chromosome is that it carries a double dose of the target gene. Translocations between the isochromosome and wheat chromosomes were induced by X-irradiation (1500 rad) of the spike just before meiosis took place in the majority of the florets. The pollen from the irradiated spike was used to pollinate standard wheat plants. Because pollen with extra chromosomal material or deficiencies is less competitive than more balanced pollen, this greatly reduces the transfer of extra chromosome material, original or structurally changed, with the resistance gene.

The progeny of 6091 plants was tested for rust resistance. If the transfer was by a reciprocal translocation, part of a wheat chromosome would be lost and most probably replaced by a non-homoeologous segment of the *Aegilops* chromosome. This can have deleterious effects and again reduces the probability of transfer through the pollen. Interchanges involving only a short distal section of a wheat chromosome are more favourable, and if the transfer is by an interstitial translocation, merely a chromosome segment is added. This then represents a relatively small duplication which is not very deleterious, and the effects of the duplication are readily compensated by selection. However, interstitial translocations are quite rare.

A total of 132 resistant plants (2.2%) were recovered and checked cytologically. Of these, 40 had an *Aegilops* isochromosome, 52 an *Aegilops* telocentric, and 28 had 43 chromosomes and formed a trivalent at meiosis, indicating that a translocation between a wheat chromosome and the *Aegilops* isochromosome has been formed. There were another 12 that did not form a trivalent and did not have 43, but 42 chromosomes. These must have been substitutions of a translocated wheat-*Aegilops* chromosome for a (partly) homologous wheat chromosome. At least one was concluded to be a small intercalary insertion with male transmission like a normal wheat chromosome. It could be made homzygous.

This approach to gene transfer has been applied in several more instances. One example is the transfer of rust resistance from *Agropyron intermedium* to wheat by Knott (1958) and Wienhues (1966, 1973). Wienhues (l.c.) compared the practical suitability of addition, substitution and translocation of chromosome segments. Surprisingly, the additions had a reasonable yield, even when not considering the positive effect of rust resistance, occasionally even equalling that of the control. The instability of the extra chromosome was the bottleneck. The substitutions were difficult to make because no monosomic series of the recipient variety was available. Preliminary data on yield were not as favourable as expected by comparison with the well-tolerated substitution of wheat chromosome 1B by rye chromosome 1R. Translocations, as in the case of Sears (1956), were the most promising, but quite laborious to isolate in sufficient numbers. There were several types, some apparently interstitial, and a few appeared practically useful.

Alien gene transfer by translocation from an addition chromosome in allopolyploids has not only been successful in wheat. Mildew resistance, for example, has been transferred from *Avena barbata* to oat (Aung and Thomas 1978) and there are more examples, including diploids (see below).

It is clear that, when the allele to be transferred is recessive and the transfer has to be monitored by selecting specific neutral closely linked markers such as RFLPs, this method of gene transfer is still very laborious, but definitely possible.

10.4.4.2.2.2 Centromere Translocations

The second method of using translocation of chromosome segments was referred to by Sears (1972). It is based on a translocation at the centromere between the donor chromosome and a homoeologous recipient chromosome spontaneously formed after centromere break during the first anaphase of meiosis when both chromosomes are univalent. The phenomenon of centromere breakage has been discussed in Section 5.5.2 as the origin of telocentrics, isochromosomes and centromere translocations. Morrison (1953, 1954) made detailed analyses of centromere breakage in wheat, and assumed restitution of the breaks into translocations. The exact course of events is still not known. Telocentrics do not appear to break at the centromere and have not reliably been reported to produce isochromosomes when univalent, nor fusions when two different isochromosomes are present together as univalents (Eizenga 1986). However, when a univalent isochromosome is combined with a bibrachial univalent, it can be involved in a centromere translocation. Davies et al. (1985) made three types of double monosomic additions of rye chromosomes to wheat. In one experiment both were bibrachial and out of 94 progeny analyzed 4 showed a centric translocation. In the second experiment one rye chromosome was bibrachial and the second telocentric. Here, among 96 progeny analyzed again 4 had a centromere translocation, or a centric fusion. When both addition chromosomes were telocentrics, and not homologous, out of 492 progeny studied none showed a fusion. This corresponds to the failure of Eizenga (1986) to find any centromere fusion between different selected telosomes of wheat.

In principle a centromere translocation can be reciprocal, but in most instances it is not. Centromere translocations have strictly defined break points. When the translocation is between homoeologous chromosomes, fully corresponding homoeologous segments are exchanged. Unlike with radiation-induced translocations, where the break points are not defined, single centromere translocation chromosomes between homoeologous chromosomes can be almost as functional as normal chromosomes because there are no duplications or deficiencies. Therefore, although most meiotic centromere translocations will be non-reciprocal, when they are formed between homoeologous chromosomes, there is a segmental homoeologous substitution and not a duplication deficiency.

In several wheat varieties into which mildew resistance has been introduced from triticale ("triticale resistance"), either an entire chromosome 1R from rye appeared to have been introduced as a substitution, usually for 1B, or the short satellited arm 1RS had been translocated to the wheat 1B long arm with the break at the centromere. In the hybrid between hexaploid wheat and octoploid triticale, there is one complete genome of rye which at meiosis forms only univalents. In the backcross to wheat, when selected for resistance, one or more single chromosomes of rye are still present, among which 1R, which is the carrier of the mildew resistance gene, occurs. In subsequent generations of selfing there is a possibility of complete substitution of the addition chromosome for a homoeologous chromosome of the backcross parent, especially when chromosome pairing is not fully regular. During the stage of monosomic substitution, there are two homoeologous univalents that have a tendency to break at anaphase I of meiosis, exchange arms and form a centromere translocation. This becomes homozygous after selfing.

Zeller (1973) concluded that there are two main origins for the several known cases of "spontaneous" substituted and translocated 1R/1B chromosomes in wheat: the plant breeding institutions in Weihen-Stephan near Munich, and Salzmünde near Halle, both in Germany. Zeller (1973) listed several varieties that had one of the mildew-resistant varieties from these institutions in their ancestry and that carried substitution or centromere translocations. Several reports describe these and similar cases.

An example of a translocation between a rye and a wheat chromosome, but this time planned, and which is probably a centromere translocation, is that of the stable transfer of Hessian fly (*Mayetiola destructor*) resistance from "Chaupon" rye to wheat via a 2BS/2RL translocation (Friebe et al. 1990). In this case, as in the cases of "triticale" resistance, the translocation was spontaneous, and not recovered after first producing a specific monosomic substitution. Wheat-rye hybrid plants were obtained after culture of scutellar embryonic tissue. The hybrids were doubled and backcrossed to wheat.

In the experiment it is possible to isolate such translocations with high frequency when the number of univalents is high. The frequency of centromere translocation can then be considerable. Lukaszewsky and Gustafson (1983) crossed hexaploid triticale (genomes AABBRR) with bread wheat (genomes AABBDD). In the hybrid 14 univalents (the R and D genomes) appeared. These univalents broke frequently at their centromeres and in addition to telocentrics and isochromosomes, many translocations were produced. Several of these R/D translocations were recovered in the progeny. In addition, a few involving chromosomes of the A and B genomes were observed. Their origin is not clear. Possibly some A and B genome chromosomes were univalent or, as in the studies of Davies et al. (1985, see above), a broken univalent had induced another chromosome to form a translocation. However, detailed analyses of meiosis were not made. In the process of establishing a complete set of individual R/D-chromosome substitutions in wheat, Friebe and Larter (1988) also recovered several rye/rye and rye/wheat (R/D) translocations. In a program of isolating rye additions to wheat, Ren et al. (1990a, b) crossed octoploid triticale (AABBDDRR) with the parental wheat (AABBDD) and selfed the hybrid (AABBDDR). In addition to rye chromosomes, which were necessarily univalent, some wheat chromosomes were lost also, apparently because they were univalent. Several translocations occurred. Among 837 progeny plants, 64 wheat/rye translocations were found and 256 rye/rye translocations. Some wheat/rye translocations were made homozygous, including a number where apparently only a small rye segment was transferred: they contained no observable rye chromatin, but still rye genes were expressed. In all cases reported by Ren et al. (1990a, b) the agronomic performance of the translocation homozygotes was less satisfactory than that of the original wheat variety. When specific wheat and rye chromosomes are made univalent to produce centromere translocations it is important to be sure that the arms involved are sufficiently homoeologous. Naranjo and Fernandez-Rueda (1991) showed that in the evolution of rye and wheat several translocations had altered the original composition of the chromosomes.

Centromere translocations between specific chromosomes can be formed in a monosomic substitution, where both the alien and the host chromosomes are univalent in meiotic anaphase. Although the frequency of translocation is much lower than when entire genomes are univalent, directed centromere translocation can be realized. King et al. (1988) made the monosomic substitution of 4S1 from *Aegilops sharonensis* (the "cuckoo" chromosome, see Sect. 6.2.1.3) for 4D (carrying the semi-dominant dwarfing gene Rht2). Four families derived from 594 Bc1/F2 plants, had 42 chromosomes and were semi-dwarf, with the "cuckoo" segment combined with the Rht2 gene in one 4S1/4D translocation chromosome. Two of these four families had formed after centromere translocation, but the other two after non-centromeric translocation. The origin of the latter remains unexplained, but is not really uncommon in such situations, although usually less frequent. The purpose of the transfer was to prevent gametes with chromosome 4D lacking to produce progeny. If they would form, they would not contain the gametocidal gene, and could not function. Plants monosomic for 4D are relatively frequent in normal commercial semi-dwarf varieties, but quite undesirable, as they appear as tall rogues. This is not acceptable from a seed certification point of view.

There is a clear variation between species with respect to centromere breakage and the formation of translocations, and, within wheat, also between genotypes (Steinitz-Sears 1973) and between individual chromosomes within varieties (Makino et al. 1977).

All three methods discussed for transferring chromosome segments from alien chromosomes to their homoeologues in the recipient species (induction of homoeologous recombination; irradiation of an addition line just prior to meiosis; centromere translocation) even when properly applied, have their advantages and disadvantages. It is not a priori possible to decide which is the best in a particular situation. This depends on the personal preference of the breeder, on the (usually only presumed) properties of the material, and previous experience with the material. In most cases when large alien chromosome segments are transferred, adjustment of the background genotype or reduction of the alien segment by homoeologous recombination or deletion are required (10.4.3). Apart from wheat the experience is limited.

There are a few cases of alien gene transfer using additions in diploids. Spontaneous transfer from an addition chromosome to the genome of the host has been reported for diploid (2n = 18) sugar beets (*Beta vulgaris*). Heijbroek et al. (1988) isolated a disomic sugar beet that was homozygous for the root nematode (*Heterodera schachtii*) resistance gene of *B. procumbens* from the progeny of an addition line. Resistant addition lines had earlier been isolated in several institutions. One of these was backcrossed repeatedly until a spontaneous recombinant was found. It arose possibly by homoeologous recombination, although the extra chromosome could almost never be seen to participate in chromosome pairing. Substitutions were not observed, and it is not known whether the transfer was the result of a homoeologous exchange or a spontaneous (non-homoeologous) translocation. Male transmission was not quite 100%, so some undersirable alien material must have been incorporated. With very limited recombination in the heterozygote, it will be difficult to "prune" the alien segment to acceptable proportions, and adjustment of the background genotype may be the only solution, or selection of a new recombinant.

The additions of wheat chromosomes to diploid rye (*Secale cereale*, 2n = 14 + 1; Schlegel et al. 1986; Schlegel and Kynast 1988) are alloplasmic because they have originated from a hybrid between triticale, which has wheat cytoplasm, and rye as the pollen parent. The genome composition of this hybrid is ABRR or ABDR if the triticale parent is octoploid triticale. In the backcross with rye the additions were recovered. They were identified by using C- and N-banding and by isozyme studies. For the transfer of chromosome segments from the wheat chromosomes into the rye genome, homoeologous recombination between the addition chromosome and the homoeologous rye chromosome is practically excluded because the two rye homologues pair preferentially. Substitutions cannot readily be constructed using primary monosomics because these are not viable. Substitution is possible only on the basis of random processes during meiosis in material with reduced chiasma frequencies. Centromere translocations might be a possibility. This would be possible in the backcross generations following the hybridization of triticale with rye.

Schlegel and Kynast (1988) report the successful transfer of a segment of chromosome 6B of wheat to rye by premeiotic irradiation with 1000 rad X-rays and using the pollen of the addition line to fertilize normal rye. This simultaneously transfers the genome to rye cytoplasm and selects the most viable translocations. The F1 was screened for alien chromatin by N-banding and chromosome pairing. Among 188 progeny 169 had $2n = 14$, 1 had one extra telocentric, 1 had one chromosome less ($2n = 13$), 1 was tetraploid and 16 had the wheat chromosome as an extra chromosome. Apparently, male transmission of the extra chromosome was not as low as expected. One plant with $2n = 14$ had the heavy N-band of 6B and had reduced awn growth, typical for this wheat chromosome.

Spontaneous somatic gene transfer from an alien chromosome into the host genome has been found to accompany the elimination of the chromosome involved. Melz and Thiele (1990) report the loss during vegetative reproduction of a wheat chromosome originally present as an addition to diploid rye, accompanied by the apparent transfer of some of the wheat genes to the rye genome. Claus and Pohler (1990) and Pohler et al. (1991) describe the mitotic elimination of rye chromosomes from barley/rye hybrids as a result of anaphase lagging and subsequent desintegration. As a consequence some of the DNA or chromatin was incorporated into the barley genome and expressed. This resembles the transfer of genes and chromosome segments from the irradiated into the untreated nucleus after asymmetric fusion (Hinnisdaels et al. 1991). The mechanism could be through activated transposable elements, or the invasion of intact chromatin by induced double strand breaks, as is possible after meiotic centromere breakage, and which may play a role in meiotic exchange.

The transfer of single genes by such methods is a sound approach for specific purposes. In case many genes from one species are to be introduced into a recipient cultivar, it may be preferred to combine the recipient and the donor into an allopolyploid. This may be a legitimate alternative when the original objectives of the breeding program are not very strict and when the transfer of a large number of alien genes, even if in principle possible by straightforward recombination, becomes problematic. This has been demonstrated by Namai et al. (1980) for the development of a fodder crop from *Brassica campestris* and *B. oleracea*, where both approaches worked: gene transfer, although laborious, resulted in promising results, as did the allopolyploid. The allopolyploid is a reconstruction of *B. napus*, which in some of its several forms is used as a fodder crop (Sect. 11.3.2.2).

Chapter 11

Manipulation of Genome Composition. B. Gene Dose: Duplication, Polyploidy and Gametic Chromosome Number

11.1 Objectives

Gene dose manipulation can go in two directions: an increasing or decreasing. Both can involve segments of chromosomes, entire chromosomes or genomes. A decrease in gene dose involving chromosome segments (deficiencies) is usually intended to remove specific genes and has been mentioned in Sections 5.1.2, 8.3.2.2.1 and 10.1. Removal of entire chromosomes results in monosomy or nullisomy which have their special applications in research and chromosome manipulation (Sects. 6.2.1 and 10.4.3). In this chapter reduction will be discussed only insofar as it concerns entire genomes, primarily in the form of the gametic chromosome number (Sect. 11.4).

The purpose of duplication of chromosome segments, chromosomes or genomes has two components: (1) *dose effects* and (2) the possibility for the stable *combination of different alleles* of the same genes in one genome. Both can be realized by the duplication of a single small or large chromosome segment, by the duplication of complete chromosomes or by the duplication of entire genomes. Single chromosome duplication for the purpose of inducing dose effects or heterosis effects is hardly ever an aim of a breeding program. Occasionally, chromosome duplication of entire (homoeologous) chromosomes in the form of a disomic (homozygous) addition has been suggested for incorporating alien genes into a cultivated species. In practice, however, this is not more than an intermediate step in gene transfer (Sect. 10.4.4.2). Entire extra chromosomes used to prevent gene transmission through the pollen with the purpose of making male sterile lines for hybrid seed production will be discussed in Section 12.4.

In a number of projects specific genes, or chromosome segments carrying specific genes, have been duplicated for the practical application of dose effects, but as yet with limited success. The duplicated segments may be identical and then have the same origin. They may also have a different origin, and then they can be heterotic. However, even when no specific heterotic allelic combination is planned, undefined, general heterozygosity has a positive effect on plant phenotypic performance, and heterozygous duplications are generally preferred over homozygous duplications, also when dose effects are the primary purpose of duplication.

When specific allelic combinations are required, the different copies have to be derived from specific different sources. Duplication of specific chromosome segments of limited size is considered in Section 11.2. Duplication of entire genomes (polyploidy) in autopolyploids, where no specific requirements for allelic combinations have been formulated, although a high general level of heterozygosity is usually necessary, is discussed in Section 11.3.1. Allopolyploids where specific allelic or even gene combinations tend to be the most important aspect of polyploidy, in addition to (or merely accepting the necessity of) genome duplication, are discussed in Section 11.3.2.

11.2 Duplications

11.2.1 Multiple Copy Genes

In nature, several eukaryotic genes are multicopy genes, either arranged in tandem (r-RNA genes at the NOR, histone genes, seed storage protein genes) or spread over the genome (t-RNA genes, some 5-S rRNA genes). This is necessary for the production of large quantities of the primary gene product which, in several of the examples mentioned, is also the end product. Amplification of normally single copy genes has been observed in cell cultures which have acquired resistance to an applied toxin. The best-analyzed example is that of methotrexate resistance in animal cell cultures (Cowell 1982). The gene for dehydrofoliase, an alternative target for methotrexate, amplifies continuously under a methotrexate regime, as long as the selective pressure continues. Amplification consists of repeated tandem duplication resulting in relatively stable, homogeneously stained megachromosomes, or in the form of small fragments (seen as "double minutes" in colchicine-treated somatic metaphase). Most of the double minutes are lost in each mitotic cycle because they lack a centromere. These small acentric fragments contain the large gene, a promotor and a regulator, an autonomous origin of replication and additional flanking sequences not necessary for gene function, but under the same amplification control system.

A comparable system in plants is glyphosate ("Roundup") tolerance in *Petunia* cell lines where elevated levels of EPSP synthase activity resulted from a 20-fold increase in gene copy number, established after step wise selection on survival in increasing concentrations of glyphosate. The amplified gene could be transferred by transformation using Ti plasmids into leaf disk cells, from which resistant plants were isolated via a callus phase (Shah et al. 1986). Other types of toxin resistance in plant cell cultures and their regenerants, which has led to resistance to certain pathogens, may also be due to gene amplification. Amplification of existing repetitive sequences in tissue culture, without selection, has been reported for instance in regenerated wheat/rye hybrids (Lapitan et al. 1988)

Spontaneous amplification of chromosome segments in the plant, not including specific functional genes has been reported for *Nicotiana* hybrids by Gerstel and Burns (1966). Telomeric heterochromatin is also known to be amplified or occasionally lost in plants (Gustafson et al. 1983). Multiple integration of concatenated genes into large blocks, or multiple integration spread over the genome is common with molecular transformation. In many cases it is a problem rather than an asset. The mechanisms behind these different forms of artificial gene or chromosome segment multiplications are not well known and cannot yet be controlled. Manipulation of gene amplification to create specific copy numbers that are most desired for specific functions has been possible only exceptionally, as in the case of the glyphosate resistance mentioned above, but this was a consequence of selection for an amplified function rather than planned production of a specific duplication. Chance successes are likely to precede any planned results.

Amplification through the generative cycle is believed to be possible by unequal crossing over when at least two extra copies are available. This is believed to occur in small duplicated segments in *Drosophila*, but it has not yet been shown to be effective in plants where even the amplification of heterochromatin is most probably a somatic phenomenon.

11.2.2 Duplication of Small Chromosome Segments

Duplications of chromosomal segments have been important in the evolution of most eukaryotes and their presence can be demonstrated not only by molecular means but also by the existence of duplicated genes and by pairing between homologous segments in haploids (Sect. 5.2.5). Several instances are known where duplicated genes have acquired new properties (Ohno 1970). It has been suggested that for plant breeding, especially for exploiting the effects of higher gene copy numbers, but also for combining different non-recombining alleles in one genome, the duplication of specific chromosome segments is a promising technique.

There are different experimental methods available for duplicating small chromosomal segments containing genes believed to make more gene product when present in multiple copies. The inevitable excess of undesired chromatin must generally be considered a disadvantage, but can be neutralized by rigorous selection for a favourable genetic background (Hagberg, pers. comm. 1985). However, every time the necessity arises to select a special genetic background for realizing conditions not directly relevant for the final objectives of the breeding program, the remaining genetic variation available for realizing these objectives is reduced. The starting material must, therefore, have a larger relevant genetic variation than otherwise considered necessary. This problem will be encountered again when in Chapter 12 the manipulation of the genetic system is considered. In addition, once a duplication has been introduced, it will be more difficult to adjust the genome to a subsequent duplication of a different segment. The simplest, direct way to produce a duplication is by asymmetric reciprocal exchange between homologous chromosome arms in the somatic tissue of a diploid plant (Fig. 5.1). Irradiation with ionizing radiation is the most appropriate technique. However, the probability of isolating the correct duplication with a dose that does not completely upset the genome is almost negligible, thus alternative methods have been developed.

The aberrant segregation of special translocations with one break point close to a chromosome end (Sybenga and Verhaar 1980), or the combination of two very similar reciprocal translocations (Gopinath and Burnham 1956 for maize; Hagberg 1965, Hagberg and Hagberg 1991 for barley in which numerous different small and large duplications have been produced) provides an opportunity to produce specific duplications. The way the duplications are produced is shown in Fig. 5.15 (Sect. 5.4.2.3). The technique is laborious when no or insufficient numbers of translocations are available, because only very specific combinations will give the desired duplications. In maize and barley, and to a more limited extent in rye and pearl millet and a few more species, such sets of translocations are available. In barley the duplication of the chromosome segment carrying the α-amylase locus has been realized with the purpose of improving the malting qualities (Hagberg 1965). Positive effects on other characteristics including yield have been reported (Hagberg and Hagberg 1991). The success was initially limited because of the negative effect of the excess of chromatin present in the duplication in addition to the gene considered, but there was a gradual improvement with selection for adapted genotypes.

A special variant of the use of translocations to duplicate chromosome segments is represented by translocations between B-chromosomes (Sects. 3.1.4.1.2.3 and 5.4.2.3) and (normal) A-chromosomes. Their use in gene localization has been mentioned in Section 8.3.3.2.2. When used to produce a chromosome segment duplication, a chromosome is added to the normal complement. Therefore, they will be discussed in Section 11.2.3.

Duplication of specific chromosome segments combining different alleles of specific genes into heterotic combinations is also of interest. This is probably so especially for self-fertilizing species which have few other ways of establishing heterotic gene combinations. It does not carry the risk of recombination which is inherent in normal disomic allelic combinations. The disadvantage of the excess of irrelevant or even deleterious material must again be overcome. Examples are not known, but there are indications that such specific heterotic duplications exist in nature in cross-fertilizing *Solanum* species (M.S. Ramanna, pers. comm. 1988). Duplications as such are very common in nature, and it is possible that in addition to creating a possibility for gene divergence and dose effects, heterosis is an important factor in their maintenance. Their detection as pairing regions in haploids, or as (groups of) morphological markers segregating as duplicate factors, or as duplicate isozyme loci in diploids, is considered in Section 6.1.1.4.

11.2.3 Addition of Entire Chromosomes

The addition of complete chromosomes as disomic additions has been mentioned as an intermediate in the transfer of small chromosome segments from one species to another (Sect. 10.4.4). As mentioned above, the direct use of a disomic addition either for dose effects, as carriers of specific genes or for heterotic effects is not promising. The partial duplication of a chromosome, with an increase in chromosome number is less deleterious, provided sufficient length is maintained for regular meiotic behaviour. Telocentrics derived from metacentric chromosomes are either too small to pair regularly or they are too large to be acceptable as a duplication. Tertiary and compensating trisomics (Khush 1973; de Vries 1984) are a theoretical possibility, because their composition can be manipulated within certain limits. If two identical extra chromosomes (or compensating sets in the case of compensating trisomics) are present, their meiotic behaviour is stabler than with only one copy (or set), but these together represent excessive amounts of extra chromatin. As a single copy (or set) they require specific balancing systems (Sect. 12.4.2.2). Instability remains a problem and is often too pronounced for practical application. Generally, when a chromosome is large enough to function in meiosis, it is too large for a duplication.

A special form of a tertiary trisomic which is large enough to pair regularly and which contains small enough amounts of active chromatin for proper functioning, is a segment of a normal chromosome translocated into a B-chromosome. The origin is a normal reciprocal A-B translocation as can be induced in any species where B-chromosomes occur or can be introduced. The main desired effect can be a gene dose effect or an interallelic interaction. Intricate combinations can incorporate specific chromosome segments of the desired size, and they can be made "homozygous" (present in a double dose) for meiotic stability (Carlson and Curtis 1986). When the duplicated A-chromosome segment is large, some multivalent formation may occur. If the accumulation properties of the B-chromosome are maintained, a higher level of duplication can be created, as shown by Shadley and Weber (1986) who duplicated an oil factor locus in maize. It is questionable whether the stability of the system will then be satisfactory. The translocation of a small specific segment of an A-chromosome into a B-chromosome is not a frequent event and practical results have not been reported. A-B translocations have been used effectively in maize for gene localization (Shadley and Weber 1986). In addition to maize (see, for instance Roman 1947; Birchler 1983, 1991; Beckett 1991; Carlson and Roseman 1991), A-B translocations have been reported for a few more plant species (*Lolium*: Evans and Macefield 1977; rye: Pohler and Schlegel 1990), but here they have not yet been used to make duplications.

11.3 Increase in the Number of Genomes: Polyploidy

Polyploidy is the most important form of duplication. The number of genomes can be manipulated during somatic development, making use of external agents or of random or systematic deviations from the normal mitotic cycle. The generative cycle can also be used to manipulate the genome number, but in quite a different way. Both approaches have been reported in Section 6.1.2. The approach chosen depends on the objectives of the program, on the possibilities the material offers and the facilities available. It appears that different methods may yield qualitatively different results. Because of the maintenance of higher levels of heterozygosity, meiotic doubling is preferred (Hermsen 1984), but usually serious technical difficulties are encountered.

The most complete review of polyploidy in plants and its evolutionary significance is given by Gottschalk (1976). Polyploidy, in general, including animals has been reviewed by Lewis (1980). The decreasing interest in polyploidy in plant breeding is the reason that the number of publications appearing since then is rather limited.

11.3.1 Autopolyploidy

11.3.1.1 Triploids

Autopolyploidy breeding primarily involves triploids and tetraploids. The greatly reduced fertility of *triploids* (Sects. 6.1.2.2.1.3 and 6.1.2.2.1.4) makes them quite valuable for fruit crops where seeds are undesired. The banana is the best-known natural example. Diploid bananas are well known in their countries of origin, but have a very limited application as the fruits are not edible because of the numerous large and hard seeds. All widely distributed cultivars are triploids and almost never produce seeds. They must be parthenocarpic in order to form fruits. Breeding bananas is obviously quite difficult and most commercial varieties have been found as spontaneous triploids (Simmonds 1976). To create new varieties two diploid varieties can be crossed, and with meiotic doubling in one of the parents, triploid progeny is formed. Finding suitable diploids, however, is a major problem. The triploids are normally too sterile to be used as parents, but a cross between a diploid and an artificial tetraploid is in principle possible.

For the same reason (pronounced sterility) artificial triploids have been proven successful in watermelon (*Citrullus*), where the seeds, although not as detrimental as in the banana, are still undesired (Kihara and Nishiyama 1951; Andrus et al. 1971). Watermelon, unlike the banana, is not vegetatively reproduced, and requires the production of triploid seeds. Using male sterile genotypes, hand-crossing a tetraploid with a diploid variety is an acceptable procedure because of the large numbers of seeds produced (if the cross-combination is sufficiently fertile) and the high production per plant. At the same time, this produces a hybrid variety.

Among ornamentals, the triploid tulip is presently a well-known example, but there are more. Sterility is an advantage because flowers that do not set fruit stand longer. In the mulberry tree, spontaneous and artificial autotriploid varieties have definite advantages over diploids (see, for instance, Basavaiah et al. 1990). The use of autotriploids in field crops is still limited, even though occasionally important. It appears that if the cross between a tetraploid and a diploid is fertile and the triploid can be made on a scale like any commercial hybrid variety, it can produce a variety which surpasses both the diploid and the tetraploid. In addition, the triploid is necessarily a hybrid, and the parents can be selected on the basis of combining ability.

Because the triploid itself is usually completely sterile or almost so, it can only be used for crops in which fruits are set parthenocarpically or in which the vegetative part is the main product. This is so, for instance, in sugar beet (*Beta vulgaris*) where the triploid has become very successful (Savitsky 1962). It is not certain whether the superiority of the triploid is primarily due to the ploidy level, or to the fact that at the same time it is a hybrid between two genetically different and highly selected strains. In sugar beets the triploid has to be reconstructed each time again, like any hybrid. Cytoplasmic male sterility (CMS) is applied to realize this. Using the tetraploid as the female, the haploid pollen of the pollinator has an advantage in pollination over the own diploid pollen of the tetraploid. This can result in relatively high levels of triploid hybrid seeds even without male sterility, but the admixture of tetraploids is still usually unacceptable. Also, using the tetraploid as the female provides only a coarse sieve against aneuploids. With the tetraploid as the male parent, aneuploid spores do not compete effectively with euploid spores, and the frequency of aneuploids is much lower, but still not negligible. Aneuploid plants tend to have a seriously reduced productivity compared with that of euploids, and they cause morphological heterogeneity which is unfavourable for modern cultivation techniques. In grass species only a fraction of the original seedlings contribute to the final stands and inferior genotypes like aneuploids are automatically eliminated. In contrast, with modern seeding techniques sugar beet seedlings immediately occupy their final position in the field, and cannot be replaced by better genotypes.

Breeding diploids, including diploid hybrids, proceeds faster than breeding the proper tetraploid-diploid combination, especially when important recessive characters have to be built in like monogerm seed and specific disease resistances. In addition, it appears to be difficult to reduce the frequency of aneuploids among the gametes of the tetraploid parent. Because of the success of diploid varieties, the interest in triploid sugar beets is decreasing. When the increase in productivity and adaptability of the diploids levels off again, there may be a new opportunity for triploids in sugar beets, starting from the improved diploids. The genetic background should be made sufficiently wide, however, because triploid breeding requires a different and probably wider variability than diploid breeding. The advantage of triploids is not just in higher productivity, but also or perhaps primarily in more specific morphological characteristics like beet shape, etc. This may contribute to giving them a new chance in the future, unless such characteristics can also be readily bred into the diploids.

It should be noted that for the production of triploids through the cross between tetraploids and diploids, incompatibility due to endosperm abortion or other developmental irregularities (triploid block: Sects. 6.1.2.2.1.1 and 11.3.1.2.1.2) may be a serious bottleneck. Although there is usually some genetic variation in the level of sterility, in species with the triploid block, satisfactory tetraploid-diploid fertility will be difficult to realize.

Triploids would be expected to be successful also in field crops that, unlike the sugar beet where the hybrid has to be reproduced every time, can be reproduced vegetatively. In potatoes (presently a tetraploid), casava (presently in practice a diploid, where promising artificial tetraploids and triploids have already been made: Hahn et al. 1990), *Allium* spp. and perhaps some more, triploidy may well be applied.

11.3.1.2 Autotetraploids

11.3.1.2.1 Induction

11.3.1.2.1.1 Somatic Induction

Although the practical interest in autopolyploidy is not restricted to tetraploidy, doubtlessly the most generally used artificial polyploid is the autotetraploid. Triploids have their special applications (Sect. 11.3.1.1), but even these often have their origin in induced autotetraploids, hybridized with diploids.

Spontaneous polyploidy as a consequence of endoreduplication or endomitosis, either systematic as part of normal development or as an incidental error, has been reported in Section 3.1.4.1.2.1. The chance that a cell, which is polyploid as a result of systematic endopolyploidization in a differentiated tissue, will develop into a shoot from which a new plant can develop, is small. Also, spontaneous polyploidy from an incidentally doubled cell in a regular growing point is a rare phenomenon. Occasionally, however, a relatively simple manipulation will enable polyploid shoots to arise from endomitotically doubled tissues. When tomato stems are cut and allowed to generate callus, among the shoots that are formed on the callus, about 10% are tetraploid (Gulcan and Sybenga 1967). It is not entirely certain whether these shoots have developed from initially tetraploid cells or from cells doubled during the callus phase. The same uncertainty exists when tetraploid regenerants are found after in vitro culture of protoplasts derived from leaf mesophyll. As discussed in Section 3.1.4.1.2.1, it can sometimes be shown that already in the first mitosis in the protoplast duplochromosomes (endoreduplication) or tetraploid cells (endomitosis) are present (Pijnacker and Ferwerda 1990). Flow cytometry of leaf mesophyll nuclei has shown considerable frequencies of tetraploidy and even higher ploidy levels. On the other hand, polyploidization during a callus phase is also common (Pijnacker et al. 1986; Ramulu 1987).

In vitro culture specifically aiming at chromosome doubling has been applied to monohaploid and dihaploid potato clones which did not respond well to other treatments (Ramulu 1987). Also, for doubling of haploids obtained from microspore cultures in anthers or in suspensions, a callus stage as an almost inevitable intermediate often results in "spontaneous" doubling. However, it is possible that in such cases the original cell had the doubled chromosome number to begin with, as a result of meiotic failure of cytokinesis. In addition to the possible technical difficulties involved in regenerating polyploids after spontaneous in vitro genome doubling in some species, there is the major disadvantage of high levels of spontaneous mutation, chromosome rearrangements and aneuploidy.

If all goes well with in vitro somatic doubling, the result is a simple duplication of the existing genomes. It is comparable to somatic genome doubling by chemical treatment as discussed below. This does not present the optimal level of heterozygosity. Chromosome doubling by fusion of genetically different protoplasts from the same species results in a much higher initial level of heterozygosity and has indeed been found a good way of inducing autopolyploidy in potato (Deimling et al. 1988), provided it is not accompanied by undesired somaclonal variation and undesired additional spontaneous doubling.

To double the chromosome number usually externally applied agents are used rather than proliferation of spontaneously doubled cells or a callus phase. By far the most generally effective agent is colchicine in an aqueous solution of about 0.05 to 0.5%. It is not an alkaloid as suggested occasionally but nevertheless very poisonous. It is found in most species of the genus *Colchicum*. It has been known as a medical drug for centuries (Eigsti and Dustin 1955). Its use for doubling chromosome numbers in plants dates from 1937 when Blakeslee and Avery (1937), and independently Nebel (1937), obtained tetraploid plants after colchicine treatment of seedlings and sprouts. In natural stands of *Colchicum*, however, no higher than normal frequencies of polyploid forms are found (Sybenga 1956).

Chromosome number doubling with the use of colchicine can be realized in principle in all tissues from which root-forming shoots can develop. Quite often germinating seeds or young seedlings are used because they grow fast and recover readily from the treatment. Roots tend to be rather sensitive and it is often favourable to protect them from the agent. The clipped young leaves of tillers of grassy species, including cereals, when at the vegetative stage, can be treated by capping them with small vials with a colchicine solution. The opening should be narrow to prevent the solution from leaking away. Treatment of axillary buds with colchicine-soaked cotton wool plugs after the buds have first been stimulated to proliferate, is also often quite successful, especially in vegetatively reproduced species like cassava (own unpublished results).

There are several more substances with effects similar to that of colchicine, but few cause so few side effects even when washed out thoroughly from the plant tissue. A successful alternative, especially for tissues in which colchicine does not produce the desired results, is N_2O (laughing gas) applied under pressure in special tanks (Zeilinga and Schouten 1968). It has been used on bulbous ornamentals just after fertilization, which results in large, doubled segments or even entire seedlings. The complete, grown plant can be treated and after the pressure is released, only very low concentrations of the gas remain. Among the more recently described mitotic spindle-disturbing substances, several of which are fungicides or herbicides, none are very suitable for chromosome doubling. An exception is oryzaline, which acts as a potent doubling agent, especially in in vitro cultures (Verhoeven, pers. comm.). For other polyploidizing treatments, the reader is referred to Gottschalk (1976).

Treatment of complex growing points leads to chimeras of doubled and non-doubled tissues: *mixoploidy*, because only cells mitotically active at the moment of treatment have a chance to have their chromosome numbers doubled. Usually, the tissue affected is thoroughly disorganized, since the microtubules on which colchicine acts do not only play a role in mitosis, but also in tissue organization (Sect. 3.1.1). The result is often a callus-like swelling of the treated plant organ, which may remain dormant for some time until sprouts arise from it. It is probable that at least some of the irregularities in chromosome number and occasionally in chromosome structure, and the formation of mutants in addition to chromosome number doubling, have their origin in this callus phase, in a way similar to abnormalities found in calluses after in vitro tissue culture. The frequency of aberrations is usually much lower than after an in vitro callus phase.

Induction of polyploidy by simply doubling the chromosome number results in two identical sets of chromosomes, implying a high degree of homozygosity. Because polyploids profit even more from a high level of heterozygosity (Bingham 1980) than diploids, and in addition require large genetic variation for selection of adapted genotypes, high levels of homozygosity are a serious drawback that can only be overcome by repeated hybridization of genetically different raw tetraploids. Doubling by somatic fusion is a potentially promising possibility in a limited number of instances (see above) and maintains all heterozygosity present. In practice, it has not been applied for this purpose, but it should receive more attention.

11.3.1.2.1.2 Meiotic Doubling

A more generally recommended technique for creating genetic diversity and maintaining heterozygosity is meiotic doubling (Skiebe et al. 1963; Hermsen 1984; Veronesi et al. 1986). Unlike somatic methods where the existing genomes are simply multiplied, with meiotic doubling both genomes of the two parents are combined, and the level of heterozygosity of both diploid parents is, at least partly, maintained (Fig. 11.1). However, from a computer simulation Watanabe et al. (1991) concluded that the advantage is not always as great as expected. Especially the advantage of FDR (first division restitution) with maintenance of most of the original heterozygosity, compared to SDR (second division restitution) where much lower levels of heterozygosity are maintained, may in practice not always be fully exploited.

Meiotic doubling is applied most readily in species with a relatively high frequency of unreduced gametes. For the direct production of tetraploids the method is usually not well suited because simultaneous occurrence of unreduced gametes in both parents is quite rare. Certain genotypes that predetermine high frequencies of unreduced gametes may result in sufficiently high frequencies to be used for direct tetraploidization. Selection for such genotypes in red clover (*Trifolium pratensis*) using tetraploid x diploid test crosses, leads to the isolation of genotypes that could produce at least some tetraploids directly (Parrott et al. 1985). In other species, too, some genetic variation in the frequency of the formation of unreduced functional gametes has been reported (for instance alfalfa: Veronesi et al. 1986). By determining the frequency of diploid pollen by flow cytometry, it is relatively simple to select genotypes that have sufficiently high frequencies of unreduced pollen (van Tuyl et al. 1989). There is a certain danger of restricting the genetic variation in the progeny when the frequency of genotypes producing sufficient numbers of functional unreduced gametes is small. It is always necessary that the tetraploids recovered are derived from a sufficiently large number of genetically different parents.

A number of specific genes are known that in a recessive condition cause high frequencies of unreduced gametes. The *ps* ("parallel spindle") gene in diploid (2n = 24) potatoes (*Solanum tuberosum*) makes the second meiotic spindles in pollen mother cells lie parallel, in contrast to the normal perpendicular orientation. Because of the absence of cell wall formation after the first meiotic division in PMCs (common in Solanaceae), this results in the fusion of the two spindles and restitution of the sporophytic chromosome number by a process which is effectively first division restitution, FDR (Peloquin et al. 1989). There is only little loss of heterozygosity as a consequence of genetic exchange. The sister chromatids of all chromosomes separate, but the chromatids derived from different chromosomes are brought together again, restoring the original allelic combinations. Only segments distal to the chiasmata can become homozygous, but only in 50%. In a diploid hybrid of potato (*Solanum tuberosum*) with *S. chacoense* recombination was reduced naturally by about 35%, resulting in almost 90% maintenance of heterozygosity (Douches and Quiros 1988). In a *ps.ps* background this resulted in highly heterotic diploid gametes formed at a high frequency. The frequency of *ps* or comparable genes with practically useful frequencies of unreduced pollen (>1%), in cultivated and related wild tuber-bearing *Solanum* species, is

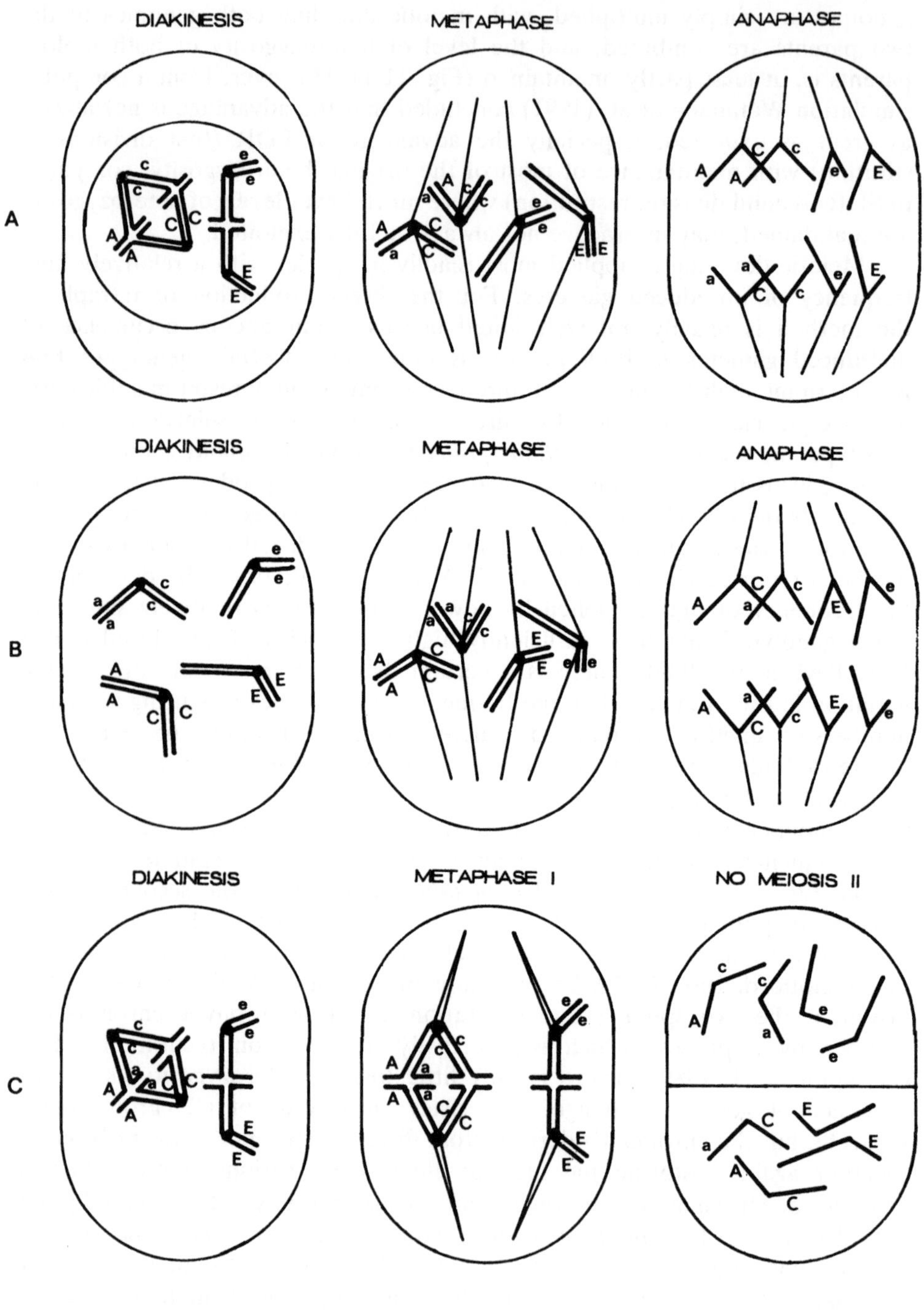
DIAKINESIS
METAPHASE
ANAPHASE
A
DIAKINESIS
METAPHASE
ANAPHASE
B
DIAKINESIS
METAPHASE I
NO MEIOSIS II
C

Fig. 11.1A–C

variable but rather high. If a Hardy-Weinberg equilibrium is assumed, the frequency can be estimated to vary between 0.4 and 0.9 (Watanabe and Peloquin 1989). The expression, however, of these genes is far from complete. An effect comparable to that of *ps* results from the action of the "elongate" (*el*) gene in maize (Rhoades and Dempsey 1966), and the "triploid inducer" (*tri*) in barley (Finch and Bennett 1979). These do not appear to have been applied in practice.

In the female, the *ps* gene does not operate, because the EMC forms a linear instead of a circular tetrad. Here, desynapsis may play a similar role, although through a different process. Using a desynaptic mutant strongly reducing recombination (almost 90% reduction) and at the same time inducing first division restitution (FDR) by lack of a sufficient number of bivalents, heterozygosity could be maintained to up to 98% (Jongedijk et al. 1991a).

The reason why some species or genotypes within species have a greater tendency to FDR, not considering specific monofactorial mutants like *ps*, is complex. Jongedijk et al. (1991b) suggest that in EMCs, where *ps* does not operate, the imbalance between different chromosomal and cellular meiotic processes plays a role in the amphitelic orientation of univalents caused by desynapsis in the potato. Together these result in complete restitution of the original diploid complement in the two nuclei formed after the aborted first division (cf. Sect. 12.5). Precocious centromere separation was also observed by Ohri and Zadoo (1986) in open (rod) bivalents in a number of PMCs in *Bougainvillea*. In telophase the chromatids of the segregating chromosomes fell apart, second division was absent, and functional unreduced spores were formed. This is equivalent to FDR. Among the progeny triploids were observed with considerable frequency, and even tetraploids, suggesting that restitution occurred also in the EMCs.

Disturbance of normal meiotic processes, due to environmental or genetic causes, including hybrid imbalance, is a major cause of restitution. Complete

Fig. 11.1 Nuclear restitution in meiosis in embryo sac mother cells. **A** First division restitution (FDR). Normal exchange recombination, but no segregation at anaphase I, which is replaced by an AII-like division separating the chromatids. The new nuclei are genetically identical to the original diploid ones except for segments distal to the point of exchange. Here, homozygosity occurs in 50% of the cells when a chiasma has formed: the level depends on the chiasma frequency. The final result is like that of *fs* (fused spindles) in PMCs. The phenomenon is infrequent in normal synaptic plants.

B First division restitution (FDR) after desynapsis (or asynapsis). Metaphase I and anaphase I are impossible and restitution follows. The chromatids separate in a meiosis II-like division. The gametes are unreduced and retain the complete genetic make-up of the diploid on which they are formed. With desynapsis following some exchange recombination, homozygosity as in **A** may result, but is limited.

C Second division restitution (SDR). The chromosomes segregate at anaphase I, but then split precociously. There is no second division, and the diploid chromosome number is restored. Only the segments distal to the chiasmata can remain heterozygous and thus considerable homozygosity results. More than one chiasma in a chromosome arm makes the situation more complicated. This is the most common type of restitution in absence of desynapsis absence of the second division, with a prolonged interphase, results directly in unreduced (SDR) spores (Werner and Peloquin 1987).

The use of specific genotypes with high levels of formation of unreduced gametes can be quite valuable (Jacobsen 1980) and has been shown to give results in several cultivated species (Peloquin et al. 1989). They may not act again at the tetraploid level, however, which would result in undesired high-level polyploids. In vegetatively reproduced crops like the potato, this is not a disadvantage. For seed-reproduced crops the use of recessive alleles, inducing a moderate level of first division restitution, is the most favourable: they are not expressed in the hybrid, and tetrasomic inheritance prevents high frequencies of homozygous recessives to appear. They may show up occasionally, however. When used for the induction of apomixis, expression at the tetraploid level (in tetraploid crops), however, is essential (Sect. 12.5).

In most cases the starting point in programs of meiotic doubling is an existing or somatically induced autotetraploid which is crossed with a good, heterozygous diploid (Jacobsen 1980). The frequency of triploids arising from such crosses is often low because of an incompatibility between the embryo, the endosperm and the maternal tissues when these do not have the standard chromosome numbers: *triploid block* (Sect. 6.1.2.2.1.1). There is very little viable progeny, but among these most are tetraploids, resulting from fertilization involving an unreduced gamete. The use of a male-sterile tetraploid as female parent of the cross simplifies large-scale hybridization, as shown by Negri and Veronesi (1989) for *Lotus corniculatus*. Here, an effective triploid block was operational. When the diploid is used as the male parent there is the disadvantage of competition between normally reduced pollen and pollen with the sporophytic chromosome number, but often the advantage of better development of the embryo and endosperm. This may compensate for the competitive disadvantage.

When there is no triploid block and triploids are recovered in large numbers, the direct use of this meiotic doubling method is limited, but triploids can, in turn, be used to produce new tetraploid hybrids. It has been observed for example in rye (own unpublished observations) that the hybrid progeny derived from the cross of a triploid and a tetraploid has chromosome numbers near the tetraploid level.

Whenever possible, meiotic induction of tetraploidy should be applied instead of colchicine treatment in order to reach as high levels of heterozygosity as possible directly from the start. However, the difficulties involved, notably the generally low frequency of success, are not negligible. Somatic fusion, if possible, is an alternative, but for most plant breeding institutions this will only very infrequently be a practical and economical alternative.

There are not many quantitative reports on the actual merits of different methods of induction of autotetraploidy. VanSanten and Casler (1990) noted good performance of 4x progeny from interploidy crosses in *Dactylis*, in forage quality and yield.

11.3.1.2.2 Effects of Autopolyploidy

11.3.1.2.2.1 Somatic Effects

Soon after its discovery as a polyploidizing agent, the use of colchicine for making tetraploids expanded tremendously, as it was assumed that the simple fact of doubling the chromosome number offered great new opportunities for improving commercially used plant species. Although several important crop species are natural autopolyploids (Gottschalk 1976), it appears difficult to produce good, new, artificial autopolyploids that can compete with their diploid progenitors. However, there are several examples of artificial tetraploids that gradually became superior in at least some important characteristics. The main distinctive somatic features of autopolyploids are:

1. Gene dosage effects, some of which are specific (including increased content of secondary metabolites), others more general (morphological changes, lateness). Some of these effects may be positive, others decidedly negative in new, still unselected autotetraploids. Reduced fertility, resulting not only from meiotic irregularities, but from gene dose imbalance, is a common effect (Bosemark 1967; Gottschalk 1976; Evans and Rahman 1990); then the correlation between irregular anaphase I segregation and fertility may not or hardly be significant.
2. Effects of changed nuclear and cell size, primarily resulting in "gigas" characteristics, but also in other changes in plant morphology.
3. Specific interallelic interactions.
4. General interallelic interactions (heterosis), which are different in polyploids because more, different alleles may be present simultaneously.

In the 1940s and 1950s many cultivated plant species had their chromosome number doubled. Quality, productivity, morphology, content of secondary metabolites and resistance to stress, pests and diseases were expected and seen to improve. Although many tests of the practical use of polyploid plants have been disappointing, there are still a sufficient number of instances with positive results to make induced polyploidy of considerable interest as a plant breeding tool, provided the set-up of the program is sufficiently broad and includes sufficient testing and scientific monitoring. The original belief that polyploidy as such, especially autopolyploidy, would have a positive, practical effect on productivity because the plants tend to be larger, was soon refuted. In practice, the expected increase in dry matter yield is disappointing and the often seriously reduced fertility is a great disadvantage, especially for crops where the seed is the economic product. Outbreeders seem to be less sensitive to the negative effects of autopolyploidy than inbreeders (see, for instance the review of Friedt 1985). The causes for this reduction in fertility, with emphasis on the meiotic factors were discussed in Section 6.1.2.2.2.4. The physiological causes of reduced fertility are often similarly negative. Meiotic irregularities further cause aneuploidy in the progeny, which may have quite undesirable

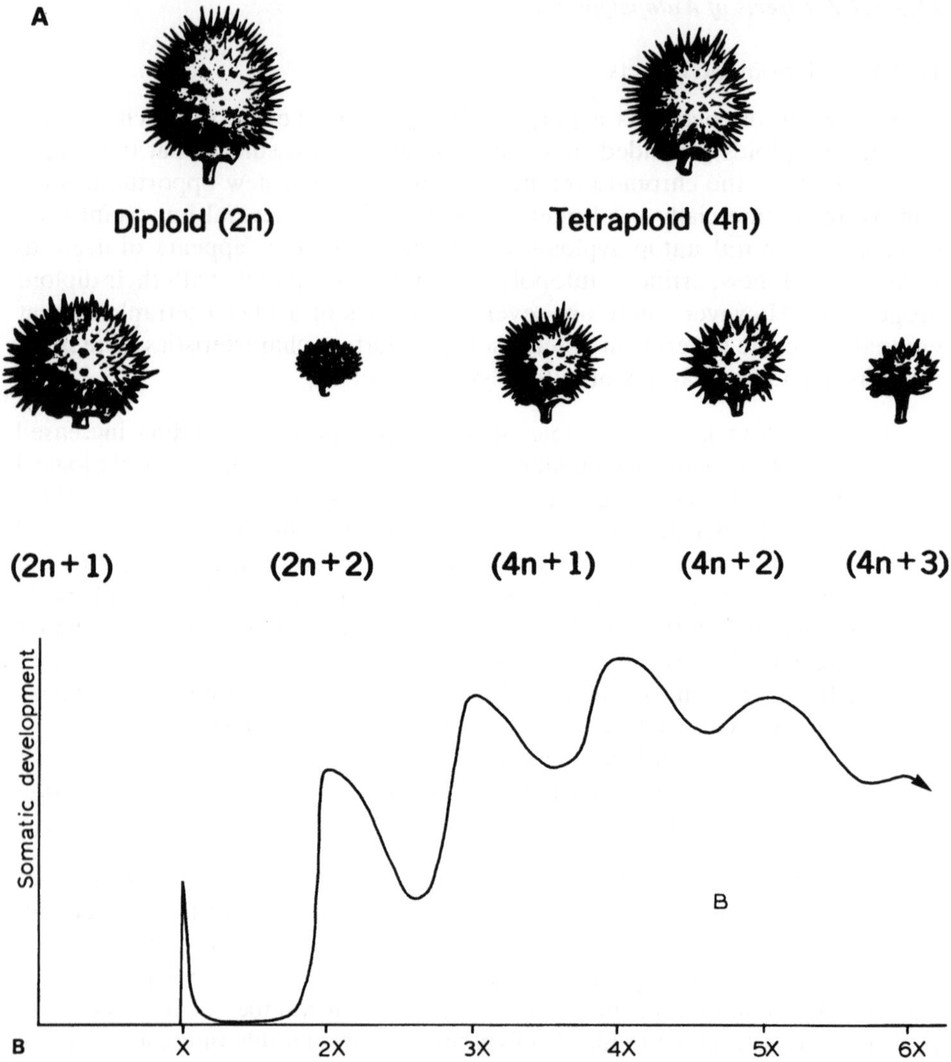

Fig. 11.2 A The effects of aneuploidy and polyploidy on the morphology of the seed capsules of *Datura stramonium* (cf. Sybenga 1972; Avery et al. 1959). **B** The combined phenotypic effects of polyploidy and aneuploidy (generalized)

consequences. Aneuploidy, however, has less effect the higher the number of genomes.

A generalized relation between the level of polyploidy and aneuploidy with vegetative vigour is given in Fig. 11.2. The sensitivity for aneuploidy is the greatest at low ploidy levels, and decreases when the number of genomes increases. According to Schwanitz (1967), Levan found for sugar beets in 1942 that aneuploidy had a very pronounced effect even around the triploid and tetraploid level. This is of considereable importance for polyploid sugar beet breeding, as discussed in Section 11.3.1.1. Above a certain optimal number of genomes, the vegetative vigour decreases again. Sugar beets, for instance are generally assumed to have their optimum at the triploid level, if not for productivity, then for specific morphological and processing reasons. It is difficult to assess this effect because most triploid sugar beet varieties profit from the hybrid vigour resulting from combining a diploid with a tetraploid strain of relatively different origin. As discussed earlier, the diploid varieties and hybrids tend to be developed faster, with the result that triploids are more and more replaced by diploids again. A similar situation is found with ryegrass (*Lolium perenne*) where tetraploid varieties have been very popular for their better palatibility, and the choice of disease-resistant initial material used for doubling, less so for higher production. Because of problems with seed production, triploid varieties are not in use with ryegrass. The less rapidly realized breeding results are a disadvantage, neutralizing the initial advantage of polyploidy. Very large-scale breeding of tetraploids might in the long run lead to results that surpass those of the successful diploids, but may not be economically justified. When diploid breeding levels off again, the most successful stocks may be tested for a possible new potential at the polyploid level. However, it is not impossible that at a certain point a high level of specialization at the diploid level makes highly bred varieties less successful as parental material for tetraploid varieties. However, little is known of such effects.

The advantage of polyploidy is usually not primarily the higher production of dry matter per unit area, although this may not be excluded. The example of tetraploid ryegrass has been mentioned above: better palatibility is more important than higher production. In sugar beets, root shape is improved in polyploids compared to diploids (Sect. 11.3.1.1). The seeds of diploid *Psathyrostachys juncea* (Russian wild rye, an important forage grass in northern America) are too small for the establishment of a good stand. Tetraploid seeds were larger and produced robuster seedlings (Wang and Berdahl 1990). In medicinal crops, the content of the active substance may be considerably increased (Gottschalk 1976). More recent examples are different species of *Hyoscyamus* (Lavania 1986; Lavania et al. 1990; Srivastava et al. 1992) where a yield increase of almost 45% was reported to result from a combination of increased dry matter production and increased alkaloid content. Many other examples could be given, but for further details on the practical aspects of polyploidy breeding other than the cytogenetic complications, the reader is referred to specialized literature (e.g. Gottschalk 1976).

11.3.1.2.2.2 Effects of the Tetrasomic Genetic System

In Section 6.1.2.2.2.4 the genetic consequences of the tetrasomic system of autopolyploids have been discussed. For plant breeding the most obvious consequences are related to the slow reaction to selection and delayed genetic stabilization, in addition to the typical physiological effects and gene interactions as discussed above. The consequences of irregularities at meiosis are not to be considered effects of the tetrasomic system as such. In their, *Quantitative Genetics and Selection in Plant Breeding*, Wricke and Weber (1986) devote a detailed section to autotetraploids. Glendinning (1989).discusses developments in autotetraploid population dynamics, including inbreeding.

11.3.1.2.3 Reproductive Stabilization of Autotetraploids

11.3.1.2.3.1 Reduced Fertility and Segregation of Aneuploids

As discussed in Section 6.1.2.2.2.3 meiosis of autotetraploids tends to be irregular, although with considerable differences between species and even between genotypes. The meiotic irregularity results in aneuploid gametes which may not function at all or which function less well than euploid gametes. This leads to reduced fertility in addition to gene balance-dependent physiological causes of reduced fertility as mentioned in Section 6.1.2.2.2.4. Both effects are highly undesired. For vegetatively reproduced species where the vegetative parts are the main commercial product, meiotic imbalance is not a major problem, but for species where the seed is the main product, sterility is serious. Even for species where the vegetative parts are the product, reproduction by seed is the usual mode of reproduction for which reduced fertility is a disadvantage. Segregation of aneuploids is almost invariably undesired because they tend to be less productive than euploids and yet occupy space in the field and all investments spent on euploids are equally spent on aneuploids. There is a difference between crops in the effect aneuploids have on overall productivity. When single plants are the basis of the productivity of the crop, occupying fixed positions in the field, as for sugar beets, potatoes, and many species of vegatables, low productivity of individual plants is directly noticeable. In a field composed of plants that are sown at an excess rate and subsequently fill the space in competition, such as most grasses, aneuploids tend to be eliminated or at least reduced in space occupied by a natural process of competition.

Klinga (1987a,b) studied the competitive interactions between euploids and aneuploids in relation to productivity of the artificially induced tetraploids of the cultivated grass species *Festuca pratensis*, *Lolium multiflorum* and *Lolium perenne* in perennial field swads. In the *F. pratensis* variety tested almost one-third of the seeds was aneuploid, 12% hypoploids and 18% hyperploids. Hypo- and hyperploids did not appear to differ statistically in behaviour and productivity. Both were eliminated rapidly with the normal, high sowing densities and thus did not affect productivity seriously after establishment of the perennial swad. In a more detailed experiment the chromosome numbers of individual plants were determined, the plants were cloned, and artificial mixtures were made in the field with somewhat lower densities than normal.

In pure aneuploid stands survival of aneuploids was as high as that of euploids, but in mixed stands the aneuploids tended again to be eliminated, but at a later stage. As a consequence, relatively more euploids, including the poorer types, survived. This resulted in a lower average productivity of the euploids as well as the aneuploids. Later, the frequency of aneuploids remained rather stable, and the difference in productivity disappeared. Not all aneuploids were apparently much less productive than the euploids. In such cases selection for low aneuploid frequencies will not be of as much importance as in sugar beets and comparable crop types.

In earlier experiments similar general conclusions on the limited role of aneuploids in dense stands of induced autotetraploids have been reported (Barcley and Armstrong 1966; Ellerström and Sjödin 1966; Norrington-Davies et al. 1981). Aneuploids produce higher aneuploid frequencies in their progeny than euploid tetraploids and a gradual increase in the frequency of aneuploids in a population may at some point become undesirable even in normally aneuploid-tolerant crops. The reason is not that aneuploids necessarily have a less regular meiosis than euploids (Schlegel et al. 1985), but simply that the chromosome involved has a much greater than average chance to be involved in aneuploidy again.

There has been great interest in the normalization of meiosis of autotetraploids by selection. Reduced fertility, insofar as it is not caused by physiological disturbances resulting from gene imbalance, and aneuploidy have their origin in the same meiotic irregularities. There are mainly two types: *reduced chiasma frequency* resulting in the formation of univalents and *unequal segregation* of the four chromosomes of a *quadrivalent*. As discussed in Sections 6.1.2.2.2.3 and 6.1.2.2.2.4, a tetrasomic quadrivalent gives a balanced segregation when two chromosomes move to each pole, independent of whether the orientation is alternate or adjacent. Adjacent orientation, however, is closely related to linear orientation which has a considerable probability of unequal segregation, and species with a high frequency of adjacent orientation often also have at least some linear orientation. Therefore, predominantly alternate orientation tends to result in much better fertility and lower aneuploid frequency in the progeny than without this preference. In fact, in species with predominantly alternate orientation fertility can even be increased by increasing the frequency of quadrivalents insofar as quadrivalent formation is dependent on chiasma frequency. Roseweir and Rees (1962) found selection for fertility in induced autotetraploid rye to be accompanied by an increase in chiasma frequency which in turn resulted in an increase in quadrivalent frequency. It should be noted that the correlation between an increase in fertility and an increase in the frequency of chiasmata has an additional reason: both react positively on a general improvement of genotype and environment. Hossain (1976), also for rye, reported an increase in the quadrivalent frequency from 6.2–7.7 per cell to 10.7, a decrease in trivalent frequency of 1.4 to 0.7, and a decrease in bivalent frequency from 18.9 to 16.9 in a period of almost 20 years. The number of chiasmata per bivalent had increased but not the number of chiasmata per quadrivalent; fertility had much improved.

There are more examples where quadrivalent frequency and fertility are not correlated. In established, natural autotetraploids of *Plantago media* (4x = 24) the frequency of quadrivalents is lower than than in the recently colchicine-doubled *P. media*, but fertility is approximately the same (van Dijk and van Delden 1990). In natural polyploids reduced seed set was compensated by a larger number of ovules, compared to recent polyploids. Wang and Berdahl (1990) found that the fertility of artificial autotetraploids of Russian wild rye (*Psathyrostachys juncea*) did not suffer from a relatively high quadrivalent frequency, and was hardly lower than that of the diploid parents.

From the various reports on the correlation between chiasma frequency and quadrivalent frequency, it appears that in general there is an increase in fertility when the number of univalents decreases. However, there is often a fertility decrease with a further increase in quadrivalent frequency because of the usually, not strictly 2:2 segregation of quadrivalents.

It is clear that if quadrivalents occur, their orientation plays an essential role in fertility. The study of multivalent orientation has advanced furthest with respect to interchange heterozygotes (Sybenga and Rickards 1987), but most of the results are directly applicable to polysomic quadrivalents, with the difference that in the latter case the point of partner exchange is variable. Chromosomal characteristics (arm ratio, chiasma frequency and localization, etc.) play a role in multivalent orientation, but also less easily definable factors, including the characteristics of the centromere, play a role. In addition to its pronounced effect on amphitelic orientation in univalents, causing meiotic doubling in female meiosis, precocious centromere separation causes malorientation of multivalents (Janicke and LaFountain 1989). Several more examples of theoretical interest could be given, but are not necessary here.

Genotype and environment not only affect fertility directly by their effect on meiotic behaviour and physiology of the reproductive process, but also indirectly by affecting the survival of aneuploids among the progeny. Ising (1967) found F1 plants of autotetraploid barley with strongly expressed hybrid vigour to be more fertile than weaker tetraploid hybrids and homozygous lines. In addition to physiological effects, this could be directly correlated with the frequency of aneuploids in the germinating seed, which was higher in the progeny of the more fertile hybrid. A tetraploid variety of rye, which had only approximately 65% seed set in Sweden, had more than 90% seed set in California, where the climate was much better for seed development. The increased seed set resulted from an increased frequency of surviving aneuploids (A. Müntzing pers. comm. 1965).

Increased fertility with a decrease in quadrivalent frequencies, not due to a reduction in the number of chiasmata per cell nor to better tolerance to aneuploidy, is more common than increased fertility accompanied by increased quadrivalent frequency. There are numerous studies on the relation between fertility and quadrivalent frequency, on the effect of selection for fertility on quadrivalent frequency and especially on the variation in quadrivalent frequency in autotetraploids. Gilles and Randolph (1951) reported a decrease from an average of 8.47 quadrivalents per cell in the first generation to 7.47 after 10 generations of selection for fertility in autotetraploid maize (4x = 40). On the other hand, Mastenbroek et al. (1982) report no effect on quadrivalent frequency of 22 generations of selection for fertility in maize. In tetraploid barley, according to Bender and Gaul (1966), an improvement in fertility is accompanied by a decrease in quadrivalent frequency. In autotetraploid *Brassica campestris* Swaminathan and Sulbha (1959) found that 19 generations of selection for fertility resulted in a significant decrease in the quadrivalent frequency and it could be shown that this was not due to a reduction in the number of chiasmata per cell. If selection is to have an effect, there should be sufficient genetic variation in the material. Another example is *Hyoscyamus albus*, where Srivastava and Lavania (1990) found a drop in the estimated multivalent pairing frequency *f* (Sybenga 1975) from the low level of 0.239 in the generation of induction of tetraploidy (C0) to 0.107 already in C3, accompanied by a marked increase in fertility. Simple chromosome doubling of a self-fertilizing diploid species like barley may not be expected to provide a sufficient basis for selection. Bender and Gaul (1966) showed that induced mutations could provide genetic variation that was suitable as a basis for selection for fertility in autotetraploid barley.

11.3.1.2.3.2 Quadrivalent Formation in Relation to Chromosome Morphology

It has been suggested that replacing metacentric chromosomes with their corresponding telocentrics (centric or Robertsonian split) should reduce the possibility of forming quadrivalents. Our unpublished results with rye showed that too often small telocentric chromosomes derived from the short arms tended to be univalent. Possibly, there is an application for centric split in species with long metacentric chromosomes. However, with high chiasma frequencies in each arm, quadrivalents will be formed with partner exchange in the middle of an arm and chiasmata on both sides of the same arm. This is not just a hypothesis, as is shown by the presence of frying-pan quadrivalents in autopolyploids with metacentric chromosomes, which have exactly this combination of chiasmata, but in addition one in the second arm. In species with acrocentric chromosomes translocation heterozygotes also form quadrivalents without a problem. Yet, there is a negative correlation between the frequency of open bivalent formation in diploids and the frequency of quadrivalents in the tetraploids derived from these diploids. Lavania (1986) and Srivastava et al. (1992) found a significant negative correlation between the frequency of open metaphase I bivalents in diploid genotypes of *Hyoscyamus muticus* and the frequency of quadrivalents in the tetraploids derived from them. The tetraploids with the lower quadrivalent frequencies were significantly more fertile.

In *Avena strigosa* Zadoo et al. (1989) could correlate the high frequency of two specific bivalents being open bivalents in the diploid to the low frequency of quadrivalent formation of the same two chromosomes in the derived tetraploid. Both were subacrocentric and there were no interstitial chiasmata which could result in quadrivalents after partner exchange in the long arm, combined with frequent failure of chiasmata in the short arm. Guignard (1986) also found low chiasma frequencies that were not randomly distributed over chromosomes in *Dactylis glomerata*, and attributed them to differences between chromosomes in a centromere location. Variation between populations within the species with respect to the distribution of quadrivalents was also explained by variation in karyotype, but the evidence was not convincing. A correlation between chromosome morphology and quadrivalent formation was not evident in *Hyoscyamus niger*, where subacrocentric and metacentric chromosomes occur together (Lavania et al. 1990). Here, the cause of low quadrivalent frequency was not simply a low chiasma frequency as shown by the low quadrivalent pairing frequencies *f* (Sybenga 1975); the explanation must be different.

A special case of low quadrivalent frequency resulting from reduced association frequency of one of the two arms of a chromosome is reported by Schlegel and Liebelt (1976) for autotetraploid rye (*Secale cereale*). Chromosome arms with large heterochromatic blocks tended to associate less readily in the tetraploid than arms without much heterochromatin. This was explained by the relatively short DNA replication period of tetraploid rye compared with the diploid and the failure of heterochromatin to replicate in time, which would affect pairing in specific chromosome arms. If this were generally true, chromosomes with one heterochromatic block would preferentially form pairs of open bivalents. In rye, however, where there is considerable polymorphism for telomeric heterochromatin, quadrivalent formation is not a serious problem in autopolyploids. They are relatively frequent, but tend to orient alternately.

11.3.1.2.3.3 Quadrivalent Distribution

In several of the cases mentioned above the distribution of quadrivalents over chromosomes was not random. In many other cases it is, in spite of morphological differences between chromosomes. Possibly, such differences have no effect, or they form a gradient, which results in only insignificant deviations from random. Simonsen (1973, 1975), in his detailed report, mentions several examples, including his own of *Lolium perenne*, and Ishiki (1985) reports on high frequencies of randomly distributed quadrivalents in African rice (*Oryza glaberrima*, 4x = 48) with almost as many quadrivalents as bivalents, close to the expected relation with random pairing.

The causes of low quadrivalent frequencies where relatively simple explanations cannot be found are generally not well understood. Several mechanisms, in addition to those just mentioned, have been proposed: preferential pairing; localized pairing initiation; shift of point of pairing partner exchange during the process of pairing and prior to chiasma formation; variable point of pairing partner exchange, interfering with chiasma formation in one arm; chiasma localization in one segment per chromosome; negative interference across the centromere and the point of pairing partner exchange.

Especially in the cases of extremely low quadrivalent frequencies, as occasionally found in natural autopolyploids that nevertheless show clear tetrasomic inheritance, the proposed explanations do not seem to be sufficient. A few, rather arbitrary examples are autohexaploid *Phleum pratense* (Nordenskiöld 1953), possibly an autoallopolyploid; several potato strains; *Vaccinium corymbosum*, where four enzyme loci showed clear tetrasomic inheritance but no quadrivalents were observed (Krebs and Hancock 1989); the very low quadrivalent frequency in artificial and natural autotetraploids of *Parthenium argentatum* (Hashemi et al. 1989) and the natural tetraploid *Heucheria grossulariafolia* (Wolf et al. 1989); low quadrivalent frequencies in *Hyoscyamus* species (Srivastava et al. 1991).

11.3.1.2.3.4 Preferential Pairing

Preferential pairing in autotetraploids can only be used as an argument for low quadrivalent frequencies when a new tetraploid is concerned, made by doubling a hybrid between different genotypes. Even then, there is little reason to assume that identical chromosomes pair preferentially over homologous, but not identical chromosomes. In a number of cases even the opposite has been observed, and this is certainly not the type of preferential pairing meant to result in low quadrivalent frequencies (Santos et al. 1983; Orellana and Santos 1985). Occasionally, within species, certain chromosome segments appear to show consistently identical rather than merely homologous preferential pairing (Benavente and Orellana 1989). However, in established autotetraploids, any initial preferential pairing that is not strong enough to result in a fuctional allopolyploid, will sooner or later recombine the preferentially pairing segments. The result is that instead of bivalents, quadrivalents will be preferentially formed until the critical segments have been redistributed. Finally, the preferential pairing segments will have been broken up. The overall level of quadrivalent pairing will not be reduced (Sybenga 1984a). The concept of segmental allopolyploidy (Stebbins 1947) does not logically apply to such situations.

11.3.1.2.3.5 Localization of Pairing Initiation and Chiasmata

Chiasma localization, other than caused by the presence of heterochromatin or as a consequence of the shortness of specific segments, has not yet been demonstrated as a cause of low quadrivalent frequencies. It would require restriction of chiasmata to only one short segment per chromosome. In interchange quadrivalents even the opposite has been observed: after quadrivalent pairing chromosomes that normally would form only one chiasma can now form one on either side of the point of pairing partner exchange, and the quadrivalent is maintained into metaphase I (Arana et al. 1987; Parker 1987).

The same is true for restriction of pairing initiation to one site per chromosome: it is in principle possible but has not been convincingly demonstrated to exist. Pairing initiation, like chiasma formation, is restricted and in a sense localized, but apparently not so extreme as to result in low quadrivalent frequencies.

11.3.1.2.3.6 Shift of Point of Pairing Partner Exchange

Probably more effective is the shift of the point of pairing partner exchange in the early stages of pairing, before the synaptonemal complex has been fixed by chromatid exchange (*Bombyx mori* females: Rasmussen and Holm 1979). It may well be a common phenomenon in any situation of competitive chromosome pairing, but the exact importance has not been established in autotetraploid plants. One reason is that the analysis on *Bombyx mori* females was done using serial sections, which is a very laborious technique. The technique presently in most general use is a spreading technique that is not particularly favourable for very early pairing stages where the SCs are not yet continuous over long stretches. Yet, Davies et al. (1990) could conclude from the analysis of SC spreads of successive early prophase stages that in *Lotus corniculatus* (possibly not an autotetraploid as previously assumed) multivalent pairing was replaced by bivalent pairing. The multivalents which were maintained until diplotene fell apart due to the lack of chiasmata.

In autotetraploids the point of pairing partner exchange is not fixed and there is not always only one point. The simple assumption that there is only one such point and that it is positioned somewhere around the middle of the chromosome or around the centromere is an oversimplification. In most autotetraploids the four chromosomes tend to be aligned rather parallel at the beginning of zygotene and pairing usually starts near the telomeres, which are often associated with the nuclear membrane. If only the ends would start pairing and do so simultaneously and if pairing would proceed at the same speed from both ends, the simple assumption of one point of partner exchange near the centromere would be correct for metacentric chromosomes. However, the four aligned chromosomes can start pairing at several different sites (review in Gillies 1989; *Allium vineale*: Loidl 1986; pentaploid *Achillea*: Loidl et al. 1990; tetrasomic B-chromosomes in *Crepis*: Jones et al. 1991).

According to Loidl (1986), any additional pairing initiation site after the first has a probability of 1/3 of involving the same two chromosomes. The number of pairs of initiation sites equals 1.5 $\times$ the number of observed partner switches + 1 (the first). This makes it possible to make an approximate estimate of the number of initiation sites, but not accurately. In *Allium vineale* there would be at least 3.1 independent sites per chromosome. This model is a simplification: for instance, when two pairing initiation points are close together they may well act in coordination, like centromeres do on the spindle. In triploid *Allium sphaerocephalon* the estimate of the number of pairing initiation sites was 6.1, almost twice as many as in tetraploid *A. vineale*, which cannot be explained by length differences alone. Although Loidl (1986) does not refer to this possibility, there may be an intrinsic difference between uneven and even numbers of pairing chromosomes. In autotriploid *Hyoscyamus muticus* Tyagi and Dubey (1989) report large numbers of branched trivalents at metaphase I, with several points of pairing partner exchange and many interstitial chiasmata. In autotetraploids of similar origin, Lavania et al. (1990) found only very infrequently more than one partner switch in the MI quadrivalents. Trivalents were very rare, but those that were found had several partner switches. Trivalent behaviour may well be different from quadrivalent behaviour. In this respect, there may be a difference between species.

Not all segments between partner switches have chiasmata. Whereas at pachytene 80% of the sets of four chromosomes forms quadrivalents in *Allium vineale*, there are only 22% at metaphase (Loidl 1986). Such extreme reductions are not often observed, partly because not all species form so many partner switches.

Around the point of pairing partner exchange there is a short stretch of chromosome where pairing is not complete or non-homologous, and where chiasmata cannot be formed. When it is normally around the middle of the chromosome, little reduction in effective chiasma formation and the resulting metaphase association in ring quadrivalents or ring bivalents is expected. When the position of this point is variable or when there are more points of pairing partner switch, extending into chromosome regions where chiasmata are normally formed, there may be considerable reduction in chiasma formation, and this may affect entire chromosome arms when chiasmata are normally concentrated in the subterminal region. In a somatic tomato (*Lycopersicon esculentum*) hybrid (4x = 48), de Jong (pers. comm. 1990) found that at pachytene occasional partner switches occur in the centromeric region (Fig. 6.4), but that they were concentrated in the subterminal euchromatic regions, with occasionally one in each arm, infrequently even more. Such a pattern would create a situation where the arm with the partner switch would often fail to have a chiasma and, consequently, open bivalents are formed instead of quadrivalents. Examples of low quadrivalent frequencies and a majority of open bivalents are the natural and induced autotetraploids of *Parthenium argentatum* (Hashemi et al. 1989) and the natural autotetraploid of *Heuchera grossulariafolia* (Wolf et al. 1989), which shows clear tetrasomic inheritance but low quadrivalent frequencies and a predominance of open bivalents (Table 11.1). The distribution of configurations fits the autotetraploid model of Jackson and Casey (1982) and Jackson and Hauber (1982), and the *f*-value of Sybenga (1975) was 0.647, quite close to the random pairing factor 0.667. In both cases ring bivalents in the diploids were infrequent, although there are no indications of pronounced acrocentric chromosome morphology. Possibly, interference acts across the centromere, as also suggested by Lavania (1986) for *Hyoscyamus*. The effect is strong enough to result in one arm always having at least one chiasma and the other quite infrequently: 0.092 in *Heuchera* (Wolf et al. 1989). In autotetraploid sunflower (*Helianthus annuus*,

Table 11.1. Meiotic chromosome associations in diploid (A), and natural (B) and induced (C) autotetraploids of *Parthenium argentatum*, x = 18, about 100 cells each (Hashemi et al. 1989), and of the natural (D) autotetraploid (45 cells) of *Heuchera grossulariafolia*, x = 7, 45 cells (Wolf et al. 1989), all tetraploids with low quadrivalent frequencies and high open bivalent frequencies, E expected for *Heuchera* when quadrivalent pairing = 0.647, one arm association frequency = 1, the other = 0.092 after quadrivalent and bivalent pairing

Type	univ[a]	obiv	rbiv	triv	rq	cq
A	–	0.826	0.174	–	–	–
B	0.003	0.728	0.083	0.007	0.180	
C	0.017	0.749	0.108	0.003	0.124	
D	–	0.878	0.043	–	0.019	0.060
E		0.854	0.032		0.005	0.108

[a] univ: univalents; obiv: open bivalents; rbiv: ring bivalents; triv: trivalents with univalent; cq: chain quadrivalents; rq: ring quadrivalents. Frequencies per set of 2 and 4 homologous chromosomes respectively.

4x = 68)) Jan et al. (1988) observed a low frequency of quadrivalents (2.59) and a higher frequency of ring bivalents (6.66) with a further 21.50 open bivalents, 0.36 trivalents and 0.85 univalents. Comparatively many of the quadrivalents (0.27) were of several different branched types, and the same was true for the trivalents in spite of the fact that the chromosomes were not large. Chain quadrivalent frequency was 1.06 and ring quadrivalent frequency 1.26. This is a typical example of occasional interstitial chiasma formation and frequent absence of chiasmata in the probably short segment distal to the point of partner exchange.

11.3.1.2.3.7 Other Effects: Interference, B-Chromosomes; Correlation with Diploid Meiotic Characteristics

In Table 11.1 the expected configuration frequencies are given for *Heuchera*, on the basis of an estimated quadrivalent pairing frequency of 0.647, and chiasmate association frequency of one arm = 1 and of the other arm = 0.092. It is seen that the observed ring bivalent and ring quadrivalent frequencies are higher, and the chain quadrivalent frequencies lower than expected, because they have formed pairs of open bivalents instead. This means that in a quadrivalent relatively too often two opposite end segments fail to have a chiasma simultaneously compared to each separately. This is the same type of *negative interference* across the point of partner exchange as observed in interchange heterozygote quadrivalents. Apparently, interference plays a role here. At present, it cannot be decided whether the arm that fails to form a chiasma is always the same arm (in that case chiasma localization could be the cause of low quadrivalent frequencies) or that both arms can fail to have a chiasma in turn (then interference and a chiasma-suppressing effect of partner switch in one arm would be the main cause).

In addition to genetic effects caused by genes in the normal chromosomes, genes in B-chromosomes of *Lolium* species have been shown to reduce quadrivalent frequency in autotetraploid *Lolium perenne* (Macefield and Evans 1976). The effect was not the same for all chromosomes: whereas in the normal autotetraploids the distribution of quadrivalents followed a binomial distribution (180 PMCs in 9 plants), this was not the case in the autotetraploids with B-chromosomes (300 PMCs in 15 plants).

In the previous sections, a correlation between diploid meiotic characteristics and quadrivalent frequency and fertility in the derived autotetraploids was mentioned in the discussion on the effect of chromosome arm length ratio and interference. The predictive value of meiosis in the diploid progenitor for quadrivalent frequency and fertility in the autotetraploid is not always clear, but in some instances can be considerable (Lavania 1991). An example is given by Srivastava et al. (1991) for *Hyoscyamus muticus*, where meiotic and fertility characteristics of autotetraploids and their diploid progenitors from 17 genetically different origins were compared. Multivalents had a negative effect on fertility in the tetraploid, and multivalent frequency was significantly correlated with the chiasmate association frequency in the diploid progenitor. Interference may have been involved, because reduced chiasma frequency was primarily expressed in an increase in open (rod) bivalents in the diploid. A regression analysis suggested that one open bivalent more in the diploid would result in almost 18% more seed set in the tetraploid at the lower fertility levels. There were several more correlations, most of less interest.

11.3.1.2.4 Autotetraploidy: Conclusion

Induced autotetraploidy has had application in several cultivated species, but the results have not always been satisfactory. One reason is that for some plant species and some applications, the disadvantages of autotetraploidy are too serious to be overcome by simple selection. Another reason is that in many cases the plant breeder has not been able to obtain sufficient genetic variation for selection on the many factors involved in autopolyploidy (meiotic regularity, fertility, other physiological adjustments to the new gene balance), in addition to the objectives proper of the breeding program. It is necessary to evaluate the importance of aneuploidy and reduced fertility for the specific requirements of the crop. If meiotic stability and fertility are important, it is useful to know to what extent it will be under the influence of the environment and general genetic background, in addition to specific genetic effects on meiosis. The increasing understanding of the processes and factors involved in tetraploid chromosome pairing and further meiotic behaviour does not immediately contribute to the improvement of tetraploid fertility. However, when simple selection for fertility does not readily have effects, knowledge of the possible causes may aid in the design of a more specific strategy for collecting the most promising genotypes and finding the most promising selection procedures. It cannot be denied that specialized cytogenetic expertise will then be necessary.

In addition to these typical cytogenetic aspects, the normal complications of polyploidy breeding require more attention than the breeding of diploids or stable allopolyploids. The typical tetrasomic genetic system makes selection less effective in the short term, and delays stabilization. The optimum level of heterozygosity is higher in polyploids than in diploids. It may be good to be aware of the fact that for these reasons the original autopolyploids have not improved as fast as the existing diploids, with the consequence that their initial advantage has not been maintained. The original diploids used to make the first generations of autopolyploids were not as thoroughly bred and were possibly more genetically variable than the more successful, present diploids. Even when these new diploids are used to enlarge the genetic variation of the autotetraplids, the progress will be relatively slow. It is even possible that the more specialized, new diploids are less favourable for polyploidization than the older varieties. As a source of genetic variation they may be quite suitable, but it may pay to use additional, possibly more exotic material for the tetraploid breeding program.

11.3.2 Artificial Allopolyploidy

11.3.2.1 Objectives and Pitfalls

There are two ways to combine the genes of two species: *introgression*, ranging from a few genes to large complexes resulting in entirely new forms (Sect. 10.4), or *allopolyploidy*. Numerous wild species and many important cultivated species are allopolyploids. In addition to the intact and stable combination of the characteristics of different specific genotypes, there are gene dose effects, complex allelic interactions and fixed heterosis. All these factors play a role when an artificial allopolyploid is made, but in most programs these are not all explicitly taken into consideration.

There are two main reasons for constructing an allopolyploid: (1) to create a new combination of genomes, not existing previously; (2) to reconstruct an existing allopolyploid with the purpose of introducing new genetic variation. In the first case difficulties may be encountered at several phases during the process of construction: sterility of the initial cross; hybrid dysgenesis; insufficient affinity differentiation; meiotic abnormalities and reduced fertility of the raw allopolyploid. In the second case these will probably be less important, but often the new allopolyploid will not be as stable as the established one and during the process of introducing genes from the new into the old allopolyploid irregularities may occur.

11.3.2.2 Construction of Allopolyploids

In principle, an allopolyploid is simply made by doubling the chromosome number of a hybrid between two species, but in practice there are several complications, even when the allopolyploid is merely a transient form made for purposes of gene transfer from one species to another or for widening the genetic base of an existing allopolyploid.

11.3.2.2.1 Adjusting Chromosome Pairing Differentiation

Effective allopolyploidy presumes effective pairing differentiation between the component genomes. If in the original hybrid between two species pairing between the chromosomes of the parents fails, the differentiation may be expected to be sufficient for the allopolyploid. In other hybrids there is considerable chiasmate chromosome association at meiosis but after doubling the differentiation may still be sufficient. An example, already used repeatedly in the previous chapters, is the hybrid between *Gossypium herbaceum* and *G. raimondii*, which shows considerable pairing, but after chromosome doubling (4x = 52) the genomes are well separated. This artificial allotetraploid is assumed to be the equivalent of the cultivated American *G. hirsutum*, upland cotton (Mursal and Endrizzi 1976; Endrizzi et al. 1985). In this case the size of the chromosomes differs considerably between the parental species, the A chromosomes being almost twice the size of the D chromosomes. This is considered a major reason for pairing differentiation. Although pairing differentiation in the polyploid is complete in both the artificial and the natural hybrid, this is not so at the dihaploid level. In the hybrid between *G. arboreum* (genome A2; related to *G. herbaceum*) and *G. raimondii* (genome D5) the pachytene bivalent frequency was about 11 per cell and in the dihaploid derived from the established allopolyploid it was 10. The decrease in chiasmate association at diakinesis and metaphase I was much more pronounced in the dihaploid than the hybrid; there were 0.80 bivalents in the dihaploid compared with 5.82 in the hybrid (Mursal and Endrizzi 1976).

In the allotetraploid (4x = 48) of *Lycopersicon esculentum* (tomato) and *L. lycopersicoides* the chromosomes do not differ much in length but still show effective pairing differentiation (22.12 II, 0.55 I, 0.74 III and 0.50 IV at metaphase I). Yet, in the diploid hybrid there is almost complete pachytene pairing and considerable bivalent formation at metaphase I (10.08 bivalents per cell and 3.84 univalents). Heterozygosity for blocks of heterochromatin does not increase preferential pairing (Menzel 1964). The length of the chromosomes of *Allium fistulosum* are about 0.7 of the length of the chromosomes of *Allium cepa*, but in the hybrid they pair well (Emsweller and Jones 1935). In the amphidiploid (4x = 32) there are consistently 16 bivalents (Emsweller and Jones 1945). Other examples could be added.

In still other cases the hybrid shows pairing between the parental species, and similar, although somewhat reduced pairing is observed in the allopolyploid. When A-genome *Triticum* species are combined with B-genome *Aegilops* species, the doubled hybrid constitutes an allotetraploid resembling tetraploid wheat (AABB), but with intergenomic chromosome pairing, and consequently genetically unstable. The real B-genome species of tetraploid (and allohexaploid) wheats is not known, and may not exist as such, but even if it could be found and hybridized with an A-genome species to form an allotetraploid, it would probably still show intergenomic pairing. Yet, the established tetraploid AABB wheat species do not normally form multivalents. The reason is that in the allopolyploid wheats, as discussed in Section 6.1.2.3, the dominant Ph (Pairing homoeologues) gene in chromosome 5B prevents intergenomic pairing. Such genes are not universally present in diploid *Aegilops* species with the B-genome, but in some genotypes they are encountered occasionally and, in addition, they are present in the B-chromosomes (not to be confused with B-genomes) found in a few B-genome species.

A gene with a comparable effect is present in the B-chromosomes of *Lolium* species, where it reduces intergenome pairing in the hybrid as well as in the amphidiploid (raw allotetraploid), for instance between *Lolium perenne* and *L. temulentum* and *L. rigidum* × *L. temulentum*. The B-chromosomes have an accumulation system, and they may also be lost. Their direct practical use for reducing homoelogous pairing is consequently limited. A-B chromosome translocations can be isolated without the accumulation mechanism and still having the homoeologous pairing restriction system (Evans and Macefield 1977). Their practical usefulness is still uncertain. The effect of the B-chromosomes can be substantially modified by the A-chromosomes genotype (Taylor and Evans 1976), and genes in A-chromosomes can apparently affect homoeologous pairing in the absence of B-chromosomes. In *Lolium multiflorum* also, genotypes have been isolated that reduce homoeologous chromosome association, and where no B-chromosomes are involved (Evans and Aung 1985). B-chromosomes reduce the quadrivalent frequency not only in tetraploid hybrids, but also in autotetraploid *Lolium perenne* (Macefield and Evans 1976; cf. Sect. 11.3.1.2.3.7). It appears that in tetraploid hybrids with a reduced as well as with a high metaphase I multivalent association frequency, quadrivalent pairing at zygotene has the same frequency. At pachytene, however, synaptonemal complex pairing is "corrected" in the low metaphase quadrivalent genotypes (Jenkins 1986). The B-chromosomes of rye do not have an effect on pairing differentiation (Roothaan and Sybenga 1976) in wheat, although the chiasma formation is affected.

The possibilities to increase homoeologous pairing differentiation by genetic means are limited, but clearly not negligible. For a review, see Evans (1988). The alternative approach, using chromosomal rearrangements to induce pairing differentiation, is discussed in Section 12.2.2 where the emphasis is on autopolyploids.

It has been suggested that allopolyploids with insufficient differentiation between the genomes could, by selection for fertility and reduced multivalent formation, be converted to true allopolyploids. According to Stebbins (1947), whose opinions on the subject have found wide acceptance, such "segmental" allopolyploids would occur in nature and be stable, without being converted into allopolyploids, nor into autopolyploids. This cannot at present be considered a real possibility (Sybenga 1984a). Whenever multivalents occur or when chromosomes of originally different genomes pair as bivalents and exchange segments, the system of differentiation will gradually disappear. The result is a genetic autopolyploid, even though segments of the chromosomes of the different species may remain intact for a long time (see Sect. 11.3.1.2.3). If the frequency of quadrivalents is reduced in subsequent generations, this must be due to factors similar to those operating in true autopolyploids. It is theoretically possible that in such polyploids some chromosomes have increased their level of differentiation, while others have become completely homologous. These would then be real segmental allopolyploids. There are superficial indications that this is the case in a few species, but the proof is hard to obtain except by using marked chromosomes or careful genetic experimentation.

The combination of autopolyploidy with allopolyploidy (autoallopolyploidy) is more realistic, although not frequent. *Senecio cambrensis* (2n = 60) is an established autoallohexaploid derived from the diploid *S. sqalidus* (2x = 20) and the relatively recent autotetraploid *S. vulgaris* (4x = 40). Ingram and Noltie (1989) conclude that the stability of the autoallopolyploid is due to the meiotic regularity of the autotetraploid *S. vulgaris*. Comparable is the experimental autoallohexaploid (SSPPPP) of *Medicago sativa* (genome S) and *M. papillosa* (genome P), which is stable, whereas autohexaploid *M. sativa* is not (McCoy 1989). From a practical point of view the autoallohexaploid looks interesting, although *M. sativa* is the cultivated species and *M. papillosa* is not. It is not clear to what extent the fact that *M. sativa* is a natural autotetraploid has contributed to the balanced behaviour of the autoallohexaploid, where it is present only twice and not four times.

11.3.2.2.2 Adjusting Mitotic and Meiotic Chromosome Behaviour: the Heterochromatin Story of Triticale

The most successful artificial allopolyploid until now, into which by far the most investment has been made, is *triticale*, the doubled hybrid between wheat and rye. The first triticale made was an octoploid based on hexaploid bread wheat and diploid rye (for reviews, see Müntzing 1979; Gupta and Reddy 1991). It has never been successful because of severe meiotic and endosperm imbalance, apparently caused mainly by the interaction between the rye (R) genome and the D-genome of wheat.

Hexaploid triticale (2n = 6x = 42), the doubled hybrid between tetraploid durum wheat (AABB) and rye (R) later appeared to have a much better potential to become a useful new small grain for regions unsuitable for wheat, and where rye did not yield sufficiently, or where a wheat-like product was desired (Müntzing 1979). In addition to combining the different qualities of wheat and rye, it has some new characteristics, including the composition of the seed protein. The hybrid polymers of the wheat and rye prolamin subunits have different properties from both parents (Field and Shewry 1987).

The differentiation between the genomes of rye and wheat is sufficient to avoid problems of genetic exchange and destabilization on account of loss of integrity of the genomes by recombination. There is some effect of the rye genome on pairing of the wheat genomes: it counteracts the effect of the *Ph* gene and causes some multivalent formation in the undoubled hybrid, but not in the allopolyploid.

The interaction between the genomes is not in all respects good, and this is expressed at two moments important for productivity: meiosis is slightly irregular, resulting in reduced fertility and relatively high aneuploid frequencies, and (early) endosperm development is often abnormal, resulting in shrivelled seed. Both phenomena have been related to differences in wheat and rye chromosomes with respect to replication time. The rye genome replicates at a lower rate than the wheat genome and it has been observed that the rye chromosomes are involved in irregularities more often than the wheat chromosomes. In addition, rye usually has considerable quantities of telomeric *heterochromatin*, and especially this delays replication (Bennett and Kaltsikes 1973). The role of heterochromatin in fertility and seed development has been the subject of many studies (reviewed by Kaltsikes and Gustafson 1985; Lukaszewski and Gustafson 1988; Schlegel et al. 1988 and others), but only a few can be mentioned here.

Merker (1976) reports a drop by one-third in univalent frequency when one large heterochromatin block in rye is absent. When it is present in one homologue, the chromosome pair concerned is either an open bivalent or univalent in the majority of cells. Hülgenhof and Schlegel (1985) describe additive effects of successive reduction of telomeric heterochromatin in different rye chromosomes on univalent frequencies. Dillé et al. (1984), on the other hand, found only a slight decrease in aneuploid frequency in the progeny when the heterochromatin blocks of chromosomes 4R and 6R were removed. Replacing chromosome 2R of rye by 2D of wheat more than compensated this removal by drastically increasing univalent frequencies.

The possible importance of rye heterochromatin for abnormalities in seed development, primarily retarded sister chromatid separation in the young endosperm, has received considerable attention (Bennett 1977). Raw triticales, especially the octoploids, but also, to a lesser extent, the hexaploids have shrivelled seed. Highly bred varieties, however, have plump seeds, almost as good as wheat, but the effect of the environment on seed development is greater than in wheat. There appears to be considerable variation in the effect of heterochromatin (Bennett and Gustafson 1982). Several studies have been carried out on the effect of the loss of the heterochromatic segment of specific chromosomes (see, for instance, Dillé and Gustafson 1990). There often is an effect, and different segments may have additive effects, but not in all genotypes.

Hülgenhof and Schlegel (1986) conclude, as do Lukaszewski and Gustafson (1988), that the role of heterochromatin in meiosis and endosperm development is still debatable. When the rye complement is relatively free of heterochromatin, the irregularities often seem to be less pronounced. However, entirely normal behaviour is observed only in highly selected varieties, and these are not necessarily free of heterochromatin. The genetic make-up is apparently more important than the absence of heterochromatin (Seal and Bennett 1985). One interesting genetic component of irregular chromosome behaviour in triticale is heterosis. According to Struss and Röbbelen (1989), hybrids between different hexaploid triticale lines, even when they have the same level of aneuploidy as the established, inbred parental lines, apparently suffer less from aneuploidy than the inbred lines. This presents a problem: selection for fertility in early generations will not contribute to meiotic stability.

When triticale is crossed with hexaploid wheat and subsequently backcrossed with triticale in order to incorporate genes from hexaploid wheat into triticale, in the F1 hybrid one single set of rye chromosomes and one set of D-genome wheat chromosomes are present. In Section 10.4.4.2.2.2 it has been mentioned that these 14 univalents have ample opportunity to form centromere translocations. When the number of backcrosses is small, these translocations, as well as complete substitutions, have a chance of being maintained. It has been proposed that these might favourably combine the good bread-making quality genes in the D-genome with the special qualities of rye. Substitutions and translocations of D-genome chromosomes for rye chromosomes continue to be the subject of research, although it has been suggested that, except for high quality bread-making, the pure rye genome, without D-genome substitutions, gives the best results (Lukaszewski and Apolinarska 1981; Kaltsikes et al. 1984). In the past, the number of R-D chromosome substitutions or translocations has repeatedly been overestimated, because loss of the rye chromosome heterochromatin made rye chromosomes appear as wheat chromosomes. Careful karyotyping has shown that the 2D-2R substitution is practically the only important rye substitution still present among commercial triticale varieties (Lukaszewski 1988). Of 35 lines studied, 29 had an unmodified rye genome, 4 had the 2D-2R substitution, and 2 had a 6D-6A substitution, not present in older triticales (cf. Gustafson et al. 1989).

Triticale requires an even broader genetic base for selection than established crops, because in addition to the normal requirements of productivity, adaptation and quality, there is the necessity of adjustment of the interaction between the not very compatible genomes of wheat and rye. There are three main approaches:

1. Producing new hexaploids from the hybrid between new rye and tetraploid (durum) wheat genotypes.
2. Crossing hexaploid triticales (AABBRR) with wheat (AABBDD) and backcrossing the hybrids (AABBRD) with triticale to recover a hexaploid triticale.

3. Producing octoploid triticales (AABBDDRR), crossing these with hexaploid triticales and selecting hexaploid triticales in the progeny. Potentially, octoploid triticales can be crossed with tetraploid triticales (A/BRR), but these give more complications (Bernard et al. 1990).

The last two procedures result in "secondary" triticales with the A- and B-genome chromosomes from bread wheat at least partly replacing those from durum wheat, which is interesting from the quality point of view, with procedure 2 and to a lesser extent procedure 3, when selfing instead of backcrossing is introduced at an early stage with the purpose of recovering as much bread wheat background as possible, there is a considerable probability of intentionally or unintentionally introducing D-genome chromosomes replacing (homoeologous) rye chromosomes, or D/R translocation chromosomes into triticale, as briefly mentioned above.

Many of the techniques used in wheat for generative transfer of alien genes (Sect. 10.4) can also be successfully applied to triticale (Gupta and Reddy 1991).

In addition to the less successful attempts to produce octoploid triticales and the more successful production of hexaploid triticale, tetraploid triticales have been produced but this has not led to practical, applicable results. The most logical approach is to cross diploid *Triticum* species with rye and to double the hybrid. This cross has appeared to be difficult and the success has been limited. Sodkiewicz (1984) reported a hybrid between *Triticum monococcum* and rye, but the induced tetraploid was sterile and could be maintained only vegetatively. It could be crossed with hexaploid triticale and with the tetraploid triticales mentioned below.

When hexaploid triticale is crossed with rye, an unstable tetraploid ABRR hybrid is formed. Selfing and selection for wheat chromosomes results in a tetraploid derivative with two complete rye genomes and a relatively balanced mixture of A- and B-genome chromosomes. Initially, some of these are homozygous, others still a pair of homoeologues (Lukaszewski et al. 1987b). At later stages all wheat chromosomes form sets of two homologues (Lukaszewski et al. 1987a). Many different A-B genome chromosome combinations have been recovered including a line with a complete B-genome (Lukaszewski et al. 1987a), but none of them is of any practical importance. They are interesting from a theoretical point of view, such as the study of the possibilities of compensation between A- and B-genome chromosomes and for the production of wheat chromosome additions to rye.

Such composite tetraploid triticales have been constructed and studied on many occasions (for reviews, see Sabeva 1984; Lukaszewski et al. 1987a,b). Attempts to regulate meiosis, fertility and endosperm development have not been successful, but there is a gradual increase in fertility. As in hexaploid triticales, large heterochromatin blocks in the rye genomes have effects, but the genetic composition is more important. The frequency of aneuploids is variable, but usually not excessively high. The process of assortment of a balanced combination of A- and B-genome chromosomes is not random. Certain combinations are more frequent than others (Dubovets et al. 1989). During the generations required for stabilization, there is some homoeologous recombination, which varies between chromosomes. It is not impossible that tetraploid triticales will gain some interest, but at present they are mainly a curiosity. In addition to those mentioned, several more studies on tetraploid triticale carried out by different groups in different countries have been reported, but wil not be discussed here.

The same tetraploid hybrid between hexaploid triticale and diploid rye from which composite tetraploid triticales are produced is also the origin of addition lines of wheat chromosomes to rye. These are obtained by simple selfing (Sects. 6.2.2.4.2 and 10.4.4.1). It should be noted that these additions are *alloplasmic*, i.e. rye in an alien (wheat) cytoplasma.

Hexaploid triticale is the most thoroughly studied and the best-selected artificial allopolyploid, but it is not the only artificial allopolyploid. Many attempts have been made to produce allopolyploids from hybrids between more or less closely related species, some with reasonable success.

Especially in fodder crops relatively successful attempts have been made to construct new allopolyploids. Here, vegetative growth is important, but when there is competition between the plants in the field, the effect of possible aneuploids is not critical for total production, and reproduction by seed, although necessary, is not as critical as in crops where the seed is the main product. Choice of parents, hybridization and chromosome doubling are part of the breeding program, but will not be discussed here further. The cytogenetic aspect primarily concerns the meiotic behaviour of the raw amphidiploids and their derivatives, as well as the segregation and cytological characterization of deviant types, and attempts to stabilize meiosis. For the analysis of the meiotic behaviour and for the study of the karyotype of deviant segregants, the reader is referred to Chapters 4 and 7.

There have been several attempts to hybridize *Brassica* species, especially *B. campestris* (2n = 20) or *Brassica oleracea* (2n = 18) with *Raphanus* species, especially *R. sativus* (2n = 18). The oldest report is by Karpechenko (1928) who produced the first amphidiploid between these species: *Raphanobrassica*. For the doubled reciprocal hybrid the name *Brassicoraphanus* is used. The practical purpose of making the hybrid is either the introduction of resistance genes or other qualitative genes from the *Raphanus* parent into the *Brassica* parent or to produce the amphidiploid (allopolyploid) in order to combine the properties of both species and to induce a gene dose effect and a heterosis effect (Dolstra 1982). The amphidiploid (genomes CCRR) between *R. sativus* (fodder radish) and *B. oleracea* (thousand-headed kale) is called "radicole" (McNaughton 1979). The amphidiploid (genomes AARR) between *B. rapa* (= *B. campestris*) and *R. sativus* is called "raparadish" (Lange et al. 1989).

Interaction between the relatively unrelated genomes results in meiotic abnormalities (Dolstra 1982; Prakash and Tsunoda 1983) and consequently to very frequent aberrant progeny and considerably reduced fertility. Tokamasu

(1976) reports yellow-flowered segregants among the progeny of artificial allopolyploid *Brassicoraphanus*, apparently the result of a deficiency, which had better fertility than the original amphidiploid. Dolstra (1982) described similar aberrant progeny. Selection for improved meiotic behaviour in *Raphanobrassica* or *Brassicoraphanus* does not have much success, partly because the population that can be effectively screened is relatively small, and partly because meiotic behaviour is a complex factor, apparently affected by many genes, requiring considerable genetic variation for recovery of the proper combination of genes. In addition, the meiotic disturbances are probably only a minor factor in sterility, which is mainly caused by physiological imbalance. There is considerable genetic variation in the level of fertility and when sufficient genetic variability is introduced, the results can be quite promising. In many instances in the development of allopolyploids, the lack of sufficient genetic variation and insufficient opportunity for recombination are important bottlenecks. Selection for fertility, which is much simpler than selection for meiotic regularity, will often have an indirect effect on the regulation of meiosis and subsequent suppression of the segregation of undesired, abnormal types. Some of the artificial allopolyploids of *Brassica* and *Raphanus* are becoming promising for practical use, especially as fodder crops (McNaughton 1979; Olsson 1986; Lange et al. 1989) even though not yet widely used.

Experience with the induction of allopolyploidy based on the grasses *Festuca pratensis* (2n = 14) and *Lolium perenne* (2n = 14) and on *F. arundinacea* (2n = 42) and *Lolium multiflorum* (diploid 2n = 2x = 14 and tetraploid 2n = 4x = 28) has been somewhat similar (Kleijer 1982), but the programs were often terminated too early to give practical results. There have been several attempts in different countries to construct *Festuca-Lolium* allopolyploids, but it appears to be difficult to obtain rapid results. The genomes of *Lolium* and *Festuca* are relatively closely related, but in the amphidiploid, pairing is predominantly in bivalents. Meiotic instability is rather common, fertility low and aneuploids are found frequently in the progeny (Essad 1962; Kleijer 1982). As in the *Brassica/Raphanus* allopolyploids, in the *Lolium/Festuca* allopolyploids, broadening the genetic background by hybridization and extensive selection programs should lead to acceptable results.

The role of cytogenetics in these programs is usually restricted to monitoring the progress, indicating the bottlenecks, and recording possible unexpected genetic characteristics such as altered recombination patterns (Morgan and Thomas 1991), i.e. providing information.

11.3.2.2.3 Reconstruction of Existing Allopolyploids

One reason to make allopolyploids is to reconstruct an existing allopolyploid and to use it to increase the genetic variation in the old type. The reconstruction may be different from the existing form when either the existing allopolyploid has an altered or a composite karyotype, or the original parental species have changed in the course of time, or the original parents are not available, but only close relatives. This may result in (sometimes considerable) meiotic disturbance in the hybrid between the established and the reconstructed allopolyploid, blocking gene transfer (Sects. 9.3 and 10.4). It is necessary, therefore, when using allopolyploid reconstructions to introduce genes into existing allopolyploids, to check meiosis of the hybrid, and if necessary, to look for opportunities for breaking recombinational blocks (Sect. 10.4).

Many allopolyploids have been reconstructed in the process of genome analysis, and a few specifically for introducing new genetic varation in crop plants. It would seem logical to attempt the reconstruction of wheat from its ancestors, but this has had no real practical consequences, and specific gene transfer from alien species as described in Section 10.4 has been much more important. One reason is that except for such specific genes, the existing genetic variability in allotetraploid and allohexaploid wheat is considerable and apparently sufficient. A second reason is that in the reconstructed allopolyploid, the absence of regulation of meiotic pairing leads to undesired intergenomic recombination, and a third reason is that the reconstructions have not been very successful, mainly because the original donor of the B-genome is apparently not available. Starting from tetraploid AABB wheats and adding the D-genome from one of the various sources is a more profitable approach. Practical breeders have not paid much attention to such possibilities mainly because it is not expected that characters can be introduced that are of sufficient interest to justify the labour of maintaining such characters in the process of reconstructing a good wheat.

Reconstruction of allotetraploid *Gossypium hirsutum* from the parental species or close relatives has been attempted on several occasions (Endrizzi et al. 1985). There is a varying difference in meiotic behaviour between the dihaploid of the established allotetraploid and the hybrid, depending on the choice of the parents (Sved 1966). The example given by Mursal and Endrizzi (1976) has been discussed above. Also discussed above is the autoallohexaploid *Senecio cambrensis* (Ingram and Noltie 1989). The reconstruction was meiotically less regular than the established polyploid, although the autotetraploid parent *S. vulgaris* used was quite stable. It is not clear whether different parental forms were involved in the artificial and the natural polyploids or natural selection in the natural species. The reconstruction of the cultivated spontaneous polyploid morning glory *Ipomoea sloteri* (2n = 59) from *I. coccinea* (2n = 29) and *I. quamoclit* (2n = 30) appeared to be quite difficult because of incompatibility between the parental species (Eckenwalder and Brown 1986). In 280000 ovules only 8 hybrid seeds were recovered and 2 of these were the result of a combination of 2 unreduced gametes. They were immediately polyploid and fertile, and a source for genetic improvement of the genetically very narrow cultivated species.

Allotetraploid *Brassica* species have been reconstructed in the course of genome analysis already in the 1920s and 1930s (Fig. 9.1). Especially *Brassica napus*, including both oil seed varieties and swedes grown for the fleshy root, has been reconstructed several times from very variable parental material. It originated from *B. oleracea* and *B. campestris*, both species with very large genetic diversity and used in many forms, including a wide variety of vegetables, but also forms producing seeds for the production of oil and others with swollen roots for animal feeding. Application of the resynthesized allopolyploids has a long history, not only for the oil seed forms. Olssen et al. (1955), for instance, succeeded in producing swedes by crossing autotetraploid kale (*B. oleracea*) with autotetraploid turnips (*B. campestris*), and these were quite useful to broaden the genetic basis of the existing swedes, even though the hybrids between the original and the resynthesized *B. napus* forms were not very fertile.

11.4 Reduction in Genome Number: Gametic Chromosome Number

11.4.1 Objectives

The term gametic chromosome number has been chosen instead of haploidy in order to avoid confusion when the starting material is polyploid. It should be understood that the way the gametic chromosome number is realized is not necessarily always through the meiotic reduction process, but exceptionally also by somatic reduction.

Plants with the gametic chromosome number have several potential theoretical and practical applications, but actual application has been limited due to the low frequency of induction. For *linkage studies* the possibility to score gametic genetic composition without the necessity of making test crosses with the proper tester genotype is of considerable importance. For a small number of genes it is possible to score the genotype of the pollen directly. In addition to starch mutants that can be distinguished by iodine staining in many plant species, in a few species of cucurbits it is even possible to perform electrophoresis of isozymes and other proteins of individual pollen grains (Mulcahy et al. 1979). When haploid callus or differentiated plants can be grown from pollen cultured in suspension or in anthers or when plants can develop by parthenogenesis from unfertilized eggs or when the chromosome number is reduced somatically (see below), the analysis can theoretically involve an unlimited number of genes, including molecular, for instance RFLP markers. The induction of haploidy may soon be sufficiently convenient for genetic analysis in some crops (Henderson and Pauls, 1992).

In *plant breeding* there is a limited, direct used for haploidy. There are a few haploids that, because of their special phenotype, have found application, especially in ornamentals. In 1967, for instance, Daker showed that the successful small *Pelargonium* cultivar Kleiner Liebling was a haploid, and there are more such exceptions, but they remain exceptions. There are primarily two applications for haploidy followed by chromosome doubling to the original chromosome number, resulting in completely homozygous plants. I does not matter how haploidy is induced, by somatic or meiotic reduction, where the latter is followed either by in-plant parthenogenesis or androgenesis, or followed by in vitro culture of pollen or eggs. The haploids are made diploid by any suitable treatment before use.

1. In *cross-breeding species* doubled haploids are the basis of completely homozygous lines to be used in hybrid breeding. The impossibility of selection in intermediate generations, which is practiced on purpose or unintended during the normal procedure of inbreeding, is partly compensated by strong natural selection for survival among the haploids at early stages of development. However, it should be noted that selection in this case is on different genotypes rather than plant selection during inbreeding. Early generation testing, of course, is not possible. Already in the 1950s haploids were isolated in maize that, after doubling, were used as inbreds for hybrid varieties. It is necessary to have a sufficient number of doubled haploids to select the few, really useful lines. During early development and during doubling, many haploids tend to be lost.

2. In *self-fertilizing species* doubled haploids can be directly used as varieties, but again, early loss and loss during doubling are great risks, and the number of lines required is large. Cost-benefit comparisons with selection during inbreeding in diploids have been caried out for instance in barley (see, for example Sitch and Snape 1986) and have shown that as yet the induction of haploidy can rarely compete with selection at the diploid level. In the near future this may change because in vitro and other techniques to produce doubled haploids continue to be improved. Doubling haploids has considerable potential. The application is typically a plant breeding subject and will not be discussed further here, where the emphasis will be on the cytogenetic aspects of induction.

11.4.2 Induction

Meiosis is the normal route for chromosome number reduction in higher plants. Meiosis has been discussed in previous chapters, and the ways to recover haploids from meiotic products will be briefly considered below, but first a few possible somatic ways of halving the chromosome number will be considered.

11.4.2.1 Somatic Reduction

Somatic reduction has two bases: genome separation ("reductional grouping") and chromosome elimination. Spontaneous separation of the genomes in somatic tissues of normal diploid plants was observed long ago on several occasions (Huskins 1948; Wilson and Cheng 1949), but is much more frequent in polyploids. Especially in the premeiotic division, reversion to diploidy can occur, sometimes accompanied by a certain degree of aneuploidy. The process of downward regulation has been described in detail for tetraploid tomato by Gottschalk (review with relevant literature in Gottschalk 1976). No pairing is involved, and it appears that the genomes or groups of genomes remain separated in the nucleus for several divisions before each forms an independent spindle, resulting in a primitive tetrad. Reduction from the diploid to the haploid level by this mechanism as part of a normal process of reduction has been described for certain insect species (Sect. 3.1.4.1.2.2), but has not been reported for plants.

A number of chemicals can induce or enhance genome separation. Agents affecting spindle polymerization, such as colchicine, vinblastine, benomyl, etc., have often been observed to induce reductional grouping but many other substances can have similar effects. Franzke and Ross (1952) attributed homozygosity of induced mutants in sorghum seedlings after colchicine treatment to reductional grouping of the genomes. Haploid groups arose with two chromatids per chromosome that stayed together in the same group, and consequently the chromosome number was immediately doubled. The new diploid nuclei were completely homozygous and any recessive mutation, induced or already present, was expressed. The phenomenon could be repeated, but was only convincing in a single variety of sorghum, Experimental III. Similar phenomena in other species were very hard to reproduce adequately. In root tips of barley "reductional grouping" can be readily observed (Sybenga 1955; Zhao and Davidson 1984). The latter authors observed a predominance of 6:8 and 4–10 segregations, followed by mitosis independently in the two groups. The arrangement of the chromosomes in the prophase nucleus, chromosome movements, and effects of colchicine on the cytoskeleton were factors to which the formation of separate chromosome groups were attributed. However, Sybenga (1955) tested abundant material in barley, involving many genotypes and heterozygous for markers in all of the seven chromosomes, but could never demonstrate induced homozygosity for more than a single marker at a time, which doubtlessly had arisen as a result of mutation.

In vitro genome separation is a subject of considerable interest because recovery of a reductional event does not depend on a few cells in a growing point. Gleba et al. (1987) reported genome separation in cell cultures of *Nicotiana* and *Atropa* after treatment with colchicine. Regeneration of cells with reduced chromosome number in plants, or homozygous regenerants were not reported. The subject is of sufficient interest to justify more extensive experimentation. An extreme case of chromosome separation, used for quite a different purpose, is realized by the use of amiprophos-methyl on dividing cells in culture (Ramulu et al. 1988). The chromosomes are spread singly or in small, random groups over the cells and remain separated at the following telophase. After removal of the cell wall micronuclei, representing individual chromosomes, can be isolated. The chance that a single, complete genome is formed and can be isolated is extremely small. Although somatic reduction of high polyploids leading of homogeneous reproductive segments of plants can be realized, somatic reduction of diploids to haploids in vivo or in vitro resulting in complete haploids is apparently very difficult and has not been realized yet, in spite of apparently frequent genome separation (Sect. 3.1.4).

More frequent and of considerable practical used is the elimination of chromosomes and ultimately entire genomes from hybrids. It is in principle possible in somatically produced, wide hybrids, but there it tends to be irregular and incomplete (Melchers et al. 1979). It can, more radically, occur after sexual hybridization, usually in a short period just after fertilization, in the developing embryo. The phenomenon has been discussed briefly in Section 3.1.4.1.2.2. It is observed in hybrids between species which are not necessarily taxonomically wide apart, but apparently have different systems of centromere regulation. The cause may be assumed to be a difference in timing of essential mitotic processes, leading to a loss of the chromosomes of the species that is the most retarded. Although observed in several different species hybrids, that involving *Hordeum bulbosum* as one of the parents is the most common. It was first observed by Kasha and Kao (1970) and Lange (1971). The *bulbosum* chromosomes are eliminated during embryo as well as endosperm development and the process has been analysed in detail (Bennett et al. 1976). It appears that not all *H. bulbosum* genotypes are equally well suitable. In the genus *Hordeum* many different hybrid combinations show early chromosome elimination. Chromosome position in the cell as well as mitotic cycle duration appear to have effects, depending on the genotype. In many institutions the system has been analyzed (see for instance Linde-Laursen and von Bothmer 1988; Jorgensen and von Bothmer 1988). For barley (*H. vulgare*) the "bulbosum technique" has become the standard method of producing haploid plants that are subsequently doubled to obtain entirely homozygous lines. The cross-compatibility between the two *Hordeum* species, however, is not sufficient for large-scale haploid production: the immature hybrids have to be rescued by in vitro culture. Culture of young pollen grains in anthers (anther culture) is becoming a successful alternative to obtain haploids in barley, and if the *bulbosum* technique is not improved, it may gradually be replaced by anther culture.

To produce wheat haploids the same approach is possible, but the rate of success of the cross between wheat (*Triticum aestivum*) and *H. bulbosum* is even lower. Yet, it is considered sufficient (after embryo rescue) to be a potentially acceptable method for constructing homozygous lines. Elimination of the *bulbosum* chromosome from the wheat × *H. bulbosum* hybrid is at least as effective as in the barley × *H. bulbosum* hybrid. Cross-compatibility is genetically determined, both by the wheat and the *H. bulbosum* parent. This means that not all wheat hybrids are equally favourable for producing homozygous lines from doubled haploids obtained with the *bulbosum* method. In wheat-maize crosses fertilization is relatively good, and early chromosome elimination of the maize chromosomes occurs. Elimination occurs also very early in the endosperm, however, and embryo rescue even in entire ovules is difficult. Although in some respects promising, this combination is not practicable on a sufficient level. Irradiation of the pollen or early embryo for transfer of chromosome segments from maize to wheat is a theoretical possibility (Laurie and Bennett 1989). As with barley, anther culture is rapidly evolving into a competitive (di)haploidization technique for wheat.

Both for wheat and barley, or any other diploid and allopolyploid crop species where the technique is applicable, the production of homozygous lines from doubled haploids is meant to replace inbreeding by selfing. In order to permit a certain degree of recombination and preliminary selection, it is presently recommended to produce the haploid from plants in F3 lines. For a review, see Sitch and Snape (1986). For autotetraploid potatoes the objectives are usually different. Here, parthenogenetic recovery of progeny with the gametic chromosome number is supposed to be more important than somatic elimination, although the latter was show to occur also (Clulow et al. 1991).

11.4.2.2 Parthenogentic Origin of Haploidy

Before the advent of the *bulbosum* technique for some material and the possibility to culture microspores on a large scale for other material, haploids were almost exclusively isolated after parthenogenetic development of unreduced eggs, occasionally sperm nuclei. For several species this is still, or could be, an important source of haploids that, after doubling, yield completely homozygous lines.

Spontaneous haploid parthenogenetic development of reduced female or male gametes occurs in nature with a variable but always low frequency except in certain apomicts, where it is part of the reproductive cycle, and is accompanied by other phenomena that assure chromosome doubling at the proper stage. The frequency is under *genetic control*, for instance in *Capsicum* (Morgan and Rappley 1954), where selection for high frequency of twin seedlings led to a twin percentage of 6% (D. Morgan, pers. comm. 1954), and in the proper genotypes, most of these contained one haploid. Although often referred to, selection of haploids among twin seedlings has not become very popular, apparently because it is cumbersome, and not successful in most cultivated species. It also restricts the genetic variation among the haploids recovered because there are few genotypes in which twins with haploids occur with sufficient frequency. Another reason is doubtlessly that other, more effective methods are available.

One comparatively successful method is relatively old. It consists of using *seedling markers* to select rare *spontaneous haploids* among large populations of diploids. Chase (1952) used a dominant coleoptile colour gene as marker in the male parent in maize. When crossed with a heterozygous female with the normal recessive allele, parthenogenetically developed haploids (and occasional diploids) are readily distinguished from the great majority of the hybrid seedlings with the dominant allele. A number of hybrids do not have sufficient expression of the dominant paternal allele, and there may also be maternal diploid parthenogentic progeny with possibly most of the maternal genotype. It is necessary, therefore, to check the chromosome number. Parthenogenetic diploids may be the result of chromosome number doubling of cells with originally the gametic chromosome number. In that case they are of similar value as haploids. If they result from first or second meiotic division restitution, they are only partly homozygous. Which mechanism predominates can be analyzed and depends on the species and to some extent on the genotype within species (Sarkar and Coe 1971).

A quite effective variant, provided the proper pollinator is available and with the potential of haploid parthenogenesis, is *pollination* with a *foreign species*. Smith (1946), when discussing several approaches to the induction of haploidy in the diploid wheat *Triticum monococcum*, reported a natural haploid frequency of 1 in about 2000. By pollination with a different, not closely related species of the Triticeae this could be increased to 1 in 50, and with delayed pollination even to 1 in 5. The frequency of seed set under such conditions is extremely low, but even then such high frequencies of haploids are exceptional. In potato (*Solanum tuberosum*), a natural autotetraploid, pollination with a foreign pollen, especially of certain strains of the related diploid *S. fureja*, has been used extensively to obtain (di)haploids (Hougas et al. 1958). With the combination of the use of a dominant seedling marker (seed spot) from *S. fureja*, the selection of (di)haploids has become very effective (Hermsen and Verdenius 1973) and from the (di)haploids complete (mono)haploids could be obtained by the same method. Not all *S. tuberosum* diploids found after pollination of tetraploid *S. tuberosum* by diploid *S. fureja* are formed parthenogenetically. Not infrequently trisomics with *S. fureja* chromosomes appear which indicates that originally triploid hybrids are formed from which the *S. fureja* chromosomes are somatically eliminated (Clulow et al. 1991; cf. Sects. 3.1.4.1.2.2 and 11.4.2.1).

The (di)haploids, or actually true diploids, have been used very extensively for genetic studies in potato, as well as for breeding at the diploid level, followed by doubling. For the latter step, protoplast fusion or meiotic doubling are the preferred approaches.

Pollination with *irradiated pollen* has occasionally led to parthenogenetic development of haploid plants, but the results are usually disappointing. When the dose is insufficient to completely prevent fertilization, diploids and aneuploids are formed with grossly aberrant phenotypes, especially in allopolyploids.

The in vitro growth of callus or embryoids from *microspores*, either in suspension or in the anther, has been successful in numerous species, and in several instances it has been possible to grow haploid plants from callus and embryoids. This has replaced other methods to produce haploids in a number of cases. In barley (and occasionally wheat, where the bulbosum technique works well) in vitro culture of haploids may soon become competitive. When in vitro culture of haploids is not yet sufficiently successful for most genotypes (rye, for instance) and other approaches do not work better, the only source of haploids is still the spontaneous origin, however, the frequency of recovery is far too limited for practical use.

Comparing the different approaches to haploid induction, a number of differences appear, apart from the technical difficulties and chances of success.

The somatic induction in established plants or tissue cultures by induced random chromosome elimination or reductional grouping leads to chromosome recombination, probably based on random assortment of chromosomes from the original set. Apart from very exceptional somatic crossing-over, this is the only form of recombination leading to segregation among the progeny. If an in vitro callus phase is part of the regeneration process, additional, usually not very desired mutational segregation can be observed. Since there are apparently differences between genotypes with respect to haploidization as well as in vitro regeneration, there is a restriction in the genetic variation among the haploids recovered. In practice the method is not yet, and may never become applicable.

Somatic elimination by the bulbosum technique is a "clean" technique, in that there is no appreciable somaclonal variation. Each individual haploid progeny plant is derived from a different gamete and between haploids derived from different, initially diploid zygotes, normal segregation is observed. There is some genetic variation in the level of success of haploid induction between parental (F1 or later generation) plants which are used to make haploids. This places a certain restriction on the total genetic variation available for selection.

There is a genetic component in the capacity of parthenogenetic haploid embryo development, both when of spontaneous origin and when induced or stimulated. Some genotypes produce haploids less readily than others. This again must restrict the total genetic variation available for selection among the haploids obtainable. There are no indications that there is selection at the level of the different haploids derived from one parent (Lashermes and Beckert 1988; Lashermes et al. 1988).

In vitro culture of haploid (micro)spores or anthers with young pollen, or unfertilized ovules, is also dependent on the genotype, but with increasing success of the technique, this will decrease. There may be selection between haploids within anthers, especially because the frequency of spores developing into embryoids is extremely low. Both types of genetic effects are more important in species where in vitro culture is generally less successful (such as wheat, rye) than where it is easier (such as in barley, many *Brassica* species and several Solanaceae). Anther culture, embryo induction, plant regeneration and frequency of green plants are among the factors involved that are genetically determined. Chromosomal constitution has an effect also (Agache et al. 1989) and can be used potentially to manipulate haploid production, thereby reducing the effects of limited genetic variation and selection. In vitro culture is known to induce mutations of different kinds, which are usually not welcome. Even after several generations of selfing, chromosomal irregularities in doubled haploids of wheat, when tested in backcrosses to their parent, are not eliminated (Youssef et al. 1989).

Chapter 12

Manipulation of the Genetic System

12.1 Natural and Artificial Variation in Generative Genetic Systems

In the present context the *generative genetic system* is understood to be the system regulating the genetic composition of the progeny of an individual or of a population through generative (or, exceptionally, vegetative) reproduction. In the case of apospory, with vegetative reproduction as an extreme form, there is no generative reproduction in the strict sense. However, in the cases where true embryos in seeds are formed, it is convenient to consider apospory as a special form of generative reproduction with a special genetic system: the genetic compsition is fundamentally identical to that of the parent, except for incidental mutations and somatic segregation.

With sexual reproduction, the genetic system has two components: the formation of gametes: *gametogenesis*, and the processes regulating *fertilization*. Both are under genetic control and can, consequently, vary quantitatively within species. Gametogenesis involves chromosome reduction in meiosis in the course of two successive cell divisions, with random chromosome assortment and crossing-over between homologous chromosomes, followed by the formation of haploid, recombined gametes (Sect. 3.2.2). Male and female gametes are subsequently combined (usually at random) into diploid zygotes during the process of fertilization, and the zygotes develop into normal organisms, capable of forming new haploid male and female gametes.

The genetic system generally accepted as the normal system appears to be no more than the most widely distributed and therefore apparently the generally most effective system. There are several natural variants: *bisexuality*; *self-incompatibility* and other systems to *enhance cross-breeding*; *absence of crossing-over* in one sex, as in male Diptera, but also in a few plant species (Sect. 3.2.3.1); *absence of random chromosome segregation* by *permanent translocation heterozygosity* (Sect. 5.4.1.4); *autopolyploidy*, which not only affects gene dose but also the segregation pattern (Sects. 6.1.2.2.1.4 and 11.3.1.2.2.2); several forms of *apomeiosis*.

There have been attempts to copy the most interesting of these in experiments with organisms where they do not naturally occur, hoping that they have advantages for practical application, under some conditions, over the system which is "normal" for the species. In addition to these natural variants, partly or even entirely artificial systems have been developed, especially for *hybrid* variety *breeding*.

In most plant breeding programs the genetic system is accepted as it is. A self-fertilizing species remains a self-fertilizer and a cross-fertilizer remains a cross-fertilizer, an autopolyploid remains an autopolyploid. A few attempts have been made to convert cross-breeders into self-fertilizers by developing inbred lines of good productivity. This has had no practical results. An attempt by Friedt (1979b) to introduce genes from the cleistogamous self-fertilizer *Secale vavilovii* into the cross-breeder *Secale cereale* (rye) was initially successful, but practical application has not been reported.

The presently most important successful change in the genetic system is the hybrid variety, where a species with a natural breeding system is forced to produce F1 or more complex hybrids exclusively. Several artificial systems, including a few that are based on typically cytogenetic approaches (discussed in Sect. 12.4), have been developed to breed and subsequently produce hybrid varieties.

In hybrid variety breeding fertilization is manipulated, i.e. heterozygotes are produced anew from homozygous parents in every generation. It is also possible in another manner to ensure heterozygosity, i.e. to maintain heterotic genotypes by limiting meiotic recombination with primarily cytogenetic techniques, and to complement this with specific systems of restriction of fertilization. One approach is the construction of *permanent translocation heterozygosity*, for which examples are available in nature: *Oenothera*, *Rhoeo* (Sect. 5.4.2.2). The possibility of artificially developing such systems will be discussed in Sect. 12.3.

The conservation of a heterotic genotype through one of the several possible variants of sexual reproduction can also be realized by *apomixis* and related systems. Apomixis is not uncommon in nature, but infrequent among cultivated species, the grass *Poa pratensis* and the blackberry *Rubus* being examples. The suppression of parts of the meiotic process as a special form of apomixis can be artificially induced and used to produce seeds which have (entirely or almost so) the genotype of the parent, even when heterozygous. These are special meiotic adaptations which result in unreduced gametes with complete or almost complete elimination of recombination. This will be briefly discussed in Section 12.5.

For still another system, no registered natural examples are available: *allopolyploidization of autopolyploids*. This will be discussed first (Sect. 12.2).

12.2 Allopolyploidization of Autopolyploids

In nature, autopolyploids are not uncommon, and several commercial plant species are natural autopolyploids. As discussed in Section 11.3.1 autopolyploidy has been induced for practical purposes in naturally diploid species, occasionally with considerable success. The possibilities and the problems involved, and the disadvantages compared to allopolyploids, have been considered: irregular meiotic segregation resulting in aneuploid progeny and reduced fertility; loss of heterozygosity with inbreeding; slow reaction to selection; delayed genetic stabilization. In order to overcome these difficulties, and especially to realize the important advantage of maintaining heterosis between different genomes, it has been attempted to convert the autopolyploid breeding system into that of an allopolyploid. To do so successfully, it is necessary to introduce an effective difference in affinity between genomes which are identical or at least fully homologous to start with.

Meiotic pairing differentiation has, basically, three components:

1. Differences in the specificities or activities of the chromosome segments responsible for the first attraction (zygomeres: Sybenga 1966; see Sect. 12.2.1).
2. Structural rearrangements, small or large, causing differences in the linear arrangement, and consequently the efficient functioning of such pairing initiation sites (Sect. 12.2.2).
3. Genes affecting pairing differentiation (Sects. 6.1.2.3 and 11.3.2). These genes can act only when there is an initial pairing differentiation system of one or both of the first two types. It is unrealistic to believe that such genes alone can introduce pairing differentiation in autotetraploids. They can act on different processes and at different times during the pairing process, for instance: decreasing or eliminating premeiotic chromosome associations; inactivating large segments of pairing sites; delaying pairing completion permitting homologous pairing to proceed at the expense of homoeologous pairing, etc. Their true action is usually not well understood, but they generally have effects on chiasma formation in addition to effects on pairing as such (Sect. 12.2.2.3).

When discussing the induction of preferential pairing in a program of allopolyploidization of autotetraploids, all three mechanisms must be considered.

12.2.1 Pairing Affinity Differentiation

Meiotic pairing affinity between genomes from different species is generally found to be weaker than affinity between genomes derived from the same species: there is pairing affinity differentiation (Sect. 11.3.2.2.1).

In some instances the differentiation is too weak to restrict pairing to completely homologous chromosomes even in the doubled hybrid, thus quadrivalents are formed. An example is the doubled hybrid between rye (*Secale cereale*) and *S. montanum* where Elçi and Sybenga (1976) observed only a limited reduction in quadrivalent formation compared with autotetraploid rye.

In such cases the enhancement of pairing differentiation by specific genetic pairing regulation systems, if available, can play a decisive role in the functioning of the allopolyploid. The best-known natural example is the case of tetraploid and hexaploid wheat (Riley and Chapman 1958; Sears and Okamoto 1958), although here the initial pairing affinity differentiation between the component genomes is stronger than in the case of *Secale*.

Estimating differences in affinity in such and more complex polyploid hybrids has been discussed in detail in Section 9.3.2. All methods are based on estimating the frequency of multivalents in comparison with the frequency of bivalents, taking into account the effect of chiasma frequency. The mechanism involved in pairing differentiation is essentially unknown, but it is not unreasonable to expect that within species, although at a much lower level, comparable chromosomally and genetically determined differentiating factors are present.

The meiotic association pattern of *marked chromosomes* gives an impression of the effect of variation in affinity, but it appears that the observations are erratic, showing that straightforward pairing preferences, if they exist, are overruled by factors of an unknown nature that can have both positive and negative effects on preferences for pairing between fully compared to not fully homologous chromosomes. Sybenga (1976) used combinations of genetically different normal and telocentric chromosomes of rye (*Secale cereale*) to see whether there were pairing preferences between recognizable chromosomes from varying genetic backgrounds. In an initial experiment it appeared that in heterozygotes apparent preferential pairing between identical chromosomes gave the expected shifts away from random pairing. In an experiment with better defined genetic relations, however, both a preference for identical chromosome pairing and a preference for non-identical pairing could be observed, even in the same genotypes. The frequency of branched configurations, implying partner exchange in the trisomic arm, varied erratically. In synaptonemal complex analyses (Qi et al. 1988) pairing partner exchange in the trisomic arm was not infrequent, and occasionally double. It was suggested that interaction between systematic variation in the location of pairing partner exchange and chiasma formation resulted in relative configuration frequencies which had nothing to do with pairing preferences. It seemed that large sectors of plants would behave similarly in this respect, and other sectors differently. The character of this apparent form of imprinting is not known.

Also in rye, Santos et al. (1983), Naranjo and Orellana (1984) and Orellana and Santos (1985) produced autotetraploids in which specific chromosomes in quadrivalents and bivalents could be recognized because of heterozygosity for specific C-bands. Here, too, it appeared that in some cases identical homologues paired preferentially, but in other cases pairing was preferentially between non-identical homologues (Sect. 11.3.1.2.3.4). Marking specific chromosome segments with the use of interstitial C-bands, Benavente and Orellana (1989) found that for the proximal segment of 6RL, homologous segments associated preferentially in metaphase I. To what extent this was due to true pairing preferences or to chiasma effects could not be determined.

Segregation of genetic markers is an alternative for measuring genetic variation in pairing affinity. The use of deviations from random segregation to estimate genome differentiation in artificial amphidiploids of cotton (Sved 1966) has been discussed in Section 9.3.2.2. There are several more cases of trisomic and tetrasomic segregations in genetically heterozygous material. An example will be given from the maize allotetraploidization program of Doyle (1979, 1982). In one of his experiments on the induction of increased differential pairing affinity, Doyle (l.c.) test-crossed trisomics for chromosome 3, simplex heterozygous (*aaA*) for the gene *A* (anthocyanin production), with diploid *aa* mothers. For trisomics of a standard line he found 1/3 *Aa* progeny, which corresponds to random pairing. Trisomic hybrids were made with the two *aa* chromosomes derived from the same standard line and the *A* chromosome from genetically different parents. These were exotic races, inbred lines, or variants of the standard line irradiated or chemically mutagenized.

Over 45000 gametes were tested from 48 different trisomic-3 plants. The frequency of recovered *A* types tended to decrease in the hybrid or mutagenized combinations, sometimes very considerably, indicating preferential pairing between the standard chromosomes with the *a* allele. However, there were always some plants with the opposite tendency where an excess of *A* chromosomes was recovered. In restituted, later-generation trisomics, these were even frequent.

For autotetraploids the gene *wx* (*waxy*) was used as a marker of chromosome 9 (Doyle 1979). It could be scored in the pollen after iodine staining, because of its effect on starch composition. In standard autotetraploids the expected frequency of 16.6% with random pairing was observed, but there was a considerable deviation in hybrids of different types and especially after irradiation of one of the parents. However, in the hybrids many cases of the opposite trend were found as well, like in the case of tetraploid rye mentioned above. Irradiation seemed to be an effective differentiating agent, and even though no gross chromosomal rearrangements were detected, the deviations were significant. However, there were still many significant deviations in the direction opposite those expected with preferential pairing, sometimes as many as there were cases of preferential pairing. In a subsequent paper Doyle (1986) reported on the positive effect of recurrent irradiation, which should in the long run result in a line sufficiently differentiated from the standard line to form a functional allopolyploid when combined with the standard in a tetraploid hybrid. The deviation from random segregation of *wx* increased every generation and the frequency of quadrivalents decreased accordingly. There were still plants, however, even in later generations, with the opposite effect. The basis of the differentiating effect of radiation is not clear, but the most reasonable explanation is the induction of small structural changes that cannot be observed in the microscope: "cryptic" rearrangements.

In apparently true autotetraploid rainbow trout, artificially produced by suppression of the first meiotic division of heterozygous diploids, 2/3 or 0.667 of the gametes are expected to be heterozygous when pairing is random. For several isozyme loci in different chromosomes, Diter et al. (1988) found an average of 0.83 heterozygous gametes. This was a significant difference, but there was a considerable variation between loci. As in the examples shown above, in some there was appreciable preferential pairing between non-identical chromosomes. Such observations on induced pairing differentiation are quite difficult to interpret. Especially when the same combination of chromosomes may behave differently in different situations, caution is required.

Cyclic irradiation of autotetraploids to increase the allopolyploid nature of autotetraploid barley, but in a more random manner, was done earlier by Gaul and coworkers (Bender and Gaul 1966). A detailed analysis of such material was published by Friedt (1978, 1979a). Reasonably fertile tetraploids could be isolated from crosses of different mutagen-treated autotetraploid varieties. Meiotic regularity was improved, with more bivalents, more chiasmata and slightly more rings among the quadrivalents. Alternate orientation was increased. The effects were apparently positive, especially regarding increased fertility, but do not represent real allopolyploidization. As in the case of the promising experiments carried out by Doyle (1979, 1986, 1990), it should be noted that as long as the differentiation is not complete and some "homoeologous" associations with genetic exchange are formed, the system is due to break down, as sets of fully homologous chromosomes will reappear. This is also the reason why in nature only very strong selection forces can counteract this tendency. Because these are usually not available, autopolyploids will remain autopolyploids even when some randomly distributed preferential pairing "floats" in the population (Sybenga 1984b). Such pairing differentiation may incidentially affect genetic ratios and multivalent frequencies, but it cannot stabilize itself. In fact, most cases of low multivalent frequencies and deviant segregation ratios have causes other than incidental preferential pairing (Sects. 6.1.2.3 and 11.3.1.2.3.4).

12.2.2 Chromosomal Rearrangements as Pairing Affinity Differentiating Factors

12.2.2.1 Translocations

The application of chromosomal rearrangements in allopolyploidization of autopolyploids has been reviewed rather extensively by Sybenga (1969, 1973). Important new developments have not been reported since.

In order to analyze the effect of rearrangements on preferential pairing as expressed in a change in the frequency of multivalents and in segregation ratios, it is necessary to first find a model for random pairing in the presence of chromosomal rearrangements to compare with the observations. Translocations and inversions are the best-analyzed rearrangements in autopolyploids. Linnert (1962), Sybenga (1973, 1975), Meister and Bretschneider (1977), and Doyle and Kimber (1983) report on the application of models predicting the expected frequency of different multivalent configurations for *translocations* in autotetraploids. The following is based on the model of Sybenga (1975; Fig. 12.1). Two of the four genomes are normal, and two carry the reciprocal translocation (duplex heterozygote); for other situations other models must be constructed. Only the end segments are considered, for many cases an acceptable simplification. Eight chromosomes are involved in the complex and there are two identical chromosomes of each type: two copies normal of one unchanged chromosome, two copies of the other unchanged chromosome, two translocated chromosomes of one type and two of the other. Each end segment is present four times, together having three different types of association. There are four different sets of four end segments, so there are $3 \times 3 \times 3 \times 3 = 81$ different combinations all together. In addition, each segment has its own probability of forming one or more chiasmata. It appears (Sybenga 1975) that when all end segments have at least one chiasma, close to 70% of the cells is expected to have a configuration larger than a quadrivalent. When one or more end segments has a smaller probability of forming at least one chiasma, the frequencies of the different configurations expected with random pairing have to be adjusted by introducing the specific chiasma frequencies into the model.

For one translocation tested in rye the chiasmate association frequencies for the eight segments were close to 1 and random pairing appeared to be realized. For two others the frequency of chiasmate association was reduced to 47 and 56% respectively for one segment, as estimated in the diploid. In both cases the frequency of large configurations, expected on the basis of random pairing in the tetraploid translocation duplex heterozygote, was somewhat higher than observed, indicating that the translocations had an effect on the pairing pattern, inducing a slight form of preferential pairing. In both cases the break was not far from the chromosome end, near the main segment where chromosome pairing is initiated.

In two-rowed spring barley, Bretschneider (1979) produced tetraploids with one to four duplex heterozygous translocations and reported some reduction in multivalent frequency and an increase in bivalent frequency in some combinations, compared to what was expected theoretically (Meister and Bretschneider 1977). The net effect of the translocations on field productivity, however, was variable and slightly negative (Scholz and Künzel 1981a).

A combination of two or more translocations in the same chromosome was found to be quite effective in a tetrasomic of barley: the rearranged chromosomes paired exclusively one with the other, as did the unchanged chromosomes (Tsuchiya 1969).

Induced tetraploids of the permanent translocation heterozygote *Rhoeo spathacea* (Walters and Gerstel 1948) could be demonstrated to show a low but significant level of preferential pairing (Sybenga 1973). In view of the limited

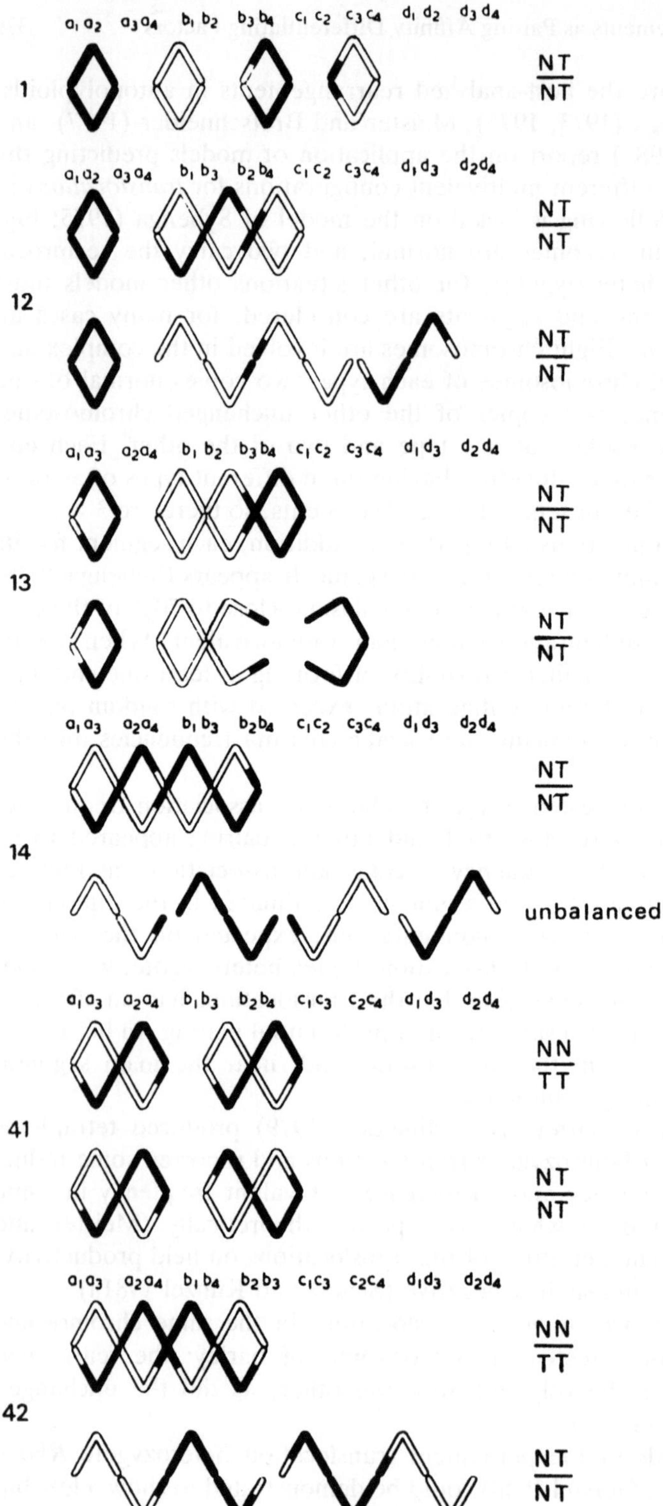
a1 a2 a3 a4 b1 b2 b3 b4 c1 c2 c3 c4 d1 d2 d3 d4
1
NT/NT
a1 a2 a3 a4 b1 b3 b2 b4 c1 c2 c3 c4 d1 d3 d2 d4
NT/NT
12
NT/NT
a1 a3 a2 a4 b1 b2 b3 b4 c1 c2 c3 c4 d1 d3 d2 d4
NT/NT
13
NT/NT
a1 a3 a2 a4 b1 b3 b2 b4 c1 c2 c3 c4 d1 d3 d2 d4
NT/NT
14
unbalanced
a1 a3 a2 a4 b1 b3 b2 b4 c1 c3 c2 c4 d1 d3 d2 d4
NN/TT
41
NT/NT
a1 a3 a2 a4 b1 b4 b2 b3 c1 c3 c2 c4 d1 d3 d2 d4
NN/TT
42
NT/NT

number of rearrangements per chromosome and the "random" location of the break points, it is not surprising that the level of preferential pairing was not high.

The effect of translocations is enhanced when they are present in interspecific tetraploid hybrids, even when without the translocation, hardly any differentiation between the parental species is detected. In the hybrid between rye (*Secale cereale*) and *S. montanum* the same rye translocations as discussed above (Sybenga 1973, 1975) had a much stronger effect on preferential pairing than in pure *S. cereale*, when studied in meiosis. In the diploid, pairing was almost like that in the species, and in the tetraploid, pairing between non-translocation chromosomes was frequently in quadrivalents, but not like that in the autotetraploid. In the same tetraploid hybrid without these but with two naturally occurring translocations, Reimann-Philipp and Eichhorn-Rohde (1970) studied the frequency of large multivalents caused by these two translocations. Basing their expectations on the simpler model of Linnert (1962), they concluded that preferential pairing was relatively strong. In the analysis of Elçi and Sybenga (1976) segregation of translocation T240W was tested in the tetraploid hybrid (Table 12.1) and found to deviate from random. In meiosis several larger configuations were observed, resulting from heterozygosity of T240W in addition to the two natural translocations, which were independent of T240W. In 28 cells analyzed, however, only 29 of such larger configurations were seen, whereas with random pairing and loss of chiasmata as in the diploid hybrid, 48 were expected. For T240W this is an additive effect of pairing differentiation between the species and a translocation effect. The duplex heterozygote of autotetraploid rye T240W does not induce noticeable preferential pairing (Sybenga 1973, 1975).

It should be noted that purely because of the characteristic meiotic behaviour of the translocation in a tetraploid, *preferential segregation* is pronounced among the balanced progeny, even without any *preferential pairing*. A frequency of heterozygous gametes (NT) higher than that of homozygous gametes (NN and TT), therefore, does not necessarily indicate preferential pairing.

Fig. 12.1 Autotetraploid duplex heterozygote for an interchange. In **1** the eight chromosomes involved (two sets of normal chromosomes and two sets of translocation chromosomes, terminal segments *a*, *b*, *c*, *d*, each in four copies, numbered *1–4*) can be seen to be able to form four homomorphic bivalents. There are $3^4 = 81$ different arm combinations, each with equal probability when pairing is random. Only one results in these four bivalents. Failure to form chiasmata reduces the size of the configurations, especially the large ones.

Six out of the 81 pairing combinations are shown, some with different combinations of chiasmata and different orientations and segregations of translocation (*T*) and normal (*N*) complexes (Sybenga 1975). The end segments associated are shown: *a–d* and *1–4*. In the second example (type **12**), for instance, the hexavalent has broken down into a quadrivalent and a bivalent as a result of chiasma failure in two segments. The two smaller configurations segregate independently

Table 12.1. Segregation of translocation T240W (3R/5R) in the progeny of the duplex heterozygous tetraploid hybrid between *Secale cereale* (rye) and the wild, close relative *S. montanum* (Elçi and Sybenga 1976)[a]

Gametic composition	Number of progeny plants Observed	Expected
NN	0	7.3
NT	92	85.4
TT	8	7.3

NN: normal karyotype for T240W
NT: heterozygous for T240W
TT: homozygous for T240W

[a] In order to reduce the effect of aneuploidy, resulting from the complex configurations in the heterozygous hybrid, the hybrid was used as the male parent: certation eliminates most of the aneuploid male gametes. The female parent was a normal rye tetraploid. There were 27 aneuploids, some possibly derived from the mother, and these were included with the euploids. The expected segregation is based on the model of Sybenga (1975) and shows that there is strong preferential segregation of the translocation, even with completely random pairing. The excess of preferential segregation is significant and due to the absence of NN gametes. It is the result of differentiation between the two species, and not primarily to the translocation, which has been shown not to affect pairing in the autotetraploid (Sybenga 1975).

12.2.2.2 Inversions

Inversions in autopolyploids follow a different pattern (Fig. 12.2). They form loops at pachytene, pretzel-shaped bivalents at diplotene, as in diploids (Sect. 5.3.4; Fig. 5.5), and the corresponding configurations at metaphase I (Figs. 5.5 and 5.6). The most striking are paracentric inversions because of the bridges they form at anaphase I or II (Sect. 5.3.4; Fig. 5.6; Dyer 1979). The frequency of cells with a bridge in duplex heterozygotes is lower than that in the corresponding diploids, because only two of the three combinations of two arms are heterozygous for the inversion. In such cells, however, two combina-

Fig. 12.2 Tetraploid simplex (**A**) and duplex (**B**) inversion heterozygotes. In the *simplex* there is always one heterozygous and one homozygous normal pair. In the *duplex* there are two possibilities: there is either a double heterozygous combination or a double homologous combination.

If the inversion is paracentric, in the duplex two bridges may be formed, in the simplex at most one. Only the pairing of the inversion segment matters, there may be partner exchange resulting in a quadrivalent in other segments of the chromosomes involved, without consequences for bridge formation

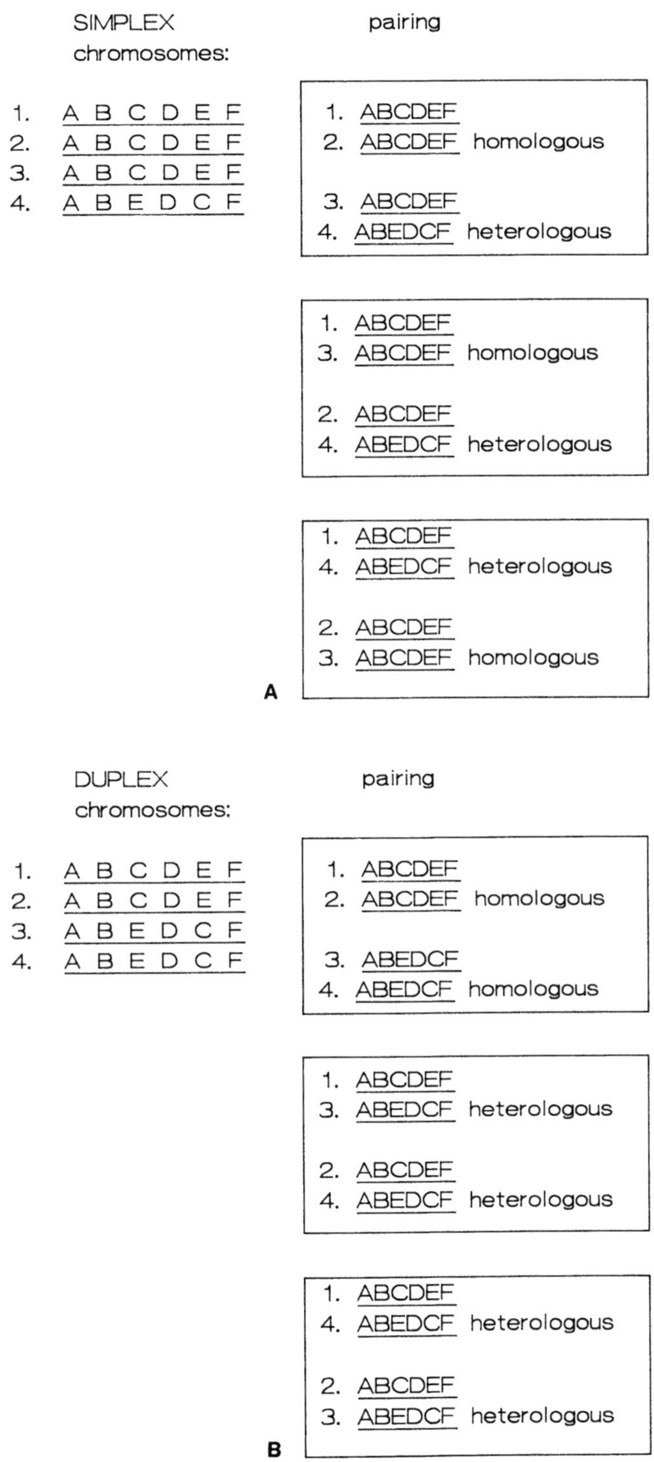
SIMPLEX
chromosomes:
pairing
1. A B C D E F
2. A B C D E F
3. A B C D E F
4. A B E D C F
1. ABCDEF
2. ABCDEF homologous
3. ABCDEF
4. ABEDCF heterologous
1. ABCDEF
3. ABCDEF homologous
2. ABCDEF
4. ABEDCF heterologous
1. ABCDEF
4. ABEDCF heterologous
2. ABCDEF
3. ABCDEF homologous
A
DUPLEX
chromosomes:
pairing
1. A B C D E F
2. A B C D E F
3. A B E D C F
4. A B E D C F
1. ABCDEF
2. ABCDEF homologous
3. ABEDCF
4. ABEDCF homologous
1. ABCDEF
3. ABEDCF heterologous
2. ABCDEF
4. ABEDCF heterologous
1. ABCDEF
4. ABEDCF heterologous
2. ABCDEF
3. ABEDCF heterologous
B

Fig. 12.2A,B

Table 12.2. Anaphase I and II bridges in simplex (InNNN) and duplex (InInNN) paracentric inversion heterozygotes in tetraploid maize (Doyle 1963). In: inversion chromosome; N: normal chromosome

Type	Anaphase I No bridge	Single bridge	Double bridge	Two bridges	Anaphase II No bridge	Bridge
Simplex	656	324	9		885	63
(%)	66.3	32.8	0.9		93.4	6.6
Duplex	754	103	6	13	460	24
(%)	86.1	11.8	0.7	1.5	95.0	5.0

Gene segregation in simplex (InNNN) and duplex (InInNN) paracentric inversion heterozygotes in tetraploid maize (Doyle 1963). Alleles marked with ′ are on inversion chromosome

Test cross	Gametes tested	%*a*
Aaaa × *aaaa*	983	54.63
A′aaa × *aaaa*	922	54.92
aaaa × *Aaaa*	2950	51.12
aaaa × *A′aaa*	1885	51.62
AAaa × *aaaa*	4413	20.83
A′A′aa × *aaaa*	3038	10.80
AAa′a′ × *aaaa*	6360	12.56

tions can form a bridge, and consequently, a significant proportion of cells must have two bridges. For bridge formation it is not relevant whether the combination of a normal and an inversion chromosome occurs in a bivalent or in a quadrivalent, as long as there is a chiasma in the pairing loop.

As shown by Doyle (1963) for maize, inversions induce preferential pairing more effectively than translocations, although probably not equally strong in all species and genotypes. Table 12.2 gives the frequencies of bridge formation as well as the segregation of a marker gene in simplex and duplex heterozygotes of maize (Doyle 1963). The simplex is a good check on the behaviour of the inversion because there is no opportunity for preferential pairing: the single inversion chromosome always pairs with a normal chromosome and the remaining two chromosomes are always normal. Again, pairing may be in bivalents or quadrivalents. Table 12.2 shows that the total bridge frequency is lower in the duplex, although the average probability of forming a heterologous (inversion/normal chromosome) combination with random pairing is the same in simplex, duplex and triplex heterozygotes (Fig. 12.2). Marker segregation showed the same deviation. The effect of marker loss as a result of bridge and fragment formation, and its effects on marker segregation and further complications have been discussed by Sybenga (1975).

Marker segregation in trisomics heterozygous for an inversion was also studied by Doyle (1979). Inversions appeared the most effective of all systems tested for inducing preferential pairing, but the effect of single inversions was still far from complete. When different inversions were compared, the nearer to the end, the greater the effect of a paracentric inversion. This corresponds to the conclusion of Burnham et al. (1972) that in maize pairing initiation is more frequently near (but not at) the chromosome ends than in more proximal regions. For pericentric inversions, the longer the inversion segment, i.e. the closer one of the breaks is to the end of the chromosomes, the more effective they are in inducing preferential pairing.

As with translocations, the effect of an inversion on preferential pairing is enhanced in tetraploid hybrids between closely related species, even when in the normal hybrid pairing differentiation is limited. This was shown for maize inversions in tetraploid maize and in tetraploid maize-teosinte hybrids (Shaver 1963).

At first sight, it would seem to be the most promising to concentrate on inversions for genome differentiation in autotetraploids, because they seem to be more effective and, in addition, involve single chromosomes, whereas translocations form complex configurations in intermediate stages in the tetraploid. When the differentiation is induced at the diploid level, translocations, although still more complex, can be handled more easily than in the tetraploid. Although there is insufficient practical experience using chromosomal rearrangements to induce pairing affinity differentiation, there is reason to expect, on the basis of the limited information provided above, that both inversions and translocations can be effective when induced at the diploid level in a series of treatments and made homozygous between treatments. In most species many more translocations are recovered than inversions, and it would be very uneconomical not to use translocations. In his later reports Doyle (1986) emphasizes "genetic" differentiation induced by mutations or "cryptic" rearrangements at the diploid level, followed by doubling. The selection of a specific recognizable rearrangement together with more cryptic differentiating agents may be an effective procedure. Large rearrangements can readily be checked for homozygosity in a systematically altered diploid line, but this is difficult for mutations and "cryptic" rearrangements. The final result will be a more or less standard altered line that can be made tetraploid and crossed with a series of normal tetraploids to produce different artificial allopolyploids from which the best can be selected. The altered line is hard to change by selection except by mutation and it will remain homozygous. Genetic variation can only be introduced through the normal tetraploids or by mutation.

As concluded by Sybenga (1969, 1973), the task of regulating pairing for each individual chromosome separately, and of completing the differentiation to a sufficient level by genetic means, is large. Yet, if allopolyploidization is ever to be successful, it may well be that a combination of chromosomal rearrangements, existing genetic diversity and mutagenization is the only way.

The possibility that the much simpler allopolyploidization of only one or a few chromosomes creates a segmental allopolyploid with specific advantages has not been studied.

12.2.2.3 Genes Enhancing Pairing Affinity Differentiation

Genes enhancing pairing affinity differentiation have been described in several allopolyploids where they are indispensable in preventing the chromosomes of the component species to pair and exchange genetic material. They have been studied in most detail in wheat. Although such genes could in principle be transferred to other Triticinae, for instance rye, this has not been attempted, and their effect on the enhancement of artificial systems of pairing affinity differentiation has not been studied. As discussed in Sections 6.1.2.3.4 and 11.3.2, in addition to wheat, a number of cases are known where genes affect pairing between homoeologous chromosomes in species hybrids and in their artificial allotetraploids. These include genes on otherwise genetically practically inactive B-chromosomes. Such genes do not have a comparable function in their natural carriers and the enhancing effect on pairing affinity differentiation must be considered a coincidence. The effect is usually insufficient for complete differentiation, except in the case of specific genotypes and B-chromosomes of *Aegilops speltoides* and *Ae. mutica*, which were found to substitute well for the *Ph* gene in wheat (Dover and Riley 1972).

The differentiation enhancing effects of the B-chromosomes of *Lolium* species in diploid and tetraploid hybrids between *L. temulentum* and *L. perenne* were not as complete as in the case of wheat. This is somewhat surprising, because the chromosomes of the two species are well differentiated with respect to size: *L. perenne* has considerably smaller chromosomes than *L. temulentum*. A similar, considerable difference in chromosome size was considered to be the main cause of the pairing differentiation between *Gossypium herbaceum* and *G. raimondii*, the presumed parental species of upland cotton, *G. hirsutum*, as discussed repeatedly before. In the hybrid the chromosomes pair well and form heteromorphic bivalents, but in the allotetraploid only homomorphic bivalents are formed (Sect. 6.1.2.3.4).

It is not certain that the mechanism of action of these genes would fit the artificial systems based on mutations, including cryptic and structural rearrangements. In view of the complexity of the transfer, and in spite of possible interesting effects, no attempts to combine pairing regulating genes with artificial differentiating systems are expected to be made in the near future, except where the genes are already available in the species involved. Both from a theoretical and a practical point of view, it is of interest that the attempts to artificially induce and reinforce pairing affinity differentiation will be continued.

12.3 Permanent Translocation Heterozygosity

In many species of animals, from insects to (less frequently) mammals, as well as in plants, chromosomal rearrangements "float" in the populations without being eliminated. It must be concluded that they have certain advantages with a stronger selection value than the obvious potential disadvantages for the fertility usually associated with them. The reduction of fertility is sometimes reduced considerably by meiotic and post-meiotic adaptations, and then a slight advantage is sufficient to cause their maintenance in spite of possible random loss. The most carefully studied are inversion polymorphisms in *Drosophila pseudoobscura* (Dobzhanski and coworkers, see for instance: Dobzhanski and Sturtevant 1938). In plants, inversion heterozygosity as a balanced polymorphism is rare. An example is *Paeonia california* where especially pericentric inversions, associated with recessive lethals, are relatively frequent in some populations (Walters 1952).

Translocation heterozygosity is more frequent, not only in animals but also in plants. In some species quite complex systems have developed (Fig. 12.3), some including sex chromosomes, especially in insects, termites and copepods. In *Rumex* Smith (1969) described translocation complexes of several chromosomes involving the sex chromosomes and a comparable system was found in *Viscum* (Barlow and Wiens 1975, 1976). As shown in Section 5.4.1.4, the chromosomes of a translocation complex form one large linkage group because there is no chromosomes recombination. Linkage between large segments of all chromosomes is often close, with crossing-over restricted to the terminal segments.

Natural, sex-linked translocation complexes must meet all requirements for alternate segregation, including terminal chiasma formation and absence of interstitial chiasmata. This reduces recombination in addition to the reduction caused by the absence of chromosome recombination. The association with the sex chromosomes ensures permanent heterozygosity in the heterogametic sex. The homogametic sex, however, does not have this advantage, and tends to be homozygous for one specific complex without much possibility to introduce variability through recombination. In theory, the complex can be enlarged indefinitely as long as the segregation remains predominantly alternate. The inevitable homozygosity of the complex in one parent probably also sets a limit to the practicable size of the complex. In the few cultivated plant species with sexual dimorphism, the introduction of a sex-linked translocation complex could be attempted.

Translocation complexes are also known in species without sex differentiation (Fig. 12.3); then there must be another system ensuring permanent heterozygosity. Single translocations may simply "float" in the population in a more or less random manner, finding an optimum frequency as a result of different random and selective forces. Single "floating" translocations have been observed in several plant species, sometimes restricted to specific populations. There they are present in relatively large numbers, whereas in other populations they have not become established (Figueiras et al. 1990). Genetically determined specific differences in multivalent behaviour could not be detected. Some of the small complexes, but especially the larger ones have acquired a special mechanism to maximize their frequency (Darlington 1965).

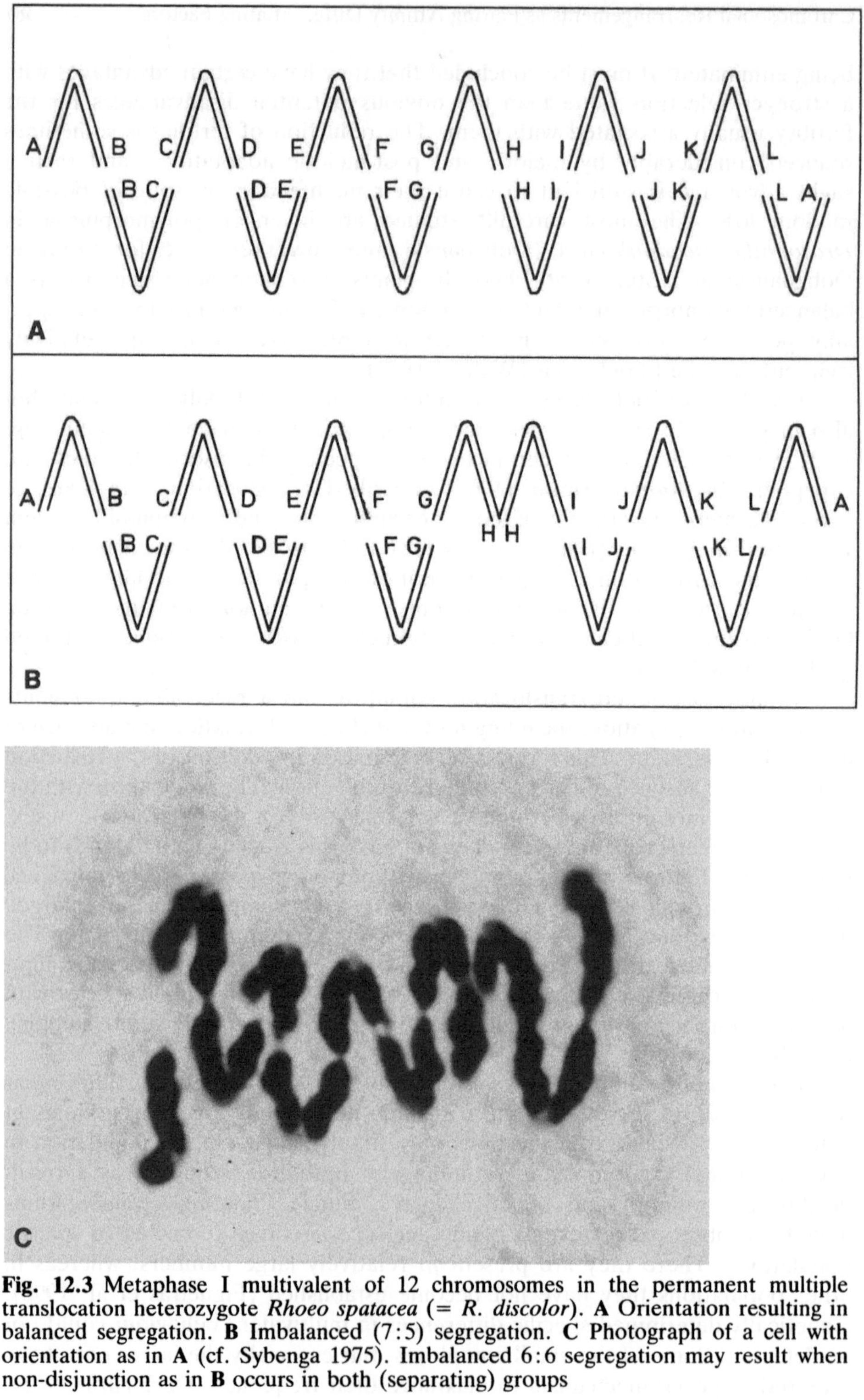

Fig. 12.3 Metaphase I multivalent of 12 chromosomes in the permanent multiple translocation heterozygote *Rhoeo spatacea* (= *R. discolor*). **A** Orientation resulting in balanced segregation. **B** Imbalanced (7:5) segregation. **C** Photograph of a cell with orientation as in **A** (cf. Sybenga 1975). Imbalanced 6:6 segregation may result when non-disjunction as in **B** occurs in both (separating) groups

In the genus *Oenothera* there is a complete range of sizes of translocation complexes. Several species have one or two linked translocations, *O. Lamarckiana* has six pairs in one complex and one chromosome pair free, whereas *O. biennis* has all 14 chromosomes in the complex (Cleland 1972). Segregation is very regular, although not always perfect.

There are two major systems which ensure that a translocation complement is always combined with a normal complement and that homozygotes of either type are not formed. One is a system of two balanced lethals, one in the normal complement and the other in the translocation complement. Homozygous embryos all die at an early stage. This implies 50% fertility reduction, compensated by a very high number of ovules. There is not necessarily a negative effect on total fertility, because early death of half of the embryos ensures good opportunities for the development of the remainder.

The second system is more complex and, if functioning properly, does not affect fertility. It is, in fact, a rather unexpected anomaly. After meiosis in the embryo sac mother cell, the four haploid meiotic products are aligned in the ovule in a linear order. The two different genomes have segregated at anaphase I, and as a consequence, in the linear tetrad two cells on one side contain a genome of one type, the two cells on the other side contain the other genome. Only a cell with one of the two genome types appears to be capable of differentiating into an embryo sac. The cell at the chalazal end is the cell which normally differentiates into the embryo sac. If this cell happens to contain the right genome, it develops normally and a female gamete is formed. If, however, the wrong genome is positioned at the chalazal end, it is not able to develop and one of the cells with the proper genome takes its place (*Renner* effect). As a consequence, the female gamete always contributes one specific genome. Pollen functions in fertilization only when it carries the other genome. Fertilization automatically restores the heterozygote.

Complete, permanent translocation heterozygosity involving all chromosomes is an exception. Besides species of the genus *Oenothera*, *Rhoeo spathacea* (syn. *R. discolor*) is an example (Fig. 12.3). There are several instances of translocation heterozygosity involving a smaller number of chromosomes and a less well perfected system ensuring heterozygosity. Apparently, in a wide variety of species and genera the potential of proper functioning of translocation heterozygosity is present and possibly in cultivated plants, too, such a system could be artificially introduced and perfected to a degree at which a desired heterozygote is reproduced automatically. When all chromosomes are involved, a complete, permanent heterozygote is produced and maintained almost like in a hybrid variety, with recombination and potential homozygosity after selfing in only small distal segments.

It would not be difficult to introduce balanced lethal genes in both genomes, but a "Renner effect" is not known to exist outside the natural, permanent translocation heterozygotes, and will be very hard to construct. This implies that reduced fertility is inevitable and it is probable that the system will function in seed crops only when the lethalities are expressed very early, and the free space in the ovary can be occupied by better developing vital embryos. This would exclude most cereals. A further complication is that lethal mutants recovered in traditional mutation programs tend to be late lethals because these are more readily detected than early lethals.

Attempts to isolate complexes of alternate orienting translocations have not yet been successful, for different reasons. Yamashita (1951) constructed a set of translocations involving all chromosomes in diploid wheat. With increasing size of the configuration fertility decreased drastically. In barley (Sisodia and Shebeski 1965) the translocations combined into a complex involving all chromosomes had originally been isolated on the basis of reduced fertility, which simplified their recognition. Reduced fertility, however, is the result of predominantly adjacent orientation, and this is exactly what is not desired and, consequently, the system was extremely infertile. Later, Scholz and Künzel (1981a) selected translocations with predominantly alternate orientation, and their combination resulted in a much better complex. In rye, where in general individual translocations have a high rate of alternate orientation, combining several translocations in one complex still resulted in considerable deviant segregation (own unpublished results), which made it unsatisfactory for application. Brar and Minocha (1982) produced a translocation complex in *Pennisetum americanum*, but this too was not functional. In diploid *Tradescantia*, which is taxonomically related to *Rhoeo*, Watanabe (1962) introduced a series of translocations sequentially by repeated irradiation in vegetative cycles and checked meiosis after each cycle of irradiation and vegetative propagation to see whether a new translocation had been added to the previous set. Clones with a larger complex were then propagated. It appeared possible to construct a complex including all chromosomes, but it had a quite unbalanced segregation. Apparently, there is still a considerable difference between the different genera of the Commelinaceae to which both *Rhoeo* and *Tradescantia* belong, with respect to centromere behaviour and chiasma localization (Sybenga and Rickards 1987).

The relation between centromere behaviour and suitability of a species or genotype for exploiting permanent translocation heterozygosity has been insufficiently studied. Several of the factors involved in the stabilization of autotetraploids (Sect. 11.3.1.2.3) may play a similar role in the establishment of permanent translocation heterozygosity. Marking centromeres by core staining and studying the timing of their separation in relation to the cell cylce, which is possible in several insect species (Janicke and LaFountain 1989; Suja et al. 1991), may soon become possible in plants, and may help in analyzing the factors involved in the meiotic behaviour of translocation heterozygotes.

Experience shows that it is not excessively difficult to produce a translocation complex involving all chromosomes, if desired. The introduction of a selective system based on early lethals should not be a major obstacle. The major problem is that the complex must function in practice. Any attempts should be restricted to species with an inherent tendency to alternate orientation and little interstitial chiasma formation. The translocations should be introduced in sequence in the same line. Combination of existing translocations into one line is possible only through interstitial crossing-over between translocations (Sects. 5.4.2.3 and 5.4.2.4). This implies interstitial chiasmata in the process of complex formation and a good probability that the end product is not free of them. These, as discussed, disturb alternate orientation. Since the genotype has an effect on orientation, sufficient genetic variation should be included in the translocation line to permit selection for regular segregation, in addition to selection for field performance characters. At each intermediate stage the meiotic behaviour and fertility of the heterozygote should be checked. Induction of new translocations should take place in the homozygote and subsequently checked in the heterozygous state, in combination with the proper tester line. When induced in the heterozygote, the new translocation added to the previous set should be checked carefully whether it is really associated with the earlier complex and does not involve homologous chromosomes of the partner genome.

It is not improbable that during the process new lethals will appear which are closely linked to the complex and that can serve as a selective factor against homozygosity. To recover early lethals, it is not sufficient to score for seedling lethals.

The complex should be combined with a number of selected partner lines in order to find the best combination, both with respect to field performance and genetic stability and fertility. There have not been enough serious attempts for a judgement of the feasability of the practical application of permanent translocation heterozygosity. In view of their widespread occurrence in nature, it is not unreasonable to expect some success in at least a few cultivated plant species. The first success may be with a permanent heterozygote involving only a limited number of chromosomes. Gradually, more chromosomes could be added. Introducing new genetic variation into permanent translocation heterozygotes, except by mutation, is not simple. As long as one of the genomes is structurally normal, any new combination with the necessarily more constant translocated genome can be made. The simplest approach is to maintain the translocated genome as a separate (homozygous) line, if necessary heterozygous for special regulating genes (lethals, for instance). Mutation and transformation are of special importance for permament translocation heterozygotes.

The special possibilities of sexually dimorphic (dioecious) species (spinach, asparagus, hemp, hops) are difficult to exploit for permanent translocation heterozygosity in practical breeding. One reason is that only one sex can be made permanently heterozygous. Another reason is that most dioecious species are gradually being converted into monoecious species, either because of higher productivity of the hermaphrodite or for simple breeding reasons. All-male asparagus could possibly be a candidate for permanent translocation heterozygosity, but it has a heterotic nature already without translocations. In hops (*Humulus*) naturally occurring, established sex-linked translocations have been observed (Darlington 1965).

12.4 Cytogenetic Approaches to Hybrid Breeding

12.4.1 Introduction

The primary aim of producing hybrid varieties is to make optimal use of heterozygosity. Hybrid breeding has this factor in common with allopolyploidization and permanent translocation heterozygosity, but the difference is that it has already had large-scale practical application. In addition to optimally exploiting heterozygosity, the advantages of hybrid varieties over open-pollinated varieties are homogeneity and the simplicity of introducing new specific genes. The species in which the hybrid variety is constructed is usually originally a random breeding cross-fertilizer, which may occasionally already have a specialized system of preferred cross-fertilization (self-incompatibility as in several *Brassicas* and related genera; mechanical means to prevent selfing as in maize). It may occasionally be a self-fertilizer, provided sufficient heterosis for yield or other production factors has been proven to be realized. Because of the higher requirement for (or capacity to exploit) heterozygosity of polyploids compared to diploids (Bingham 1980), self-pollinating allopolyploids like wheat are more successful as hybrids than self-pollinating diploids like barley, even though allopolyploids already have some built-in permanent heterozygosity.

In order to produce a hybrid variety it is necessary to make a large-scale cross between two homozygous (or almost homozygous) lines. The most commonly applied approach is to plant rows of an all-female line in the field, with a smaller number of rows of the pollinator at regular intervals. The production and maintenance of the all-female line are the primary stumbling blocks in hybrid variety breeding. Several systems have been proposed, but the most universally adopted is the use of male sterility conditioned by the cytoplasm (cytoplasmic male sterility: *CMS*). It has been widely applied in maize in hybrid varieties until the most generally used cytoplasm (the so-called Texas cytoplasm) appeared to carry extreme susceptibility to the southern corn leaf blight *Helminthosporium maydis* (Tatum 1971). Although the search for new male-sterile cytoplasms with the proper restorers has not been unsuccessful, male sterility has to a large extent been replaced by the old method of mechanical removal of the tassels.

In some horticultural crops where the price of individual seeds is high and each emasculation and pollination produce a large number of seeds, hand emasculation is feasable. In similar situations segregating recessive genetic male steriles can be selected and used for hand or mass pollination but this again is possible only on a restricted scale.

A very promising system that has not come up to expectation is the use of *chemical gametocides*. They appear to present a certain health hazard to the farmer, and are specific for specific crops, and there is even a difference between varieties in effectivity. They have not found the extensive use originally expected.

In most cultivated species where hybrid varieties are bred and cytoplasmic male sterility (CMS) does not present serious problems, it is the system used or planned. However, all apparently functional cytoplasms have a few (potential) disadvantages. One is the (probably rare) possibility of sensitivity to new races of pathogens, another is the difficulty to find sufficiently effective restorer genes. These must be present in the pollinator in order to restore the fertility of the plants in the production field when the seed is the end product. In cases where the vegetative parts are the product (onion, for instance) pollen fertility of the hybrid is not a requirement.

These disadvantages have long been known and have, several years ago, prompted a search for methods which apply the much more common, recessive nuclear gene male sterility. The advantage of nuclear gene male sterility is not only that there are many genes in many chromosomes that can mutate to male sterility of a useful type, but also that each normal line is automatically a perfect restorer. The great problem with nuclear gene male sterility is the difficulty in maintaining and increasing all-female lines for large-scale hybridization in the field.

Most systems exploiting nuclear gene male sterility are based on recessive alleles (*ms*) of dominant, wild-type (*Ms*) fertility factors (Rao et al. 1990). Dominant male sterility, however, has special possibilities, although typically cytogenetic approaches have not yet been proposed. Dominant sterility is very rare. It has been found as a spontaneous mutant in Chinese cabbage (*Brassica rapa*, van der Meer 1987), and mutagenically induced in in vitro culture of broccoli (*Brassica oleracea*) by N-nitroso-N-methylurea (NMU) (Dunemann and Grunewaldt 1991). Genetically stable dominant male sterility in tobacco and oilseed rape resulted from the transformation with constructed chimaeric ribonuclease genes based on cDNA clones derived from anther mRNAs (Mariani et al. 1990). These genes are organ-specific and block normal anther development. A method to exploit dominant male sterility has been proposed by Dunemann and Grunewalt (1991).

In addition to a few purely genetic systems (Rao et al. 1990), a number of typically cytogenetic systems have been developed that can (in principle) cope with the problem of maintenance and increase of all male-sterile (all female) stocks. Basically, fertile plants heterozygous for the male sterility gene (*Ms.ms*) are made to produce functional pollen of *ms* genotype only. *Ms* pollen, if formed, does not participate in fertilization because its genetic make-up makes it non-competitive or even non-functional. *Ms* egg cells, however, are fully functional. Consequently, heterozygous (*Ms.ms*) plants when selfed produce 50% male-sterile (*ms.ms*) and 50% male-fertile heterozygous progeny (*Ms.ms*). A homozygous male sterile *ms.ms* plant, pollinated by the pollen of the heterozygote, gives 100% male-sterile progeny when *Ms* pollen is not functional. With sufficient pollen production, there are no serious fertility problems. The simplest solution would be to have a *pollen killer* gene in the homologue of the chromosome carrying the *ms* allele, and closely linked (in repulsion) with the the *ms* gene (Lehman 1981; Rao et al. 1990). Such genes exist, but have not yet been found suitable for exploitation in crops where the construction of hybrid varieties based on nuclear gene male sterility genes has been attempted.

Comparable to pollen killing genes but slightly different, with some potential but also serious defects, are *gametocidal alien genes* introduced from specific chromosomes of some *Aegilops* species into wheat (Maan 1975; Endo 1982; Tsujimoto and Tsunewaki 1988, for example). These genes permit the microspores in which they occur to function, but microspores in which they do not occur do not develop into functional pollen. It looks as if the anther has become unsuitable for pollen development, except for pollen that carries the particular gene. Originally, the effect was observed when entire alien chromosomes of this particular type were present ("cuckoo" chromosomes), but later it appeared possible to recombine the genes responsible for this effect into the homoeologous wheat chromosomes. Several such genes and different alleles are presently known. Tsujimoto and Noda (1989) report a high frequency of mutations in hexaploid wheat into which the gametocidal genes Gc1a and Gc1b had been introduced by backcrossing from *Aegilops speltoides*. In their analysis speltoid mutants were emphasized, as these are readily recognized: the hemizygous expression of the Q-gene in chromosome 5A results in the appearance of the speltoid phenotype. In a total of 25 speltoid mutants the size of chromosome 5A was studied and 22 had an observable deletion of the long arm. There were indications that a transposable element was not involved. Such highly unfavourable side effects do not make the use of these genes attractive, but the principle remains interesting. A gametocidal fragment of a *Thinopyrum distychum* chromosome introduced into chromosome arm 7DL of wheat (Marais 1990) had less detrimental effects on the plant. It disturbed the segregation of the genes linked to it, but it was sufficient for a possible application in only a few genotypes. Both sexes were affected, but male transmission was affected most.

The review of Rao et al. (1990) discusses several applications of genic male sterility in plant breeding.

A very interesting development not yet discussed by Rao et al. (1990) is the present availability for transformation into the genome of Brassica spp. and possibly other crops, of a construct combining male sterility (cf. Mariani et al. 1990, see above) with resistance against herbicides with short persistence in the field.

12.4.2 Chromosomal Self-Regulating Systems for the Maintenance of Homozygous ms.ms Lines

In earlier Sections (5.2.5; 6.2.2.1.4; 6.2.2.4) the failure of extra chromosomal material (large duplications and extra chromosomes) to be transmitted through the pollen was mentioned. This phenomenon is the basis of a number of techniques which prevent pollen with an *Ms* allele to function. In principle, a plant with two normal chromosomes with *ms* alleles is made male-fertile by giving it an extra chromosome segment or entire chromosome with an *Ms* allele. This *ms.ms.Ms* plant is male-fertile, but pollen with the extra material carrying the *Ms* allele does not function in fertilization. The extra segment or chromosome with the *Ms* allele can be carried over to the next generation through the egg. Selfing the *ms.ms.Ms* stock results in a segregating progeny of male-sterile (*ms.ms*) and male-fertile (*ms.ms.Ms*) plants, which serves to reproduce the male-fertile stocks. To produce an all female stock the fertile *ms.ms.Ms* line is ued as the pollinator of a male-sterile (*ms.ms*) line. The male-fertile stocks are planted in rows, alternating with male-sterile plants in the field, the frequency of the fertile rows depending on pollen production. This combination reproduces the male-sterile stock, but the male-sterile plants must be harvested separately. Their seed can be used in the field as the male-sterile parent for producing the final hybrid variety.

There are a number of complications:

1. The reproductive capacity of the male-fertile stock carrying the trisomy or duplication is usually low. The transmission of the extra chromosomal material through the egg is at most 50% and usually much lower. The use of markers permitting early elimination of male-steriles without the extra material is possible, but complicates the system.
2. Complete absence of male transmission of the extra chromosomal material carrying the *Ms* allele is not guaranteed when the material is small enough for acceptable female fertility and field performance. Especially with low pollination rates there may be insufficient competition in fertilization to cause exclusive fertilization by euploid *ms* gametes (cf. Janse 1987). Low pollination rates are expected when it is attempted to maximize seed production by reducing the number of fertile rows compared with the number of male-sterile rows in the male-sterile propagation field. However, a small number of male-fertile escapes is not very detrimental. The fertile plants may be distinguished due to the morphological effects of the extra chromosomal material, and then removed. If this is not practicable, a few functional fertile plants among the steriles in the field where the hybrid seed is produced at most results in a limited number of "selfs" of the maternal line. These are less productive in the farmer's field, but when not too frequent this is tolerable. There may be a *problem with certification* when the selfs are morphologically very distinct.
3. More serious is recombination between the extra material and the standard chromosomes, bringing the *Ms* allele into a normal chromosome that is regularly transmitted through the pollen. This will result in increasing introgression of male fertility into the male-sterile stocks, which may ultimately be disastrous for the entire system.

Three different systems of chromosomal regulation of male *ms* transmission, and some variants, will be discussed.

12.4.2.1 Duplications

A system using a large duplication, although not the first proposed, is the simplest system and theoretically promising, because it has the least complications. Patterson (1973) proposed using a large duplication in maize, derived from a practically terminal translocation (Sects. 5.2.1 and 5.4.1.4). The normal chromosome carries the *ms* allele, the large duplicated segment the *Ms* allele (Fig. 12.4). The pollen transmission of the extra segment with the *Ms* allele is usually very low and the few resulting progeny can be removed from the field when their special morphology makes them recognizable at an early stage of development. Female transmission is high because there is, in principle, 50% probability of inclusion of the duplicated segment with the *Ms* allele in a functional egg cell. Production of male-sterile stocks as well as maintenance of the male-fertile duplication stock seem guaranteed. There are, however, several complications. Meiosis is not entirely regular in the presence of the large duplicated segment, and it results in the segregation of some abnormal progeny. Pollen production of the male-fertile duplication stock is irregular and not always sufficient, whereas male transmission of the duplication is occasionally too high and not always readily recognized in the field. Finally, the duplication, in order to be effective in limiting male transmission, may be large enough for recombination with the homologous segments in the normal chromosomes. This not only leads to meiotic irregularity and abnormal progeny, but also to the risk of transferring the *Ms* allele to normal chromosomes.

In such systems the major problem is probably to find small chromosome segments with a dominant allele of a functional *ms* gene and specific genotypes which, in combination, give sufficient reduction of male transmission and at the same time sufficently normal plant development and normal meiosis without recombination in the critical segment. It is not impossible that such combinations are found when sufficiently large material is analyzed. Menzel and Dougherty (1987) report that in cotton the male transmission of duplication-deficiency chromosomes is not primarily dependent on the size of the segments involved, but apparently mainly on their specific genetic composition. Selection of the proper duplication and the genetic background are clearly essential.

Patterson's (1973) system has not been pursued, in spite of the fact that the number of translocations available to choose from was very large, and it seemed probable that at least one derived duplication would be suitable. Finding the proper genetic background may be one of the bottlenecks.

Fig. 12.4 The balanced duplication system of Patterson (1973). **A** The segment with the dominant *Ms* allele is present as a (terminal) duplication in a chromosome other than that in which the recessive alleles are located. There are two types of gametes: one has the duplication (not male transmissible) with the dominant allele, combined with a recessive allele in another chromosome that is not expressed. The other gametic type has only the recessive allele, which is expressed after fertilization of a similar female gamete (below). **B** Selfing of the duplication heterozygote results in recovery of the fertile duplication types as well as homozygous recessive male steriles

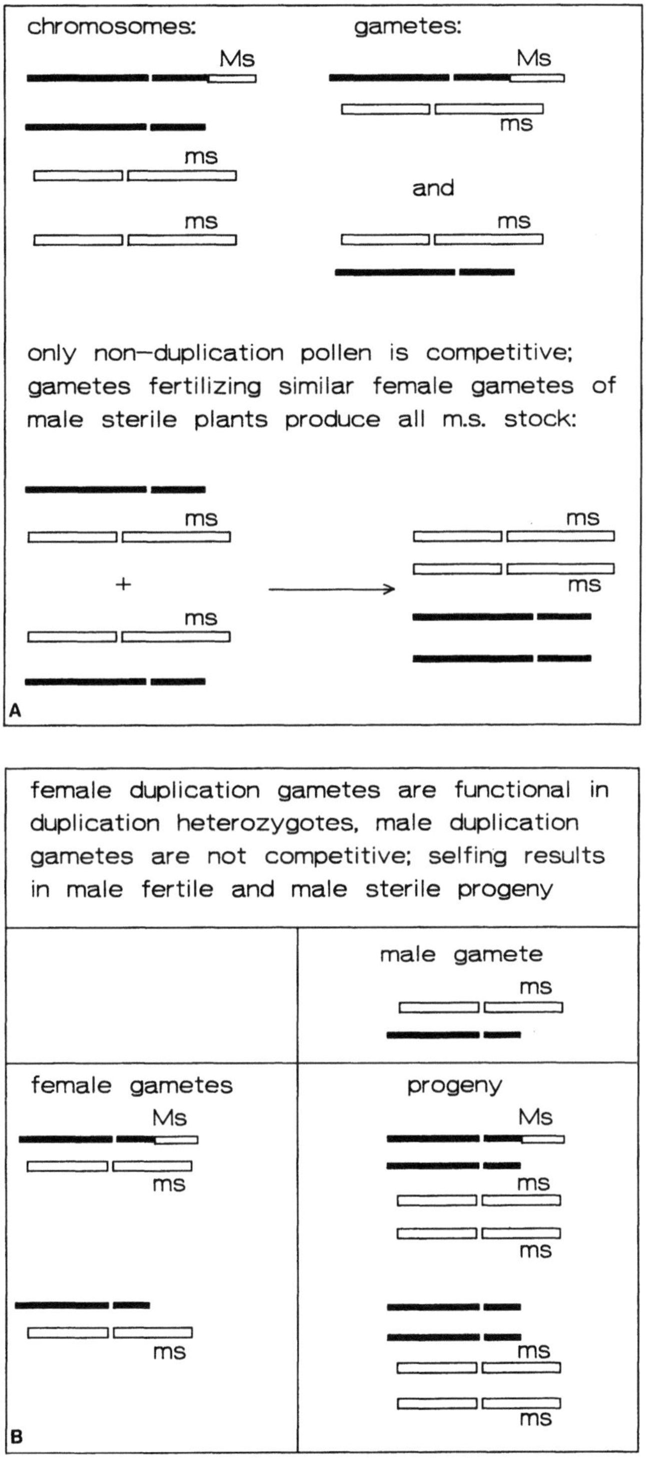
chromosomes:
gametes:
Ms
Ms
ms
ms
ms
and
ms
only non–duplication pollen is competitive; gametes fertilizing similar female gametes of male sterile plants produce all m.s. stock:
ms
ms
+
ms
ms
A
female duplication gametes are functional in duplication heterozygotes, male duplication gametes are not competitive; selfing results in male fertile and male sterile progeny
male gamete
ms
female gametes
Ms
ms
ms
progeny
Ms
ms
ms
ms
ms
B

Fig. 12.4A,B

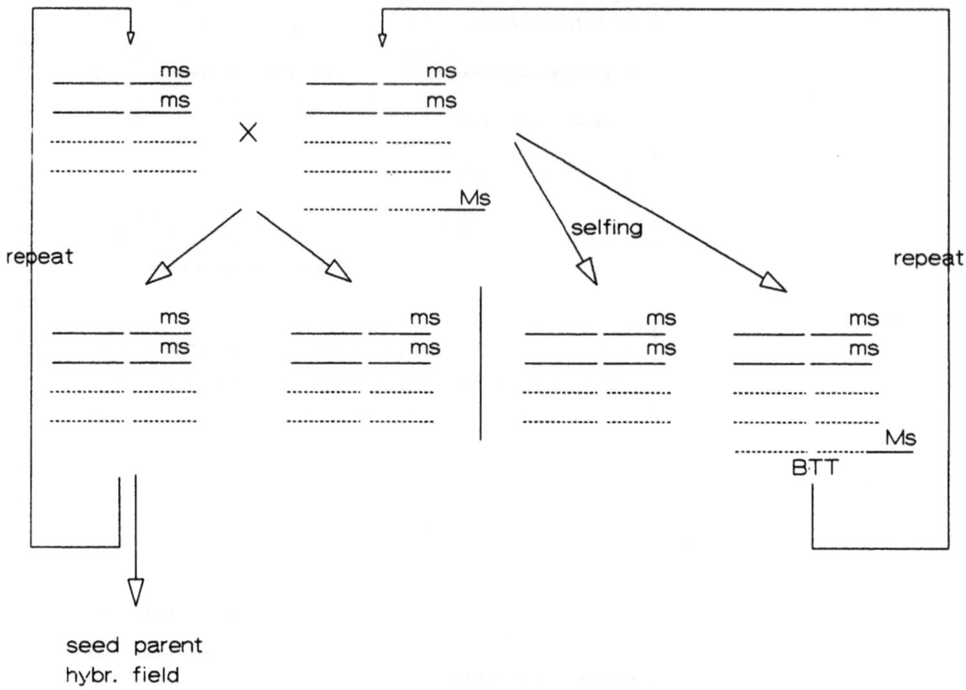

Fig. 12.5 The balanced tertiary trisomic system of Ramage (1983). When a male-sterile disomic is crossed as the female with a male-fertile tertiary trisomic, only male-sterile disomics result, because pollen with the extra chromosome, carrying the dominant allele, does not function. These male-steriles are used to produce the hybrid seed in the seed production field, with normal male-fertile partners as pollinators.

When the male-fertile tertiary trisomic is selfed, part of the progeny is male-fertile trisomic again, part is male-sterile. Recovery of the male-fertiles can be simplified when the trisomics can be made recognizable or when the male-steriles are eliminated by the presence of (conditional) lethal genes

12.4.2.2 Balanced Tertiary Trisomy and Variants

The first system proposed as a self-regulating chromosomal system for the propagation of nuclear gene male-sterile stocks was one involving balanced tertiary trisomy. It was developed for barley (Ramage 1965, 1983, 1991) and has attracted attention in most centers where large collections of translocations in barley are available (Lehman 1981; Scholz and Künzel 1986). This is important because the tertiary trisomics that are functional in some climates are not necessarily so in others (Scholz and Künzel 1986).

The principle of the application of balanced tertiary trisomy (BTT) for the reproduction of male-sterile (all female) lines is shown in Fig. 12.5. The extra chromosome is a translocation chromosome, which cannot replace any of the normal chromosomes. Its transmission through the male must be sufficiently low, so it must have a certain minimum size. For a minimal effect on plant development and female transmission, it must not be too large, however, and apparently the construction of the chromosome and the genotype of the carrier are both important factors influencing the success of the system. The meiotic behaviour of tertiary trisomics is such that it pairs relatively frequently with another chromosome and can then orientate sufficiently regularly to be included in one of the daughter cells. Nevertheless, occasionally a new segregational product, for instance a primary trisomic, is formed (de Vries 1984).

Because of the rearrangement, recombination in the neighbourhood of the break and in most of the interstitial segment between centromere and break is eliminated. The dominant *Ms* allele will find a safe place here. It balances the tertiary trisomy in the sense that the trisomics are necessary for reproduction of the population and, therefore, will be maintained automatically. Selfing tertiary trisomics results in a minority (about 30% in favourable cases, theoretical maximum 44.4%, de Vries 1984) of progeny with the extra chromosome.

In order to increase the frequency of male-fertile trisomics originally a DDT sensitivity gene was introduced, closely linked with the *ms* alleles in the normal chromosomes. Spraying the field with a DDT solution removed the male-steriles. After DDT had to be abandoned, selective lethal markers were introduced that automatially eliminate male-steriles among the progeny of selfed fertile trisomics. The disadvantage is that the trisomics inevitably carry the lethal marker in their normal chromosomes and when they are used for the propagation of the all male-sterile stocks, these markers are transferred to these stocks. Consequently, after one generation these all female stocks are heterozygous for the lethal factor and 50% of their progeny will die. This level will be maintained because all survivors are heterozygous for the lethality gene. This places an extra burden on the reproduction of male-steriles which increases the cost. In practice, this possibility has not been not used. In the reproduction field of the trisomics, seeding is sparse (5–7 kg/ha) and the recognizable disomic seedlings are removed by hand; the productivity of the trisomics is then optimal. In the male-sterile rows of the hybrid production field, seeding is dense, eliminating the less competitive trisomics. Scholz and Künzel (1986) in Germany, however, are not pessimistic regarding the use of seedling lethals.

In spite of some initial successes the BTT system has not yet been a commercial success in barley. There are several reasons. One is that the variation in the material tested has necessarily been limited because of the restrictions in the size of the programs. Another reason is that sufficient pollen production of the fertile lines remains a problem in this self-fertilizing, often almost cleistogamous species. Some lines produce more pollen than others, but probably the introduction of special genes from related wild relatives (*Hordeum spontaneum*, for instance), a complicated task, will be necessary for sufficient pollen production. The possibility of recombination between the *ms* locus and the rearrangement can be important, but apparently in the material tested this was not a serious problem. It is often assumed that genes in interstitial segments in translocation heterozygotes will not recombine, and for barley this may often be true. In other material sufficient reduction in crossing-over in the interstitial segment of translocation chromosomes cannot be relied upon (de Vries 1983). In addition, recombination patterns may be quite different in trisomics than they are in translocation heterozygotes and, on the basis of the recombinational behaviour of a translocation, it cannot be concluded that it will produce a useful tertiary trisomic. Absence of transmission of the extra chromosome through the pollen cannot be guaranteed, as mentioned above (Janse 1987), when the rate of pollination is low. Nevertheless, if sufficient input is available, it should be possible to construct functional BTT systems. It may be questioned, however, whether barley is the best crop for this purpose. In a diploid self-fertilizer, heterosis, although definitely present in heterozygotes, does not reach the same level as in polyploids and between selected inbred lines of cross-fertilizers.

A variant of tertiary trisomy in chromosomal self-regulation of the reproduction of male-sterile stocks proposed by Sybenga (1982) and worked out by de Vries (1985) for rye, is the use of *compensating trisomics* (cf. Khush 1973; Fig. 6.13). These trisomics will, in principle, be closer to the maximum of 50% trisomics in the progeny of selfed trisomics than tertiary trisomics because the extra chromosome is, like a duplication, an intrinsic part of the genome in half of the gametes. In addition, there are more possibilities to incorporate specific, different chromosome segments into the system. The meiotic complications are serious but may be overcome, especially with telocentric compensating trisomics.

12.4.2.3 XYZ System in Wheat

The third type of chromosomal self-regulating system of reproducing male-sterile lines was developed for wheat by Driscoll (1972, 1985). Polyploids present problems in the use of recessive male sterility genes. If a locus in one genome mutates from dominant to recessive, it may still not be expressed in the homozygous state because of the possible presence of homoeologous dominant alleles in the other genomes. Whereas in barley almost 30 male sterility genes are known, the number is quite small in wheat. They are alleles of loci on a small number of chromosomes which do not have dominant alleles in homoeologues. One is the "cornerstone" male-sterile mutant. Isolation of mutants of such specific genes is facilitated by using monosomics which allow the detection of hemizygous mutants when pollinated with pollen from irradiated spikes (cf. Sect. 8.3.2.1.1). Using this procedure Driscoll and Barlow (1976) isolated mutants on chromosome 4A which were very promising for application.

The XYZ system makes use of one of the available recessive male sterility factors in wheat and an alien addition chromosome carrying a dominant *Ms* allele which suppresses the expression of the recessive (*ms*) male sterility allele. The alien addition can be derived for instance from rye (*Secale cereale*, Hossain and Driscoll 1983), or from *Aegilops umbellulata*, *Agropyrum elongatum*, or *Triticum monococcum*. Even a barley chromosome might function. The "cornerstone" mutant on 4A could be compensated by the short arm of chromosome 4R of rye, but, surprisingly, 2Rs was even better. However, the entire chromosome 2R was less effective. Apparently, antifertility genes in the addition chromosomes are one of the factors that can disturb the system (Hossain and Driscoll 1983). Stability of the disomic addition line is important for the functioning of the system, and additions from very distant species are not always acceptable in this respect.

Basically, the system functions as follows. As the name indicates, there are three lines: all three are homozygous *ms.ms* and contain two (X-line), one (Y-line) and zero (Z-line) alien chromosomes respectively. The euploid 21 chromosome pollen of the Y-line has a competitive advantage over the 22 chromosome pollen with the alien addition chromosome. It has an almost complete monopoly on fertilization. Therefore, in practice, the Y-line contributes only *ms* alleles to the progeny when used as the male parent. The X-line is necessary because the Y-line, unlike tertiary trisomics, produces insuffficent trisomic progeny. The X-line is relatively stable, although the alien chromosome, which is not required for functional haplo- and diplophases, is not infrequently lost during generative reproduction.

The Z-line plants are male-sterile and may be obtained from the Y-line, which itself produces about 75% Z-progeny. For the production of a Y-population, rows of Z-plants are alternated with rows of X-plants and the seed produced on the Z-plants is Y-type, with some admixtures of Z, especially when the X-line is somewhat unstable. To increase the male-sterile Z-population, Z-rows are grown alternating with Y-rows and the seeds harvested on the Z-plants are (practically) all male-sterile. These are used to produce the hybrid seed with a selected normal pollinator as the male parent.

Later, the system was made simpler and equally effective by omitting the X-line and using the selfed progeny of the Y-line instead (Driscoll 1985). It contains, depending on the addition chromosome, approximately 72% Z-plants, 27% Y-plants and 1% X-plants. This simplifcation permits the use of an isochromosome, which usually cannot be made disomic but has the advantages of a double male fertility dose, very low male transmission and good stability as a univalent. The admixture of Z-plants reduces the pollen density in the field and larger pollinator blocks must be used. Isochromosomes of chromosomes 4 of *Triticum urartu* and possibly of *Hordeum vulgare* are good addition chromosomes.

There has been some practical application, but CMS and later the use of gametocides, when these still gave the impression of being very promising, have for some time reduced the interest in the conceptually more complex BTT and XYZ systems. It is well possible that in the near future this interest will increase again.

Table 12.3. Terminology of apomixis. (Rutishauser 1967)

1. *Development of the gametophyte*
Apomeiosis: Development of a female gametophyte without reduction in chromosome number. The degree of apomeiosis is the percent unreduced female gametes formed.
Apospory: Formation of an unreduced embryo sac from a vegetative cell of the ovule.
Diplospory: Formation of an unreduced embryo sac from a generative cell (archespore cells) of the ovule.

2. *Development of the sporophyte*
Parthenogenesis: Autonomous development of an embryo from an unfertilized egg cell. Parthenogenesis is haploid when the egg is reduced, diploid when unreduced.
Pseudogamy: Parthenogenetic development of an egg cell under the stimulus of pollen.
Adventitious or *nucellar embryony*: Development of an embryo from one or more somatic cells of the nucellus of the ovule. It is *autonomous* when no induction is necessary; it is *induced* when pollination is required.
Apogamety: Development of an embryo from a vegatative cell of the embryo sac (synergid, for instance). Infrequent.

12.5 Apomixis and Related Systems

The most faithful reproduction of specific heterozygous genotypes by seed is through apomixis. There are several forms of apomixis (Table 12.3; cf. Rutishauser 1967). Some leave out meiosis, and diploid embryo sacs or even embryos are formed directly from diploid tissues of the reproductive organs, for instance the integuments (*apomeiosis*, *apospory*). Others involve a modified but still mechanically functional meiosis without reduction, or a completely normal meiosis after premeiotic chromosome doubling. In both cases meiosis must be followed by *parthenogenetic development* of embryos from one of the meiotic products.

One problem with the practical application of apomictic reproduction is that it is not readily induced in normal, generatively reproducing plants. Yet, for plant breeding a generative phase preceding the final genetic fixation by apomixis is a prerequisite. The solution is to introduce genes for apomixis into the breeding material and to select for both apomixis and a desired genotype in the later breeding phases. This is in principle possible for both types of apomixis, but it requires the availability of some generative reproduction or generative segregants in an apomictic population from which apomixis genes can be introduced into a normal generative population. Among the several attempts reported, those by Murty et al. (1984) and by Dujardin and Hanna (1986) are relatively recent examples of the introduction of apomixis genes into sorghum and pearl millet respectively.

Transfer of apomixis genes and their exploitation require a thorough understanding of the type and the genetics of apomixis in the species concerned. In his review on reproduction and meiosis of apomictic flowering plants, Rutishauser (1967) discusses the genetics of the different types of apomicts (see also Asker 1980). A somewhat older, but very extensive review of apomixis in plants, also considering genetics and meiosis, is that of Gustafsson (1947). The genetics of apomictic development will not be further discussed here. The emphasis is on cytogenetic aspects of induction and maintenance of apomixis.

It has been suggested on several occasions that interspecific hybridity and polyploidy play a role in apomixis (Ernst 1918; Gustafsson 1947; Rutishauser 1967; Gottschalk 1976). One possible reason why several apomicts are hybrids and/or polyploids is that their sexual fertility is often insufficient to guarantee reproduction at the necessary level and that only those types that happen to have a tendency to apomixis can maintain themselves. This does not exclude the possibility that polyploidy does stimulate parthenogenetic development (Gröber and Zacharias 1983) by slightly deregulating the subtle timing of processes involved in generative reproduction. Asker (1980), on the other hand, reports that diploid facultative apomicts can give rise to totally sexual autotetraploids.

Another example where apomixis occurred in polyploids without polyploidy being the cause of apomixis is that of Hashemi et al. (1989), who compared the natural diplosporous and pseudogamous apomictic autotetraploid ($4x = 72$) *Parthenium argentatum* (the rubber-producing guayule) with the artificial autotetraploid ($4x = 72$) made from the natural diploid *Parthenium argentatum* ($2x = 36$). They observed that the artificial tetraploid, after crossing with the highly compatible, related but morphologically distinct, diploid *P. rollinsianum* ($2x = 36$), always formed triploid $3x = 54$ progeny. The natural tetraploid apomict formed $3x = 54$ triploids in only 15 out of 179 progeny (8.38%). The remainder was matromorphic and tetraploid, apparently produced by apomixis. Apparently, polyploidy as such was not the reason for apomixis, nor was the natural apomictic tetraploid an interspecific hybrid, since in meiosis it behaved almost exactly like the artificial autotetraploid. It is also clear that apomixis was not complete in this experiment, as had been observed earlier.

A strong tendency to parthenogenetic development must be the basis of apomixis in plants where the main reproductive cycle has remained intact. For the artificial induction of apomixis such a tendency is a necessary starting point. In addition to spontaneous parthenogenetically developed haploids (Sects. 6.1.1.1 and 11.4.2.2), diploid spontaneous "matromorphs" have been found in several species with varying frequencies. The genus *Brassica* is well known for this tendency (Eenink 1974a) and both heterozygous and homozygous diploids have been observed (Eenink 1974b; Röbbelen 1966). Starting from a heterozygous mother plant, heterozygous matromorphous progeny are formed from unreduced embryo sac mother cells and homozygous (partly) matromorphs from reduced and subsequently doubled, unfertilized eggs. Prickle pollination with own or even foreign pollen is sometimes necessary to stimulate parthenogenetic development. In several cases fertilization of

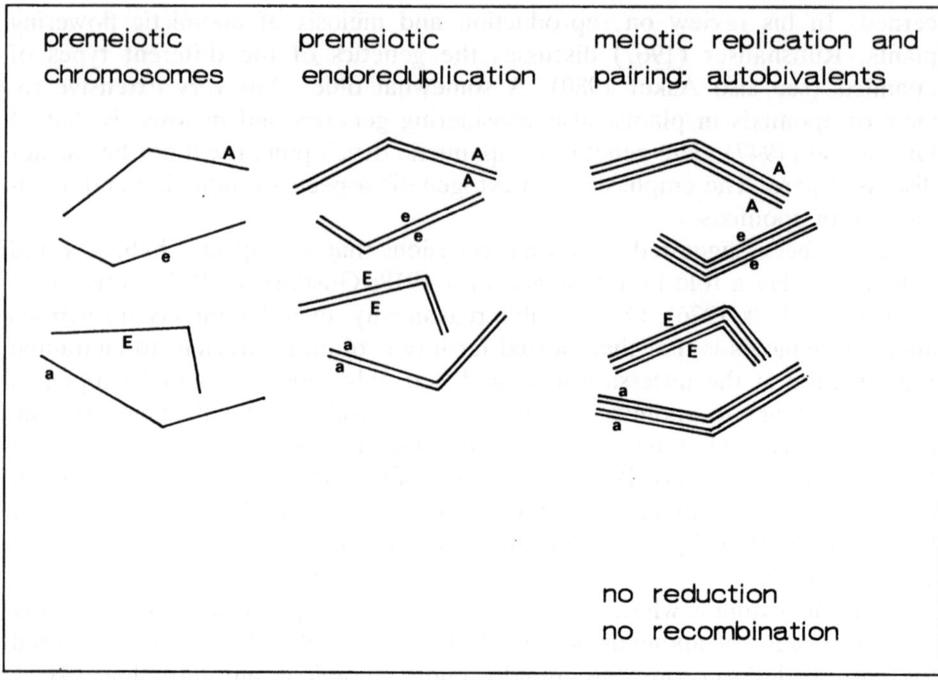

Fig. 12.6 Autobivalent formation. After premeiotic endoreduplication two complete sister chromosomes are formed, each with two chromatids. Because of close proximity, they consistently pair with each other with exclusion of other homologous chromosomes in the same cell. They form chiasmata and segregate normally, but since the chromosomes involved in the bivalents are identical, there is no gene segregation. The resulting spores have the doubled chromosome number and are genetically identical to the parent on which they are formed

the secondary pole nucleus is required for endosperm development, which in turn is a prerequisite for proper embryo development (pseudogamy).

Incidental matromorphy and other forms of incompletely expressed apomictic tendencies have been observed in several more species, such as *Solanum* (Hermsen 1980), *Pisum* (Virk and Gupta 1984), cereals (Matzk 1982), including maize (Sarkar and Coe 1971), sorghum (Murty et al. 1984), *Hordeum – Triticum* hybrids (Mujeeb-Kazi 1981) and several more, some mentioned earlier. Apospory, haploid parthenogenesis and homozygous matromorphy resulting from the doubling of reduced gametes will not be discussed here further.

On the basis of a capacity for parthenogenetic development, mainly two processes can lead to the induction of functional apomixis. One is *premeiotic endoreduplication* or *endomitosis*, which results in two identical chromosomes very close together, in parallel alignment, both with two chromatids. These pair at zygotene to form *autobivalents* (Fig. 12.6), with the exclusion of the other homologues, which are similarly double. At anaphase I two identical *diploid daughter nuclei* are formed resulting in a tetrad of four diploid cells with the parental genotype. When the egg cell develops parthenogenetically into an embryo, the apomictic cycle is complete. It may require fertilization of the secondary polar cell (pseudogamy) to ensure endosperm development. The phenomenon occurs in parthenogenetic insects (Pijnacker and Ferwerda 1982) where pseudogamy is not essential, but in plants the phenomenon is rare. Examples have been given by Håkanson and Levan (1957) for *Allium nutans* (2n = 40) and autotetraploid *A. odorum* (2n = 32), and by Gohil and Kaul (1981) for the autotetraploid Indian *A. tuberosum* (Chinese chives). In the latter case, which has regional commercial importance, related sexual forms are available for the transfer of the premeiotic doubling genes to sexual forms, and possibly to other related species. It is interesting to note that all *Allium* species recorded having this particular system of apomictic reproduction are polyploids. If it does appear possible to transfer the critical genes from one *Allium* species to another, an interesting form of permanent heterosis can be created. In other species occasional premeiotic doubling has been observed, but a systematic search for mutants with consistent premeiotic doubling has not been made.

The second type of apomixis, based primarily on a meiotic abnormality combined with a tendency to parthenogenetic development with proven potential for manipulation, is *first division restitution* (*FDR*; Fig. 11.1). Potato cytogeneticists (see Hermsen 1980; Hermsen et al. 1985) have proposed using FDR to produce potato seeds with all or at least most of the (heterozygous) maternal genotype intact.

There are two main reasons to replace vegetative reproduction of the potato by reproduction by seed. One is a phytosanitary reason: virus infection is a serious threat to potato production. With strict sanitary regimes it can be kept at a low level, but this is not possible under all conditions, especially in tropical countries. Reproduction by seed eliminates the virus. The second reason is that, again especially in tropical countries, tubers are much more difficult to keep from one season to the next than seeds. A disadvantage of seeds is the longer juvenile period. There are, however, genotypes with rapid seedling development and good tuber formation.

With normal bivalent (or quadrivalent) formation at meiosis there is at least some recombination between homologues and when, after FDR, the chromatids separate, there is a good chance that in the same daughter nucleus identical alleles are combined in a process somewhat resembling double reduction in autopolyploids (Fig. 11.1A). In several tuberous *Solanum* species FDR occurs as a spontaneous deviation with variable frequency, apparently under the influence of the genotype. Second division restitution (SDR) is more common but does not have the same conserving effect on the genotype (Fig. 11.1C). By the action of the parallel spindle (*ps*) mutant, the frequency of SDR is greatly enhanced. These phenomena have been discussed in Section 11.3.1.2.1.2.

For apomixis it is important that genes for desynapsis can be introduced (Ramanna 1983; Hermsen et al. 1985), which reduce the level of recombination and stimulate FDR. Most experiments have been carried out with diploid potatoes, which are easier to handle and to make homozygous for synaptic mutants. Douches and Quiros (1988) reported that in the presence of a homozygous recessive desynaptic mutant in a diploid hybrid between *Solanum tuberosum* and *S. chacoense*, most chromosomes were present as univalents at diakinesis. In a half-tetrad analysis using the 4x × 2x cross, a strong reduction in recombination was found. Heterozygosity in the unreduced gametes was maintained at a very high level. It has appeared possible to isolate desynaptic mutants with even lower bivalent frequencies and strong but variable reduction in recombination (Jongedijk et al. 1991a) and practically complete maintenance of heterozygosity. FDR is stimulated by desynapsis, especially in genotypes with a strong tendency to early centromere activation. This is expressed as complete amphitelic univalent orientation at anaphase I in the presence of desynapsis, resulting in systematic FDR. In absence of desynapsis, early centromere activation is expressed as centromere split during interkinesis, which results in SDR (Jongedijk 1991). A number of desynapsis genes have been isolated in tuber-bearing *Solanums*, and their meiotic behaviour in the EMCs has been described (Jongedijk 1986; Jongedijk et al. 1991a). With the combination of the proper desynapsis genes and specific meiotic background genotypes, which are already available, it is in principle possible to produce genetically homogeneous and constantly heterotic true seed potatoes.

When at the diploid level occasional functional gametes are formed, it is possible to cross two selected desynaptic diploids to produce desynaptic tetraploids directly. Assuming that at the tetraploid level parthenogenetic development is even stronger than at the diploid level, tetraploid seed potatoes breeding true to seed can then be obtained. Since potato normally reproduces vegetatively, the original, completely heterozygous form can be maintained vegetatively on a limited scale as elite material and, if necessary, used as a source for new seeds of the original composition. Especially with incomplete transmission of heterozygosity, vegetative maintenance may be an advantage. Even if the obstacles against the introduction of true seed potatoes mentioned above could be overcome, several practical problems still have to be solved. One is the limited amount of fruits with seeds formed in numerous varieties.

Although the same principle can be applied to other cultivated plant species, for instance the genus *Brassica*, the developments have nowhere gone as far as in the potato.

In addition to the types of apomixis mentioned, and other special systems for conserving heterozygosity (permanent translocation heterozygosity: Sect. 12.3), a number of natural systems, sometimes rather bizarre, exist that have the same capacity. Insects have developed special cytogenetic systems more effectively than plants, but even in plants complex chromosome behaviour at meiosis may be established as reproductive innovations (Darlington 1965) with specific, apparently useful consequences. Although, in principle, these could be constructed artificially in cultivated plants, for the time being these are merely interesting as a curiosity.

None of these typically cytogenetic systems of manipulating the genetic system of plants has really been able to find an important place in crop production, although some have come close to it. The theoretical basis for a directed search for genotypes suitable for their construction is apparently still lacking. There has been constant progress in the cell biology and genetics of generative reproduction, including meiosis, and sooner or later some of the proposed approaches will come within reach of practical application, especially when in their development use is made of molecular genetic manipulation and molecular methods of analysis. They certainly deserve continued attention.

References

Agache S, Bachelier B, Buyer J de, Henry Y, Snape J (1989) Genetic analysis of anther culture response in wheat using aneuploid, chromosome substitution and translocation lines. Theor Appl Genet 77:7–11

Alberts B, Bray D, Lewis J, Raff M, Roberts K, Watson JD (1989) Molecular biology of the cell, 2nd edn. Garland, New York

Albini SM, Jones GH (1990) Synaptonemal complex spreading in *Allium cepa* and *Allium fistulosum*. III. The F1 hybrid. Genome 33:854–866

Alonso LC, Kimber G (1981) The analysis of meiosis in hybrids II. Triploid hybrids. Can J Genet Cytol 23:221–234

Alonso LC, Kimber G (1983) A study of genome relationships in wheat based on telocentric chromosome pairing II. Z Pflanzenzuecht 90:273–284

Amato Fd' (1977) Nuclear cytology in relation to development. Cambridge University Press, Cambridge

Ambros PF, Matzke MA, Matzke AJM (1986) Detection of a 17kb unique sequence (T-DNA) in plant chromosomes by in situ hybridization. Chromosoma 94:11–18

Anderson E (1949) Introgressive hybridization. Wiley and Sons, New York

Andrus CF, Seshadri VS, Grimball PC (1971) Production of seedless watermelons. US Dep Agric Tech Bull 1425, 12 pp

Appels R, McIntyre CL (1985) Cereal genome organization as revealed by molecular probes. Oxf Surv Plant Mol Cell Biol 2:235–252

Appels R, Driscoll C, Peacock WJ (1978) Heterochromatin and highly repeated DNA sequences in rye (*Secale cereale*). Chromosoma 70:67–89

Arana P, Santos JL, Henriques-Gil N (1987) Interference relationships in grasshopper reciprocal translocation heterozygotes. Heredity 59:85–93

Armstrong KC, Craig IL, Merrit C (1987) Hordeum chilense (2n = 14) computer-assisted Giemsa karyotypes. Genome 29:683–688

Ashmore SE, Shapcott AS (1989) Cytogenetic studies of *Haplopappus gracilis* in both callus and suspension cultures. Theor Appl Genet 78:249–259

Asker S (1980) Gametophytic apomixis: elements and genetic regulation. Hereditas 93:277–293

Aung T, Thomas H (1978) The structure and breeding behaviour of a translocation involving the transfer of mildew resistance from *Avena barbata* Pott. into the cultivated oat. Euphytica 27:731–739

Avery AG, Satina S, Rietsema J (1959) Blakeslee: the genus *Datura*. Ronald Press, New York

Avivi L, Feldman M, Brown M (1982) An ordered arrangement of chromosomes in the somatic nucleus of common wheat. Chromosoma 86:1–26

Bak P, Bak AL, Zeuthen J (1979) Characterization of human chromosomal unit fibres. Chromosoma 73:301–315

Banks P (1984) A new diploid chromosome number for tomato (*Lycopersicon esculentum*)? Can J Genet Cytol 26:636–639

Barcley PC, Armstrong JM (1966) Certain aspects of chromosome number and seed size in induced tetraploid pasture plants. In: Hill AG (ed), Proc 10th Int Grassl Congr Helsinki, Finn Grassl Ass, pp 667–671

Barlow BA, Wiens D (1975) Permanent translocation heterozygosity in *Viscum hildebrandtii* Engl. and *V. engleri* Tiegh. (Viscaceae) in East Africa. Chromosoma 53:265–272

Barlow BA, Wiens D (1976) Translocation heterozygosity and sex ratio in *Viscum fischeri*. Heredity 37:27–40

Basavaiah M, Rajan V, Dandin SB, Suryanarayana N, Sengupta K (1990) Chromosomal association and meiotic behaviour of four triploid varieties of mulberry (*Morus* sp.). Cytologia 55:327–333

Beckett JB (1991) Cytogenetic, genetic and plant breeding applications of B-A translocations in maize. In: Gupta PK, Tsuchiya T (eds) Chromosome engineering in plants: genetics, breeding, evolution, Part A. Elsevier, Amsterdam, pp 493–529

Bedbrook JR, Jones J, O'Dell M, Thompson RD, Flavell RB (1980) A molecular description of telomeric heterochromatin in *Secale* species. Cell 19:545–560

Belfield AM, Riley R (1969) The relationships of the genomes of allohexaploid wheat. Chromosomes Today 2:5–11

Belling J, Blakeslee AF (1924) The configurations and sizes of the chromosomes in the trivalents of 25-chromosome Daturas. Proc Natl Acad Sci USA 10:116–120

Belling J, Blakeslee AF (1927) The assortment of chromosomes in haploid Daturas. Cellule 37:353–365

Benavente E, Orellana J (1989) Pairing competition between identical and homologous chromosomes in autotetraploid rye heterozygous for interstitial C-bands. Chromosoma 98:225–232

Benavente R (1982) Holocentric chromosomes of Arachnids: presence of kinetochore plates during meiotic divisions. Genetica 59:23–27

Bender K, Gaul H (1966) Zur Frage der Diploidisiering autotetraploider Gerste. Z Pflanzenzuecht 56:164–183

Bennett MD (1971) The duration of meiosis. Proc R Soc Lond B 178:277–299

Bennett MD (1976) The cell in sporogenesis and spore development. In: Yeoman M (ed) Cell division in higher plants. Academic Press, New York, pp 161–198

Bennett MD (1977) Heterochromatin, aberrant endosperm nuclei and grain shriveling in wheat-rye genotypes. Heredity 39:411–419

Bennett MD (1982) Nucleotypic basis of spatial ordering of chromosomes in eukaryotes and the implications of the order for genome evolution and phenotypic variation. In: Dover GA, Flavell RB (eds) Genome evolution. Academic Press, London, pp 240–261

Bennett MD (1984) The genome, the natural karyotype, and biosystematics. In: Grant WF (ed) Plant biosystematics. Academic Press, Toronto, pp 41–66

Bennett MD (1988) Parental genome separation in F1 hybrids between grass species. In: Brandham PE (ed) Kew chromosome conference III. HMSO, London, pp 195–208

Bennett MD, Gustafson JP (1982) The effect of telomeric heterochromatin from *Secale cereale* on triticale (X Triticosecale). II. The presence or absence of blocks of heterochromatin in isogenic backgrounds. Can J Genet Cytol 24:93–100

Bennett MD, Kaltsikes PJ (1973) The duration of meiosis in diploid rye, a tetraploid wheat and the hexaploid triticale derived from them. Can J Genet Cytol 15:671–679

Bennett MD, Smith JB (1975) Confirmation of the identification of the rye chromosome in 1B/1R wheat-rye chromosome substitution and translocation lines. Can J Genet Cytol 17:117–120

Bennett MD, Smith JB (1976) Nuclear DNA amounts in angiosperms. Philos Trans R Soc Lond 274:227–274

Bennett MD, Stern H, Woodward M (1974) Chromatin attachment to nuclear membrane of wheat pollen mother cells. Nature 252:395–396

Bennett MD, Finch RA, Barclay IR (1976) The time rate and mechanism of chromosome elimination in *Hordeum* hybrids. Chromosoma 54:175–200

Bernard M, Bernard S, Joudrier P, Autran JC (1990) Biochemical characterization of

lines descended from 8x triticale × 4x triticale cross. Theor Appl Genet 79:646–653
Bernards A, Michels PAM, Lincke CR, Borst P (1983) Growth of chromosome ends in multiplying trypanosomes. Nature 303:592–597
Bernatzky R, Tanksley SD (1986) Toward a saturated linkage map in tomato based on isozymes and random cDNA sequences. Genetics 112:887–898
Bianchi NO, Bianchi MS, Cleaver JE, Wolff S (1985) The pattern of restriction-enzyme-induced banding in the chromosomes of chimpanzee, gorilla and orangutan and its evolutionary significance. J Mol Evol 22:323–333
Bingham ET (1980) Maximum heterozygosity in autopolyploids. In: Lewis WH (ed) Polyploidy. Biological relevance. Plenum Press, New York, pp 471–490
Birchler JA (1983) Chromosomal manipulation in maize. In: Swaminathan MS, Gupta PK, Sinha U (eds) Cytogenetics of crop plants. MacMillan India, Delhi, pp 379–403
Birchler JA (1991) Chromosome manipulations in maize. In: Gupta PK, Tsuchiya T (eds) Chromosome engineering in plants: genetics, breeding, evolution, Part A. Elsevier, Amsterdam, pp 531–559
Blakeslee AF, Avery AG (1937) Methods of inducing doubling of chromosomes in plants. J Hered 28:393–411
Blakeslee AF, Avery AG (1938) Fifteen-year breeding records of 2n + 1 types in *Datura stramonium*. Carnegie Inst Wash Publ 501:315–351
Blakeslee AF, Belling J, Farnham M (1923) Inheritance in tetraploid Daturas. Bot Gaz 76:329–373
Bosemark NO (1967) Post-meiotic nuclear fragmentation and pollen sterility in autotetraploid sugarbeet. Hereditas 57:217–238
Bostock CJ, Sumner AT (1978) The eukaryotic chromosome. North Holland, Amsterdam
Brandham PE (1990) Meiotic crossing-over between sites on opposite sides of the centromeres of homoeologues is frequent in hybrid Aloeaceae. Genome 33:170–176
Brar DS, Minocha JL (1982) Multiple chromosomal interchanges in pearl millet. Theor Appl Genet 219:689–693
Bretschneider H (1979) Zum Einfluss von Translokationen auf meiotische Chromosomenpaarung und Fertilität bei autotetraploider Gerste. Tagungsber Akad Landwirtsch-Wiss DDR 175:209–216
Brinkley BR, Tousson A, Valdivia MM (1985) The kinetochore of mammalian chromosomes: structure and function in normal mitosis and aneuploidy. In: Dellarco VL, Voytek PE, Hollaender A (eds) Aneuploidy. Etiology and mechanisms. Plenum Press, New York, pp 243–267
Brown SW, Zohari D (1955) The relationship of chiasmata and crossing-over in *Lilium formosanum*. Genetics 40:850–873
Brown WV, Stack SM (1968) Somatic pairing as a regular preliminary of meiosis. Bull Torrey Bot Club 95: 369–378
Burk LG, Tso TC (1960) Genetic tumors of Nicotiana associated with chromosome loss. J Genet 51:184–187
Burnham CR (1954) Tester set of translocations. Maize Gent Coop Newsl 28:58–66
Burnham CR (1956) Chromosomal interchanges in plants. Bot Rev 22:419–552
Burnham CR (1962) Discussions in cytogenetics. Burgess, Minneapolis
Burnham CR, Stout JT, Weinheimer WH, Kowles RV, Philips RL (1972) Chromosome pairing in maize. Genetics 71:111–126
Burr B, Burr FA, Thompson KH, Albertson MC, Stuber CW (1988) Gene mapping with recombinant inbreds in maize. Genetics 118:519–526
Cai Q, Bullen MR (1991) Characterization of genomes in timothy (*Phleum pratense* L.). I. Karyotypes and C-banding patterns in cultivated timothy and two wild relatives. Genome 34:52–58
Callan HG (1986) Lampbrush chromosomes. Springer, Berlin Heidelberg New York

Callan HG, Montalenti G (1947) Chiasma interference in mosquitoes. J Genet 48:119–134
Callow RS (1985) Comments on Bennett's model on chromosome disposition. Heredity 54:171–177
Carlson WR (1988) The cytogenetics of corn. In: Sprague GF, Dudley JW (eds) Corn and corn improvement. "Agronomy" Madison, WI, pp 259–343
Carlson WR, Curtis C (1986) A new method for producing homozygous duplications in maize. Can J Genet Cytol 28:1034–1040
Carlson WR, Roseman RR (1991) Segmental duplication of distal chromosomal regions in maize. Genome 34:537–542
Caspersson T, Zech L (1970) Fluorescent labelling of chromosomal DNA: superiority of quinacrine mustard to quinacrine. Science 170:762
Caspersson T, Farber S, Foley GE, Kudynowski J, Modest EJ, Simonsson E, Wagh U, Zech L (1968) Chemical differentiation along metaphase chromosomes. Exp Cell Res 49:219–222
Cavalier-Smith T (1983) Cloning chromosome ends. Nature 301:112–113
Chao S, Sharp PJ, Worland AJ, Warham EJ, Koebner RMD, Gale MD (1989) RFLP-based genetic maps of wheat homoeologous group 7 chromosomes. Theor Appl Genet 78:495–504
Chapman CGD, Kimber G (1992) Developments in the meiotic analysis of hybrids. V. Second order models for tetraploids and pentaploids. Heredity 68:205–210
Chase SS (1952) Monoploids in maize. In: Gowen JD (ed) Heterosis. Iowa State Coll Press, Ames, Iowa, pp 389–399
Clarke L, Carbon J (1985) The structure and function of yeast centromeres. Annu Rev Genet 19:29–56
Claus E, Pohler W (1990) Ergebnise der Bastardierung von Hordeum vulgare L. mit Secale-Arten. Biol Zent bl 109:489–498
Cleland RE (1972) Oenothera – cytogenetics and evolution. Academic Press, London
Clulow SA, Wilkinson MJ, Waugh R, Baird E, DeMaine MJ, Powell W (1991) Cytological and molecular obervations on *Solanum phureja*-induced dihaploid potatoes. Theor Appl Genet 82:545–551
Coe EH, Hoisington DA, Neuffer MG (1990) Linkage map of corn (maize) (Zea mays L.) (2N = 20) (Aug. 1989). In: Stephen J O'Brien (ed) Genetic maps. Locus maps of complex genomes. 5th Edition. Cold Spring Harb. Lab Press, pp 6.39–6.67
Conger AD (1953) Culture of pollen tubes for chromosomal analysis at the pollen tube division. Stain Technol 28:289–293
Cortés F, Escalza P (1986) Analysis of different banding patterns and late replicating regions in chromosomes of *Allium cepa*, *A. sativum* and *A. nigrum*. Genetica 71:39–46
Cortés F, Gonzales-Gil G, Lopez-Saez JF (1980) Differential staining of late replicating DNA-rich regions in *Allium cepa* chromosomes. Caryologia 33:193–202
Cowell JK (1982) Double minutes and homogeneously staining regions: gene amplification in mammalian cells. Annu Rev Genet 16:21–59
Crane CF, Sleper DA (1989a) A model of meiotic chromosome association in triploids. Genome 32:82–98
Crane CF, Sleper DA (1989b) A model of meiotic chromosome association in tetraploids. Genome 32:691–707
Creighton HB, McClintock B (1931) A correlation of cytological and genetical crossing-over in *Zea mays*. Proc Natl Acad Sci USA 17:492–497
Cuadrado MC, Romero C (1988) Different genetic systems in rye affecting homoeologous pairing in wheat-rye combinations. Geome 30:793–796
Curtis C, Feldman M (1988) Increased proximal recombination frequency in common wheat by premeiotic colchicine treatment. In: Miller TE, Koebner RMD (eds) Proc 7th Int Wheat Genet Symp. IPSR, Cambridge, pp 243–248
Curtis CA, Lukaszewski AJ (1991) Genetic linkage between C-bands and storage protein genes in chromosome 1B of tetraploid wheat. Theor Appl Genet 81:245–252

Curtis CA, Lukaszewski AJ, Chrzastek M (1991) Metaphase I pairing of deficient chromosomes and genetic mapping of deficiency break points in common wheat. Genome 34:553–560
Daker MG (1967) Cytological studies on a haploid cultivar of *Pelargonium* and its colchicine induced diploids. Chromosoma 21:250–271
Darlington CD (1965) Cytology. J and A Churchill, London
Darlington CD, LaCour L (1940) Nucleic acid starvation in chromosomes in *Trillium*. J Genet 40:185–213
Darlington CD, LaCour LF (1976) The handling of chromosomes, 6th edn. George Allen and Unwin, London
Davies A, Jenkins G, Rees H (1990) Diploidisation of *Lotus corniculatus* L. (Fabaceae) by elimination of multivalents. Chromosoma 99:289–295
Davies DR, Cullis CA (1982) A simple plant polytene chromosome system, and its use for in situ hybridization. Plant Mol Biol 1:301–304
Davies PA, Pallota MA, Driscoll, CJ (1985) Centric fusion between nonhomologous rye chromosomes in wheat. Can J Genet Cytol 27:627–632
Deimling S, Zitzlsperger J, Wenzel G (1988) Somatic fusion for breeding of tetraploid potatoes. Plant Breed 201:181–189
Demarly Y (1963) Genétique des tétraploides et amélioration des plantes. Ann Amélior Plant 13:307–400
Dewey DR (1984) The genomic system of classification as a guide to intergeneric hybridization with the perennial Triticeae. In: Gustafson JP (ed) Gene manipulation in plant improvement. 16th Stadler Genetics Symp, Columbia, MO, pp 209–279
Dijk P van, Delden W van (1990) Evidence for autotetraploidy in Plantago media and comparisons between natural and artificial cytotypes concerning cell size and fertility. Heredity 65:349–357
Dillé JE, Gustafson JP (1990) Influence of telomeric heterochromatin loss on growth and development of the early embryo and endosperm in triticale. Plant Breed 104:34–39
Dillé JE, Gustafson JP, Bennett MD (1984) The effect of telomeric heterochromatin on the frequency of aneuploids in triticale. Can J Genet Cytol 26:46–49
Diter A, Guyomard R, Chourrout D (1988) Gene segregation in induced tetraploid rainbow-trout: genetic evidence of preferential pairing of homologous chromosomes. Genome 30:547–553
Dobzhanski T, Sturtevant AH (1938) Inversions in the chromosomes of *Drosophila pseudoobscura*. Genetics 23:28–64
Dolstra O (1982) Synthesis and fertility of × *Brassicoraphanus* and ways of transferring *Raphanus* characters to *Brassica*. Agric Res Rep (Wageningen) 917:1–90
Dorninger D, Timischl W (1987) Geometrical contraints on Bennett's predictions of chromosome order. Heredity 58:321–325
Douches DS, Quiros CF (1988) Genetic recombination in a diploid synaptic mutant of a *Solanum tuberosum* × *S. chacoense* diploid hybrid. Heredity 60:183–191
Dover GA, Riley R (1972) Prevention of pairing of homoeologous meiotic chromosomes of wheat by an activity of supernumerary chromosomes of *Aegilops*. Nature 240:159–161
Doyle GG (1963) Preferential pairing in structural heterozygotes of *Zea mays*. Genetics 48:1011–1027
Doyle GG (1979) The allotetraploidization of maize. Part 1: The physical basis – differential pairing affinity. Theor Appl Genet 54:103–112
Doyle GG (1982) The allotetraploidization of maize. 3. Gene segregation in trisomic heterozygotes. Theor Appl Genet 61:81–89
Doyle GG (1986) The allotetraploidization of maize. 4. Cytological and genetic evidence indicative of substantial progress. Theor Appl Genet 71:585–594
Doyle GG (1990) The allotetraploidization of maize. 5. The utilization of zygomeres. Theor Appl Genet 79:145–152

Doyle GG, Kimber G (1983) A computer simulation of the behavior of reciprocal translocations in autotetraploids. Theor Appl Genet 65:53–60
Driscoll CJ (1972) XYZ system of producing hybrid wheat. Crop Sci 12:516–517
Driscoll CJ (1983) The relationships of chromosome 4A of hexaploid wheat to its homoeologues. In: Reddi VR, Rao MK (eds) Cytogenetics of higher plants: chromosome structure and behaviour. Proc Meet Andhra Univ India pp 19–22
Driscoll CJ (1985) Modified XYZ system of producing hybrid wheat. Crop Sci 25:1115–1116
Driscoll CJ Barlow KK (1976) Male sterility in plants. Induction, isolation and utilization. Induced mutations in cross-breeding. Proc Adv Group FAO/IAEA Vienna, 1975, pp 123–131
Driscoll CJ, Darvey NL, Barber HN (1967) Effect of colchicine on meiosis of hexaploid wheat. Nature 216:687–688
Driscoll CJ, Bielig LM, Darvey NL (1979) An analysis of frequencies of chromosome configurations in wheat and wheat hybrids. Genetics 91:755–767
Dubovets NI, Badaev NS, Bolsheva NL, Badaeva ED, Shcherbakova AM, Bormotov VE, Zelenin AV (1989) Regularities of karyotype formation in tetraploid triticale. Cereal Res Commun 17:253–257
Dudits D, Maroy E, Praznovszky T, Olah Z, Gyorgyey J, Cella R (1987) Transfer of resistance traits from carrot into tobacco by asymmetric somatic hybridization. Proc Natl Acad Sci USA 84:8434–8438
Dujardin M, Hanna W (1986) An apomictic polyhaploid obtained from a pearl millet × *Pennisetum squamulatum* apomictic interspecific hybrid. Theor Appl Genet 72:33–36
Dunemann F, Grunewaldt J (1991) Identification of a monogenic dominant male sterility mutant in broccoli (*Brassica oleracea* var. *italica* Plenck). Plant Breed 106:161–163
Dvorak J (1977) Effect of rye on homoeologous chromosome pairing in wheat × rye hybrids. Can J Genet Cytol 19:549–556
Dvorak J (1983) The origin of wheat chromosomes 4A and 4B and their genome reallocation. Can J Genet Cytol 18:371–377
Dvorak J, Appels R (1986) Investigation of homologous crossing-over and sister chromatid exchange in the wheat NOR-B2 locus coding for rRNA and Gli-B2 locus coding for gliadins. Genetics 113:1037–1056
Dyer AF (1979) Investigating chromosomes. Edward Arnold, London
Earnshaw WC, Migeon BR (1985) Three related centromere proteins are absent from the inactive centromere of a stable isodicentric chromosome. Chromosoma 92:290–296
Eckenwalder JE, Brown BP (1986) Polyploid speciation in hybrid morning glories of *Ipomoea* L. sect. Quamoclit Griseb. Can J Genet Cytol 28:17–20
Eenink AH (1974a) Matromorphy in *Brassica oleracea* L. 4. Formation of homozygous and heterozygous diploid products of gametogenesis and qualitative genetical research on matromorphic plants. Euphytica 23:719–724
Eenink AH (1974b) Matromorphy in *Brassica oleracea* L. 1. Terminology, parthenogenesis in Cruciferae and the formation and usability of matromorphic plants. Euphytica 23:429–433
Ehrendorfer F (1986) Chromosomal differentiation and evolution of angiosperm groups. In: Iwatsuki K, Raven PH, Bock WJ (eds) Modern aspects of species. Univ Tokyo Press, Tokyo pp 59–86
Eigsti OJ, Dustin P (1955) Colchicine – in agriculture, medicine, biology and chemistry. Iowa State College Press, Ames
Eizenga GC (1986) Failure of centric fusion of selected wheat telosomes. Cereal Res Commun 14:211–214
Eizenga GC (1987) Locating the Agropyron segment in wheat – Agropyron transfer no. 12. Genome 29:365–366

Elçi S, Sybenga J (1976) Incomplete preferential pairing in a tetraploid *Secale* hybrid carrying translocations: multivalent configuration frequencies and marker segregation. Genetica 46:177–182
Ellerström S, Sjödin J (1966) Frequency and vitality of aneuploids in a population of tetraploid red clover. Hereditas 55:166–182
Emsweller SL, Jones HA (1935) Meiosis in *Allium fistulosum*, *Allium cepa* and their hybrid. Hilgardia 9:277–294
Emsweller SL, Jones HA (1945) Further studies on the chiasmata of the *Allium cepa* × *A. fistulosum* hybrid and its derivatives. Am J Bot 32:370–379
Endo TR (1982) Gametocidal chromosomes of three Aegilops species in common wheat. Can J Genet Cytol 24:201–206
Endo TR, Gill BS (1983) Identification of wheat chromosomes by N-banding. In: Sakanioto S (ed) Proc 6th Int Wheat Genet Symp Kyoto, Kyoto Univ, pp 355–359
Endrizzi J, Kohel RJ (1966) Use of telosomes in mapping three chromosomes of cotton. Genetics 54:535–550
Endrizzi JE, Turcotte EL, Kohel RJ (1985) Genetics, cytology and evolution of *Gossypium*. Adv Genet 23:271–375
Ernst A (1918) Bastardierung als Ursache der Apogamie im Pflanzenreich. Gustav Fischer, Jena
Espinasse A, Kimber G (1981) The analysis of meiosis in hybrids IV. Pentaploid hybrids. Can J Genet Cytol 23:623–638
Essad S (1962) Etude génétique et cytogénétique des espèces *Lolium perenne* L., *Festuca pratensis* Huds. et de leur hybrides. Ann Amélior Plant 12:1–103
Essad S, Arnoux J, Maia N (1966) Controle de validité des caryogrammes. Application au caryotype de *Lolium perenne* L. Chromosoma 20:202–220
Esser K, Kuenen R (1965) Genetik de Pilze. Springer, Berlin Heidelberg New York
Evans GM (1988) Genetic control of chromosome pairing in polyploids. In: Brandham P (ed) Proc Kew Chrom Conf II. HMSO, London, pp 253–260
Evans GM, Aung T (1985) Identification of a diploidizing genotype of *Lolium multiflorum*. Can J Genet Cytol 27:498–505
Evans GM, Macefield AJ (1972) Suppression of homoeologous pairing by B-chromosomes in a *Lolium* species hybrid. Nat New Biol 236:110–111
Evans GM, Macefield AJ (1977) Induced translocation of an A and a B chromosome in *Lolium perenne*. Chromosoma 61:257–266
Evans GM, Rahman MM (1990) The basis of low grain yield and infertility in autotetraploid barley (*Hordeum vulgare*). Heredity 64:305–313
Evans HJ, Savage JRK (1963) The relation between DNA synthesis and chromosome structure as resolved by X-ray damage. J Cell Biol 18:525–539
Evans HJ, Buckton KE, Sumner AT (1971) Cytological mapping of human chromosomes: results obstained with quinacrine fluorescence and the acetic-saline-Giemsa techniques. Chromosoma 35:310–325
Evola SV, Burr FA, Burr B (1986) The suitability of restriction fragment length polymorphisms as genetic markers in maize. Theor Appl Genet 71:765–771
Feldman M (1988) Cytogenetic and molecular approaches to alien gene transfer in wheat. In: Miller TE, Koebner RMD (eds) Proc 7th Int Wheat Genet Symp Cambridge. IPSR, Cambridge, pp 23–32
Field JM, Shewry PR (1987) Prolamin subunit interactions in hexaploid and octaploid triticales and their wheat and rye parents. J Cereal Sci 6:199–210
Figueiras AM, González-Jaén MT, Candela M (1990) Chromosomal identification and meiotic behavior of reciprocal translocations in a rye polymorphic population. Evolutionary implications. Theor Appl Genet 79:686–692
Filion WG, Walden DB (1973) Karyotype analysis: the detection of chromosomal alterations in the somatic karyotype of *Zea mays* L. Chromosoma 41:183–194
Finch RA, Bennett MD (1979) Action of triploid inducer (tri) on meiosis in barley (*Hordeum vulgare* L.). Heredity 43:87–93

Finch RA, Smith JB, Bennett MD (1981) Hordeum and Secale mitotic genomes lie apart in a hybrid. J Cell Sci 52:391–403
Fitzgerald-Hayes M, Clarke L, Carbon J (1982) Nucleotide sequence comparisons and functional analysis of yeast centromere DNAs. Cell 29:235–244
Flach M (1966) Nutmeg cultivation and its sex problems. Dissertation, agricultural University Wageningen
Flavell RB (1981) The analysis of plant genes and chromosomes by using DNA cloned in bacteria. Philas Trans R Soc Lond B 292:579–588
Flavell RB, Rimpau J, Smith DB, O'Dell M, Bedbrook JR (1979) The evolution of plant genome structure. In: Leaver CJ (ed) Genome organization and expression in plants. Plenum Press, New York, pp 35–47
Ford JH, Correll T (1989) Chromosome displacement and aneuploidy. In: Mechanisms of chromosome distribution and aneuploidy. Alan R Liss, New York, pp 231–240
Franzke CJ, Ross, JG (1952) Colchicine induced variants in *Sorghum*. J Hered 43:106–115
Friebe B, Larter EN (1988) Identification of a complete set of isogenic wheat/rye D-genome substitution lines by means of Giemsa C-banding. Theor Appl Genet 76:473–479
Friebe B, Hatchett JH, Sears RG, Gill BS (1990) Transfer of Hessian fly resistance from "Chaupan" rye to hexaploid wheat via a 2BS/2RL wheat – rye chromosome translocation. Theor Appl Genet 79:385–398
Friedt W (1978) Untersuchungen an autotetraploiden Gersten unter besonderer Berüchsichtigung der Diploidisierung. I. Fertilität, Vitalität und Kornertrag. Z Pflanzenzuecht 81:118–139
Friedt W (1979a) Untersuchungen an autotetraploiden Gersten unter besonderer Berücksichtigung der Diploidisierung. II. Meiosemerkmale. Z Pflanzenzuecht 82:311–339
Friedt W (1979b) The use of *Secale vavilovii* in rye breeding. In: Proc Conf Broadening Genet Base of Crops. Pudoc Wageningen, pp 221–224
Friedt W (1985) The present status of breeding autotetraploid cereals. In: Horn W, Jensen CJ (eds) Genetic manipulation and plant breeding. Proc Eucarpia Congr Berlin, pp 123–128
Fu TK, Sears ER (1973) The relationship between chiasmata and crossing-over in *Triticum aestivum*. Genetics 75:231–246
Fukada I (1984) Chromosome banding and biosystematics. In: Grant WF (ed) Plant biosystematics. Academic Press, Toronto, pp 97–116
Fukui K (1986) Standardization of karyotyping plant chromosomes by a newly developed chromosome image analyzing system (CHIAS). Theor Appl Gent 72:27–32
Funaki K, Matsui S, Sasaki M (1975) Location of nucleolar organizers in animal and plant chromosomes by means of an improved N-banding technique. Chromosoma 49:357–370
Fussell CP (1987) The Rabl orientation: a prelude to synapsis. In: Moens PB (ed) Meiosis. Academic Press Toronto, pp 275–299
Gall JG, Pardue ML (1969) Formation and detection of RNA-DNA hybrid molecules in cytological preparations. Proc Natl Acad Sci USA 63:378–383
Ganner E, Evans JH (1971) The relationship between patterns of DNA replication and of quinacrine fluorescence in the human chromosome complement. Chromosoma 35:326–341
Gasser CS, Fraley RT (1989) Genetically engineering plants for crop improvement. Science 249:1293–1299
Gaul H (1954) Asynapsis und ihre Bedeutung für die Genomanalyse. Z Induct Abstammungs Vererbungsl 86:69–100
Gaul H (1958) A critical survey of genome analysis In: Jenkins BC, Proc 1st Int Wheat Genet Symp Winnipeg, Univ Manitoba, pp 194–206
Geitler L (1939) Die Entstehung der polyploiden Somakerne der Heteropteren durch Chromosomenteilung ohne Kernteilung. Chromosoma 1:1–22

Gerstel DU (1945) Inheritance in *Nicotiana tabacum*. XX. The addition of *Nicotiana glutinosa* chromosomes to tobacco. J Hered 36:197–206
Gerstel DU, Burns JA (1966) Chromosomes of unusual length in hybrids between two species of *Nicotiana*. Chromosomes Today 1:41–56
Gilles A, Randolph LG (1951) Reduction of quadrivalent frequency in autotetraploid maize during a period of ten years. Am J Bot 38:12–17
Gillies CB (1984) The synaptonemal complex in higher plants. CRC Crit Rev Plant Sci 2:81–116
Gillies CB (1989) Chromosome pairing and fertility in polyploids. In: Gillies CB (ed) Fertility and chromosome pairing: recent studied in plants and animals. CRC Press, Boca Raton, pp 137–176
Gleba YY, Parakonny A, Kotov V, Negrutiu I, Momot V (1987) Spatial separation of parental genomes in hybrids of somatic plant cells. Proc Natl Acad Sci USA 84:3709–3713
Glendinning DR (1989) Some aspects of autotetraploid population dynamics. Theor Appl Genet 78:233–242
Gohil RN, Kaul R (1981) Studies on male and female meiosis in Indian *Allium*. 2. Autotetraphoid *Allium tuberosum*. Chromosoma 82:735–739
Goodpasture C, Bloom SE (1975) Visualization of nucleolar organizing regions in mammalian chromosomes using silver staining. Chromosoma 53:37–50
Goodspeed TH, Clausen RE (1928) Interspecific hybridization in *Nicotiana* VIII. The sylvestris-tomentosa-tabacum hybrid triangle and its bearing on the origin of tabacum. Univ Calif Publ Bot 11:245–256
Gopinath DM, Burnham CR (1956) A cytogenetics study in maize of duplication-deficiency produced by crossing interchanges involving the same chromosomes. Genetics 41:382–395
Gottschalk W (1976) Die Bedeutung der Polyploidie für die Evolution der Pflanzen. Gustav Fischer, Stuttgart
Grant WF (ed) (1984) Plant biosystematics. Academic Press, Toronto
Green DK, Carmeron J (1972) Metaphase cell finding by machine. Cytogenetics 11:475–487
Greilhuber J (1974) Hy-banding: a new quick technique for heterochromatin staining in plant chromosomes. Naturwissenschaften 61(4):170–171
Greilhuber J (1975) Heterogeneity of heterochromatin in plants: comparison of Hy- and C-bands in *Vicia faba*. Plant Syst Evol 124:139–156
Grell RF (1964) Chromosome size at distributive pairing in *Drosophila melanogaster* females. Genetics 50:151–166
Grell RF, Oakberg EF, Generoso EE (1980) Synaptonemal complexes at premeiotic interphase in the mouse spermatocyte. Proc Natl Acad Sci USA 77:6720–6723
Griesbach RJ, Malmberg RL, Carlson PS (1982) Uptake of isolated Lilly chromosomes by tobacco protoplasts. J Hered 73:151–152
Gröber K, Zacharias M (1983) über Nutzungsmöglichkeiten der apomiktischen Samenausbildung bei ausgewählten Kulturpflanzen. Kulturpflanze 31:145–163
Guignard G (1986) Comportement meiotique de populations tetraploides du genre *Dactylis*. Can J Genet Cytol 28:899–905
Gulcan R Sybenga J (1967) Relative rate of development of aneuploid cells in an euploid environment. Genetica 38:167–170
Gupta PK, Fedak G (1987) Segregation in the pollen of F2 rye (*Secale cereale*) plants for induction of homoeologous chromosome pairing in hybrids with wheat. Genome 29:888–891
Gupta PK, Reddy, VRK (1991) Cytogenetics of triticale – a man-made cereal. In: Gupta PK, Tsuchiya T (eds) Chromosome engineering in plants: genetics, breeding, evolution, Part A. Elsevier, Amsterdam, pp 335–359
Gupta PP, Schieder O, Gupta M (1984) Intergeneric nuclear gene transfer between somatically and sexually incompatible plants through asymmetric protoplast fusion. Mol Gen Genet 197:30–35

Gustafson JP, Dillé JE (1989) Wheat substitutions in hexaploid triticale. Plant Breed 102:109–112
Gustafson JP, Lukaszewski AJ, Bennett MD (1983) Somatic deletion and/or redistribution of telomeric heterochromatin in the genus *Secale* and Triticale. Chromosoma 88:293–298
Gustafson JP, Butler E, McIntyre CL (1990) Physical mapping of low-copy DNA sequence in rye (*Secale cereale* L.). Proc Natl Acad Sci USA 87:1899–1902
Gustaffson Å (1947) Apomixis in higher plants. Lunds Universitets Årsskrift NF Avd 2 Bd 42, 43
Hagberg A (1965) Use of induced translocations in directed production of duplications. In: The use of induced mutations in plant breeding. FAO/IAEA Rome. Suppl Rad Bot Pergamon Press, Oxford, pp 741–752
Hagberg A, Hagberg P (1991) Production and analysis of chromosome duplications in barley. In: Gupta PK, Tsuchiya T (eds) Chromosome engineering in plants: genetics, breeding, evolution, Part A. Elsevier, Amsterdam, pp 401–410
Hahn SK, Bai KV, Asiedu R (1990) Tetraploids, triploids, and 2n pollen from diploid interspecific crosses with cassava. Theor Appl Genet 79:433–439
Håkanson A, Levan A (1957) Endoduplicational meiosis in *Allium odorum*. Hereditas 43:179–200
Haldane JBS (1919) The combination of linkage values, and the calculation of the distance between the loci of linked factors. J Genet 8:299–309
Harlan JR, DeWet JMJ (1971) Toward a rational classification of cultivated plants. Taxon 20:509–517
Harlan JR, DeWet JMJ (1975) On Ö. Winge and a prayer: the origins of polyploidy. Bot Rev 41:361–390
Harris JE, Nasjleti CE, Kowalski CJ (1973) Discrimination between groups of chromosomes and individual chromosomes in the normal human karyotype. Chromosoma 40:269–284
Hashemi A, Estilai A, Waines JG (1989) Cytogenetics and reproductive behaviour of induced and natural tetraploid guayule (*Parthenium argentatum* Gray). Genome 32:1100–1104
Hearne EM, Huskins CL (1935) Chromosome pairing in *Melanoplus femur-rubrum*. Cytologia 6:123–147
Heijbroek W, Roelands AJ, Jong JH de, Hulst C van, Schoone AHL, Munning RC (1988) Sugar beets homozygous for resistance to beet cyst nematode (*Heterodera schachtii* Schm.), developed from monosomic additions of *Beta procumbens* to *B. vulgaris*. Euphytica 38:121–131
Hein T, Przewozny T, Schieder O (1983) Culture and selection of somatic hybrids using an auxotrophic cell line. Theor Appl Genet 64:119–122
Heitz E (1929) Heterochromatin, Chromozentren, Chromomeren. Ber Dtsch Bot Ges 47:274–284
Helentjaris T (1987) A genetic linkage map for maize based on RFLPs. Trends Genet 3(8):217–221
Helentjaris T, Weber D, Wright S (1988) Identification of the genome locations of duplicate nucleotide sequences in maize by analysis of restriction fragment length polymorphism. Genetics 118:353–363
Henderson CAP, Pauls KP (1992) The use of haploidy to develop plants that express several recessive traits using light-seeded canola (Brassica napus) as an example. Theor Appl Genet 83:476–479
Hermsen JG Th (1980) Breeding for apomixis in potato: persuing a utopian scheme? Euphytica 29:595–607
Hermsen JGT (1984) The potential of meiotic polyploidization in breeding allogamous crops. Iowa State J Res 58:435–448
Hermsen, JGT, Verdenius J (1973) Selection from *Solanum tuberosum* group Fureja of genotypes combining high frequency haploid induction with homozygosity for embryo spot. Euphytica 22:244–259

Hermsen JG Th, Ramanna MS, Jongedijk E (1985) Apomictic approach to introduce uniformity and vigour into progenies from true potato seed (TBS). Rep 26 Planning Conf Present and future strategies for potato breeding and improvement, 1983. CIP, Lima, Peru, pp 99–114

Heslop-Harrison JS, Leitch AR, Schwarzacher T, Anamthawat-Jonsson K (1990) Detection and characterization of 1B/1R translocations in hexaploid wheat. Heredity 65:385–392

Hinnisdaels, Bariller L, Mouras A, Sidorov V, Del-Favero J, Veuskens J, Negrutiu I, Jacobs M (1991) Highly asymmetric intergeneric nuclear hybrids between Nicotiana and Petunia: evidence for recombinogenic and translocation events in somatic hybrid plants after "gamma" fusion. Theor Appl Genet 82:609–614

Holliday R (1964) A mechanism for gene conversion in fungi. Genet Res 5:282–304

Holliday R (1967) Altered recombination frequencies in radiation sensitive strains of *Ustilago*. Mutat Res 4:275–288

Holm PB (1988) Chromosome pairing and synaptonemal complex formation in hexaploid wheat, nullisomic for chromosome 5B. Carlsberg Res Commun 53: 91–110

Holm PB, Rasmussen SW (1981) Chromosome pairing, crossing-over, chiasma formation and disjunction as revealed by three-dimensional reconstructions. In: Schweiger HG (ed) Int Cell Biol 1980–1981. Springer, Berlin Heidelberg New York, pp 195–204

Hossain MA, Driscoll CJ (1983) Fertility compensation of Cornerstone male sterility of wheat by rye. Genetics 104:181–189

Hossain MG (1976) The significance of chromosome association in an advanced population of tetraploid rye. Can J Genet Cytol 18:601–609

Houben A, Schlegel R, Ahne R, Huhn P (1990) C-banding und computer gestützte Chromosomenanalyse bei Weizen (*Triticum aestivum* L. cv. "Chinese Spring"). Arch Züchtungsf 20:133–142

Hougas RW, Peloquin SJ, Ross RW (1958) Haploids of the common potato. J Hered 47:103–107

Hsam SLK, Zeller FJ, Huber W (1982) Genetic control of 6-phosphogluconate dehydrogenase (6-PGD) isozymes in cultivated wheat and rye. Theor Appl Genet 62:317–320

Hu J, Quiros CF (1991) Molecular and cytological evidence of deletions in alien chromosomes for two monosomic addition lines of *Brassica campestris – oleracea*. Theor Appl Genet 81:221–226

Hülgenhof E, Schlegel R (1985) Veränderungen der Chromosomenstruktur beim hexaploiden Triticale und ihren Wirkung auf cytologische sowie ertragsbeeinflussende Merkmale II. Experimentelle Befunde zum cytologischeen Verhalten einiger heterochromatindefizienter Biol Zentralbl 104:245–259

Hülgenhof E, Schlegel R (1986) Veränderungen der Chromosomenstruktur beim hexaploiden Triticale und ihre Wirkung auf cytologische sowie ertragsbeeinflussende Merkmale I. Ein Literaturüberblick. Biol Zentralbl 105:347–359

Huskins CL (1948) Segregation and reduction in somatic tissues. J Hered 39:311–325

Ingram R, Noltie HJ (1989) Early adjustment of patterns of metaphase association in the evolution of polyploid species. Genetica 78:21–24

Ishiki K (1985) Cytogenetical studies on African rice, *Oryza glaberrima* Steud. 1. Colchicine-induced autotetraploid. Jpn J Genet 60:225–234

Ising G (1967) Fertility and heterosis in autotetraploid barley. Hereditas 57:327–332

Islam AKMR, Shepherd KW (1991) Alien genetic variation in wheat improvement. In: Gupta PK, Tsuchiya T (eds) Chromosome engineering in plants: genetics, breeding, evolution, Part A. Elsevier, Amsterdam, pp 291–312

Jackson RC, Casey J (1982) Cytogenetic analyses of autopolyploids: models and methods for triploids to octoploids. Am J Bot 69:487–501

Jackson RC, Hauber DP (1982) Autotriploid and autotetraploid analyses: correction coefficients for proposed binomial models. Am J Bot 69:644–646

Jacobsen E (1980) Increase of diplandroid formation and seed set in 4x x 2x-crosses in potatoes by genetical manipulation of dihaploids and some theoretical consequences. Z Pflanzenzuecht 85:110–121
Jahier J, Chèvre AM, Tanguy AM, Eber F (1989) Extraction of disomic addition lines of *Brassica napus- B. nigra*. Genome 32:408–413
Jain HK, Basak SL (1963) Genetic interpretation of chiasmata in *Delphinium*. Genetics 48:329–339
Jan CC, Chandler JM, Wagner SA (1988) Induced tetraploidy and trisomic production of *Helianthus annuus*. Genome 30:647–651
Janicke MA, LaFountain JR (1989) Centromere dots in Crane-fly spermatocytes: meiotic maturation and malorientation. Chromosoma 98:358–367
Janse J (1987) Certation between euploid and aneuploid pollen grains from a tertiary trisomic of rye, *Secale cereale* L. Genome 29:353–356
Jauhar PP (1975) Genetic regulation of diploid-like chromosome pairing in the hexaploid species *Festuca arundinacea* Schreb. and *F. rubra* L. (Gramineae). Chromosoma 52:363–382
Jena KK, Khush GS (1989) Monosomic alien addition lines of rice: production, morphology, cytology, and breeding behavior. Genome 32:449–455
Jenkins G (1986) Synaptonemal complex formation in hybrids between *Lolium temulentum* x *Lolium perenne* (L.) III. Tetraploid. Chromosoma 93:413–419
Jenkins G, White J (1990) Elimination of synaptonemal complex irregularities in a *Lolium* hybrid. Heredity 64:45–53
John B (1990) Meiosis. Cambridge University Press, Cambridge
John B, Lewis KR (1965) The meiotic system. Protoplasmatologia VI, F1. Springer, Wien
Johnson BL (1967) Confirmation of the genome donors of *Aegilops cylindrica*. Nature 216:859–862
Jonson NL (1962) The folded normal distribution: accuracy by maximum likelihood. Technometrics 4:249–256
Johnson RT, Rao PN (1970) Mammalian cell fusion: Induction of premature chromosome condensation in interphase nuclei. Nature 226:717–722
Jones DF (1937) Somatic segregation and its relation to atypical growth. Genetics 22:484–522
Jones GH (1967) The control of chiasma distribution in rye. Chromosoma 22:69–90
Jones GH (1969) Further correlations between chiasmata and U-type exchanges in rye meiosis. Chromosoma 26:105–118
Jones GH (1978) Giemsa C-banding of rye meiotic chromosomes and the nature of "terminal" chiasmata. Chromosoma 66:45–57
Jones GH (1984) The control of chiasma distribution. In: Evans CW, Dickinson HG (eds) Controlling events in meiosis. Symp Soc Exp Biologists 38; Comp Biologists Cambridge, pp 293–320
Jones GH (1987) Chiasmata. In: Moens PB (ed) Meiosis. Academic Press Toronto, pp 213–244
Jones GH, Albini SM, Whitehorn JAF (1991) Ultrastructure of meiotic pairing in B chromosomes of *Crepis capillaris* II. 4B pollen mother cells. Chromosoma 100:193–202
Jones K (1974) Chromosome evolution by Robertsonian translocation in *Gibasis* (Commelinaceae). Chromosoma 45:353–368
Jones K (1976) Multiple Robertsonian fusions in the evolution of a plant genus. In: Jones K, Brandham PE (eds) Current chromosome research. Proc 1st Kew Chrom Conf. North-Holland, Amsterdam, pp 220–221
Jones RN (1975) B chromosome systems in flowering plants and animal species. Int Rev Cytol 40:1–100
Jones RN (1983) Cytogenetic evolution in the genus *Allium*. In: Swaminathan MS, Gupta PK, Sinha U (eds) Cytogenetics of crop plants. MacMillan India, Delhi, pp 555–582

Jones RN, Rees H (1968) Nuclear DNA variation in *Allium*. Heredity 23:591–605
Jones RN, Rees H (1982) B-chromosomes. Academic Press, London
Jong JH de, Speckmann GJ, Bock Th SM de, Voorst A van (1985) Monosomic additions with resistance to beet cyst nematode obtained from hybrids of *Beta vulgaris* and wild *Beta* species of the section patellarea. II. Comparative analysis of the alien chromosomes. Z. Pflanzenzuecht 95:84–94
Jong JH de, Eden, J van, Sybenga J (1989) Synaptonemal complex formation and metaphase I configuration patterns in a translocation heterozygote of rye (*Secale cereale*). Genome 32:72–81
Jong JH de, Havekes F, Roca A, Naranjo T (1991) Synapsis and chiasma formation in a ditelo-substituted haploid of rye. Genome 34:109–120
Jongedijk E (1986) The present stae of research into the induction of apomixis in potato, In: Beekman AGB, Louwes KM, Dellaert LMW, Neele AEF (eds) Potato research of tomorrow, Proc Int Seminar, Wageningen, The Netherlands, Oct 1985. Pudoc, Wageningen, pp 120–123
Jongedijk E (1987) A rapid methyl salicylate clearing technique for routine phase-contrast observations on female meiosis in *Solanum*. J Microsc (Oxf) 146:157–162
Jongedijk E, Hutten RCB, Wolk JMASA vander, Schuurmans-Stekhoven SEJ (1991a) Synaptic mutants in potato, *Solanum tuberosum* L. III. Effect of the Ds-1/ds-1 locus (desynapsis) on genetic recombination in male and female meiosis. Genome 34:121–130
Jongedijk E, Ramanna MS, Sawor Z, Hermsen JGT (1991b) Formation of first division restitution (FDR) 2n-gametes through pseudo-homotypic division in ds-1 (desynapsis) mutants of diploid potato: routine production of tetraploid progeny from 2xFDR-2xFDR crosses. Theor Appl Genet 82:645–656
Jorgensen RB, Bothmer R von (1988) Haploids of *Hordeum vulgare* and *H. marinum* from crosses between the two species. Hereditas 108:207–212
Kaltsikes PJ, Gustafson JP (1985) The heterochromatin story in triticale. In: Bernard M, Bernard J (eds) Genetics and breeding of triticale. Proc Eucarpia Meet on Triticale, Clermont-Ferrand 1984, INRA, pp 5–13
Kaltsikes PJ, Gustafson JP, Lukaszewski AJ (1984) Chromosome engineering in triticale. Can J Genet Cytol 26:105–110
Karpechenko G (1928) Polyploid hybrids of *Raphanus sativus* L. x *Brassica oleracea*. Z Indukt Abstummungs Vererbungsl 48:1–85
Kasha KJ, Kao KN (1970) High frequency haploid production in barley (*Hordeum vulgare* L.). Nature 225:874–875
Kavander T, Viinika Y (1987) Neocentric activity in open pollinated cultivars of rye. Hereditas 107:229–233
Kessous A, Caussinus H, Jaylet A, Beetschen JC (1968) Essay d'analyse biométrique du caryotype de l'amphibien Urodèle *Salamandra salamandra* L. Chromosoma 23:324–332
Khush GS (1973) Cytogenetics of aneuploids. Academic Press, New York
Khush GS (1984) IRRI breeding program and its worldwide impact on increasing rice production. In: Gustafson JP (ed) Gene manipulation and plant improvement. 16th Stadler Genetics Symp, Columbia, MO, pp 61–94
Khush GS, Rick CM (1966) The origin, identification and cytogenetic behavior of tomato monosomics. Chromosoma 18:407–420
Khush GS, Rick CM (1968) Cytogenetics analysis of the tomato genome by means of induced deficiencies. Chromosoma 23:452–484
Khush GS, Singh RJ (1991) Chromosome architecture and aneuploidy in rice. In: Gupta PK, Tsuchiya T (eds) Chromosome engineering in plants: genetics, breeding, evolution, Part A. Elsevier, Amsterdam, pp 577–598
Kihara H (1930) Genomanalyse bei *Triticum* and *Aegilops*. Cytologia 1:263–270
Kihara H, Nishiyama I (1951) An application of sterility of autotriploid to the breeding of seedless watermelon. Seiken Ziho 3:5–15

Kimber G (1983) The B-genome of wheat: the present status. In: Swaminathan MS, Gupta PK, Sinha U (eds) Cytogenetics of crop plants. MacMillan India, Delhi, pp 214–224
Kimber G (1984) Technique selection for the introduction of alien variation in wheat. Z Pflanzenzuecht 92:15–21
Kimber G, Alonso LG (1981) The analysis of meiosis in hybrids. III. Tetraploid hybrids. Can J Genet Cytol 23:235–254
Kimber G, Riley R (1963) Haploid angiosperms. Bot Rev 29:480–531
Kimber G, Yen Y (1989) Hybrids involving wheat relatives and autotetraploid *Triticum umbellulatum*. Genome 32:1–5
Kimber G, Alonso LG, Sallee PJ (1981) The analysis of meiosis in hybrids. I. Aneuploid hybrids. Can J Genet Cytol 23:209–219
King IP, Reader SM, Miller TE (1988) Exploitation of the "Cuckoo" chromosome (4Sl) of *Aegilops sharonensis* for eliminating segregation for height in semi-dwarf Rht2 bread wheat cultivars. In: Miller TE, Koebner RMD (eds) Proc 7th Int Wheat Genet Symp Cambridge. IPSR, Cambridge, pp 337–341
Klasterska I (1977) The concept of the prophase of meiosis. Hereditas 86:205–210
Kleijer G (1982) Cytogenetic studies of *Lolium multiflorum* Lam., *Festuca arundinacea* Schreb., their hybrids and amphidiploids. Dissertation, Agricultural University, Wageningen
Klinga K (1987a) Aneuploidy in induced autotetraploid populations of *Festuca pratensis*, *Lolium multiflorum*, and *Lolium perenne*. III. Performance of plants at different sowing densities in a perennial swad of tetraploid *Featuca pratensis*. Hereditas 106:277–287
Klinga K (1987b) Aneuploidy in induced autotetraploid populations of *Festuca pratensis*, *Lolium multiflorum* and *Lolium perenne*, IV, Competitive interactions between euploid and aneuploid plants in a simulated perennial swad of tetraploid *Festuca pratensis*. Hereditas 107:249–256
Knott DR (1958) The effect on wheat of an *Agropyron* chromosome carrying rust resistance. In: Proc 10th Int Congr Genet, Montreal 2:148
Knott DR (1988) The chromosome location for four recombinants between Agropyron chromosome 7el2 and a wheat chromosome. Genome 30:97–98
Koebner RMD, Shepherd KW (1986) Controlled introgression to wheat of genes from rye chromosome arm 1RS by induction of allosyndesis. 1. Isolation of recombinants. Theor Appl Genet 73:197–208
Koebner RMD, Shepherd KW (1987) Allosyndetic recombination between a chromosome of *Aegilops umbellulata* and wheat chromosomes. Heredity 59:33–45
Koebner RMD, Shepherd KW (1988) Isolation and agronomic assessment of allosyndetic recombinants derived from wheat/rye translocation 1DL.1RS, carrying reduced amounts of rye chromatin. In: Miller TE, Koebner RMD (eds) Proc 7th Int Wheat Genet Symp. IPSR, Cambridge, pp 343–348
Koller DL, Zeller FJ (1976) The homoeologous relationships of rye chromosomes 4R and 7R with wheat chromosomes. Genet Res 28:177–188
Koornneef M, Veen JH van der (1983) Trisomics in *Arabidopsis thaliana* and the location of linkage groups. Genetica 61:41–46
Kornberg RD, Klug A (1981) The nucleosome. Sci Am 244, 2:48–61
Kosambi DD (1944) The estimation of map distance from recombination values. Ann Eugen 12:172–175
Kota RS, Dvorak J (1988) Genome instability in wheat induced by chromosome 6Bs of *Triticum speltoides*. Genetics 120:1085–1094
Kourilsky P, Chambon P (1978) The ovalbumin gene: an amazing gene in eight pieces. TIBS Nov 1978:244–247
Kramer HH (1954) Recombination in selfed interchange heterozygotes. In: Kempthorne O et al. (eds) Statistics and mathematics in biology. Iowa State College Press, Ames, Iowa, pp 511–522

Krebs SL, Hancock JF (1989) Tetrasomic inheritance of isozyme markers in the highbush blueberry, *Vaccinium corymbosum* L. Heredity 63:11–18
Kumar H, Mercykutty VC, Srivastava CP (1990) Chiasma formation in induced autotetraploids of pea (*Pisum sativum* L.) – an efficient method of scoring. Cytologia 55:287–292
Kwack BH, Kim I-H (1967) An improved method for culturing *Tradescantia* pollen tubes for chromosome analysis. Cytologia 32:1–5
Laat AAM de, Blaas J (1984) Flow-cytometric characterization and sorting of plant chromosomes. Theor Appl Genet 67:463–467
Laat AMM de, Verhoeven HA, Ramulu KS, Dijkhuis P (1987) Efficient induction by amiprophos-methyl and flow cytometric sorting of micronuclei in *Nicotiana plumbaginifolia*. Planta 172:473–478
LaCour LF (1952) The Luzula system analysed by X-rays. Heredity (Suppl) 6:77–81
Lammerts WE (1934) On the nature of chromosome association in *N. tabacum* haploids. Cytologia 6:38–50
Lange W (1971) Crosses between *Hordeum vulgare* L. and *H. bulbosum* L. II. Elimination of chromosomes in hybrid tissues. Euphytica 20:181–194
Lange W, Toxopeus H, Lubberts, JH, Dolstra O, Harrewijn JL (1989) The development of Raparadish (×*Brassicoraphanus*, 2n = 38), a new crop in agriculture. Euphytica 40:1–14
Lapitan NLV, Sears RG, Gill BS (1984) Translocations and other karyotypic structural changes in wheat × rye hybrids regenerated from tissue culture. Theor Appl Genet 68:547–554
Lapitan NLV, Sears RG, Rayburn AL, Gill BS (1986) Wheat-rye translocations. Detection of chromosome breakpoints by in situ hybridization with a biotin-labelled DNA probe. J Hered 77:415–419
Lapitan NLV, Sears RG, Gill BS (1988) Amplification of repeated sequences in wheat × rye hybrids regenerated from tissue culture. Theor Appl Genet 75:381–388
Larkin PJ, Scowcroft WR (1981) Somaclonal variation – a novel source of variability from cell cultures. Theor Appl Genet 60:197–214
Larsen J, Kimber G (1973) The arm ratio of chromosome 5B from *Triticum aestivum* var. Chinese Spring at mitosis and meiosis. Hereditas 73:316–318
Lashermes P, Beckert M (1988) Genetic control of maternal haploidy in maize (*Zea mays* L.) and selection of haploid inducing lines. Theor Appl Genet 76:405–410
Lashermes P, Gaillard A, Beckert M (1988) Gynogenetic haploid plants analysis for agronomic and enzymatic markers in maize (*Zea mays* L.). Theor Appl Genet 76:570–572
Laurie DA, Bennett MD (1989) The timing of chromosome elimination in hexaploid wheat × maize crosses. Genome 32:953–961
Lavania UC (1986) Genetic improvement of Egyptian henbane, *Hyoscyamus muticus* L. through induced tetraploidy. Theor Appl Genet 73:292–298
Lavania UC (1991) Polyploid breeding: meiosis in the diploid progenitor and its preductive value for fertility in the autotetraploid. Proc Indian Natl Sci Acad B57:83–90
Lavania UC, Srivastava S, Sybenga J (1991) Cytogenetics of fertility improvement in artificial autotetraploids of *Hyoscyamus niger* L. Genome 34:190–194
Law CN (1966) Biometrical analysis using chromosome substitutions within a species. In: Riley R, Lewis KR (eds) Chromosome manipulations and plant genetics Oliver & Boyd, Edinburgh, pp 59–85
Lawrence CW (1961) The efffect of the irradiation of different stages in microsporogenesis in chiasma frequency. Heredity 16:83–89
Lawrence CW (1965) Influence of non-lethal doses of radiation on recombination in *Chlamydomonas reinhardi*. Nature 206:789–791
Lawrence GJ, Appels R (1986) Mapping the nucleolar organizing region, seed protein loci and isozyme loci on chromosome 1R in rye. Theor Appl Genet 71:742–749

Le HT, Armstrong KC, Miki B (1989) Detection of rye DNA in wheat rye hybrids and wheat translocation stocks using total genomic DNA as a probe. Plant Mol Biol Rep 7:150–158

Lehmann LC (1981) Where is hybrid barley today? Barley Genetics 4. 4th Barley Genet Symp Edinburgh, University Press, Edinburgh, pp 772–777

Lenz EM, Crane CF, Sleper DA, Loegering WQ (1983) An assessment of preferential chromosome pairing at meiosis in *Dactylis glomerata*. Can J Genet Cytol 25:222–232

Levan A (1940) Meiosis in *Allium porrum*, a tetraploid species with chiasma localization. Hereditas 26:454–462

Levan A (1941) The cytology of the species hybrid *Allium cepa* × *fistulosum* and its polyploid derivatives. Hereditas 27:253–272

Levan A (1942) Studies on the meiotic mechanism of haploid rye. Hereditas 28:127–211

Levan A, Fredga K, Sandberg AA (1964) Nomenclature for centromeric position on chromosomes. Hereditas 52:201–220

Lewis WH (ed) (1980) Polyploidy. Biological relevance. Plenum Press, New York

Lilienfeld FA (1951) H. Kihara: Genome analysis in *Triticum* and *Aegilops*. X. Concluding review. Cytologia 16:101–123

Lima-de-Faria A (1956) The role of the kinetochore in chromosome organization. Hereditas 42:85–160

Lima-de-Faria A (1983) Molecular evolution and organization of the chromosome. Elsevier, Amsterdam

Lima-de-Faria A, Jaworska H (1968) Late DNA-synthesis in heterochromatin. Nature 217:138–142

Lima-de-Faria A, Isaksson M, Olsson E (1980) Action of restriction endonucleases on the DNA and the chromosomes of *Muntiacus muntjak*. Hereditas 92:267–273

Lin B-Y (1982) Association of endosperm reduction with parental imprinting in maize. Genetics 100:475–486

Lin B-Y (1984) Ploidy barrier to endosperm development in maize. Genetics 107:103–115

Lin H-PP, Church K (1982) Meiosis in *Drosophila melanogaster*. III. The effect of orientation disruptor (ord) on gonial mitotic and the meiotic divisions in males. Genetics 102:751–770

Linde-Laursen I, Bothmer R von (1988) Elimination and duplication of particular *Hordeum vulgare* chromosomes in aneuploid interspecific *Hordeum* hybrids. Theor Appl Genet 76:897–908

Linnert G (1962) Untersuchungen an hemiploiden Nachkommen Autotetraploider 1. Der Verteilungsmodus einer reziproken Translokation und die daraus folgende Erhöhung der Heterozygotenfrequenz in der Nachkommenschaft Duplex-Heterozygoten Z Vererbungsl 93:389–398

Loidl J (1986) Synaptonemal complex spreading in *Allium*. II. Tetraploid *Allium vineale*. Genome 28:754–761

Loidl J (1987) Heterochromatin and differential chromosome staining in meiosis. Biol Zentralbl 106:641–662

Loidl J (1990) The initiation of meiotic chromosome pairing: the cytological view. Genome 33:759–778

Loidl J, Jones GH (1986) Synaptonemal complex spreading in *Allium*. I Triploid *A. sphaerocephalon*. Chromosoma 93:420–428

Loidl J, Ehrendorfer F, Schweizer D (1990) EM analysis of meiotic chromosome pairing in a pentaploid *Achillea* hybrid. Heredity 65:11–20

Lukaszewski AJ (1988) Chromosome constitution of hexaploid triticale lines in the recent international yield trials. Plant Breed 100:268–272

Lukaszewski AJ, Apolinarska B (1981) Chromosome constitution of hexaploid winter triticale. Can J Genet Cytol 23:281–285

Lukaszewsky AJ, Gustafson JP (1983) Translocations and modifications of chromosomes in triticale × wheat hybrids. Theor Appl Genet 64:299–348

Lukaszewski AJ, Gustafson JP (1988) Cytogenetics of triticale. Plant Breed Rev 5:41–94
Lukaszewski AJ, Apolinarska B, Gustafson JP, Krolow K-D (1987a) Chromosome pairing and aneuploidy in tetraploid triticale. I. Stabilized karyotypes. Genome 29:554–561
Lukaszewski AJ, Apolinarska B, Gustafson JP, Krolow KD (1987b) Chromosome pairing and aneuploidy in tetraploid triticale. II. Unstabilized karyotypes. Genome 29:562–569
Lundsteen C, Piper J (1989) Automation of cytogenetics. Springer, Berlin Heidelberg New York
Luykx P (1970) Cellular mechanisms of chromosome distribution. Academic Press, New York
Lyon MF (1963) Attempts to test the inactive-X theory of dosage compensation in mammals. Genet Res 4:93–103
Maan SS (1975) Exclusive preferential transmission of alien chromosomes in common wheat. Crop Sci 15:278–292
Macefield AJ, Evans GM (1976) The effect of B chromosomes on meiosis in autotetraploid *Lolium perenne*. Heredity 36:393–397
Magoon ML, Khanna KR (1963) Haploids. Caryologia 6:191–235
Makino T, Sasaki M, Morris R (1977) Misdivision of homoeologous group 5 univalent chromosomes in hexaploid wheat I. Univalents derived from Japanese cultivars. Cytologia 42:73–83
Marais GF (1990) Preferential transmission in bread wheat of a chromosome segment derived from *Thinopyrum distichum* (Thunb.) Löve. Plant Breed 104:152–159
Mariani C, DeBeuckeleer M, Truettner J, Leemans J, Goldberg RB (1990) Induction of male sterility in plants by a chimaeric ribonuclease gene. Nature 347:737–741
Mariani C, DeBeuckeleer M, Truettner J, Leemans J, Goldberg RB (1990) Induction of male sterility in plants by a chimaeric ribonuclease gene. Nature 347:737–741
Marks GE, Schweizer D (1974) Giemsa banding: karyotype differences in some species of *Anemone* and *Hepatica notilis*. Chromosoma 44:217–224
Mastenbroek I, DeWet JMJ, Lu C-Y (1982) Chromosome behaviour in early and advanced generations of tetraploid maize. Caryologia 35:463–470
Matérn B, Simak M (1968) Statistical problems in karyotype analysis. Hereditas 59:280–288
Matzk F (1982) Vorstellungen über potentielle Wege zur Apomixis bei Getreide. Arch Züchtungsf Berl 12(3):183–195
McClintock B (1931) Cytological observations of deficiencies involving known genes, translocations and an inversion in *Zea mays*. Mo Agric Exp Stn Res Bull 163:1–30
McClintock B (1938) The fusion of broken ends of sister half chromatids following chromatid breakage at meiotic anaphases. Mo Agric Exp Stn Res Bull 290:1–48
McCoy TJ (1989) A potential solution to the chromosome instability problem in hexaploid alfalfa, *Medicago sativa*. Genome 32:302–306
McFadden ES, Sears ER (1944) The artificial synthesis of *Triticum spelta*. Genetics 30:14
McIntosh JR (1985) Spindle structure and the mechanisms of chromosome movement. In: Dellarco VL, Voytek PE, Hollaender A (eds) Aneuploidy. Etiology and mechanisms. Plenum Press, New York, pp 197–229
McLeish J, Snoad B (1958 etc) Looking at chromosomes. Macmillan, London
McNaughton IH (1979) The current position and problems in the breeding of Raphanobrassica (Radicole) as a forage crop. In: Marrewijk NPH van, Toxopeus H (eds) Cruciferae. Proc Eucarpia Conf, Wageningen. pp 22–28
Meer van der QP (1987) Chromosomal monogenic dominant male sterility in Chinese cabbage (*Brassica rapa* subsp. *pekinensis* [Lour] Hanelt). Euphytica 36:927–931
Meister A, Bretschneider H (1977) Präferentielle Paarung der Chromosomen autotetraploider Gerste – mathematisches Modell. Kulturpflanze 25:297–311
Melchers G, Sacristan MD, Holder AA (1979) Somatic hybrid plants of potato and tomato regenerated from fused protoplasts. Carlsberg Res Commun 43:203–218

Melz G, Thiele V (1990) Spontaneous somatic transfer of a segment from a wheat addition chromosome into the rye genome. Genome 33:794–797

Melz G, Winkel A (1986) Evidence for monotelodisomics in rye (*Secale cereale* L.). Plant Breed 97:368–370

Melz G, Schlegel R, Sybenga J (1988) Identification of the chromosomes in the "Esto" set of rye trisomics. Plant Breed 100:169–172

Menzel MY, Dougherty BJ (1987) Transmission of duplication-deficiencies from cotton translocations is unrelated to map lengths of the unbalanced segments. Genetics 116:321–330

Menzel Y (1964) Preferential chromosome pairing in allotetraploid *Lycopersicum esculentum – Solanum lycopersicoides*. Genetics 50:855–862

Merker A (1976) The cytogenetic effect of heterochromatin in hexaploid triticale. Hereditas 83:215–222

Michie D, Wallace ME (1953) Affinity: a new genetic phenomenon in the house mouse. Nature (Lond) 171:26–27

Miller TE, Reader SM (1985) The effect of increased dosage of wheat chromosomes on chromosome pairing and an analysis of chiasma frequencies of individual wheat bivalents. Can J Genet Cytol 27:421–425

Minocha JL (1991) Pearl millet cytogenetics. In: Gupta PK, Tsuchiya T (eds) Chromosome engineering in plants: genetics, breeding, evolution, Part A. Elsevier, Amsterdam, pp 599–611

Moens PB, Earnshaw WC (1989) Anti-topoisomerase II recognizes meiotic chromosome cores. Chromosoma 98:317–322

Moens PB, Pearlman RE (1990) Telomere and centromere DNA are associated with the cores of meiotic prophase chromosomes. Chromosoma 100:8–14

Moore R, Gregory G (1963) Biometrics of the karyotype of *Protemnodon bicolor*, with reference to the limitations in accuracy of determining human chromosomes. Nature 200:234–237

Morgan DT, Rappley R (1954) A cytogenetics study of the origin of multiple seedlings of *Capsicum frutescens*. Am J Bot 41:576–586

Morgan WG, Thomas H (1991) A study of chromosome association and chiasma formation in the amphidiploid between *Lolium multiflorum* and *Festuca drymeja*. Heredity 67:241–245

Morinaga T (1934) Interspecific hybridization in *Brassica*. VI. The cytology of F1 hybrids between *B. juncea* and *B. nigra*. Cytologia 6:62–67

Morrison JW (1953) Chromosome behaviour in wheat monosomics. Heredity 7:203–217

Morrison JW (1954) Chromosome interchange by misdivision in *Triticum*. Can J Bot 32:281–284

Mouras A, Salesses G, Lutz A (1978) Sur l'utilisation des protoplastes en cytologie: amélioration d'une méthode recente en vue de l' identification des chromosomes mitotiques des genres *Nicotiana* et *Prunus*. Caryologia 31:117–127

Mouras A, Wildenstein C, Salesses G (1986) Analysis of karyotype and C-banding pattern of *Nicotiana plumbaginifolia* using two techniques. Genetica 68:197–202

Mouras A, Saul MW, Essad S, Potrykus I (1987) Localization by in situ hybridization of a low copy chimaeric resistance gene introduced into plants by direct gene transfer. Mol Gen Genet 207:204–209

Mujeeb-Kazi A (1981) Apomictic progeny derived from intergeneric *Hordeum-Triticum* hybrids. J Hered 72:284–285

Mulcahy DL, Mulcahy GB, Robinson RW (1979) Evidence for postmeiotic genetic activity in pollen of *Cucurbita* species. J Hered 70:365–368

Müntzing A (1979) Triticale. Results and problems. Fortschr Pflanzenzuecht 10:1–103

Mursal IEM, Endrizzi JE (1976) A reexamination of the diploidlike meiotic behavior of polyploid cotton. Theor Appl Genet 47:171–178

Murty ER, Kirti PB, Bharathi M, Rao NGP (1984) The nature of apomixis and its utilization in the production of hybrids ("Vybrids") in *Sorghum bicolor* (L.). Z Pflanzenzuecht 92:30–39

Nagl W (1969) Banded polytene chromosomes in the legume *Phaseolus vulgaris*. Nature 221:70–71
Nagl W (1978) Endopolyploidy and polyteny in differentiation and evolution. North-Holland, Amsterdam
Namai H, Sarashima N, Hosoda T (1980) Interspecific and intergeneric hybridization breeding in Japan. In: Tsumoda S, Hinata H, Gómez-Campo G (eds) *Brassica* crops and wild allies. Biology and breeding. Jpn Sci Soc Press, Tokyo, pp 191–203
Naranjo CA, Poggio L, Brandham PE (1983) A practical method of chromosome classification on the basis of centromere position. Genetica 62:51–53
Naranjo T, Fernandez-Rueda P (1991) Homoeology of rye chromosome arm to wheat. Theor Appl Genet 82:577–586
Naranjo T, Orellana J (1984) Meiotic behaviour of chromosomes 1R, 2R and 5R in autotetraploid rye. Chromosoma 89:143–150
Naranjo T, Fernandez-Rueda P, Goicoechea PG, Roca A, Giraldez R (1989) Homoeologous pairing and recombination between the long arms of group 1 chromosomes in wheat × rye hybrids. Genome 32:293–301
Nebel BR (1937) Mechanism of polyploids through colchicine. Nature 140:1101
Negri V, Veronesi F (1989) Evidence for the existence of 2n gametes in *Lotus tenuis* Wald. et Kit. (2n = 2x = 12): their relevance in evolution and breeding of *Lotus corniculatus* L. (2n = 4x = 24). Theor Appl Genet 78:400–404
Neijzing MG (1982) Chiasma formation in duplicated segments of the haploid rye genome. Chromosoma 85:287–298
Neijzing MG (1985) Number and localization of sets of homologous segments in the genome of rye on the basis of Giemsa banded metaphase I chromosome associations in a haploid. Genetica 66:41–52
Nishiyama I, Yabuno T, Taira T (1989) Genomic affinity relations in the genus *Avena*. Plant Breed 102:22–30
Noda S (1968) Achiasmate bivalent formation by parallel pairing in PMCs of *Fritillaria amabilis*. Bot Mag Tokyo 81:344–345
Nordenskiöld H (1953) A genetical study in the mode of segregation in hexaploid *Phleum pratense*. Hereditas 39:469–488
Nordenskiöld H (1962) Studies of meiosis in *Luzula purpurea*. Hereditas 34:104–112
Norrington-Davies J, Moore K, Blijenburg JG (1981) The yield and competitive ability of euploid and aneuploid genotypes of tetraploid *Lolium perenne* in two seasons at two locations. Euphytica 30:675–686
Ockey CH (1969) The position of chromosomes in human fiboblasts and their DNA synthesis behaviour. Chromosoma 27:308–320
Ohno S (1970) Evolution by gene duplication. Springer, Berlin Heidelberg New York
Ohri D, Zadoo SN (1986) Cytogenetics of cultivated Bougainvilleas. IX. Precocious centromere division and origin of polyploid taxa. Plant Breed 97:227–231
Olins AL, Olins DE (1974) Spheroid chromatin units (nu-bodies). Science 183:330–332
Olsson G (1986) Allopolyploids in *Brassica*. In: Olsson G (ed) Research and results in plant breeding. LTs, Stockholm, pp 114–119
Olsson G, Josefsson A, Hagberg A, Ellerström S (1955) Synthesis of the ssp. *rapifera* of *Brassica napus*. Hereditas 41:241–249
Orellana J (1985) Most of the homoeologous pairing at metaphase I in wheat-rye hybrids is not chiasmatic. Genetics 111:917–931
Orellana J, Santos JL (1985) Pairing competition between identical and homologous chromosomes in autotetraploid rye. I. Submetacentric chromosomes. Genetics 111:933–944
Östergren G (1951) The mechanism of coorientation in bivalents and multivalents. The theory of orientation by pulling. Hereditas 37:85–156
Oud JL, Reutlinger AHH (1981) The chromosome behaviour during early meiotic prophase of mouse primary spermatocytes. Chromosoma 83:395–407
Oud JL, Jong JH de, Rooij DG de (1979) A sequential analysis of meiosis in the male mouse using a restricted spermatocyte population obtained by a hydroxyurea/triaziquone treatment. Chromosoma 71:237–248

Oud JL, Kakes P, Jong JH de (1987) Computerized analysis of chromosome parameters in karyotype studies. Theor Appl Genet 73:630–634

Oud JL, Mans A, Brakenhoff GJ, Voort HTM vander, Spronsen EA van, Nanninga N (1989) 3D chromosome arrangement of *Crepis capillaris* in mitotic prophase and anaphase as studied by convocal scanning laser microscopy. J Cell Sci 92:329–339

Painter TS (1934) Salivary chromosomes and the attack on the gene. J Hered 25:465–476

Pardue ML, Gall JG (1969) Molecular hybridization of radioactive DNA to the DNA of cytological proparations. Proc Natl Acad Sci USA 64:600–604

Parker JS (1975) Chromosome specific desynapsis in *Hypochoeris radicata*. Chromosoma 49:391–406

Parker JS (1987) Increased chiasma frequency as a result of chromosome rearrangement. Heredity 58:87–94

Parker JS, Jones GH, Edgar LA, Whitehouse C (1990) The population cytogenetics of *Crepis capillaris*. III. B-chromosome effects on meiosis. Heredity 64:377–385

Parrott WA, Smith RR, Smith MM (1985) Bilateral sexual tetraploidization in red clover. Can J Genet Cytol 27:64–68

Patterson EB (1973) Genic male sterility and hybrid maize production. In: Proc 7th Meet Maize and Sorghum section Eucarpia, Zagreb, Vugoslavia

Peloquin SJ, Yerk GL, Werner JE, Darmo E (1989) Potato breeding with haploids and 2n gametes. Genome 31:1000–1004

Person C (1956) Some aspects of monosomic wheat breeding. Can J Bot 34:60–70

Phillips LL (1964) Segregation in new allopolyploids of *Gossypium*. V. Multivalent formation in New World × Asiatic and New World × wild American hexaploids. Am J Bot 51:324–329

Pijnacker LP (1966) The maturation divisions of the parthenogenetic stick insect *Carausius morosus* Br. (Orthoptera, Phasmidae). Chromosoma 19:99–112

Pijnacker LP, Ferwerda MA (1982) Normal and BrdC-substituted chromosomes during spermatogenesis with an endomeiotic chromosome reduplication in *Carausius morosus*. Chromosoma 85:67–82

Pijnacker LP, Ferwerda MA (1990) Orientation of diplochromosomes in mitoses of potato (*Solanum tuberosum* L.) callus cultures. Genome 33:328–332

Pijnacker LP, Walch K, Ferwerda MA (1986) Behaviour of chromosomes in potato leaf tissue cultured in vitro as studied by BrdC-Giemsa labelling. Theor Appl Genet 72:833–839

Pilch J, Hesemann CU (1986) Variation of telomeric heterochromatin among inbred lines in diploid rye (*Secale cereale*). Plant Breed 97:371–374

Piper J, Lundsteen C (1987) Human chromosome analysis by machine. Trends Genet 3:309–313

Pirrie A, Power JB (1986) The production of fertile, triploid somatic hybrid plants (*Nicotiana glutinosa*(n) + *N. tabacum* (2n)) via gametic somatic protoplast fusion. Theor Appl Genet 72:48–52

Pohler W, Claus E (1985) Genomallozyklie verursacht zytologische Störungen bei *Hordeum-Secale*-Bastarden. Biol Zentralbl 104:145–154

Pohler W, Clauss E, Senula A (1991) Cytology of somatically unstable Hordeum × Secale hybrids and of barley derivatives. Proc Meeting Cereal Section Eucarpia, Schwerin 1991; Vorträge Pflanzenzüchtung 20:121–126

Pohler W, Schlegel R (1990) A rye plant with frequent A–B chromosome pairing. Hereditas 112:217–220

Prakash S, Tsunoda S (1983) Cytogenetics ot *Brassica*. In: Swaminathan MS, Gupta PK, Sinha U (eds) Cytogenetics of crop plants. MacMillan India, Delhi, pp 482–513

Prakken R (1943) Studies of asynapsis in rye. Hereditas 29:475–495

Prakken R, Müntzing A (1942) A meiotic peculiarity in rye, simulating a terminal centromere. Hereditas 28:441–482

Puertas MJ, Peña A de la, Estades B, Merino F (1984) Early sensitivity to colchicine in developing anthers of rye. Chromosoma 89:121–126

Qi L-L, Sybenga J, Jong JH de (1988) Synaptonemal complex pairing and metaphase I association in a telo-substituted telotrisomic of rye (*Secale cereale* L.). Chromosoma 97:88–95

Rajhathy T (1983) The chromosomal architecture of *Avena*. In: Swaminathan MS, Gupta PK, Sinha U (eds) Cytogenetics of crop plants. MacMillan India, Delhi, pp 283–307

Rajhathy T (1991) The chromosomes of *Avena*. In: Gupta PK, Tsuchiya T (eds) Chromosome engineering in plants: genetics, breeding, evolution, Part A. Elsevier, Amsterdam, pp 449–467

Ramage RT (1965) Balanced tertiary trisomics for use in hybrid seed production. Crop Sci 5:177–178

Ramage RT (1983) Heterosis and hybrid seed production in barley. In: Frankel R (ed) Heterosis – reappraisal of theory and practice. Monogr Theor Appl Genet 6. Springer, Berlin Heidelberg New York, pp 71–93

Ramage RT (1991) Chromosome manipulations in barley breeding. In: Gupta PK, Tsuchiya T (eds) Chromosome engineering in plants: genetics, breeding, evolution, Part A. Elsevier, Amsterdam, pp 385–400

Ramage RT, Burnham CR, Hagberg A (1961) A summary of translocation studies in barley. Crop Sci 1:277–279

Ramanna MS (1983) First division restitution gametes through fertile desynaptic mutants of potato. Euphytica 32:337–350

Ramanna MS, Wagenvoort M (1976) Identification of the trisomic series in diploid *Solanum tuberosum* L., group tuberosum. I. Chromosome identification. Euphytica 25:233–240

Ramulu KS (1987) Genetic instability during plant regeneration in potato: origin and implications. Plant Physiol (Life Sci Adv) 6:211–218

Ramulu KS, Verhoeven HA, Dijkhuis P (1988) Mitotic dynamics of micronuclei induced by amiprophos-methyl and prospects for chromosome mediated gene transfer in plants. Theor Appl Genet 75:575–584

Rao MK, Devi KU, Arundhati A (1990) Applications of genic male sterility in plant breeding. Plant Breed 105:1–25

Rasmussen SW, Holm PB (1979) Chromosome pairing in autotetraploid *Bombyx mori* females. Mechanism for exclusive bivalent formation. Carlsberg Res Commun 44:101–125

Rattner JB, Lin CC (1987) The higher order structure of the centromere. Genome 29:588–593

Rayburn AL, Auger JA (1990) Genome size variation in *Zea mays* ssp. *mays* adapted to different altitudes. Theor Appl Genet 79:470–474

Rayburn AL, Gill BS (1985) Use of biotin labeled probes to map specific DNA sequences on wheat chromosomes. J Hered 76:78–81

Rees H, Jones GH (1967) Chromosome evolution in *Lolium*. Heredity 22:1–18

Rees H, Jones RN (1977) Chromosome genetics. Edward Arnold, London

Rees H, Thompson JB (1956) Genotypic control of chromosome behaviour in rye. III. Chiasma frequency in homozygotes and heterozygotes. Heredity 3:409–424

Reimann-Philip R, Eichhorn-Rohde H (1970) Studies on the relationship of fertility and mode of chromosome pairing in tetraploid F1 species hybrids of rye (*Secale cereale* × *Secale montanum*). Theor Appl Genet 40:99–105

Ren ZL, Lelley T, Röbbelen G (1990a) The use of monosomic rye addition lines for transferring rye chromatin into bread wheat. I. The occurrence of translocations. Plant Breed 105:257–264

Ren ZL, Lelley T, Röbbelen G (1990b) The use of monosmic rye addition lines for transferring rye chromatin into bread wheat. II. Breeding value of homozygous wheat/rye translocations. Plant Breed 105:265–270

Rhoades MM (1936) A cytogenetic study of a chromosome fragment in maize. Genetics 21:491–502

Rhoades MM (1947) Crossover chromosomes in unreduced gametes of asynaptic maize. Genetics 32:101

Rhoades MM (1952) Preferential segregation in maize. In: Gowen JW (ed) Heterosis. Iowa State College Press, Ames, pp 66–80
Rhoades MM (1968) Studies on the cytological basis of crossing-over. In; Peacock WJ, Brock RD (eds) Replication and recombination of genetic material. Austr Acad Sci, pp 229–241
Rhoades MM, Dempsey E (1953) Cytogenetic studies of deficient-duplicate chromosomes derived from inversion heterozygotes in maize. Am J Bot 40:405–424
Rhoades MM, Dempsey E (1966) Induction of chromosome doubling at meiosis by the elongate gene in maize. Genetics 54:505–522
Richards EJ, Ausubel FM (1988) Isolation of a higher eukaryotic telomere from *Arabidopsis thaliana*. Cell 53:127–136
Rickards GK (1983) Orientation behaviour of chromosome multiples of interchange (reciprocal translocation) heterozygotes. Annu Rev Genet 17:443–498
Rickards GK (1988) The spatial distribution of chromosomes in the pre-meiotic nucleus as indicated by asynapsis and chromosome stretching in an interchange quadrivalent. In: Brandham P (ed) Proc Kew Chrom Conf III, HMSO, pp 219–227
Riley R (1966) Genotype-environmental interaction affecting chiasma frequency in *Triticum aestivum*. Chromosomes Today 1:57–65
Riley R, Chapman V (1958) Genetic control of the cytologically diploid behaviour of hexaploid wheat. Nature 182:713–715
Riley R, Chapman V (1966) Estimates of the homoeology of wheat chromosomes by measurements of differential affinity at meiosis. In: Riley R, Lewis KR (eds) Chromosome manipulations and plant genetics. Oliver & Boyd, Edinburgh, pp 46–58
Riley R, Kimber G (1961) Aneuploids, and the cytogenetic structure of wheat varietal populations. Heredity 16:275–290
Riley R, Law C (1984) Chromosome manipulation in plant breeding: progress and prospects. In: Gustafson JP (ed) Genetic manipulation in plant improvement. 16th Stadler Genetics Symp, pp 301–322
Riley R, Chapman V, Johnson R (1968) The incorporation of alien disease resistance in wheat by genetic interference with the regulation of meiotic chromosome synapsis. Genet Res 12:199–219
Röbbelen G (1966) Beobachtungen bei interspezifischen *Brassica*-Kreuzungen, insbesondere über die Entstehung matromorpher F1-Pflanzen. Angew Bot 39:205–221
Robertson WRB (1916) Chromosome studies I. Taxonomic relationships shown in the chromosomes of Tettigidae and Acrididae: V-shaped chromosomes and their significance in Acrididae, Locustidae, and the Gryllidae: chromosomes and variation. J Morphol 27:179–331
Roeder GS (1990) Chromosome synapsis and genetic recombination: their roles in meiotic chromosome segregation. Trends Genet 6:385–389
Rogowsky PM, Guidet FLY, Langridge P, Shepherd KW, Koebner RMD (1991) Isolation and characterization of wheat-rye recombinants involving chromosome arm 1DS of wheat. Theor Appl Genet 82:537–544
Roman H (1947) Mitotic non-disjunction in the case of interchanges involving the B-type chromosome in maize. Genetics 32:391–409
Romera F, Jiménez MM, Puertas MJ (1991) Genetic control of the rate of transmission of rye B-chromosomes. I. Effects in 2B × 0B crosses. Heredity 66:61–65
Roothaan M, Sybenga J (1976) No 5B compensation by rye B-chromosomes. Theor Appl Genet 48:63–66
Roseweir J, Rees H (1962) Fertility and chromosome pairing in autotetraploid rye. Nature 195:203–204
Rufas JS, Gimenez-Abian J, Suja JA, Garcia de la Vega C (1987) Chromosome organization in meiosis revealed by light microscope analysis of silver stained cores. Genome 29:706–712

Rutishauser A (1967) Fortpflanzungsmodus und Meiose apomiktischer Blütenpflanzen. Protoplasmatologia VI F3. Springer, Berlin Heidelberg New York
Sabeva Z (1984) Obtaining and investigation of tetraploid triticale forms. Cereal Res Commun 13:71–76
Säll T, Flink J, Bengtsson BO (1990) Genetic control of recombination in barley. I. Variation in recombination frequency measured with inversion heterozygotes. Hereditas 112:157–170
Santos JL, Orellana J, Giraldez R (1983) Pairing competition between identical and homologous chromosomes in rye and grasshoppers. Genetics 104:677–684
Santos JL, Ciprés G, Lacadena JR (1987) Metaphase I chiasmata in silver stained cores of bivalents in grasshopper spermatocytes. Genome 29:235–238
Sarkar KR, Coe EH (1971) Origin of parthenogenetic diploids in maize and its implications for the production of homozygous lines. Crop Sci 11:543–544
Sarvella P, Meyer JR, Owings A (1961) A scotch tape method for counting or measuring stomata. Crop Sci 1:81–82
Savitsky VF (1962) Sucrose and weight of root in tetraploid monogerm sugar beet populations under different mating systems. J Am Soc Sugar Beet Technol 11:676–711
Schlegel R (1982) First evidence for rye-wheat additions. Biol Zentralbl 101:641–646
Schlegel R, Kynast R (1988) Wheat chromosome 6B compensates genetical information of diploid rye, *Secale cereale* L. In: Miller TE, Koebner RM (eds) Proc 7th Int Wheat Genet Symp Cambridge, IPRS Cambridge, pp 421–426
Schlegel R, Liebelt H (1976) Untersuchungen zum meiotischen Paarungsverhalten Giemsa-gefärbter Chromosomen des autotetraploiden Roggens (*Secale cereale* L.). Biol Zentralbl 95:725–734
Schlegel R, Scholz I, Fischer K (1985) Aneusomie als Ursache für Meiosestörungen beim tetraploidem Roggen? Biol Zentralbl 104:375–384
Schlegel R, Kynast R, Schmidt JC (1986) Alien chromosome transfer from wheat into rye. In: Horn W, Jensen CJ, Odenbach W, Schieder O (eds) Genetic manipulation in plant breeding. De Gruyter, Berlin, pp 129–136
Schlegel R, Pohler W, Hülgenhof E, Meister A (1988) Karyotype modifocations in triticale and their utilization. Biol Zentralbl 107:395–402
Scholz F, Künzel G (1981a) Induced chromosome and gene mutations for specific goals in barley genetics and breeding. In: Induced mutations – a tool in plant research. Proc IAEA/FAO Symp Vienna, 1981, pp 23–35
Scholz F, Künzel G (1981b) Progress and problems with hybrid barley. In: Asher MJC, Ellis RP, Hayter AM, Whitehouse RNH (eds) Barley genetics 4. Proc 4th Int Barley Genet Symp Edinburgh, Edinb Univ Press, pp 758–765
Scholz F, Künzel G (1986) Genetisch-Züchterische Probleme bei Hybridgerste, insbesondere Fragen der Entwicklung genetischer Systeme für die Hybridsaatguterzeugung. In: Arbeitstagung Bundesanst alpenl Landw Gumpenstein, Irdning, pp 153–164
Schubert I, Rieger R, Döbel P (1984) G and/or C-bands in plant chromosomes? J Cell Sci 71:111–120
Schubert V, Blüthner W-D, Schlegel R (1987) N-banding and Feulgen karyogram of *Aegilops markgrafii* (Greuter) Hammer var. *markgrafii*. Cereal Res Commun 15:317–320
Schwanitz F (1967) Die Evolution der Kulturpflanzen. Bay. Landw. Verlag, München
Schweizer D (1976) Reverse fluorescent chromosome banding with chromomycin and DAPI. Chromosoma 58:307–324
Seabright MA (1972) The use of proteolytic enzymes for the mapping of structural re-arrangements in the chromosomes of man. Chromosoma 36:204–210
Seal AG, Bennett MD (1985) Effect of parental genotype on early seed development in wheat-rye hybrids. In: Bernard M, Bernard S (eds), Genetics and breeding of triticale. Proc Eucarpia Meet Clermont-Ferrand, 1984, pp 167–168
Sears ER (1954) The aneuploids of common wheat. Mo Agric Exp Stn Res Bul 572

Sears ER (1956) The transfer of leaf rust resistance from *Aegilops umbellulata* to wheat. In: Genetics in plant breeding. Brookhaven Symp Biol. BNL, Brookhaven, 9, pp 1–22
Sears ER (1969) Wheat cytogenetics. Annu Rev Genet 3:451–468
Sears ER (1972) Chromosome engineering in wheat. In: Stadler Symp Univ Missouri, Columbia, MO, pp 23–38
Sears ER (1976) Genetic control of chromosome pairing in wheat. Annu Rev Genet 10:31–51
Sears ER (1984) Mutations in wheat that raise the level of meiotic chromosome pairing. In: Gustafson JP (ed) Genetic manipulation in plant improvement. 16th Stadler Genet Symp, Columbia, MO, pp 295–300
Sears ER, Okamoto M (1958) Intergenomic chromosome relationships of non-homologous chromosomes in wheat. In: Proc 10th Int Congr Genet, Montreal, 2:258–259
Sficas AG (1963) Statistical analysis of chromosome distribution to the poles in interspecific hybrids with variable chromosome pairing. Genet Res 4:266–275
Shadley J, Weber D (1986) Location of chromosomal regions controlling fatty acid composition of embryo oil in *Zea mays* L. Can J Genet Cytol 28:260–265
Shah DM et al. (1986) Engineering herbicide tolerance in transgenic plants. Science 233:478–481
Sharma AK, Sharma A (1980) Chromosome techniques. Theory and practice, 3rd edn. Butterworths, London
Shaver DL (1963) The effect of structural heterozygosity on the degree of preferential pairing in allotetraploids of *Zea*. Genetics 48:515–524
Sheldon J, Wilson C, Dickinson HG (1988) Interaction between nucleus and cytoskeleton during the pairing stages of male meiosis in flowering plants. In: Brandham P (ed) Proc Kew Chrom Conf III HMSO London, pp 27–35
Simmonds NW (1976) Evolution of crop plants. Longman, London
Simonsen O (1973) Cytogenetic investigations in diploid and autotetraploid populations of *Lolium perenne* L. Hereditas 75:157–188
Simonsen O (1975) Cytogenetic investigations in diploid and tetraploid populations of *Festuca pratensis*. Hereditas 79:73–108
Singh NK, Shepherd KW, McIntosh RA (1990) Linkage mapping of genes for resistance to leaf, stem and stripe rusts and Ω-secalins on the short arm of rye chromosome 1R. Theor Appl Genet 80:609–616
Sisodia NS, Shebeski LH (1965) Synthesis of complete interchange stocks in barley (*Hordeum vulgare* L.) Can J Genet Cytol 7:164–170
Sitch LA, Snape JW (1986) Doubled haploid production in winter wheat and triticale genotypes, using the *Hordeum bulbosum* system. Euphytica 35:1045–1051
Skiebe K, Jahr W, Stein M (1963) Bedeutung von Valenzkreuzungen für die Polyploidiezüchtung. Z Pflanzenzuecht 50:26–33
Slizynska H (1938) Salivary gland analysis of the white-facet region of *Drosophila melanogaster*. Genetics 23:291–299
Smith BW (1969) Evolution of sex-determining mechanisms in *Rumex*. Chromosomes Today 2:172–182
Smith FH (1946) Haploidy in einkorn. J Agric Res 73:291–301
Sodkiewicz W (1984) Amphidiploid *Triticum monococcum* L. x *Secale cereale* L. (AARR) – a new form of tetraploid triticale. Cereal Res Commun 12:35–40
Srivastava S, Lavania UC (1990) Meiotic regularization, restauration of seed fertility and alkaloid content in induced autotetraploids of *Hyoscyamus albus* L. Plant Breed 104:160–166
Srivastava S, Lavania UC (1991) Enhanced productivity of tropane alkaloids and fertility in artificial autotetraploids of *Hyoscyamus niger*. Euphytica (in press)
Srivastava S, Lavania UC, Sybenga J (1992) Genetic variation in meiotic behaviour and fertility in tetraploid *Hyoscyamus muticus*: correlation with diploid meiosis. Heredity 68:231–239

Stack S, Anderson L (1987) Hypotonic bursting method for spreading synaptonemal complexes of *Zea mays*. J Hered 78:178–182
Stack SM, Anderson LK, Sherman JD (1989) Chiasmata and recombination nodules in *Lilium longiflorum*. Genome 32:486–498
Stam P (1977) Selection response under random mating and under selfing in the progeny of a cross of homozygous parents. Euphytica 26:169–184
Stebbins GL (1947) Types of polypoids: their classification and significance. Adv Genet 1:403–429
Stebbins GL (1971) Chromosomal evolution in higher plants. Edward Arnold, London
Steinitz-Sears LM (1973) Misdivision of five different 3B monosomes in Chinese Spring wheat. In: Proc 4th Int Wheat Genet Symp Missouri Agric Exp Stn, Columbia, MO, pp 739–743
Stern H, Hotta Y (1987) The biochemistry of meiosis. In: Moens PB (ed) Meiosis. Academic Press Toronto, pp 303–331
Struhl K (1983) The new yeast genetics. Nature 305:391–397
Struss D, Röbbelen G (1989) Cytological irregularities masked by heterozygosity in triticale. Plant Breed 102:265–272
Stubblefield E (1975) Analysis of the replication pattern of Chinese hamster chromosomes using 5-bromodeoxyuridine suppression of 33258 Hoechst fluorescence. Chromosoma 53:209–221
Suja JA, Torre J de-la, Giménez-Albián JF, Garcia-de-la-Vega JF, Rufas JS (1991) Meiotic chromosome structure. Kinetochores and chromatid cores in standard and B chromosomes of *Acryptera fusca* (Orthoptera) revealed by silver staining. Genome 34:19–27
Sved JA (1966) Telomere attachment of chromosomes. Some genetical and cytological consequences. Genetics 53:747–756
Swaminathan MS, Sulbha K (1959) Multivalent frequency and seed fertility in raw and evolved tetraploids of *Brassica campestris* var. *toria*. Z Vererbungsl 90:385–392
Swaminathan MS, Gupta PK, Sinha U (1983) Cytogenetics of crop plants. MacMillan India, Delhi
Swanson CP (1957) Cytology and cytogenetics. Prentice Hall, Englewood Cliffs, NJ
Swanson CP, Merz T, Young WJ (1981) Cytogenetics, the chromosome in division, inheritance and evolution, 2nd edn. Prentice Hall, Englewood Cliffs, NJ
Sybenga J (1955) Variation in *Hordeum vulgare* L. and *Trifolium repens* L. after colchicine treatment. Thesis, University of Marland, College Park, MD, 60 pp
Sybenga J (1956) The significance of colchicine from *Colchicum autumnale* L. for the induction of polyploidy in nature. Genetica 28:217–222
Sybenga J (1958) Inbreeding effects in rye. Z Vererbungsl 89:338–354
Sybenga J (1959) Some sources of error in the determination of chromosome length. Chromosoma 10:355–364
Sybenga J (1964) Accurate chromosome measurements with the camera lucida. Z Wiss Mikrosk Mikrosk Tech 65:437–439
Sybenga J (1965) The quantitative analysis of chromosome pairing and chiasma formation based on the relative frequencies of MI configurations. II. Primary trisomics. Genetica 36:339–350
Sybenga J (1966) The zygomere as hypothetical unit of chromosome pairing initiation. Genetica 37:186–198
Sybenga J (1969) Allopolyploidization of autopolyploids. I. Possibilities and limitations. Euphytica 18:355–371
Sybenga J (1972) General cytogenetics. North Holland/American Elsevier, Amsterdam
Sybenga J (1973) Allopolyploidization of autopolyploids. 2. Manipulation of the chromosome pairing system. Euphytica 22:433–444
Sybenga J (1975) Meiotic configurations. Springer Berlin Heidelberg New York
Sybenga J (1976) Quantitative variation in chromosome pairing affinities within a species, *Secale cereale*. In: Jones K, Brandham PE, (eds) Current chromosome research. Elsevier/North Holland Biomedical Press, Amsterdam, pp 143–150

Sybenga J (1981) Specialization in the behaviour of chromosomes on the meiotic spindle. Genetica 57:143–151

Sybenga J (1982) Cytogenetic systems and chromosomal male sterility genes in hybrid rye breading Tagungsber Akad Landwirtschaftl Wiss DDR 198:161–163

Sybenga J (1983a) Genetic manipulation in plant breeding: somatic versus generative. Theor Appl Genet 66:179–201

Sybenga J (1983b) Rye chromosome nomenclature and homoeology relationships. Workshop report. Z Pflanzenzuecht 90:297–304

Sybenga J (1984a) Preferential pairing in *Dactylis glomerata*? Can J Genet Cytol 26:640

Sybenga J (1984b) The taxonomy of multivalent orientation: six modes of alternate or one? Can J Genet Cytol 26:389–392

Sybenga J (1987) Aneuploidy resulting from complex meiotic configurations. In: Vig BK, Sandberg AA (eds) Aneuploidy, Part A. Incidence and etiology. Alan R Liss, New York, pp 297–329

Sybenga J (1988) Mathematical models for estimating preferential pairing and recombination in triploid hybrids. Genome 30:745–757

Sybenga J (1989) Genetic manipulation: generative versus somatic. In: Bajaj YPS (eds) Biotechnology in agriculture and forestry 9. Plant protoplasts and genetic engineering II. Springer, Berlin Heidelberg New York. pp 26–53

Sybenga J, Mastenbroek I (1980) Combined genetic and cytological analysis of positive and negative interference in an interchange heterozygote of rye (*Secale cereale* L.). Heredity 44:83–92

Sybenga J, Prakken R (1962) Gene analysis in rye. Genetica 33:95–105

Sybenga J, Rickards GK (1987) The orientation of multivalents at meiotic metaphase I: a workshop report. Genome 29:612–620

Sybenga J, Verhaar HM (1980) Chiasma patterns in a translocation derived duplication heterozygote of rye. Chromosoma 80:147–162

Sybenga J Wolters AHG (1972) The classification of the chromosomes of rye (*Secale cereale* L.): a translocation tester set. Genetica 43:453–464

Sybenga J, Parmar S, Eden J van, Shewry P (1990) Mapping seed storage protein loci Sec-1 and Sec-3 in relation to five chromosomal rearrangements in rye (*Secale cereale* L.). Theor Appl Genet 79:201–208

Szabados L, Hadlaczky G, Dudits D (1981) Uptake of isolated plant chromosomes by plant protoplasts. Planta 151:141–151

Tanksley SD, Miller JC, Paterson A, Bernatzki R (1988) Molecular mapping of plant chromosomes. In: Gustafson JP, Appels R (eds) Chromosome structure and function. Plenum Press, New York, pp 157–173

Tatum LA (1971) The Southern corn leaf blight epidemic. Science 171:1113–1116

Taylor IB, Evans, GM (1976) The effect of B-chromosomes on homoeologous pairing in species hybrids. III. Interspecific variation. Chromosoma 57:25–32

Taylor JH, Woods PS, Hughes WL (1957) The organization and duplication of chromosomes as revealed by autoradiographic studies using tritium labelled thymidine. Proc Natl Acad Sci USA 43:122–128

Tease C, Jones GH (1976) Chromosome-specific control of chiasma formation in *Crepis capillaris*. Chromosoma 57:33–49

Thomas H (1968) The addition of single chromosomes of *Avena hirtula* to the cultivated hexaploid oat *A. sativa*. Can J Genet Cytol 10:551–563

Thomas H (1981) Interspecific manipulation of chromosomes. Philos Trans R Soc Lond B 292:519–527

Thomas JB, Kaltsikes PJ (1977) The effect of colchicine on chromosome pairing. Can J Genet Cytol 19:321–249

Tokamasu S (1976) The increase of seed fertility of *Brassicoraphanus* through cytological irregularity. Euphytica 25:463–470

Tsuchiya T (1969) Cytogenetics of 16 chromosome plants in barley. Chromosoma 26:130–139

Tsuchiya T (1991) Chromosome mapping by means of aneuploid analysis in barley. In: Gupta PK, Tsuchiya T (eds) Chromosome engineering in plants: genetics, breeding, evolution, Part A. Elsevier, Amsterdam, pp 361–384
Tsujimoto H Noda K (1989) Structure of chromosome 5A of wheat speltoid mutants induced by the gametocidal genes of *Aegilops speltoides*. Genome 32:1085–1090
Tsujimoto H, Tsunewaki K (1988) Gametocidal genes in wheat and its relatives. III. Chromosome location and effects of two *Aegilops speltoides*-derived gametocidal genes in common wheat. Genome 30:239–244
Tuyl JM van, Vries JN de, Bino RJ, Kwakkenbos TAM (1989) Identification of 2n-pollen producing interspecific hybrids of *Lilium* using flow cytometry. Cytologia 54:737–745
Tyagi BR, Dubey R (1989) Meiotic chromosome behaviour and fertility in autotriploids of Egyptian henbane, *Hyoscyamus muticus* L. Caryologia 42:153–163
U N (1935) Genome analysis in *Brassica* with special reference to the experimental formation of *B. napus* and peculiar mode of fertilization. Jpn J Genet 7:389–452
Unrau J (1950) The use of monosomes and nullisomes in cytogenetic studies of common wheat. Sci Agric 30:66–89
VanSanten E, Casler MD (1990) Forage quality and yield in 4x progeny from interploidy crosses in *Dactylis*. Crop Sci 30:35–39
Varghese BM (1971) Chromosome constitution and nucleolar behaviour in the endosperm of *Curcurbita pepo*. Genetics 42:231–238
Veen JH van der, Blankestijn-de Vries H (1973) Double reduction in tetraploid *Arabidopsis thaliana* studied by means of a chlorophyll mutant with a distinct simplex phenotype. Arabidopsis Inf Serv 10:11–12
Veronesi F, Mariani A, Bingham ET (1986) Unreduced gametes in diploid *Medicago* and their importance in alfalfa breeding. Theor Appl Genet 72:37–41
Vig BK (1983) Sequence of centromere separation: occurrence, possible significance, and control. Cancer Genet Cytogenet 8:249–274
Virk DS, Gupta K (1984) Matromorphy in *Pisum sativum* L. Theor Appl Genet 68:207–211
Vosa CG, Marchi P (1972) On the quinacrine fluorescence and Giemsa staining patterns of the chromosomes of *Vicia faba*. G Bot Ital 106:151–159
Vosselman L (1979) Sex determination of the onion fly *Hylemya antiqua* (Meigen). II. Sex ratio distortion by unstable somatic behaviour of chromosome Y2, and inheritance of a nonfunctional Y2(Ym). Chromosoma 75:353–367
Vosselman L (1981) Meiotic segregation of five different reciprocal translocations in the onion fly, *Hylemya antiqua* (Meigen). Chromosoma 81:727–738
Vries JN de (1983) High recombination between the breakpoint of a reciprocal translocation in rye (*Secale cereale* L.) and an interstitially located gene. Theor Appl Genet 66:329–339
Vries JN de (1984) Meiotic behaviour, stability, transmission and recombination in four balanced tertiary trisomics of rye (*Secale cereale* L.). Chromosoma 89:24–32
Vries JN de (1985) Isolation of telotertiary compensating trisomics from telocentric translocation trisomics and telo-substituted translocation heterozygotes of rye (*Secale cereale* L.). Genetica 68:47–58
Vries JN de (1989) De cytogenetica van *Allium* soortshybriden. Annu Rep Inst Hortic Plant Breed, pp 51–53 Wageningen
Vries JN de, Sybenga J (1984) Chromosomal location of 17 monogenically inherited morphological markers in rye (*Secale cereale* L.) using the translocation tester set. Z Pflanzenzuecht 92:117–139
Vries JN de, Zevenbergen MJ, Oostveen BDM, Huibers W, Jongerius MC (1991) Monosomic additions of *Allium cepa* L. in a diploid *A. fistulosum* background. In: Proc Int Symp: The Genus *Allium*, Gatersleben, June 1991 (in press)
Wagenaar EB (1966) High mitotic synchronization induced by 5-aminouracyl in root tip cells of *Allium cepa* L. Exp Cell Res 43:184–190

Wagenaar EB (1969) End-to-end chromosome attachments in mitotic interphase and their possible significance to meiotic chromosome pairing. Chromosoma 26:410–426
Walters JL (1952) Heteromorphic chromosome pairs in *Paeonia californica*. Am J Bot 39(2):145–151
Walters MS, Gerstel DU (1948) A cytological investigation of a tetraploid *Rhoeo discolor*. Am J Bot 35:141–150
Wang HC, Kao KN (1988) G-banding in plant chromosomes. Genome 30:48–51
Wang RR-C (1989) An assessment of genome analysis based on chromosome pairing in hybrids of perennial Triticeae. Genome 32:179–189
Wang RR-C, Berdahl JD (1990) Meiotic association at metaphase I in diploid, triploid, and tetraploid Russian wild rye (*Psatyrostachys juncea* (Fisch.) Nevski). Cytologia 55:639–643
Watanabe H (1962) An X-ray induced strain of ring of 12 chromosomes in *Tradescantia paludosa*. Nature 193:603
Watanabe K, Peloquin SJ (1989) Occurrence of 2n pollen and *ps* gene frequencies in cultivated groups and their related wild species in tuber-bearing Solanums. Theor Appl Genet 78:329–336
Watanabe K, Peloquin SJ, Endo M (1991) Genetic significance of mode of polyploidization: somatic doubling or 2n gametes? Genome 34:28–34
Watson JD, Hopkins NH, Roberts JW, Steitz JA, Weiner AM (1987) Molecular biology of the gene, vol 1. General principles, vol 2, special aspects, 4th edn. Benjamin/Cummings, Menlo Park, CA
Watson WAF (1969) Studies on a recombination-deficient mutant of *Drosophila*. I. Dominant lethals. Mutat Res 8:91–100
Weber DF (1983) Monosomic analysis in diploid crop plants. In: Swaminathan MS, Gupta PK, Sinha U (eds) Cytogenetics of crop plants. MacMillan India, Delhi, pp 351–378
Weber DF (1991) Monosomic analysis in maize and other diploid crop plants. In: Gupta PK, Tsuchiya T (eds) Chromosome engineering in plants: genetics, breeding, evolution, Part A. Elsevier, Amsterdam, pp 181–209
Werner JE, Peloquin SJ (1987) Frequency and mechanisms of 2n egg formation in haploid *Tuberosum*-wild species F1 hybrids. Am Potato J 64:641–654
Wettstein D von, Rasmussen SW, Holm PB (1984) The synaptonemal complex in genetic segregation. Annu Rev Genet 18:331–413
White J, Rees H (1987) Chromosome weights and measures in *Petunia*. Heredity 58:139–143
White MJD (1973) Animal cytology and evolution. Cambridge University Press, Cambridge
Whitehouse HLK (1963) A theory of crossing-over by means of hybrid deoxyribonucleic acid. Nature 199:1034–1040
Wienhues A (1966) Transfer of rust resistance of *Agropyron* to wheat by addition, substitution and translocation. In: Proc 2nd Int Wheat Genet Symp Lund, Suppl Hereditas, pp 328–341
Wienhues A (1973) Translocation between wheat chromosomes and an *Agropyron* chromosome conditioning rust resistance. In: Proc 4th Int Wheat Genet Symp, Columbia, MO, pp 201–207
Wijbrandi J, Posthuma A, Kok JM, Rijken R, Vos JGM, Koornneef M (1989) Asymmetric somatic hybrids between *Lycopersicon esculentum* and irradiated *Lycopersicon peruvianum*. I. Cytogenetics and morphology. Dissertation, Agricultural University, Wageningen, pp 41–55
Wilson EB (1925) The cell in development and heredity, 3rd edn. Macmillan, New York
Wilson GB, Cheng KG (1949) Segregation and reduction in somatic tissues II. The separation of homologous chromosomes in *Trillium* sp. J Hered 40:3–6

Wolf PG, Soltis PS, Soltis DE (1989) Tetrasomic inheritance and chromosome pairing behaviour in the naturally occurring autotetraploid *Heuchera grossulariifolia* (Saxifragaceae). Genome 32:655–659
Wolff S, Luippold HE (1956) Obtaining large numbers of metaphases in barley root tips. Stain Technol 31:201–205
Worland AJ, Law CN (1985) Aneuploidy in semi-dwarf wheat varieties. Euphytica 34:317–327
Wricke G, Weber WE (1986) Quantitative genetics and selection in plant breeding. Walter de Gruyter, Berlin
Yacobi YZ, Mello-Sampayo T, Feldman M (1982) Genetic induction of bivalent interlocking in common wheat. Chromosoma 87:165–175
Yamashita K (1951) Studies on X-ray induced reciprocal translocations in Einkorn wheats. III. A newly synthesized ring of 14 chromosomes in a complex heterozygote, *Aegilopoides monococcum*. Cytologia 16:164–176
Yamashita Y, Terada R, Nishibayashi S, Shimamoto K (1989) Asymmetric somatic hybrids of *Brassica*: partial transfer of *B. campestris* genome into *B. oleracea* by cell fusion. Theor Appl Genet 77:189–194
Yang SJ (1965) Numerical chromosome instability in *Nicotiana hybrids*. II. Intra-plant variation. Can J Genet Cytol 7:112–119
Yang X-F, Zhang Z-L (1988) In situ induction of G-bands and macrocoils in plant chromosomes. Hereditas 109:45–51
Yen Y, Kimber G (1989) A triploid hybrid between autotetraploid *Triticum longissimum* and *T. speltoides*. Cereal Res Comm 17:259–264
Yen Y, Kimber G (1990a) Production and meiotic analysis of autotriploid *Triticum speltoides* and *T. bicorne*. Theor Appl Genet 79:525–528
Yen Y, Kimber G (1990b) Genomic relationships of *Triticum sharonense* with other S-genome diploid *Triticum* species. Plant Breed 104:53–57
Young ND, Tanksley SD (1989) RFLP analysis of the size of chromosomal segments retained around the Tm-2 locus of tomato during backcross breeding. Theor Appl Genet 77:353–359
Youssef SS, Morris R, Baenziger PS, Papa CM (1989) Cytogenetic studies of progenies from crosses between "Centurk" wheat and its doubled haploids derived from anther culture. Genome 32:622–628
Zadoo SN, Choubey RN, Gupta SK, Sybenga J (1989) Meiotic chromosome association in diploid and tetraploid *Avena strigosa* Schrel. Genome 32:972–977
Zeilinga AE, Schouten HP (1968) The production of tetraploids. Euphytica 17:303–310
Zeller FJ (1973) 1B/1R wheat-rye chromosome substitutions and translocations. In: Proc 4th Int Wheat Genet Symp, Columbia, MO, pp 209–221
Zen S (1961) Chiasma studies in structural hybrids VI. Heteromorphic bivalent and reciprocal translocation in *Allium fistulosum*. Cytologia 26:67–73
Zhao J-P, Davidson D (1984) Distribution of chromosomes into discrete groups in colchicine-induced C-metaphases of barley. Caryologia 37:331–342

Subject Index

Very common terms have been indexed only if used in a special context: chiasma, chromosome, configuration, division, homologous, karyotype, linkage, locus, segregation, for example. The same is true for terms not of immediate interest for the subject of the book: e.g. RNA, translation, or such terms have not been listed at all: meiocyte, pistil. Similarly, meiotic and mitotic stages (leptotene, zygotene, pachytene, prometaphase I, metaphase I, anaphase I, telophase I, interkinesis, prophase II, metaphase II, anaphase II etc.), even when not frequently referred to in the text (leptotene, prophase II), are given only in special contexts. For details on these stages the reader is referred to Chapter 3.

MIX
Papier aus verantwortungsvollen Quellen
Paper from responsible sources
FSC® C105338

If you have any concerns about our products,
you can contact us on
ProductSafety@springernature.com

In case Publisher is established outside the EU,
the EU authorized representative is:
Springer Nature Customer Service Center GmbH
Europaplatz 3, 69115 Heidelberg, Germany

Printed by Libri Plureos GmbH
in Hamburg, Germany